Student's Solutions Manual

COLLEGE ALGEBRA

SECOND EDITION

Student's Solutions Manual

College Algebra

SECOND EDITION

Bittinger/Beecher

Judith A. Penna

 ADDISON-WESLEY

An imprint of Addison Wesley Longman, Inc.

Reading, Massachusetts • Menlo Park, California • New York • Harlow, England
Dan Mills, Ontario • Sydney • Mexico City • Madrid • Amsterdam

Reproduced by Addison-Wesley from camera-ready copy supplied by the author.

Copyright © 1994 Addison Wesley Longman, Inc.

ISBN: 0-201-52585-2

10 11 12 CRW 020100

TABLE OF CONTENTS

Special thanks are extended to Patsy Hammond for her
excellent typing and to Pam Smith for her careful
proofreading. Their patience, efficiency, and good
humor made the author's work much easier.

Exercise Set 1.1

1. – 6. Consider the numbers

-6, 0, 3, $-\frac{1}{2}$, $\sqrt{3}$, -2, $-\sqrt{7}$, $\sqrt[3]{2}$, $\frac{5}{8}$, 14, $-\frac{9}{4}$,

8.53, $9\frac{1}{2}$, $\sqrt{16}$, $-\sqrt[3]{8}$

1. Natural numbers: 3, 14, $\sqrt{16}$ ($\sqrt{16}$ = 4)

2. Whole numbers: 0, 3, 14, $\sqrt{16}$

3. Irrational numbers: $\sqrt{3}$, $-\sqrt{7}$, $\sqrt[3]{2}$

4. Rational numbers: -6, 0, 3, $-\frac{1}{2}$, -2, $\frac{5}{8}$, 14, $-\frac{9}{4}$,

8.53, $9\frac{1}{2}$, $\sqrt{16}$, $-\sqrt[3]{8}$

5. Integers: -6, 0, 3, -2, 14, $\sqrt{16}$,

$-\sqrt[3]{8}$ ($-\sqrt[3]{8}$ = -2)

6. Real numbers: All of them

7. $-\frac{6}{5}$ can be expressed as $\frac{-6}{5}$, a quotient of integers. Thus, $-\frac{6}{5}$ is <u>rational</u>.

8. Rational

9. -9.032 is an ending decimal. Thus, -9.032 is <u>rational</u>.

10. Rational

11. 4.$\overline{516}$ is a repeating decimal. Thus, 4.$\overline{516}$ is <u>rational</u>.

12. Rational

13. 4.303003000300003... has a pattern but not a pattern formed by a repeating block of digits. The numeral is an unending nonrepeating decimal and thus <u>irrational</u>.

14. Irrational

15. $\sqrt{6}$ is <u>irrational</u> because 6 is not a perfect square.

16. Irrational

17. $-\sqrt{14}$ is <u>irrational</u> because 14 is not a perfect square.

18. Irrational

19. $\sqrt{49}$ is <u>rational</u> because 49 is a perfect square.

$\sqrt{49} = \sqrt{7^2} = 7 = \frac{7}{1}$ (Quotient of integers)

20. Rational

21. $\sqrt[3]{5}$ is <u>irrational</u> because 5 is not a perfect cube.

22. Irrational

23. "-x" means "find the opposite of x." If x = -7, then we are finding the opposite of -7.

-(-7) = 7   (The opposite of -7 is 7.)

"-1·x" means "find negative one times x." If x = -7, then we are finding -1 times -7.

-1·(-7) = 7   (The product of two negative numbers is positive.)

24. $\frac{10}{3}$, $\frac{10}{3}$

25. "-x" means "find the opposite of x." If x = 57, then we are finding the opposite of 57.

-(57) = -57   (The opposite of 57 is -57.)

"-1·x" means "find negative one times x." If x = 57, then we are finding -1 times 57.

-1·57 = -57   (The product of a negative number and a positive number is negative.)

26. $-\frac{13}{14}$, $-\frac{13}{14}$

27. $-(x^2 - 5x + 3)$

$= -(12^2 - 5·12 + 3)$ (Substituting 12 for x)

$= -(144 - 60 + 3)$

$= -87$

28. -107

29. $-(7 - y)$

$= -[7 - (-9)]$ (Substituting -9 for y)

$= -[7 + 9]$

$= -16$

30. 12

31. -3.1 + (-7.2)

The sum of two negative real numbers is a negative real number.

-3.1 + (-7.2) = -10.3

32. -416

33. $\frac{9}{2} + \left(-\frac{3}{5}\right)$

The positive number $\left(\frac{9}{2}\right)$ is farther from 0 on a number line than $-\frac{3}{5}$ is, so the answer is positive.

$\frac{9}{2} + \left(-\frac{3}{5}\right) = \frac{39}{10}$

34. -20

35. $-7(-4) = 28$

    The product of two negative numbers is positive.

36. 12

37. $(-8.2) \times 6 = -49.2$

    The product of a negative number and a positive number is negative.

38. -48

39. $\quad -7(-2)(-3)(-5)$

    $= 14 \cdot 15 \qquad$ (The product of two negative numbers is positive.)

    $= 210$

40. 16.33

41. $\quad -\dfrac{14}{3}\left(-\dfrac{17}{5}\right)\left(-\dfrac{21}{2}\right)$

    $= -\dfrac{17}{5}\left[\left(-\dfrac{14}{3}\right)\left(-\dfrac{21}{2}\right)\right] \qquad$ (Using associativity and commutativity)

    $= -\dfrac{17}{5} \cdot \dfrac{294}{6} \qquad$ (The product of two negative numbers is positive.)

    $= -\dfrac{17}{5} \cdot 49$

    $= -\dfrac{833}{5} \qquad$ (The product of a negative number and a positive number is negative.)

42. $\dfrac{598}{5}$

43. $\dfrac{-20}{-4} = 5 \qquad$ (The quotient of two negative numbers is positive.)

44. -7

45. $\quad \dfrac{-10}{70}$

    $= -\dfrac{10}{70} \qquad$ (The quotient of a negative number and a positive number is negative.)

    $= -\dfrac{1}{7} \qquad$ (Simplifying)

46. -5

47. $\quad \dfrac{2}{7} \div \left(-\dfrac{14}{3}\right)$

    $= \dfrac{2}{7} \cdot \left(-\dfrac{3}{14}\right) \qquad$ (Multiplying by the reciprocal)

    $= -\dfrac{6}{98}$

    $= -\dfrac{3}{49}$

48. $\dfrac{7}{10}$

49. $\quad -\dfrac{10}{3} \div \left(-\dfrac{2}{15}\right)$

    $= -\dfrac{10}{3} \cdot \left(-\dfrac{15}{2}\right) \qquad$ (Multiplying by the reciprocal)

    $= \dfrac{150}{6}$

    $= 25$

50. 8

51. $\quad 11 - 15$

    $= 11 + (-15) \qquad [a - b = a + (-b)]$

    $= -4$

52. -29

53. $\quad 12 - (-6)$

    $= 12 + 6 \qquad [a - b = a + (-b)]$

    $= 18$

54. -9

55. $\quad 15.8 - 27.4$

    $= 15.8 + (-27.4) \qquad [a - b = a + (-b)]$

    $= -11.6$

56. -34.8

57. $\quad -\dfrac{21}{4} - \left(-\dfrac{7}{8}\right)$

    $= -\dfrac{21}{4} + \dfrac{7}{8} \qquad [a - b = a + (-b)]$

    $= -\dfrac{42}{8} + \dfrac{7}{8}$

    $= -\dfrac{35}{8}$

58. $\dfrac{59}{12}$

59. a) $\quad (1.4)^2 = 1.96$

        $(1.41)^2 = 1.9881$

        $(1.414)^2 = 1.999396$

        $(1.4142)^2 = 1.999962$

        $(1.41421)^2 = 1.999990$

    b) $\sqrt{2}$

60. a) 9.261000, 9.938375, 9.993948, 9.999517, 9.999935

    b) $\sqrt[3]{10}$

61. $k + 0 = k \qquad$ Additive identity property

62. Commutative law of multiplication

63. $-1(x + y) = (-1x) + (-1y)$

    Distributive law of multiplication over addition

64. Associative law of addition

65. $c + d = d + c$ Commutative law of addition

66. Multiplicative identity property

67. $4(xy) = (4x)y$ Associative law of multiplication

68. Distributive law of multiplication over addition

69. $y\left(\dfrac{1}{y}\right) = 1$, $y \neq 0$ Multiplicative inverse property

70. Additive inverse property

71. $a + (b + c) = a + (c + b)$
Commutative law of addition

72. Commutative law of multiplication

73. $a(b + c) = a(c + b)$
Commutative law of addition

74. Distributive law of multiplication over addition

75. $7 - 5 \neq 5 - 7$
$7 - 5 = 2$
$5 - 7 = -2$
$2 \neq -2$

76. $16 \div 8 \neq 8 \div 16$
$16 \div 8 = 2$
$8 \div 16 = \dfrac{1}{2}$
$2 \neq \dfrac{1}{2}$

77. $16 \div (4 \div 2) \neq (16 \div 4) \div 2$
$16 \div (4 \div 2) = 16 \div 2 = 8$
$(16 \div 4) \div 2 = 4 \div 2 = 2$
$8 \neq 2$

78. $7 - (5 - 2) \neq (7 - 5) - 2$
$7 - (5 - 2) = 7 - 3 = 4$
$(7 - 5) - 2 = 2 - 2 = 0$
$4 \neq 0$

79. $\pi = 3.1415926535...$
$3.14 = 3.14$
$\dfrac{22}{7} = 3.\overline{142857}$
$\dfrac{3927}{1250} = 3.1416$

Of these three, $\dfrac{3927}{1250}$ is the best approximation to π.

80. At the third decimal place

81. Let $n = 0.\overline{9} = 0.999...$

Then $10n = 9.999...$
$\underline{n = 0.999}$
$9n = 9$ Subtracting
$n = 1.$

82. $\dfrac{371}{99}$

83. Let $n = 18.3\overline{245} = 18.3245245...$

Then $10,000n = 183,245.245...$
$\underline{10n = 183.245...}$
$9990n = 183,062$ Subtracting
$n = \dfrac{183,062}{9990}.$

84. $\dfrac{12,346,418}{999,900}$

85. Prove: $(b + c)a = ba + ca$

Proof: $(b + c)a = a(b + c)$ Commutative law of multiplication
$= ab + ac$ Distributive law
$= ba + ca$ Commutative law of multiplication, used twice

86. Let a be a positive number. Then by the definition of subtraction,
$a - 0 = c$ if and only if $0 + c = a$.
Now $0 + c = c$ (identity property of addition), so $c = a$ and $a - 0 = a$ where a is a positive number. Then $a - 0$ is positive, so
$0 < a$ or $a > 0$.

Exercise Set 1.2

1. $2^3 \cdot 2^{-4} = 2^{3+(-4)} = 2^{-1} = \dfrac{1}{2}$

2. $\dfrac{1}{3}$

3. $b^2 \cdot b^{-2} = b^{2+(-2)} = b^0 = 1$

4. 1

5. $4^2 \cdot 4^{-5} \cdot 4^6 = 4^{2+(-5)+6} = 4^3$

6. 5^3

7. $2x^3 \cdot 3x^2 = 2 \cdot 3 \cdot x^{3+2} = 6x^5$

8. $12y^7$

9. $(5a^2b)(3a^{-3}b^4) = 5 \cdot 3 \cdot a^{2+(-3)} \cdot b^{1+4} = 15a^{-1}b^5$
$= \dfrac{15b^5}{a}$

10. $\dfrac{12y^7}{x^3}$

11. $(2x)^3(3x)^2 = 2^3x^3 \cdot 3^2x^2 = 8 \cdot 9 \cdot x^{3+2} = 72x^5$

12. $432y^5$

13. $(6x^5y^{-2}z^3)(-3x^2y^3z^{-2})$
$= 6 \cdot (-3) \cdot x^{5+2}y^{-2+3}z^{3+(-2)}$
$= -18x^7yz$

14. $-10x^6yz$

15. $\dfrac{b^{40}}{b^{37}} = b^{40-37} = b^3$

16. a^7

17. $\dfrac{x^2y^{-2}}{x^{-1}y} = x^{2-(-1)}y^{-2-1} = x^3y^{-3} = \dfrac{x^3}{y^3}$

18. $\dfrac{x^4}{y^5}$

19. $\dfrac{9a^2}{(-3a)^2} = \dfrac{9a^2}{9a^2} = 1$

20. 1

21. $\dfrac{24a^5b^3}{8a^4b} = 3a^{5-4}b^{3-1} = 3ab^2$

22. $6x^3y^2$

23. $\dfrac{12x^2y^3z^{-2}}{21xy^2z^3} = \dfrac{12}{21}x^{2-1}y^{3-2}z^{-2-3} = \dfrac{4}{7}xyz^{-5} = \dfrac{4xy}{7z^5}$

24. $\dfrac{x^2y^3}{3z^8}$

25. $(2ab^2)^3 = 2^3a^3b^{2 \cdot 3} = 8a^3b^6$

26. $16x^2y^6$

27. $(-2x^3)^4 = (-2)^4x^{3 \cdot 4} = 16x^{12}$

28. $81x^8$

29. $-(2x^3)^4 = -2^4x^{3 \cdot 4} = -16x^{12}$

30. $-81x^8$

31. $(6a^2b^3c)^2 = 6^2a^{2 \cdot 2}b^{3 \cdot 2}c^2 = 36a^4b^6c^2$

32. $25x^6y^4z^2$

33. $(-5c^{-1}d^{-2})^{-2} = (-5)^{-2}c^{-1(-2)}d^{-2(-2)} = \dfrac{1}{25}c^2d^4$

$\left[(-5)^{-2} = \dfrac{1}{(-5)^2} = \dfrac{1}{25}\right]$

34. $\dfrac{1}{16}x^2z^4$

35. $\dfrac{4^{-2} + 2^{-4}}{8^{-1}} = \dfrac{\frac{1}{4^2} + \frac{1}{2^4}}{\frac{1}{8}} = \dfrac{\frac{1}{16} + \frac{1}{16}}{\frac{1}{8}} = \dfrac{\frac{2}{16}}{\frac{1}{8}} =$

$\dfrac{\frac{1}{8}}{\frac{1}{8}} = 1$

36. $\dfrac{119}{72}$

37. $\dfrac{(-2)^4 + (-4)^2}{(-1)^8} = \dfrac{16 + 16}{1} = \dfrac{32}{1} = 32$

38. 25

39. $\dfrac{(3a^2b^{-2}c^4)^3}{(2a^{-1}b^2c^{-3})^2}$

$= \dfrac{3^3a^6b^{-6}c^{12}}{2^2a^{-2}b^4c^{-6}}$

$= \dfrac{27}{4}a^{6-(-2)}b^{-6-4}c^{12-(-6)}$

$= \dfrac{27}{4}a^8b^{-10}c^{18}$

$= \dfrac{27a^8c^{18}}{4b^{10}}$

40. $\dfrac{8a^{11}c^{19}}{9b^3}$

41. $\dfrac{6^{-2}x^{-3}y^2}{3^{-3}x^{-4}y} = \dfrac{3^3}{6^2}x^{-3-(-4)}y^{2-1} = \dfrac{27}{36}xy = \dfrac{3}{4}xy$

42. $\dfrac{8}{25}xy^2$

43. $\left[\dfrac{24a^{10}b^{-8}c^7}{3a^6b^{-3}c^5}\right]^5 = \left[\dfrac{8a^4c^2}{b^5}\right]^5 = \dfrac{8^5a^{20}c^{10}}{b^{25}} =$

$\dfrac{32,768a^{20}c^{10}}{b^{25}}$

44. $\dfrac{q^{80}}{625p^{16}r^{148}}$

45. When $x = 5$, $-x^2 = -5^2 = -25$
and $(-x)^2 = (-5)^2 = 25$

46. $-49, 49$

47. When $x = -1.08$, $-x^2 = -(-1.08)^2 = -1.1664$
and $(-x)^2 = [-(-1.08)]^2 = (1.08)^2 = 1.1664$

48. $-3, 3$

49. $5.8,000,000.$

7 moves

Since the number to be converted is large, the exponent is positive.

$58,000,000 = 5.8 \times 10^7$

50. 2.7×10^4

51.   3.65,000.

    5 moves

Since the number to be converted is large, the exponent is positive.

$365,000 = 3.65 \times 10^5$

52. 3.645×10^3

53.   0.000002.7

    6 moves

Since the number to be converted is small, the exponent is negative.

$0.0000027 = 2.7 \times 10^{-6}$

54. 6.58×10^{-5}

55.   0.02.7

    2 moves

Since the number to be converted is small, the exponent is negative.

$0.027 = 2.7 \times 10^{-2}$

56. 3.8×10^{-3}

57.   0.0000000000000000000000000009.11

    28 moves

Since the number to be converted is small, the exponent is negative.

$0.0000000000000000000000000000911 = 9.11 \times 10^{-28}$

58. 9.3×10^7

59.   3.664,000,000.

    9 moves

Since the number to be converted is large the exponent is positive.

$3,664,000,000 = 3.664 \times 10^9$

60. 10^{-9}

61. $4 \times 10^5 = 400,000$

62.   0.0005

63. $6.2 \times 10^{-3} = 0.0062$

64.   7,800,000

65. $7.69 \times 10^{12} = 7,690,000,000,000$

66.   0.000000854

67. $5.67 \text{ E} - 7 = 0.000000567$

68.   1,314,000,000,000

69. $9.46 \times 10^{12} = 9,460,000,000,000$

70.   0.000066

71. $7.69 \text{ E} - 8 = 0.0000000769$

72.   0.0000000008603

73. $2.567 \text{ E } 8 = 256,700,000$

74.   111,300,000,000

75. $(3.1 \times 10^5)(4.5 \times 10^{-3})$

    $= (3.1 \times 4.5) \times (10^5 \times 10^{-3})$

    $= 13.95 \times 10^2$

    $= (1.395 \times 10) \times 10^2$

    $= 1.395 \times 10^3$

76. 7.462×10^{-13}

77. $\dfrac{6.4 \times 10^{-7}}{8.0 \times 10^6} = \dfrac{6.4}{8.0} \times \dfrac{10^{-7}}{10^6}$

        $= 0.8 \times 10^{-13}$

        $= (8 \times 10^{-1}) \times 10^{-13}$

        $= 8 \times 10^{-14}$

78. 5.5×10^{30}

79.   The amount of water discharged in one hour is the amount discharged per second times the number of seconds in one hour.

    $4,200,000 \times 3600$

    $= (4.2 \times 10^6) \times (3.6 \times 10^3)$ Using scientific notation

    $= (4.2 \times 3.6) \times (10^6 \times 10^3)$

    $= 15.12 \times 10^9$ Multiplying

    $= (1.512 \times 10) \times 10^9$

    $= 1.512 \times 10^{10}$

In one hour the discharge of water is 1.512×10^{10} cubic feet.

80. 1.095×10^9 gallons

81.   We begin by expressing one billionth as a fraction.

    $\dfrac{1}{1,000,000,000} = \dfrac{1}{10^9}$ Using scientific notation

                $= 10^{-9}$

Scientific notation for 1 nanosecond is 10^{-9} sec.

82. $5.8404 \quad 10^8$ mi

83. The distance light travels in 13 weeks is the portion of a year in 13 weeks times the number of miles that light travels in one year. (See Example 23.) We will use 52 weeks for 1 year. Then 13 weeks is $\frac{13}{52}$, or 0.25 of a year.

$$0.25 \times (5.88 \times 10^{12})$$
$$= (2.5 \times 10^{-1}) \times (5.88 \times 10^{12}) \quad \text{Using scientific notation}$$
$$= (2.5 \times 5.88) \times (10^{-1} \times 10^{12})$$
$$= 14.7 \times 10^{11}$$
$$= (1.47 \times 10) \times 10^{11}$$
$$= 1.47 \times 10^{12}$$

Light travels 1.47×10^{12} miles in 13 weeks.

84. 2.4×10^{-2} m³, or 2.4×10^{7} mm³

85. $3 \cdot 2 + 4 \cdot 2^2 - 6(3 - 1)$

$= 3 \cdot 2 + 4 \cdot 2^2 - 6 \cdot 2$ Working inside parentheses

$= 3 \cdot 2 + 4 \cdot 4 - 6 \cdot 2$ Evaluating 2^2

$= 6 + 16 - 12$ Multiplying

$= 22 - 12$ Adding in order

$= 10$ from left to right

86. 18

87. $\dfrac{4(8 - 6)^2 + 4 \cdot 3 - 2 \cdot 8}{3^1 + 19^0}$

$= \dfrac{4 \cdot 2^2 + 4 \cdot 3 - 2 \cdot 8}{3 + 1}$ Calculating in the numerator and in the denominator

$= \dfrac{4 \cdot 4 + 4 \cdot 3 - 2 \cdot 8}{4}$

$= \dfrac{16 + 12 - 16}{4}$

$= \dfrac{28 - 16}{4}$

$= \dfrac{12}{4}$

$= 3$

88. -5

89. $16 \div 4 \cdot 4 \div 2 \cdot 256$

$= 4 \cdot 4 \div 2 \cdot 256$ Multiplying and dividing in order from left to right

$= 16 \div 2 \cdot 256$

$= 8 \cdot 256$

$= 2048$

90. 2

91. $\left[\dfrac{5^2}{8} + 5(5) - \dfrac{5^3}{12}\right] - \left[\dfrac{(-4)^2}{8} + 5(-4) - \dfrac{(-4)^3}{12}\right]$

$= \left[\dfrac{25}{8} + 5(5) - \dfrac{125}{12}\right] - \left[\dfrac{16}{8} + 5(-4) - \dfrac{-64}{12}\right]$

$= \left[\dfrac{25}{8} + 25 - \dfrac{125}{12}\right] - \left[2 - 20 + \dfrac{16}{3}\right]$

$= \dfrac{425}{24} - \left[-\dfrac{38}{3}\right]$

$= \dfrac{729}{24}$

$= \dfrac{243}{8}$

92. $\dfrac{9}{8}$

93. The distance of 12 from 0 is 12, so $|12| = 12$.

94. 2.56

95. The distance of -47 from 0 is 47, so $|-47| = 47$.

96. 0

97. $|-7a| = |-7| \cdot |a| = 7|a|$

98. $10|m| |n|$

99. $|-8x^6| = |-8| \cdot |x^6| = 8x^6$

100. $5x^4 y^8$

101. $|9xy| = |9| \cdot |xy| = 9|xy|$ or $9|x| |y|$

102. y^4

103. $|3a^2 b| = |3| \cdot |a^2| \cdot |b| = 3a^2 |b|$

104. $\dfrac{4|a|}{b^2}$

105. $(x^t \cdot x^{3t})^2 = (x^{4t})^2 = x^{4t \cdot 2} = x^{8t}$

106. 1

107. $(t^{a+x} \cdot t^{x-a})^4 = (t^{2x})^4 = t^{2x \cdot 4} = t^{8x}$

108. $m^2 yt$

109. $(x^a y^b \cdot x^b y^a)^c = (x^{a+b} y^{a+b})^c = x^{ac+bc} y^{ac+bc}$
$$= (xy)^{ac+bc}$$

110. $(mn)^{x^2}$

111. $\left[\dfrac{(3x^a y^b)^3}{(-3x^a y^b)^2}\right]^2 = \left[\dfrac{27x^{3a} y^{3b}}{9x^{2a} y^{2b}}\right]^2$

$$= \left[3x^a y^b\right]^2$$

$$= 9x^{2a} y^{2b}$$

112. $\dfrac{x^{6r}}{y^{18t}}$

113. P = \$92,000 - \$14,000 = \$78,000, n = 12·25 = 300

$$M = P\left[\frac{\frac{i}{12}\left(1 + \frac{i}{12}\right)^n}{\left(1 + \frac{i}{12}\right)^n - 1}\right]$$

$$M = 78,000\left[\frac{\frac{0.1075}{12}\left(1 + \frac{0.1075}{12}\right)^{300}}{\left(1 + \frac{0.1075}{12}\right)^{300} - 1}\right]\text{(Substituting)}$$

$$= \$750.43 \qquad \text{(Using a calculator)}$$

114. \$791.88, \$728.12

115. $x^4(x^3)^2 = x^4 \cdot x^5 = x^9$

Exponents were added instead of multiplied.

$x^4(x^3)^2 = x^4 \cdot x^6 = x^{10}$ (Correct)

116. Exponents were added instead of subtracted; $\dfrac{x^6}{y^{12}}$

117. $(2x^{-4}\,y^6\,z^3)^3 = 6x^{-1}\,y^3\,z^6$
①②③④   ①②③④

Four errors were made.

① $2 \cdot 3$ instead of 2^3

② Exponents were added instead of multiplied.

③ Exponents were subtracted instead of multiplied.

④ Exponents were added instead of multiplied.

$(2x^{-4}y^6z^3)^3 = 2^3x^{-4 \cdot 3}y^{6 \cdot 3}z^{3 \cdot 3} = 8x^{-12}y^{18}z^9 =$

$$\frac{8y^{18}z^9}{x^{12}} \quad \text{(Correct)}$$

Exercise Set 1.3

1. $-11x^4 - x^3 + x^2 + 3x - 9$

Term	Degree	
$-11x^4$	4	
$-x^3$	3	
x^2	2	
$3x$	1	$(3x = 3x^1)$
-9	0	$(-9 = -9x^0)$

The degree of a polynomial is the same as that of its term of highest degree.

The term of highest degree is $-11x^4$. Thus, the degree of the polynomial is 4.

2. 3, 2, 1, 0; 3

3. $y^3 + 2y^6 + x^2y^4 - 8$

Term	Degree	
y^3	3	
$2y^6$	6	
x^2y^4	6	$(2 + 4 = 6)$
-8	0	$(-8 = -8x^0)$

The terms of highest degree are $2y^6$ and x^2y^4. Thus, the degree of the polynomial is 6.

4. 2, 5, 7, 0; 7

5. $a^5 + 4a^2b^4 + 6ab + 4a - 3$

Term	Degree	
a^5	5	
$4a^2b^4$	6	$(2 + 4 = 6)$
$6ab$	2	$(1 + 1 = 2)$
$4a$	1	$(4a = 4a^1)$
-3	0	$(-3 = -3a^0)$

The term of highest degree is $4a^2b^4$. Thus, the degree of the polynomial is 6.

6. 6, 8, 4, 2, 0; 8

7. $(5x^2y - 2xy^2 + 3xy - 5)+(-2x^2y - 3xy^2 + 4xy + 7)$
$= (5 - 2)x^2y + (-2 - 3)xy^2 + (3 + 4)xy + (-5 + 7)$
$= 3x^2y - 5xy^2 + 7xy + 2$

8. $2x^2y - 7xy^2 + 8xy + 5$

9. $(-3pq^2 - 5p^2q + 4pq + 3)+(-7pq^2 + 3pq - 4p + 2q)$
$= (-3 - 7)pq^2 - 5p^2q + (4 + 3)pq - 4p + 2q + 3$
$= -10pq^2 - 5p^2q + 7pq - 4p + 2q + 3$

10. $-9pq^2 - 3p^2q + 11pq - 6p + 4q + 5$

11. $(2x+3y+z-7) + (4x-2y-z+8) + (-3x+y-2z-4)$
$= (2+4-3)x + (3-2+1)y + (1-1-2)z + (-7+8-4)$
$= 3x + 2y - 2z - 3$

12. $7x^2 + 12xy - 2x - y - 9$

13. $\left[7x\sqrt{y} - 3y\sqrt{x} + \dfrac{1}{5}\right] + \left[-2x\sqrt{y} - y\sqrt{x} - \dfrac{3}{5}\right]$
$= (7 - 2)x\sqrt{y} + (-3 - 1)y\sqrt{x} + \left[\dfrac{1}{5} - \dfrac{3}{5}\right]$
$= 5x\sqrt{y} - 4y\sqrt{x} - \dfrac{2}{5}$

14. $7x\sqrt{y} - 5y\sqrt{x} + 1$

15. The opposite of $5x^3 - 7x^2 + 3x - 6$ can be symbolized as

$-(5x^3 - 7x^2 + 3x - 6)$ (With parentheses)

or

$-5x^3 + 7x^2 - 3x + 6$ (Without parentheses)

16. $-(-4y^4 + 7y^2 - 2y - 1)$, $4y^4 - 7y^2 + 2y + 1$

17. $(3x^2 - 2x - x^3 + 2) - (5x^2 - 8x - x^3 + 4)$

$= (3x^2 - 2x - x^3 + 2) + (-5x^2 + 8x + x^3 - 4)$

$= (3 - 5)x^2 + (-2 + 8)x + (-1 + 1)x^3 + (2 - 4)$

$= -2x^2 + 6x - 2$

18. $-4x^2 + 8xy - 5y^2 + 3$

19. $(4a - 2b - c + 3d) - (-2a + 3b + c - d)$

$= (4a - 2b - c + 3d) + (2a - 3b - c + d)$

$= (4 + 2)a + (-2 - 3)b + (-1 - 1)c + (3 + 1)d$

$= 6a - 5b - 2c + 4d$

20. $8a - 8b - 2c + 6d$

21. $(x^4 - 3x^2 + 4x) - (3x^3 + x^2 - 5x + 3)$

$= (x^4 - 3x^2 + 4x) + (-3x^3 - x^2 + 5x - 3)$

$= x^4 - 3x^3 + (-3 - 1)x^2 + (4 + 5)x - 3$

$= x^4 - 3x^3 - 4x^2 + 9x - 3$

22. $2x^4 - 5x^3 - 7x^2 + 10x - 5$

23. $(7x\sqrt{y} - 4y\sqrt{x} + 7.5) - (-2x\sqrt{y} - y\sqrt{x} - 1.6)$

$= (7x\sqrt{y} - 4y\sqrt{x} + 7.5) + (2x\sqrt{y} + y\sqrt{x} + 1.6)$

$= (7 + 2)x\sqrt{y} + (-4 + 1)y\sqrt{x} + (7.5 + 1.6)$

$= 9x\sqrt{y} - 3y\sqrt{x} + 9.1$

24. $13x\sqrt{y} - 5y\sqrt{x} + \dfrac{5}{3}$

25. $(0.565p^2q - 2.167pq^2 + 16.02pq - 17.1)$

$\quad + (-1.612p^2q - 0.312pq^2 - 7.141pq - 87.044)$

$= (0.565 - 1.612)p^2q + (-2.167 - 0.312)pq^2 +$

$\quad (16.02 - 7.141)pq + (-17.1 - 87.044)$

$= -1.047p^2q - 2.479pq^2 + 8.879pq - 104.144$

26. $2859.6xy^{-2} - 6153.8xy + 7243.4\sqrt{xy} - 10{,}259.12$

Exercise Set 1.4

1.
$$
\begin{array}{r}
2x^2 + 4x + 16 \\
3x - 4 \\
\hline
-8x^2 - 16x - 64 \\
6x^3 + 12x^2 + 48x \qquad\quad \\
\hline
6x^3 + 4x^2 + 32x - 64
\end{array}
$$

(Multiplying by -4)
(Multiplying by $3x$)
(Adding)

2. $6y^3 + 3y^2 + 9y + 27$

3.
$$
\begin{array}{r}
4a^2b - 2ab + 3b^2 \\
ab - 2b + 1 \\
\hline
4a^2b - 2ab + 3b^2 \\
-8a^2b^2 + 4ab^2 - 6b^3 \qquad\quad \\
4a^3b^2 - 2a^2b^2 \qquad\qquad\qquad\qquad\quad +3ab^3 \\
\hline
4a^3b^2 - 10a^2b^2 + 4ab^2 - 6b^3 + 4a^2b - 2ab + 3b^2 + 3ab^3
\end{array}
$$

4. $2x^4 - 2y^4 - 4x^3y + 3xy^3 - x^2y^2$

5.
$$
\begin{array}{r}
a^2 + ab + b^2 \\
a - b \\
\hline
-a^2b - ab^2 - b^3 \\
a^3 + a^2b + ab^2 \qquad\qquad \\
\hline
a^3 \qquad\qquad\qquad - b^3
\end{array}
$$

The product is $a^3 - b^3$.

6. $t^3 + 1$

7. $(2x + 3y)(2x + y)$

$= 4x^2 + 2xy + 6xy + 3y^2$ (FOIL)

$= 4x^2 + 8xy + 3y^2$

8. $4a^2 - 8ab + 3b^2$

9. $\left(4x^2 - \dfrac{1}{2}y\right)\left(3x + \dfrac{1}{4}y\right)$

$= 12x^3 + x^2y - \dfrac{3}{2}xy - \dfrac{1}{8}y^2$ (FOIL)

10. $6y^4 - \dfrac{1}{2}xy^3 + \dfrac{3}{5}xy - \dfrac{1}{20}x^2$

11. $(2p^2q^3 - r^2)(5pq - 2r)$

$= 10p^3q^4 - 4p^2q^3r - 5pqr^2 + 2r^3$ (FOIL)

12. $9y^3 - 3xy^2 - 6y + 2x$

13. $(2x + 3y)^2$

$= (2x)^2 + 2(2x)(3y) + (3y)^2$

$\qquad [(A + B)^2 = A^2 + 2AB + B^2]$

$= 4x^2 + 12xy + 9y^2$

14. $25x^2 + 20xy + 4y^2$

15. $(2x^2 - 3y)^2$

$= (2x^2)^2 - 2(2x^2)(3y) + (3y)^2$

$\qquad [(A - B)^2 = A^2 - 2AB + B^2]$

$= 4x^4 - 12x^2y + 9y^2$

16. $16x^4 - 40x^2y + 25y^2$

17. $(2x^3 + 3y^2)^2$

$= (2x^3)^2 + 2(2x^3)(3y^2) + (3y^2)^2$

$\qquad [(A + B)^2 = A^2 + 2AB + B^2]$

$= 4x^6 + 12x^3y^2 + 9y^4$

18. $25x^6 + 20x^3y^2 + 4y^4$

19. $\left[\frac{1}{2}x^2 - \frac{3}{5}y\right]^2$

$= \left[\frac{1}{2}x^2\right]^2 - 2\left[\frac{1}{2}x^2\right]\left[\frac{3}{5}y\right] + \left[\frac{3}{5}y\right]^2$

$\qquad\qquad [(A - B)^2 = A^2 - 2AB + B^2]$

$= \frac{1}{4}x^4 - \frac{3}{5}x^2y + \frac{9}{25}y^2$

20. $\frac{1}{16}x^4 - \frac{1}{3}x^2y + \frac{4}{9}y^2$

21. $(0.5x + 0.7y^2)^2$

$= (0.5x)^2 + 2(0.5x)(0.7y^2) + (0.7y^2)^2$

$\qquad\qquad [(A + B)^2 = A^2 + 2AB + B^2]$

$= 0.25x^2 + 0.7xy^2 + 0.49y^4$

22. $0.09x^2 + 0.48xy^2 + 0.64y^4$

23. $(3x - 2y)(3x + 2y)$

$= (3x)^2 - (2y)^2 \qquad [(A - B)(A + B) = A^2 - B^2]$

$= 9x^2 - 4y^2$

24. $9x^2 - 25y^2$

25. $(x^2 + yz)(x^2 - yz)$

$= (x^2)^2 - (yz)^2 \qquad [(A + B)(A - B) = A^2 - B^2]$

$= x^4 - y^2z^2$

26. $4x^4 - 25x^2y^2$

27. $(3x^2 - \sqrt{2})(3x^2 + \sqrt{2})$

$= (3x^2)^2 - (\sqrt{2})^2 \qquad [(A - B)(A + B) = A^2 - B^2]$

$= 9x^4 - 2$

28. $25x^4 - 3$

29. $(2x + 3y + 4)(2x + 3y - 4)$

$= (2x + 3y)^2 - 4^2 \qquad [(A + B)(A - B) = A^2 - B^2]$

$= 4x^2 + 12xy + 9y^2 - 16$

30. $25x^2 + 20xy + 4y^2 - 9$

31. $(x^2 + 3y + y^2)(x^2 + 3y - y^2)$

$= (x^2 + 3y)^2 - (y^2)^2 \qquad [(A + B)(A - B) = A^2 - B^2]$

$= x^4 + 6x^2y + 9y^2 - y^4$

32. $4x^4 + 4x^2y + y^2 - y^4$

33. $(x + 1)(x - 1)(x^2 + 1)$

$= (x^2 - 1)(x^2 + 1)$

$= x^4 - 1$

34. $y^4 - 16$

35. $(2x + y)(2x - y)(4x^2 + y^2)$

$= (4x^2 - y^2)(4x^2 + y^2)$

$= 16x^4 - y^4$

36. $625x^4 - y^4$

37. $(0.051x + 0.04y)^2$

$= (0.051x)^2 + 2(0.051x)(0.04y) + (0.04y)^2$

$= 0.002601x^2 + 0.00408xy + 0.0016y^2$

38. $1.065024x^2 - 5.184768xy + 6.310144y^2$

39. $(37.86x + 1.42)(65.03x - 27.4)$

$= 2462.0358x^2 - 1037.364x + 92.3426x - 38.908$

$= 2462.0358x^2 - 945.0214x - 38.908$

40. $169.625105x^2 - 711.87827x - 546.525$

41. $(y + 5)^3$

$= y^3 + 3 \cdot y^2 \cdot 5 + 3 \cdot y \cdot 5^2 + 5^3$

$= y^3 + 15y^2 + 75y + 125$

42. $t^3 - 21t^2 + 147t - 343$

43. $(m^2 - 2n)^3$

$= [m^2 + (-2n)]^3$

$= (m^2)^3 + 3(m^2)^2(-2n) + 3(m^2)(-2n)^2 + (-2n)^3$

$= m^6 - 6m^4n + 12m^2n^2 - 8n^3$

44. $27t^6 + 108t^4 + 144t^2 + 64$

45. $(\sqrt{2}x^2 - y^2)(\sqrt{2}x - 2y)$

$= 2x^3 - 2\sqrt{2}x^2y - \sqrt{2}xy^2 + 2y^3$ (FOIL)

46. $3y^3 - \sqrt{3}xy^2 - 2\sqrt{3}y + 2x$

47. $(a^n + b^n)(a^n - b^n)$

$= (a^n)^2 - (b^n)^2$

$= a^{2n} - b^{2n}$

48. $t^{2a} - 3t^a - 28$

49. $(x^m - t^n)^3$

$= [x^m + (-t^n)]^3$

$= (x^m)^3 + 3(x^m)^2(-t^n) + 3(x^m)(-t^n)^2 + (-t^n)^3$

$= x^{3m} - 3x^{2m}t^n + 3x^mt^{2n} - t^{3n}$

50. $y^{3n+3}z^{n+3} - 4y^4z^{3n}$

51. $(x - 1)(x^2 + x + 1)(x^3 + 1)$

$= (x^3 - 1)(x^3 + 1)$

$= (x^3)^2 - 1^2$

$= x^6 - 1$

52. $a^{2n} + 2a^nb^n + b^{2n}$

9

53. $[(2x - 1)^2 - 1]^2$

    $= [4x^2 - 4x + 1 - 1]^2$

    $= [4x^2 - 4x]^2$

    $= (4x^2)^2 - 2(4x^2)(4x) + (4x)^2$

    $= 16x^4 - 32x^3 + 16x^2$

54. $-a^4 - 2a^3b + 25a^2 + 2ab^3 + b^4 - 25b^2$

55. $(x^{a-b})^{a+b}$

    $= x^{(a-b)(a+b)}$

    $= x^{a^2-b^2}$

56. $t^{2m^2+2n^2}$

57. $(a + b + c)^2$

    $= (a + b + c)(a + b + c)$

    $= a^2 + ab + ac + ba + b^2 + bc + ca + cb + c^2$

    $= a^2 + b^2 + c^2 + 2ab + 2ac + 2bc$

58. $a^3 + 3a^2b + 3ab^2 + b^3 + 3a^2c + 6abc + 3b^2c +$
                              $3ac^2 + 3bc^2 + c^3$

59. $(a + b)^4$

    $= (a + b)(a + b)^3$

    $= (a + b)(a^3 + 3a^2b + 3ab^2 + b^3)$

    $= a^4 + 3a^3b + 3a^2b^2 + ab^3 + a^3b + 3a^2b^2 + 3ab^3 + b^4$

    $= a^4 + 4a^3b + 6a^2b^2 + 4ab^3 + b^4$

60. $x^5 - y^5$

61. $(m + t)(m^4 - m^3t + m^2t^2 - mt^3 + t^4)$

    $= m^5 - m^4t + m^3t^2 - m^2t^3 + mt^4 +$

          $m^4t - m^3t^2 + m^2t^3 - mt^4 + t^5$

    $= m^5 + t^5$

62. $a^8 - b^8$

63. $(3a + b)^2 = 3a^2 + b^2$

    The first term is $9a^2$, not $3a^2$.
    The middle term, $2 \cdot 3a \cdot b$, is missing.
    $(3a + b)^2 = (3a)^2 + 2 \cdot 3a \cdot b + b^2$

                $= 9a^2 + 6ab + b^2$ (Correct)

64.   The middle term, $-12xy$, is missing; $-9y^2$ should
    be $+9y^2$; correct answer is $4x^2 - 12xy + 9y^2$.

65. $2x(x + 3) + 4(x^2 - 3)$

    $= 2x^2 + 3x + 4x^2 - 3$ (1)

    The 3x should be 6x; the 2 and 3 were not
    multiplied. Similarly, -3 should be -12; the
    4 and -3 were not multiplied.

    $= 6x^2 + x$

    3x - 3 is not x since they are not like terms.

    $2x(x + 3) + 4(x^2 - 3)$

    $= 2x^2 + 6x + 4x^2 - 12$

    $= 6x^2 + 6x - 12$ (Correct)

66.   In (1) the middle term, $-5ab$, is missing; in going
    from (1) to (2) the factor 6 has been omitted;
    correct answer is $6a^2 - 5ab - 6b^2$.

67.   Using the results of Exercises 60 and 62 we might
    conclude that the product is $x^n - y^n$. We can
    confirm this by multiplying as follows:

$$x^{n-1} + x^{n-2}y + x^{n-3}y^2 + \cdots + x^2y^{n-3} + xy^{n-2} + y^{n-1}$$
$$\underline{\qquad\qquad\qquad\qquad\qquad\qquad\qquad\qquad x - y}$$
$$\underline{-x^{n-1}y - x^{n-2}y^2 - \cdots - x^2y^{n-2} - xy^{n-1} - y^n}$$
$$\underline{x^n + x^{n-1}y + x^{n-2}y^2 + \cdots + x^2y^{n-2} + xy^{n-1}}$$
$$x^n \qquad\qquad\qquad\qquad\qquad\qquad\qquad\qquad -y^n$$

    The product is $x^n - y^n$.

Exercise Set 1.5

1. $p^2 + 6p + 8$
    We look for factors of 8 whose sum is 6. By
    trial we determine the factors to be 4 and 2.
    $p^2 + 6p + 8 = (p + 4)(p + 2)$

2. $(w - 5)(w - 2)$

3. $2n^2 + 9n - 56$
    We look for binomials $an + b$ and $cn + d$ for which
    the product of the first terms is $2n^2$. The
    product of the last terms must be -56. When we
    multiply the inside terms, then the outside terms,
    and add, we must have 9n. By trial, we determine
    the factorization to be

        $(2n - 7)(n + 8)$.

4. $(3y - 5)(y + 4)$

5. $y^4 - 4y^2 - 21$
    Think of this polynomial as $u^2 - 4u - 21$, where
    we have mentally substituted u for y^2. Then we
    look for factors of -21 whose sum is -4. By trial
    we determine the factors to be -7 and 3, so

        $u^2 - 4u - 21 = (u - 7)(u + 3)$.

    Then, substituting y^2 for u, we obtain the
    factorization of the original trinomial.

        $y^4 - 4y^2 - 21 = (y^2 - 7)(y^2 + 3)$

6. $(m^2 - 10)(m^2 + 9)$

7. $18a^2b - 15ab^2$
   $= 3ab \cdot 6a - 3ab \cdot 5b$
   $= 3ab(6a - 5b)$

8. $4xy(x + 3y)$

9. $a(b - 2) + c(b - 2)$
   $= (a + c)(b - 2)$

10. $(a - 2)(x^2 - 3)$

11. $x^3 + 3x^2 + 6x + 18$
    $= x^2(x + 3) + 6(x + 3)$
    $= (x^2 + 6)(x + 3)$

12. $(x^2 - 6)(3x + 1)$

13. $y^3 - 3y^2 - 4y + 12$
    $= y^2(y - 3) - 4(y - 3)$
    $= (y^2 - 4)(y - 3)$
    $= (y + 2)(y - 2)(y - 3)$

14. $(p + 3)(p - 3)(p - 2)$

15. $9x^2 - 25$
    $= (3x)^2 - 5^2$
    $= (3x + 5)(3x - 5)$

16. $(4x - 3)(4x + 3)$

17. $4xy^4 - 4xz^2$
    $= 4x(y^4 - z^2)$
    $= 4x[(y^2)^2 - z^2]$
    $= 4x(y^2 + z)(y^2 - z)$

18. $5xy^4 - 5xz^4$
    $= 5x(y^4 - z^4)$
    $= 5x(y^2 + z^2)(y^2 - z^2)$
    $= 5x(y^2 + z^2)(y + z)(y - z)$

19. $y^2 - 6y + 9$
    $= y^2 - 2 \cdot y \cdot 3 + 3^2$
    $= (y - 3)^2$

20. $(x + 4)^2$

21. $1 - 8x + 16x^2$
    $= 1^2 - 2 \cdot 1 \cdot 4x + (4x)^2$
    $= (1 - 4x)^2$

22. $(1 + 5x)^2$

23. $4x^2 - 5$
    $= (2x)^2 - (\sqrt{5})^2$
    $= (2x + \sqrt{5})(2x - \sqrt{5})$

24. $(4x - \sqrt{7})(4x + \sqrt{7})$

25. $x^2y^2 - 14xy + 49$
    $= (xy)^2 - 2 \cdot xy \cdot 7 + 7^2$
    $= (xy - 7)^2$

26. $(xy - 8)^2$

27. $4ax^2 + 20ax - 56a$
    $= 4a(x^2 + 5x - 14)$
    $= 4a(x + 7)(x - 2)$

28. $y(7x - 4)(3x + 2)$

29. $a^2 + 2ab + b^2 - c^2$
    $= (a + b)^2 - c^2$
    $= [(a + b) + c][(a + b) - c]$
    $= (a + b + c)(a + b - c)$

30. $(x - y + z)(x - y - z)$

31. $x^2 + 2xy + y^2 - a^2 - 2ab - b^2$
    $= (x^2 + 2xy + y^2) - (a^2 + 2ab + b^2)$
    $= (x + y)^2 - (a + b)^2$
    $= [(x + y) + (a + b)][(x + y) - (a + b)]$
    $= (x + y + a + b)(x + y - a - b)$

32. $(r + s + t - v)(r + s - t + v)$

33. $5y^4 - 80x^4$
    $= 5(y^4 - 16x^4)$
    $= 5(y^2 + 4x^2)(y^2 - 4x^2)$
    $= 5(y^2 + 4x^2)(y + 2x)(y - 2x)$

34. $6(y^2 + 4x^2)(y + 2x)(y - 2x)$

35. $x^3 + 8$
    $= x^3 + 2^3$
    $= (x + 2)(x^2 - 2x + 4)$

36. $(y - 4)(y^2 + 4y + 16)$

37. $3x^3 - \dfrac{3}{8}$
    $= 3\left[x^3 - \dfrac{1}{8}\right]$
    $= 3\left[x^3 - \left[\dfrac{1}{2}\right]^3\right]$
    $= 3\left[x - \dfrac{1}{2}\right]\left[x^2 + \dfrac{1}{2}x + \dfrac{1}{4}\right]$

38. $5\left[y + \dfrac{1}{3}\right]\left[y^2 - \dfrac{1}{3}y + \dfrac{1}{9}\right]$

39. $x^3 + 0.001$
    $= x^3 + (0.1)^3$
    $= (x + 0.1)(x^2 - 0.1x + 0.01)$

40. $(y - 0.5)(y^2 + 0.5y + 0.25)$

41. $3z^3 - 24$
    $= 3(z^3 - 8)$
    $= 3(z^3 - 2^3)$
    $= 3(z - 2)(z^2 + 2z + 4)$

42. $4(t + 3)(t^2 - 3t + 9)$

43. $a^6 - t^6$
    $= (a^3)^2 - (t^3)^2$
    $= (a^3 + t^3)(a^3 - t^3)$
    $= (a + t)(a^2 - at + t^2)(a - t)(a^2 + at + t^2)$

44. $(4m^2 + y^2)(16m^4 - 4m^2y^2 + y^4)$

45. $16a^7b + 54ab^7$
    $= 2ab(8a^6 + 27b^6)$
    $= 2ab[(2a^2)^3 + (3b^2)^3]$
    $= 2ab(2a^2 + 3b^2)(4a^4 - 6a^2b^2 + 9b^4)$

46. $3a^2x(2x - 5a^2)(4x^2 + 10a^2x + 25a^4)$

47. $x^2 - 17.6$
    $= x^2 - (\sqrt{17.6})^2$
    $= (x + \sqrt{17.6})(x - \sqrt{17.6})$
    $= (x + 4.19524)(x - 4.19524)$

48. $(x + 2.8337)(x - 2.8337)$

49. $37x^2 - 14.5y^2$
    $= 37(x^2 - 0.391892y^2)$
    $= 37[x^2 - (0.626y)^2]$
    $= 37(x + 0.626y)(x - 0.626y)$

50. $1.96(x + 2.980y)(x - 2.980y)$

51. $(x + 3)^2 - 2(x + 3) - 35$
    Think of this polynomial as $u^2 - 2u - 35$, where we have mentally substituted u for x + 3. Then we look for factors of -35 whose sum is -2. By trial we determine the factors to be -7 and 5, so
    $$u^2 - 2u - 35 = (u - 7)(u + 5).$$
    Then, substituting x + 3 for u, we obtain the factorization of the original trinomial.
    $(x + 3)^2 - 2(x + 3) - 35 = (x + 3 - 7)(x + 3 + 5)$
    $\qquad\qquad\qquad\qquad = (x - 4)(x + 8)$

52. $(y + 4)(y - 7)$

53. $3(a - b)^2 + 10(a - b) - 8$
    Think of this polynomial as $3u^2 + 10u - 8$, where we have mentally substituted u for a - b. Then we look for binomials ax + b and cx + d for which the product of the first terms is $3u^2$ and the product of the last terms is -8. When we multiply the inside terms, then the outside terms, and add, the sum must be $10u$. By trial, we determine the factorization to be
    $$(3u - 2)(u + 4).$$
    Then, substituting a - b for u, we obtain the factorization of the original trinomial.
    $\quad 3(a - b)^2 + 10(a - b) - 8$
    $= [3(a - b) - 2](a - b + 4)$
    $= (3a - 3b - 2)(a - b + 4)$

54. $(6p + 3q + 5)(4p + 2q - 5)$

55. $(x + h)^3 - x^3$
    $= [(x + h) - x][(x + h)^2 + x(x + h) + x^2]$
    $= (x + h - x)(x^2 + 2xh + h^2 + x^2 + xh + x^2)$
    $= h(3x^2 + 3xh + h^2)$

56. $0.02(x + 0.005)$

57. $y^4 - 84 + 5y^2$
    $= y^4 + 5y^2 - 84$
    $= u^2 + 5u - 84$ $\qquad$ Substituting u for y^2
    $= (u + 12)(u - 7)$
    $= (y^2 + 12)(y^2 - 7)$ $\qquad$ Substituting y^2 for u

58. $(x^2 + 16)(x^2 - 5)$, or
    $(x^2 + 16)(x + \sqrt{5})(x - \sqrt{5})$

59. $y^2 - \dfrac{8}{49} + \dfrac{2}{7}y$
    $= y^2 + \dfrac{2}{7}y - \dfrac{8}{49}$
    $= \left[y + \dfrac{4}{7}\right]\left[y - \dfrac{2}{7}\right]$

60. $\left[x + \dfrac{4}{5}\right]\left[x - \dfrac{1}{5}\right]$, or $\dfrac{1}{25}(5x + 4)(5x - 1)$

61. $t^2 - 0.27 + 0.6t$
    $= t^2 + 0.6t - 0.27$
    $= (t + 0.9)(t - 0.3)$

62. $(m - 0.1)(m + 0.5)$

63. $x^{2n} + 5x^n - 24$
    $= (x^n)^2 + 5x^n - 24$
    $= (x^n + 8)(x^n - 3)$

64. $(2x^n - 3)(2x^n + 1)$

65. $x^2 + ax + bx + ab$

= $x(x + a) + b(x + a)$

= $(x + b)(x + a)$

66. $(by + a)(dy + c)$

67. $\frac{1}{4} t^2 - \frac{2}{5} t + \frac{4}{25}$

= $\left[\frac{1}{2} t\right]^2 - 2 \cdot \frac{1}{2} t \cdot \frac{2}{5} + \left[\frac{2}{5}\right]^2$

= $\left[\frac{1}{2} t - \frac{2}{5}\right]^2$

68. $\frac{1}{3}\left[\frac{2}{3}r + \frac{1}{2}s\right]^2$; other forms of the answer are
$\frac{1}{108}(4r + 3s)^2$ and $\left[\frac{1}{9}r + \frac{1}{12}s\right]\left[\frac{4}{3}r + s\right]$.

69. $25y^{2m} - (x^{2n} - 2x^n + 1)$

= $(5y^m)^2 - (x^n - 1)^2$

= $[5y^m + (x^n - 1)][5y^m - (x^n - 1)]$

= $(5y^m + x^n - 1)(5y^m - x^n + 1)$

70. $2(x^{2a} + 3)(2x^{2a} + 5)$

71. $3x^{3n} - 24y^{3m}$

= $3(x^{3n} - 8y^{3m})$

= $3[(x^n)^3 - (2y^m)^3]$

= $3(x^n - 2y^m)(x^{2n} + 2x^ny^m + 4y^{2m})$

72. $(x^{2a} - t^b)(x^{4a} + x^{2a}t^b + t^{2b})$

73. $(y - 1)^4 - (y - 1)^2$

= $(y - 1)^2[(y - 1)^2 - 1]$

= $(y - 1)^2[y^2 - 2y + 1 - 1]$

= $(y - 1)^2(y^2 - 2y)$

= $y(y - 1)^2(y - 2)$

74. $(x^2 + 1)(x + 1)(x - 1)^3$

75. $5x - 9 = 5 \cdot x - 5 \cdot \frac{9}{5}$

= $5\left[x - \frac{9}{5}\right]$

76. $\frac{2}{3}\left[x - \frac{21}{2}\right]$

77.   a)

$$
\begin{array}{r}
x^2 - x + 1 \\
x^3 + x^2 - 1 \\
\hline
-x^2 + x - 1 \\
x^4 - x^3 + x^2 \\
x^5 - x^4 + x^3 \\
\hline
x^5 \qquad\qquad + x - 1
\end{array}
$$

The product is $x^5 + x - 1$.

b) We use the result in part (a).

$x^5 + x - 1 = (x^2 - x + 1)(x^3 + x^2 - 1)$

Exercise Set 1.6

1. $\frac{3x - 3}{x(x - 1)}$

Since division by zero is not defined, any number that makes the denominator zero is not a meaningful replacement. When x is replaced by 0 or 1, the denominator is 0. Thus, all real numbers except 0 and 1 are meaningful replacements.

2. All real numbers except -2, 1, and -1

3. We first factor the denominator completely.

$$\frac{7x^2 - 28x + 28}{(x^2 - 4)(x^2 + 3x - 10)} = \frac{7x^2 - 28x + 28}{(x+2)(x-2)(x+5)(x-2)}$$

We see that $x + 2 = 0$ when $x = -2$, $x - 2 = 0$ when $x = 2$, and $x + 5 = 0$ when $x = -5$. Thus, all real numbers except -2, 2, and -5 are meaningful replacements.

4. All real numbers except 0, 3, and -2

5. $\frac{3x - 3}{x(x - 1)} = \frac{3\cancel{(x - 1)}}{x\cancel{(x - 1)}}$

= $\frac{3}{x}$

All real numbers except 0 are meaningful replacements in the simplified expression.

6. All real numbers except 1

7. $\frac{7x^2 - 28x + 28}{(x^2 - 4)(x^2 + 3x - 10)} = \frac{7(x^2 - 4x + 4)}{(x+2)(x-2)(x+5)(x-2)}$

= $\frac{7\cancel{(x - 2)^2}}{(x+2)(x+5)\cancel{(x-2)^2}}$

= $\frac{7}{(x + 2)(x + 5)}$

All real numbers except -2 and -5 are meaningful replacements in the simplified expression.

8. $\frac{7x - 3}{x(x - 3)}$; all real numbers except 0 and 3

9. $\frac{25x^2y^2}{10xy^2} = \frac{\cancel{5xy^2} \cdot 5x}{\cancel{5xy^2} \cdot 2}$

= $\frac{5x}{2}$

All real numbers are meaningful replacements in the simplified expression.

10. $\frac{x - 2}{x + 3}$; all real numbers except -3

11. $\dfrac{x^2 - 3x + 2}{x^2 + x - 2} = \dfrac{(x - 2)\cancel{(x - 1)}}{(x + 2)\cancel{(x - 1)}}$

$\qquad = \dfrac{x - 2}{x + 2}$

All real numbers except −2 are meaningful replacements in the simplified expression.

12. $\dfrac{1}{(a - 2)(a + 5)}$; all real numbers except 2 and −5

13. $\dfrac{x^2 - y^2}{(x - y)^2} \cdot \dfrac{1}{x + y}$

$\qquad = \dfrac{(x^2 - y^2) \cdot 1}{(x - y)^2 (x + y)}$

$\qquad = \dfrac{\cancel{(x + y)}(x - y) \cdot 1}{(x - y)(x - y)\cancel{(x + y)}}$

$\qquad = \dfrac{1}{x - y}$

14. 1

15. $\dfrac{x^2 - 2x - 35}{2x^3 - 3x^2} \cdot \dfrac{4x^3 - 9x}{7x - 49}$

$\qquad = \dfrac{\cancel{(x - 7)}(x + 5)\cancel{(x)}(2x + 3)\cancel{(2x - 3)}}{x \cdot x\cancel{(2x - 3)}(7)\cancel{(x - 7)}}$

$\qquad = \dfrac{(x + 5)(2x + 3)}{7x}$

16. $\dfrac{(x - 5)(3x + 2)}{7x}$

17. $\dfrac{a^2 - a - 6}{a^2 - 7a + 12} \cdot \dfrac{a^2 - 2a - 8}{a^2 - 3a - 10}$

$\qquad = \dfrac{\cancel{(a - 3)}(a + 2)\cancel{(a - 4)}(a + 2)}{\cancel{(a - 4)}\cancel{(a - 3)}(a - 5)\cancel{(a + 2)}}$

$\qquad = \dfrac{a + 2}{a - 5}$

18. $\dfrac{(a + 3)^2}{(a - 6)(a + 4)}$

19. $\dfrac{m^2 - n^2}{r + s} \div \dfrac{m - n}{r + s}$

$\qquad = \dfrac{m^2 - n^2}{r + s} \cdot \dfrac{r + s}{m - n}$

$\qquad = \dfrac{(m + n)\cancel{(m - n)}\cancel{(r + s)}}{\cancel{(r + s)}\cancel{(m - n)}}$

$\qquad = m + n$

20. $a - b$

21. $\dfrac{3x + 12}{2x - 8} \div \dfrac{(x + 4)^2}{(x - 4)^2}$

$\qquad = \dfrac{3x + 12}{2x - 8} \cdot \dfrac{(x - 4)^2}{(x + 4)^2}$

$\qquad = \dfrac{3\cancel{(x + 4)}\cancel{(x - 4)}(x - 4)}{2\cancel{(x - 4)}\cancel{(x + 4)}(x + 4)}$

$\qquad = \dfrac{3(x - 4)}{2(x + 4)}$

22. $\dfrac{a + 1}{a - 3}$

23. $\dfrac{x^2 - y^2}{x^3 - y^3} \cdot \dfrac{x^2 + xy + y^2}{x^2 + 2xy + y^2}$

$\qquad = \dfrac{\cancel{(x + y)}\cancel{(x - y)}\cancel{(x^2 + xy + y^2)} \cdot 1}{\cancel{(x - y)}\cancel{(x^2 + xy + y^2)}(x + y)(x + y)}$

$\qquad = \dfrac{1}{x + y}$

24. $c - 2$

25. $\dfrac{(x - y)^2 - z^2}{(x + y)^2 - z^2} \div \dfrac{x - y + z}{x + y - z}$

$\qquad = \dfrac{(x - y)^2 - z^2}{(x + y)^2 - z^2} \cdot \dfrac{x + y - z}{x - y + z}$

$\qquad = \dfrac{\cancel{(x - y + z)}(x - y - z)\cancel{(x + y - z)}}{(x + y + z)\cancel{(x + y - z)}\cancel{(x - y + z)}}$

$\qquad = \dfrac{x - y - z}{x + y + z}$

26. $\dfrac{a + b - 3}{a - b + 3}$

27. $\dfrac{3}{2a + 3} + \dfrac{2a}{2a + 3}$

$\qquad = \dfrac{3 + 2a}{2a + 3}$

$\qquad = 1$

28. 2

29. $\dfrac{y}{y - 1} + \dfrac{2}{1 - y}$

$\qquad = \dfrac{y}{y - 1} + \dfrac{-1}{-1} \cdot \dfrac{2}{1 - y}$

$\qquad = \dfrac{y}{y - 1} + \dfrac{-2}{y - 1}$

$\qquad = \dfrac{y - 2}{y - 1}$

30. 1

31. $\dfrac{x}{2x - 3y} - \dfrac{y}{3y - 2x}$

$\qquad = \dfrac{x}{2x - 3y} - \dfrac{-1}{-1} \cdot \dfrac{y}{3y - 2x}$

$\qquad = \dfrac{x}{2x - 3y} - \dfrac{-y}{2x - 3y}$

$\qquad = \dfrac{x + y}{2x - 3y} \qquad [x - (-y) = x + y]$

32. $\dfrac{5a}{3a - 2b}$

33. $\dfrac{3}{x + 2} + \dfrac{2}{x^2 - 4}$

$\qquad = \dfrac{3}{x + 2} + \dfrac{2}{(x + 2)(x - 2)}$, $\quad$ LCD is $(x + 2)(x - 2)$

$\qquad = \dfrac{3}{x + 2} \cdot \dfrac{x - 2}{x - 2} + \dfrac{2}{(x + 2)(x - 2)}$

$\qquad = \dfrac{3x - 6}{(x + 2)(x - 2)} + \dfrac{2}{(x + 2)(x - 2)}$

$\qquad = \dfrac{3x - 4}{(x + 2)(x - 2)}$

34. $\dfrac{5a + 13}{(a + 3)(a - 3)}$

35. $\dfrac{y}{y^2 - y - 20} + \dfrac{2}{y + 4}$

   $= \dfrac{y}{(y + 4)(y - 5)} + \dfrac{2}{y + 4}$,   LCD is $(y + 4)(y - 5)$

   $= \dfrac{y}{(y + 4)(y - 5)} + \dfrac{2}{y + 4} \cdot \dfrac{y - 5}{y - 5}$

   $= \dfrac{y}{(y + 4)(y - 5)} + \dfrac{2y - 10}{(y + 4)(y - 5)}$

   $= \dfrac{3y - 10}{(y + 4)(y - 5)}$

36. $\dfrac{-5y - 9}{(y + 3)^2}$

37. $\dfrac{3}{x + y} + \dfrac{x - 5y}{x^2 - y^2}$

   $= \dfrac{3}{x + y} + \dfrac{x - 5y}{(x + y)(x - y)}$,   LCD is $(x + y)(x - y)$

   $= \dfrac{3}{x + y} \cdot \dfrac{x - y}{x - y} + \dfrac{x - 5y}{(x + y)(x - y)}$

   $= \dfrac{3x - 3y}{(x + y)(x - y)} + \dfrac{x - 5y}{(x + y)(x - y)}$

   $= \dfrac{4x - 8y}{(x + y)(x - y)}$

38. $\dfrac{2a}{(a + 1)(a - 1)}$

39. $\dfrac{9x + 2}{3x^2 - 2x - 8} + \dfrac{7}{3x^2 + x - 4}$

   $= \dfrac{9x + 2}{(3x + 4)(x - 2)} + \dfrac{7}{(3x + 4)(x - 1)}$,

   LCD is $(3x + 4)(x - 2)(x - 1)$

   $= \dfrac{9x + 2}{(3x + 4)(x - 2)} \cdot \dfrac{x - 1}{x - 1} + \dfrac{7}{(3x + 4)(x - 1)} \cdot \dfrac{x - 2}{x - 2}$

   $= \dfrac{9x^2 - 7x - 2}{(3x + 4)(x - 2)(x - 1)} + \dfrac{7x - 14}{(3x + 4)(x - 1)(x - 2)}$

   $= \dfrac{9x^2 - 16}{(3x + 4)(x - 2)(x - 1)}$

   $= \dfrac{\cancel{(3x + 4)}(3x - 4)}{\cancel{(3x + 4)}(x - 2)(x - 1)}$

   $= \dfrac{3x - 4}{(x - 2)(x - 1)}$

40. $\dfrac{y}{(y - 2)(y - 3)}$

41. $\dfrac{5a}{a - b} + \dfrac{ab}{a^2 - b^2} + \dfrac{4b}{a + b}$

   $= \dfrac{5a}{a - b} + \dfrac{ab}{(a + b)(a - b)} + \dfrac{4b}{a + b}$,

   LCD is $(a + b)(a - b)$

   $= \dfrac{5a}{a - b} \cdot \dfrac{a + b}{a + b} + \dfrac{ab}{(a + b)(a - b)} + \dfrac{4b}{a + b} \cdot \dfrac{a - b}{a - b}$

   $= \dfrac{5a^2 + 5ab}{(a + b)(a - b)} + \dfrac{ab}{(a + b)(a - b)} + \dfrac{4ab - 4b^2}{(a + b)(a - b)}$

   $= \dfrac{5a^2 + 10ab - 4b^2}{(a + b)(a - b)}$

42. $\dfrac{6a^2 + 9ab + 3b^2 + 5}{(a + b)(a - b)}$

43. $\dfrac{7}{x + 2} - \dfrac{x + 8}{4 - x^2} + \dfrac{3x - 2}{4 - 4x + x^2}$

   $= \dfrac{7}{x + 2} - \dfrac{x + 8}{(2 + x)(2 - x)} + \dfrac{3x - 2}{(2 - x)^2}$

   LCD is $(2 + x)(2 - x)^2$

   $= \dfrac{7}{2 + x} \cdot \dfrac{(2 - x)^2}{(2 - x)^2} - \dfrac{x + 8}{(2 + x)(2 - x)} \cdot \dfrac{2 - x}{2 - x} +$

   $\qquad\qquad \dfrac{3x - 2}{(2 - x)^2} \cdot \dfrac{2 + x}{2 + x}$

   $= \dfrac{28 - 28x + 7x^2 - (16 - 6x - x^2) + 3x^2 + 4x - 4}{(2 + x)(2 - x)^2}$

   $= \dfrac{28 - 28x + 7x^2 - 16 + 6x + x^2 + 3x^2 + 4x - 4}{(2 + x)(2 - x)^2}$

   $= \dfrac{11x^2 - 18x + 8}{(2 + x)(2 - x)^2}$

44. $\dfrac{33 - 32x + 9x^2}{(3 + x)(3 - x)^2}$

45. $\dfrac{1}{x + 1} + \dfrac{x}{2 - x} + \dfrac{x^2 + 2}{x^2 - x - 2}$

   $= \dfrac{1}{x + 1} + \dfrac{x}{2 - x} + \dfrac{x^2 + 2}{(x + 1)(x - 2)}$

   $= \dfrac{1}{x + 1} + \dfrac{-1}{-1} \cdot \dfrac{x}{2 - x} + \dfrac{x^2 + 2}{(x + 1)(x - 2)}$

   $= \dfrac{1}{x + 1} + \dfrac{-x}{x - 2} + \dfrac{x^2 + 2}{(x + 1)(x - 2)}$,

   LCD is $(x + 1)(x - 2)$

   $= \dfrac{1}{x + 1} \cdot \dfrac{x - 2}{x - 2} + \dfrac{-x}{x - 2} \cdot \dfrac{x + 1}{x + 1} + \dfrac{x^2 + 2}{(x + 1)(x - 2)}$

   $= \dfrac{x - 2}{(x + 1)(x - 2)} + \dfrac{-x^2 - x}{(x + 1)(x - 2)} + \dfrac{x^2 + 2}{(x + 1)(x - 2)}$

   $= \dfrac{x - 2 - x^2 - x + x^2 + 2}{(x + 1)(x - 2)}$

   $= \dfrac{0}{(x + 1)(x - 2)}$

   $= 0$

46. $\dfrac{3}{2 + x}$

47. $\dfrac{\dfrac{x^2 - y^2}{xy}}{\dfrac{x - y}{y}} = \dfrac{x^2 - y^2}{xy} \cdot \dfrac{y}{x - y}$

   $= \dfrac{(x + y)\cancel{(x - y)}\cancel{y}}{x\cancel{y}\cancel{(x - y)}}$

   $= \dfrac{x + y}{x}$

48. $\dfrac{a}{a + b}$

49. $\dfrac{a - a^{-1}}{a + a^{-1}} = \dfrac{a - \dfrac{1}{a}}{a + \dfrac{1}{a}} = \dfrac{a \cdot \dfrac{a}{a} - \dfrac{1}{a}}{a \cdot \dfrac{a}{a} + \dfrac{1}{a}}$

$$= \dfrac{\dfrac{a^2 - 1}{a}}{\dfrac{a^2 + 1}{a}}$$

$$= \dfrac{a^2 - 1}{a} \cdot \dfrac{a}{a^2 + 1}$$

$$= \dfrac{a^2 - 1}{a^2 + 1}$$

50. $\dfrac{a^2(b - 1)}{b^2(a - 1)}$

51. $\dfrac{c + \dfrac{8}{c^2}}{1 + \dfrac{2}{c}} = \dfrac{c \cdot \dfrac{c^2}{c^2} + \dfrac{8}{c^2}}{1 \cdot \dfrac{c}{c} + \dfrac{2}{c}}$

$$= \dfrac{\dfrac{c^3 + 8}{c^2}}{\dfrac{c + 2}{c}}$$

$$= \dfrac{c^3 + 8}{c^2} \cdot \dfrac{c}{c + 2}$$

$$= \dfrac{(c + 2)(c^2 - 2c + 4)c}{c \cdot c(c + 2)}$$

$$= \dfrac{c^2 - 2c + 4}{c}$$

52. $\dfrac{x^2y^2}{y^2 - yx + x^2}$

53. $\dfrac{x^2 + xy + y^2}{\dfrac{x^2}{y} - \dfrac{y^2}{x}} = \dfrac{x^2 + xy + y^2}{\dfrac{x^2}{y} \cdot \dfrac{x}{x} - \dfrac{y^2}{x} \cdot \dfrac{y}{y}}$

$$= \dfrac{x^2 + xy + y^2}{\dfrac{x^3 - y^3}{xy}}$$

$$= (x^2 + xy + y^2) \cdot \dfrac{xy}{x^3 - y^3}$$

$$= \dfrac{(x^2 + xy + y^2)xy}{(x - y)(x^2 + xy + y^2)}$$

$$= \dfrac{xy}{x - y}$$

54. $\dfrac{a + b}{ab}$

55. $\dfrac{\dfrac{x}{y} - \dfrac{y}{x}}{\dfrac{1}{y} + \dfrac{1}{x}} = \dfrac{\dfrac{x}{y} - \dfrac{y}{x}}{\dfrac{1}{y} + \dfrac{1}{x}} \cdot \dfrac{xy}{xy}$, LCM is xy

$$= \dfrac{\left(\dfrac{x}{y} - \dfrac{y}{x}\right)(xy)}{\left(\dfrac{1}{y} + \dfrac{1}{x}\right)(xy)}$$

$$= \dfrac{x^2 - y^2}{x + y}$$

$$= \dfrac{(x + y)(x - y)}{(x + y) \cdot 1}$$

$$= x - y$$

56. $-a - b$

57. $\dfrac{x^2y^{-2} - y^2x^{-2}}{xy^{-1} + yx^{-1}} = \dfrac{\dfrac{x^2}{y^2} - \dfrac{y^2}{x^2}}{\dfrac{x}{y} + \dfrac{y}{x}}$, LCM is x^2y^2

$$= \dfrac{\dfrac{x^2}{y^2} - \dfrac{y^2}{x^2}}{\dfrac{x}{y} + \dfrac{y}{x}} \cdot \dfrac{x^2y^2}{x^2y^2}$$

$$= \dfrac{x^4 - y^4}{x^3y + xy^3}$$

$$= \dfrac{(x^2 + y^2)(x + y)(x - y)}{xy(x^2 + y^2)}$$

$$= \dfrac{(x + y)(x - y)}{xy}$$

58. $\dfrac{a^2 + b^2}{ab}$

59. $\dfrac{\dfrac{a}{1 - a} + \dfrac{1 + a}{a}}{\dfrac{1 - a}{a} + \dfrac{a}{1 + a}} = \dfrac{\dfrac{a}{1 - a} \cdot \dfrac{a}{a} + \dfrac{1 + a}{a} \cdot \dfrac{1 - a}{1 - a}}{\dfrac{1 - a}{a} \cdot \dfrac{1 + a}{1 + a} + \dfrac{a}{1 + a} \cdot \dfrac{a}{a}}$

$$= \dfrac{\dfrac{a^2 + (1 - a^2)}{a(1 - a)}}{\dfrac{(1 - a^2) + a^2}{a(1 + a)}}$$

$$= \dfrac{1}{a(1 - a)} \cdot \dfrac{a(1 + a)}{1}$$

$$= \dfrac{1 + a}{1 - a}$$

60. $\dfrac{1 - x}{1 + x}$

61. $\dfrac{\dfrac{1}{a^2} + \dfrac{2}{ab} + \dfrac{1}{b^2}}{\dfrac{1}{a^2} - \dfrac{1}{b^2}} = \dfrac{\dfrac{1}{a^2} + \dfrac{2}{ab} + \dfrac{1}{b^2}}{\dfrac{1}{a^2} - \dfrac{1}{b^2}} \cdot \dfrac{a^2b^2}{a^2b^2}$, LCM is a^2b^2

$$= \dfrac{b^2 + 2ab + a^2}{b^2 - a^2}$$

$$= \dfrac{(b + a)(b + a)}{(b + a)(b - a)}$$

$$= \dfrac{b + a}{b - a}$$

62. $\dfrac{y + x}{y - x}$

63. $\dfrac{(x + h)^2 - x^2}{h} = \dfrac{x^2 + 2xh + h^2 - x^2}{h}$

$$= \dfrac{2xh + h^2}{h}$$

$$= \dfrac{h(2x + h)}{h \cdot 1}$$

$$= 2x + h$$

64. $\dfrac{-1}{x(x + h)}$

65. $\dfrac{(x+h)^3 - x^3}{h} = \dfrac{x^3 + 3x^2h + 3xh^2 + h^3 - x^3}{h}$

$= \dfrac{3x^2h + 3xh^2 + h^3}{h}$

$= \dfrac{\cancel{h}(3x^2 + 3xh + h^2)}{\cancel{h} \cdot 1}$

$= 3x^2 + 3xh + h^2$

66. $\dfrac{-2x - h}{x^2(x+h)^2}$

67. $\left[\dfrac{\frac{x+1}{x-1} + 1}{\frac{x+1}{x-1} - 1}\right]^5 = \left[\dfrac{\frac{(x+1)+(x-1)}{x-1}}{\frac{(x+1)-(x-1)}{x-1}}\right]^5$

$= \left[\dfrac{2x}{x-1} \cdot \dfrac{x-1}{2}\right]^5$

$= \left[\dfrac{2x\cancel{(x-1)}}{1 \cdot 2\cancel{(x-1)}}\right]^5$

$= x^5$

68. $\dfrac{5x+3}{3x+2}$

69. $\dfrac{a}{b} \div \left(\dfrac{a}{3} + \dfrac{b}{4}\right)$

$= \dfrac{a}{b} \cdot \left(\dfrac{3}{a} + \dfrac{4}{b}\right)$ (1) Step (1) uses the wrong reciprocal. The reciprocal of a sum is not the sum of the reciprocals.

$= \dfrac{a}{b} \cdot \left(\dfrac{3b + 4a}{ab}\right)$ (2) Step (2) would be correct if step (1) had been correct.

$= \dfrac{a(3b + 4a)}{ab^2}$ (3) Step (3) would be correct if steps (1) and (2) had been correct.

$= \dfrac{4a + 3b}{b}$ (4) Step (4) has an improper simplification of the b in the denominator.

$\dfrac{a}{b} \div \left(\dfrac{a}{3} + \dfrac{b}{4}\right) = \dfrac{a}{b} \div \left(\dfrac{4a + 3b}{12}\right)$

$= \dfrac{a}{b} \cdot \dfrac{12}{4a + 3b}$

$= \dfrac{12a}{b(4a + 3b)}$ (Correct answer)

70. In step (1) the $\dfrac{5x^2}{4xy}$ should be $\dfrac{10x^2}{4xy}$.

Step (2) would be correct if step (1) had been.

In going from step (2) to step (3) the quantity x is <u>not</u> a factor of <u>both</u> terms in the numerator of (2).

In going from step (3) to step (4) the quantity y is not a factor of <u>both</u> terms in the numerator of (3).

The correct answer is $\dfrac{10x^2 + 3y^2}{4xy}$.

71. $\dfrac{n(n+1)(n+2)}{2 \cdot 3} + \dfrac{(n+1)(n+2)}{2}$

$= \dfrac{n(n+1)(n+2)}{2 \cdot 3} + \dfrac{(n+1)(n+2)}{2} \cdot \dfrac{3}{3}$, LCD is $2 \cdot 3$

$= \dfrac{n(n+1)(n+2) + 3(n+1)(n+2)}{2 \cdot 3}$

$= \dfrac{(n+1)(n+2)(n+3)}{2 \cdot 3}$ (Factoring the numerator by grouping)

72. $\dfrac{(n+1)(n+2)(n+3)(n+4)}{2 \cdot 3 \cdot 4}$

Exercise Set 1.7

1. $\sqrt{x-3}$

We substitute -2 for x in x - 3: -2 - 3 = -5. Since the radicand is negative, -2 is not a meaningful replacement.

We substitute 5 for x in x - 3: 5 - 3 = 2. Since the radicand is not negative, 5 is a meaningful replacement.

2. Yes, no

3. $\sqrt{3 - 4x}$

We substitute -1 for x in 3 - 4x: 3 - 4(-1) = 7. Since the radicand is not negative, -1 is a meaningful replacement.

We substitute 1 for x in 3 - 4x: 3 - 4·1 = -1. Since the radicand is negative, 1 is not a meaningful replacement.

4. Yes, yes

5. $\sqrt{1 - x^2}$

We substitute 1 for x in 1 - x²: 1 - 1² = 0. Since the radicand is not negative, 1 is a meaningful replacement.

We substitute 3 for x in 1 - x²: 1 - 3² = -8. Since the radicand is negative, 3 is not a meaningful replacement.

6. Yes, yes

7. $\sqrt[3]{2x + 7}$

Since every real number, positive, negative, or zero, has a cube root, -4 and 5 are both meaningful replacements.

8. No, no

9. $\sqrt{(-11)^2} = |-11| = 11$

10. 1

11. $\sqrt{16x^2} = \sqrt{(4x)^2} = |4x| = 4|x|$

12. $6|t|$

13. $\sqrt{(b+1)^2} = |b+1|$

14. $|2c - 3|$

15. $\sqrt[3]{-27x^3} = \sqrt[3]{(-3x)^3} = -3x$

16. $-2y$

17. $\sqrt{x^2 - 4x + 4} = \sqrt{(x-2)^2} = |x - 2|$

18. $|y + 8|$

19. $\sqrt[5]{32} = \sqrt[5]{2^5} = 2$

20. -2

21. $\sqrt{180} = \sqrt{36 \cdot 5} = \sqrt{36} \cdot \sqrt{5} = 6\sqrt{5}$

22. $4\sqrt{3}$

23. $\sqrt[3]{54} = \sqrt[3]{27 \cdot 2} = \sqrt[3]{27} \cdot \sqrt[3]{2} = 3\sqrt[3]{2}$

24. $3\sqrt[3]{5}$

25. $\sqrt{128c^2d^4} = \sqrt{64c^2d^4 \cdot 2} = |8cd^2|\sqrt{2} = 8|c|d^2\sqrt{2}$

26. $9c^2|d^3|\sqrt{2}$

27. $\sqrt{3}\sqrt{6} = \sqrt{18} = \sqrt{9 \cdot 2} = 3\sqrt{2}$

28. $4\sqrt{3}$

29. $\sqrt{2x^3y}\sqrt{12xy} = \sqrt{24x^4y^2} = \sqrt{4x^4y^2 \cdot 6} = 2x^2y\sqrt{6}$

30. $2y^2z\sqrt{15}$

31. $\sqrt[3]{3x^2y}\,\sqrt[3]{36x} = \sqrt[3]{108x^3y} = \sqrt[3]{27x^3 \cdot 4y} = 3x\sqrt[3]{4y}$

32. $2xy\sqrt[5]{x^2}$

33. $\sqrt[3]{2(x+4)}\,\sqrt[3]{4(x+4)^4} = \sqrt[3]{8(x+4)^5}$
$$= \sqrt[3]{8(x+4)^3 \cdot (x+4)^2}$$
$$= 2(x+4)\sqrt[3]{(x+4)^2}$$

34. $2(x+1)\sqrt[3]{9(x+1)}$

35. $\dfrac{\sqrt{21ab^2}}{\sqrt{3ab}} = \sqrt{\dfrac{21ab^2}{3ab}} = \sqrt{7b}$

36. $\dfrac{2\sqrt{2ab}}{a}$

37. $\dfrac{\sqrt[3]{40m}}{\sqrt[3]{5m}} = \sqrt[3]{\dfrac{40m}{5m}} = \sqrt[3]{8} = 2$

38. $\sqrt{5y}$

39. $\dfrac{\sqrt[3]{3x^2}}{\sqrt[3]{24x^5}} = \sqrt[3]{\dfrac{3x^2}{24x^5}} = \sqrt[3]{\dfrac{1}{8x^3}} = \dfrac{1}{2x}$

40. $y\sqrt[3]{5}$

41. $\dfrac{\sqrt{a^2-b^2}}{\sqrt{a-b}} = \sqrt{\dfrac{a^2-b^2}{a-b}} = \sqrt{\dfrac{(a+b)(a-b)}{a-b}}$
$$= \sqrt{a+b}$$

42. $\sqrt{x^2 + xy + y^2}$

43. $\sqrt{\dfrac{9a^2}{8b}} = \sqrt{\dfrac{9a^2}{8b} \cdot \dfrac{2b}{2b}} = \sqrt{\dfrac{9a^2}{16b^2} \cdot 2b} = \dfrac{3a\sqrt{2b}}{4b}$

44. $\dfrac{b\sqrt{15a}}{6a}$

45. $\sqrt[3]{\dfrac{2x^2 2y^3}{25z^4}} = \sqrt[3]{\dfrac{2x^2 2y^3}{25z^4} \cdot \dfrac{5z^2}{5z^2}} = \sqrt[3]{\dfrac{y^3}{125z^6} \cdot 20x^2z^2}$
$$= \dfrac{y\sqrt[3]{20x^2z^2}}{5z^2}$$

46. $\dfrac{2x}{y}$

47. $\dfrac{\left(\sqrt[3]{32x^4y}\right)^2}{\left(\sqrt[3]{xy}\right)^2} = \left(\sqrt[3]{\dfrac{32x^4y}{xy}}\right)^2 = \left(\sqrt[3]{32x^3}\right)^2$
$$= \left(\sqrt[3]{2^5x^3}\right)^2$$
$$= \sqrt[3]{(2^5x^3)^2}$$
$$= \sqrt[3]{2^{10}x^6}$$
$$= \sqrt[3]{2^9x^6 \cdot 2}$$
$$= 2^3x^2\sqrt[3]{2}$$
$$= 8x^2\sqrt[3]{2}$$

48. $4\sqrt[3]{4x^2}$

49. $\dfrac{3\sqrt{a^2b^2}\,\sqrt{4xy}}{2\sqrt{a^{-1}b^{-2}}\,\sqrt{9x^{-3}y^{-1}}} = \dfrac{3}{2}\sqrt{\dfrac{a^2b^2}{a^{-1}b^{-2}}}\sqrt{\dfrac{4xy}{9x^{-3}y^{-1}}}$
$$= \dfrac{3}{2}\sqrt{a^3b^4}\sqrt{\dfrac{4}{9}x^4y^2}$$
$$= \dfrac{3}{2} \cdot ab^2\sqrt{a} \cdot \dfrac{2}{3}x^2y$$
$$= ab^2x^2y\sqrt{a}$$

50. xy^2a^3b

51. $9\sqrt{50} + 6\sqrt{2} = 9\sqrt{25 \cdot 2} + 6\sqrt{2}$

$\qquad = 9 \cdot 5\sqrt{2} + 6\sqrt{2}$

$\qquad = 45\sqrt{2} + 6\sqrt{2}$

$\qquad = (45 + 6)\sqrt{2}$

$\qquad = 51\sqrt{2}$

52. $29\sqrt{3}$

53. $\quad 8\sqrt{2} - 6\sqrt{20} - 5\sqrt{8}$

$\quad = 8\sqrt{2} - 6\sqrt{4 \cdot 5} - 5\sqrt{4 \cdot 2}$

$\quad = 8\sqrt{2} - 6 \cdot 2\sqrt{5} - 5 \cdot 2\sqrt{2}$

$\quad = 8\sqrt{2} - 12\sqrt{5} - 10\sqrt{2}$

$\quad = -2\sqrt{2} - 12\sqrt{5}$

54. $4\sqrt{3}$

55. $\quad 2\sqrt[3]{8x^2} + 5\sqrt[3]{27x^2} - 3\sqrt[3]{x^3}$

$\quad = 2 \cdot 2\sqrt[3]{x^2} + 5 \cdot 3\sqrt[3]{x^2} - 3x$

$\quad = 4\sqrt[3]{x^2} + 15\sqrt[3]{x^2} - 3x$

$\quad = 19\sqrt[3]{x^2} - 3x$

56. $(5a^2 - 3b^2)\sqrt{a + b}$

57. $\quad 3\sqrt{3y^2} - \dfrac{y\sqrt{48}}{\sqrt{2}} + \sqrt{\dfrac{12}{4y^{-2}}}$

$\quad = 3\sqrt{3y^2} - y\sqrt{24} + \sqrt{3y^2}$

$\quad = 3y\sqrt{3} - 2y\sqrt{6} + y\sqrt{3}$

$\quad = 4y\sqrt{3} - 2y\sqrt{6}$

58. $(3x - 2)\sqrt[3]{x^2}$

59. $\quad (\sqrt{3} - \sqrt{2})(\sqrt{3} + \sqrt{2})$

$\quad = (\sqrt{3})^2 - (\sqrt{2})^2$

$\quad = 3 - 2$

$\quad = 1$

60. -12

61. $(1 + \sqrt{3})^2 = 1^2 + 2 \cdot 1 \cdot \sqrt{3} + (\sqrt{3})^2$

$\qquad = 1 + 2\sqrt{3} + 3$

$\qquad = 4 + 2\sqrt{3}$

62. $27 - 10\sqrt{2}$

63. $\quad (\sqrt{t} - x)^2$

$\quad = (\sqrt{t})^2 - 2 \cdot \sqrt{t} \cdot x + x^2$

$\quad = t - 2x\sqrt{t} + x^2$

64. $a + 2 + \dfrac{1}{a}$, or $a + 2 + a^{-1}$, or $\dfrac{(a + 1)^2}{a}$

65. $\quad 5\sqrt{7} + \dfrac{35}{\sqrt{7}}$

$\quad = 5\sqrt{7} + \dfrac{35}{\sqrt{7}} \cdot \dfrac{\sqrt{7}}{\sqrt{7}}$

$\quad = 5\sqrt{7} + \dfrac{35\sqrt{7}}{7}$

$\quad = 5\sqrt{7} + 5\sqrt{7}$

$\quad = 10\sqrt{7}$

66. $2a^2b + 5a\sqrt{by} - 3y$

67. $\quad (\sqrt{x + 3} - \sqrt{3})(\sqrt{x + 3} + \sqrt{3})$

$\quad = (\sqrt{x + 3})^2 - (\sqrt{3})^2$

$\quad = (x + 3) - 3$

$\quad = x$

68. h

69. We substitute 90 for L in the formula.

$\quad r = 2\sqrt{5L}$

$\qquad = 2\sqrt{5 \cdot 90}$

$\qquad = 2\sqrt{450}$

$\qquad \approx 42.43 \qquad$ Using a calculator

The speed of the car was about 42.43 mph.

70. 46.90 mph

71. We use the Pythagorean theorem to find b, the airplane's horizontal distance from the airport. We have a = 3700 and c = 14,200.

$\qquad c^2 = a^2 + b^2$

$\quad 14,200^2 = 3700^2 + b^2$

$201,640,000 = 13,690,000 + b^2$

$187,950,000 = b^2$

$\quad 13,709.5 \approx b \qquad$ Using a calculator

The airplane is about 13,709.5 ft horizontally from the airport.

72. 102.8 ft

73. a) $h^2 + \left(\dfrac{a}{2}\right)^2 = a^2$ (Pythagorean theorem)

$$h^2 + \dfrac{a^2}{4} = a^2$$

$$h^2 = \dfrac{3a^2}{4}$$

$$h = \sqrt{\dfrac{3a^2}{4}}$$

$$h = \dfrac{a}{2}\sqrt{3}$$

b) Using the result of part (a) we have

$$A = \dfrac{1}{2} \cdot base \cdot height$$

$$A = \dfrac{1}{2}a \cdot \dfrac{a}{2}\sqrt{3} \qquad \left(\dfrac{a}{2} + \dfrac{a}{2} = a\right)$$

$$A = \dfrac{a^2}{4}\sqrt{3}$$

74. $c = s\sqrt{2}$

75.

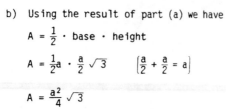

$x^2 + x^2 = (8\sqrt{2})^2$ (Pythagorean theorem)

$$2x^2 = 128$$
$$x^2 = 64$$
$$x = 8$$

76. 50 ft²

77. $\dfrac{6}{3 + \sqrt{5}} = \dfrac{6}{3 + \sqrt{5}} \cdot \dfrac{3 - \sqrt{5}}{3 - \sqrt{5}}$

$$= \dfrac{6\left(3 - \sqrt{5}\right)}{9 - 5}$$

$$= \dfrac{6\left(3 - \sqrt{5}\right)}{4}$$

$$= \dfrac{3\left(3 - \sqrt{5}\right)}{2}$$

78. $\sqrt{3} + 1$

79. $\sqrt[3]{\dfrac{16}{9}} = \sqrt[3]{\dfrac{16}{9} \cdot \dfrac{3}{3}} = \sqrt[3]{\dfrac{48}{27}} = \sqrt[3]{\dfrac{8}{27} \cdot 6} = \dfrac{2\sqrt[3]{6}}{3}$

80. $\dfrac{\sqrt[3]{4}}{2}$

81. $\dfrac{4\sqrt{x} - 3\sqrt{xy}}{2\sqrt{x} + 5\sqrt{y}} = \dfrac{4\sqrt{x} - 3\sqrt{xy}}{2\sqrt{x} + 5\sqrt{y}} \cdot \dfrac{2\sqrt{x} - 5\sqrt{y}}{2\sqrt{x} - 5\sqrt{y}}$

$$= \dfrac{8x - 20\sqrt{xy} - 6x\sqrt{y} + 15y\sqrt{x}}{4x - 25y}$$

82. $\dfrac{15x + 10\sqrt{xy} + 6x\sqrt{y} + 4y\sqrt{x}}{9x - 4y}$

83. $\dfrac{8\sqrt{2} + 5\sqrt{3}}{5\sqrt{3} - 7\sqrt{2}} = \dfrac{8\sqrt{2} + 5\sqrt{3}}{5\sqrt{3} - 7\sqrt{2}} \cdot \dfrac{5\sqrt{3} + 7\sqrt{2}}{5\sqrt{3} + 7\sqrt{2}}$

$$= \dfrac{40\sqrt{6} + 56 \cdot 2 + 25 \cdot 3 + 35\sqrt{6}}{25 \cdot 3 - 49 \cdot 2}$$

$$= \dfrac{40\sqrt{6} + 112 + 75 + 35\sqrt{6}}{75 - 98}$$

$$= \dfrac{187 + 75\sqrt{6}}{-23}$$

84. $\dfrac{-66 + 43\sqrt{2}}{-94}$, or $\dfrac{66 - 43\sqrt{2}}{94}$

85. $\dfrac{p - q\sqrt{s}}{q + \sqrt{s}} = \dfrac{p - q\sqrt{s}}{q + \sqrt{s}} \cdot \dfrac{q - \sqrt{s}}{q - \sqrt{s}}$

$$= \dfrac{pq - p\sqrt{s} - q^2\sqrt{s} + qs}{q^2 - s}$$

86. $\dfrac{c^2\sqrt{t} + cdt + cd + d^2\sqrt{t}}{c^2 - d^2 t}$

87. $\dfrac{\sqrt{2} + \sqrt{5a}}{6} = \dfrac{\sqrt{2} + \sqrt{5a}}{6} \cdot \dfrac{\sqrt{2} - \sqrt{5a}}{\sqrt{2} - \sqrt{5a}}$

$$= \dfrac{2 - 5a}{6(\sqrt{2} - \sqrt{5a})}$$

88. $\dfrac{3 - 5y}{4(\sqrt{3} - \sqrt{5y})}$

89. $\dfrac{\sqrt{x + 1} + 1}{\sqrt{x + 1} - 1} = \dfrac{\sqrt{x + 1} + 1}{\sqrt{x + 1} - 1} \cdot \dfrac{\sqrt{x + 1} - 1}{\sqrt{x + 1} - 1}$

$$= \dfrac{(x + 1) - 1}{(x + 1) - 2\sqrt{x + 1} + 1}$$

$$= \dfrac{x}{x - 2\sqrt{x + 1} + 2}$$

90. $\dfrac{x}{x + 8 + 4\sqrt{x + 4}}$

91. $\dfrac{\sqrt{a + 3} - \sqrt{3}}{3} = \dfrac{\sqrt{a + 3} - \sqrt{3}}{3} \cdot \dfrac{\sqrt{a + 3} + \sqrt{3}}{\sqrt{a + 3} + \sqrt{3}}$

$$= \dfrac{(a + 3) - 3}{3(\sqrt{a + 3} + \sqrt{3})}$$

$$= \dfrac{a}{3(\sqrt{a + 3} + \sqrt{3})}$$

92. $\dfrac{1}{\sqrt{a + h} + \sqrt{a}}$

93. $\dfrac{\sqrt{x+h} - \sqrt{x}}{h} = \dfrac{\sqrt{x+h} - \sqrt{x}}{h} \cdot \dfrac{\sqrt{x+h} + \sqrt{x}}{\sqrt{x+h} + \sqrt{x}}$

$= \dfrac{x + h - x}{h(\sqrt{x+h} + \sqrt{x})}$

$= \dfrac{h \cdot 1}{h(\sqrt{x+h} + \sqrt{x})}$

$= \dfrac{1}{\sqrt{x+h} + \sqrt{x}}$

94. $\dfrac{1}{\sqrt{p+7} + \sqrt{p}}$

95. $\dfrac{\sqrt{x} + \sqrt{y}}{\sqrt{x} - \sqrt{y}} = \dfrac{\sqrt{x} + \sqrt{y}}{\sqrt{x} - \sqrt{y}} \cdot \dfrac{\sqrt{x} - \sqrt{y}}{\sqrt{x} - \sqrt{y}}$

$= \dfrac{x - y}{x - \sqrt{xy} - \sqrt{xy} + y}$

$= \dfrac{x - y}{x - 2\sqrt{xy} + y}$

96. $\dfrac{a - b}{a + 2\sqrt{ab} + b}$

97. $\sqrt{1 + x^2} + \dfrac{1}{\sqrt{1 + x^2}}$

$= \sqrt{1 + x^2} \cdot \dfrac{1 + x^2}{1 + x^2} + \dfrac{1}{\sqrt{1 + x^2}} \quad \dfrac{\sqrt{1 + x^2}}{\sqrt{1 + x^2}}$

$= \dfrac{(1 + x^2)\sqrt{1 + x^2}}{1 + x^2} + \dfrac{\sqrt{1 + x^2}}{1 + x^2}$

$= \dfrac{(2 + x^2)\left(\sqrt{1 + x^2}\right)}{1 + x^2}$

98. $\dfrac{(2 - 3x^2)\sqrt{1 - x^2}}{2(1 - x^2)}$

99. Let $a = 16$ and $b = 9$.

Then $\sqrt{a + b} = \sqrt{16 + 9} = \sqrt{25} = 5$
and $\sqrt{a} + \sqrt{b} = \sqrt{16} + \sqrt{9} = 4 + 3 = 7$.

100. $5 + 2\sqrt{6}$

101. We substitute 0.1 for h, 0.03 for d, and 0.6 for v_0 in the formula.

$w = d\sqrt{\dfrac{v_0}{\sqrt{v_0{}^2 + 19.6h}}}$

$= 0.03\sqrt{\dfrac{0.6}{\sqrt{(0.6)^2 + 19.6(0.1)}}}$

$= 0.03\sqrt{\dfrac{0.6}{\sqrt{0.36 + 1.96}}}$

$= 0.03\sqrt{\dfrac{0.6}{\sqrt{2.32}}}$

$\approx 0.0188 \qquad$ (Using a calculator)

The width of the stream is about 0.0188 m.

102. $51 - 10\sqrt{26}$

103. a) We substitute 6.672×10^{-11} for G, 5.97×10^{24} for M, 6.37×10^6 for R in the formula.

$V_0 = \sqrt{\dfrac{2GM}{R}}$

$= \sqrt{\dfrac{2 \times 6.672 \times 10^{-11} \times 5.97 \times 10^{24}}{6.37 \times 10^6}}$

$\approx 11{,}183 \qquad$ (Using a calculator)

The escape velocity is about 11,183 m/sec.

b) We substitute 6.672×10^{-11} for G, 6.27×10^{23} for M, and 3.3917×10^6 for R in the formula.

$V_0 = \sqrt{\dfrac{2GM}{R}}$

$= \sqrt{\dfrac{2 \times 6.672 \times 10^{-11} \times 6.27 \times 10^{23}}{3.3917 \times 10^6}}$

$\approx 4967 \qquad$ (Using a calculator)

The escape velocity is about 4967 m/sec.

Exercise Set 1.8

---

1. $x^{3/4} = \sqrt[4]{x^3}$

2. $\sqrt[5]{y^2}$

3. $16^{3/4} = (16^{1/4})^3 = \left(\sqrt[4]{16}\right)^3 = 2^3 = 8$

4. 128

5. $125^{-1/3} = \dfrac{1}{125^{1/3}} = \dfrac{1}{\sqrt[3]{125}} = \dfrac{1}{5}$

6. $\dfrac{1}{16}$

7. $a^{5/4}\, b^{-3/4} = \dfrac{a^{5/4}}{b^{3/4}} = \dfrac{\sqrt[4]{a^5}}{\sqrt[4]{b^3}} = \dfrac{a\sqrt[4]{a}}{\sqrt[4]{b^3}}$, or $\dfrac{a\sqrt[4]{ab}}{b}$

8. $\dfrac{\sqrt[5]{x^2y^4}}{y}$

9. $\sqrt[3]{20^2} = 20^{2/3}$

10. $17^{3/5}$

11. $\left(\sqrt[4]{13}\right)^5 = \sqrt[4]{13^5} = 13^{5/4}$, or $\sqrt[4]{13^5} = 13\sqrt[4]{13}$

12. $12^{4/5}$

13. $\sqrt[3]{\sqrt{11}} = \left(\sqrt{11}\right)^{1/3} = (11^{1/2})^{1/3} = 11^{1/6}$

14. $7^{1/12}$

15. $\sqrt{5}\,\sqrt[3]{5} = 5^{1/2} \cdot 5^{1/3} = 5^{1/2 + 1/3} = 5^{5/6}$

16. $2^{5/6}$

17. $\sqrt[5]{32^2} = 32^{2/5} = (32^{1/5})^2 = 2^2 = 4$

18. $\dfrac{1}{16}$

19. $\sqrt[3]{8y^6} = (8y^6)^{1/3} = (2^3y^6)^{1/3} = 2^{3/3}y^{6/3} = 2y^2$

20. $2c^2d^3$

21. $\sqrt[3]{a^2 + b^2} = (a^2 + b^2)^{1/3}$

22. $(a^3 - b^3)^{1/4}$

23. $\sqrt[3]{27a^3b^9} = (3^3a^3b^9)^{1/3} = 3^{3/3}a^{3/3}b^{9/3} = 3ab^3$

24. $3x^2y^2$

25. $\sqrt[6]{\dfrac{m^{12}n^{24}}{64}} = \left(\dfrac{m^{12}n^{24}}{2^6}\right)^{1/6} = \dfrac{m^{12/6}n^{24/6}}{2^{6/6}} = \dfrac{m^2n^4}{2}$

26. $\dfrac{m^2n^3}{2}$

27. $(2a^{3/2})(4a^{1/2}) = 8a^{3/2 + 1/2} = 8a^2$

28. $24a\sqrt{a}$

29. $\left(\dfrac{x^6}{9b^{-4}}\right)^{-1/2} = \left(\dfrac{x^6}{3^2b^{-4}}\right)^{-1/2} = \dfrac{x^{-3}}{3^{-1}b^2} = \dfrac{3}{x^3b^2}$

30. $\dfrac{2\sqrt[3]{x^2}}{xy}$

31. $\dfrac{x^{2/3}y^{5/6}}{x^{-1/3}y^{1/2}} = x^{2/3 - (-1/3)}y^{5/6 - 1/2} = xy^{1/3} = x\sqrt[3]{y}$

32. $\sqrt[4]{ab}$

33. $\sqrt[3]{6}\,\sqrt{2} = 6^{1/3}2^{1/2} = 6^{2/6}2^{3/6}$
$= (6^2 2^3)^{1/6}$
$= \sqrt[6]{36 \cdot 8}$
$= \sqrt[6]{288}$

34. $2\sqrt[4]{2}$

35. $\sqrt[4]{xy}\,\sqrt[3]{x^2y} = (xy)^{1/4}(x^2y)^{1/3} = (xy)^{3/12}(x^2y)^{4/12}$
$= \left[(xy)^3(x^2y)^4\right]^{1/12}$
$= [x^3y^3x^8y^4]^{1/12}$
$= \sqrt[12]{x^{11}y^7}$

36. $b\sqrt[6]{a^5b}$

37. $\sqrt[3]{a^4\sqrt{a^3}} = \left(a^4\sqrt{a^3}\right)^{1/3} = (a^4a^{3/2})^{1/3}$
$= (a^{11/2})^{1/3}$
$= a^{11/6}$
$= \sqrt[6]{a^{11}}$
$= a\sqrt[6]{a^5}$

38. $a\sqrt[6]{a^5}$

39. $\dfrac{\sqrt{(a+x)^3}\,\sqrt[3]{(a+x)^2}}{\sqrt[4]{a+x}} = \dfrac{(a+x)^{3/2}(a+x)^{2/3}}{(a+x)^{1/4}}$
$= \dfrac{(a+x)^{26/12}}{(a+x)^{3/12}}$
$= (a+x)^{23/12}$
$= \sqrt[12]{(a+x)^{23}}$
$= (a+x)\sqrt[12]{(a+x)^{11}}$

40. $\dfrac{\sqrt[3]{x+y}}{x+y}$

41. $\left(\sqrt[4]{13}\right)^5 = 13^{5/4} = 13^{1.25} \approx 24.685$

42. 8.372

43. $12.3^{3/2} = 12.3^{1.5} \approx 43.138$

44. 2.098

45. $105.6^{3/4} = 105.6^{0.75} \approx 32.942$

46. 11.671

47. $L = \dfrac{0.000169d^{2.27}}{h}$
$= \dfrac{0.000169(180)^{2.27}}{4}$
≈ 5.56 ft

48. 1.46 ft

49. $L = \dfrac{0.000169d^{2.27}}{h}$

$L = \dfrac{0.000169(200)^{2.27}}{4}$

≈ 7.07 ft

50. 17.74 ft

51. $T = 34\,x^{-0.41}$

When x = 1:   When x = 6:

$T = 34(1)^{-0.41}$ $T = 34(6)^{-0.41}$

$= 34$ hr ≈ 16.3 hr

When x = 8:   When x = 10:

$T = 34(8)^{-0.41}$ $T = 34(10)^{-0.41}$

≈ 14.5 hr ≈ 13.2 hr

When x = 32:   When x = 64:

$T = 34(32)^{-0.41}$ $T = 34(64)^{-0.41}$

≈ 8.2 hr ≈ 6.2 hr

52. 25.6 hr, 21.7 hr, 17.6 hr, 13.8 hr, 12.7 hr, 10.2 hr, 5.1 hr

53. $a^{-2}b^5 - a^3b^{-5}$

a) The variables common to both terms involve powers of a and b.

b) a^{-2} has a smaller exponent than a^3.
b^{-5} has a smaller exponent than b^5.

c) $a^{-2}b^{-5}$ factored out of $a^{-2}b^5$ leaves
$a^{-2-(-2)}b^{5-(-5)} = a^0b^{10} = b^{10}$.
$a^{-2}b^{-5}$ factored out of a^3b^{-5} leaves
$a^{3-(-2)}b^{-5-(-5)} = a^5b^0 = a^5$.

d) Thus,
$a^{-2}b^5 - a^3b^{-5} = a^{-2}b^{-5}(b^{10} - a^5) = \dfrac{b^{10} - a^5}{a^2b^5}$.

54. $\dfrac{p^{13} + q^6}{q^2p^5}$

55. $5a^{2/3}b^{-1/2} + 2a^{-1/3}b^{1/2}$

a) The variables common to both terms involve powers of a and b.

b) $a^{-1/3}$ has a smaller exponent than $a^{2/3}$.
$b^{-1/2}$ has a smaller exponent than $b^{1/2}$.

c) $a^{-1/3}b^{-1/2}$ factored out of $5a^{2/3}b^{-1/2}$ leaves
$5a^{2/3-(-1/3)}b^{-1/2-(-1/2)} = 5a^1b^0 = 5a$.
$a^{-1/3}b^{-1/2}$ factored out of $2a^{-1/3}b^{1/2}$ leaves
$2a^{-1/3-(-1/3)}b^{1/2-(-1/2)} = 2a^0b^1 = 2b$.

d) Thus,
$5a^{2/3}b^{-1/2} + 2a^{-1/3}b^{1/2} =$
$a^{-1/3}b^{-1/2}(5a + 2b) = \dfrac{5a + 2b}{a^{1/3}b^{1/2}}$.

56. $\dfrac{p + 2q^4}{p^{1/5}q^2}$

57. $x^{-1/3}y^{3/4} - x^{2/3}y^{-1/4}$

a) The variables common to both terms involve powers of x and y.

b) $x^{-1/3}$ has a smaller exponent than $x^{2/3}$.
$y^{-1/4}$ has a smaller exponent than $y^{3/4}$.

c) $x^{-1/3}y^{-1/4}$ factored out of $x^{-1/3}y^{3/4}$ leaves
$x^{-1/3-(-1/3)}y^{3/4-(-1/4)} = x^0y^1 = y$.
$x^{-1/3}y^{-1/4}$ factored out of $x^{2/3}y^{-1/4}$ leaves
$x^{2/3-(-1/3)}y^{-1/4-(-1/4)} = x^1y^0 = x$.

d) Thus,
$x^{-1/3}y^{3/4} - x^{2/3}y^{-1/4} = x^{-1/3}y^{-1/4}(y - x) = \dfrac{y - x}{x^{1/3}y^{1/4}}$.

58. $\dfrac{2(2a - 3b)}{a^{1/2}b^{3/4}}$

59. $(2x - 3)^{-3}(x + 1)^{5/4} + (2x - 3)^{-2}(x + 1)^{1/4}$
$= (2x - 3)^{-3}(x + 1)^{1/4}[(x + 1) + (2x - 3)]$
                                Factoring
$= (2x - 3)^{-3}(x + 1)^{1/4}(3x - 2)$ Simplifying
$= \dfrac{(x + 1)^{1/4}(3x - 2)}{(2x - 3)^3}$

60. $\dfrac{x(13x + 6)}{(5x + 3)^{1/3}}$

61. $2(x+1)^{1/2}(3x+4)^{-3/4} - 10(x+1)^{-1/2}(3x+4)^{1/4}$
$= 2(x + 1)^{-1/2}(3x + 4)^{-3/4}[(x + 1) - 5(3x + 4)]$
                                Factoring
$= 2(x + 1)^{-1/2}(3x + 4)^{-3/4}(x + 1 - 15x - 20)$
                                Simplifying
$= 2(x + 1)^{-1/2}(3x + 4)^{-3/4}(-14x - 19)$
$= \dfrac{2(-14x - 19)}{(x + 1)^{1/2}(3x + 4)^{3/4}}$, or $\dfrac{-2(14x + 19)}{(x + 1)^{1/2}(3x + 4)^{3/4}}$

62. $\dfrac{-4(-x + 11)}{(2x - 5)^3(3x + 1)^{2/3}}$, or $\dfrac{4(x - 11)}{(2x - 5)^3(3x + 1)^{2/3}}$

63. $3(x^2 + 1)^3 + 3(3x - 5)(x^2 + 1)^2(2x)$
$= 3(x^2 + 1)^2[(x^2 + 1) + (3x - 5)(2x)]$ Factoring
$= 3(x^2 + 1)^2(x^2 + 1 + 6x^2 - 10x)$ Simplifying
$= 3(x^2 + 1)^2(7x^2 - 10x + 1)$

64. $x^2(x - 1)^3(7x - 3)$

65. $$\frac{x^3(2x) - (x^2 + 1)(3x^2)}{x^6}$$

$$= \frac{2x^4 - 3x^4 - 3x^2}{x^6}$$

$$= \frac{-x^4 - 3x^2}{x^6} = \frac{x^2(-x^2 - 3)}{x^2 \cdot x^4}$$

$$= \frac{-x^2 - 3}{x^4}$$

66. $\frac{7x^3 - 4}{2x^{1/2}}$

67. $$\frac{x^2(x^2 + 1)^{-1/2}(x) - (2x)(x^2 + 1)^{1/2}}{x^4}$$

$$= \frac{x(x^2 + 1)^{-1/2}[x^2 - 2(x^2 + 1)]}{x^4}$$

$$= \frac{x(x^2 + 1)^{-1/2}(x^2 - 2x^2 - 2)}{x^4}$$

$$= \frac{x(x^2 + 1)^{-1/2}(-x^2 - 2)}{x \cdot x^3}$$

$$= \frac{-x^2 - 2}{x^3(x^2 + 1)^{1/2}}$$

68. $\frac{x - 3}{2(x - 1)^{3/2}}$

69. $\left[\sqrt{a^{\sqrt{a}}}\right]^{\sqrt{a}} = \left(a^{\sqrt{a}/2}\right)^{\sqrt{a}} = a^{a/2}$

70. $48 \cdot 2^{1/3} \cdot a^{34/3} \cdot b^{47/9} \cdot c^{34/35}$

Exercise Set 1.9

1. $36 \text{ ft} \cdot \frac{1 \text{ yd}}{3 \text{ ft}}$

$= \frac{36}{3} \cdot \frac{\text{ft}}{\text{ft}} \cdot \text{yd}$

$= 12 \text{ yd}$

2. 96 oz

3. $6 \text{ kg} \cdot 8 \frac{\text{hr}}{\text{kg}}$

$= 6 \cdot 8 \cdot \frac{\text{kg}}{\text{kg}} \cdot \text{hr}$

$= 48 \text{ hr}$

4. 27 km

5. $3 \text{ cm} \cdot \frac{2g}{2 \text{ cm}}$

$= \frac{3 \cdot 2}{2} \cdot \frac{\text{cm}}{\text{cm}} \cdot g$

$= 3 \text{ g}$

6. 18 km

7. $6m + 2m$

$= (6 + 2)m$

$= 8 \text{ m}$

8. 16 tons

9. $5 \text{ ft}^3 + 7 \text{ ft}^3$

$= (5 + 7) \text{ ft}^3$

$= 12 \text{ ft}^3$

10. 27 yd³

11. $\frac{3 \text{ kg}}{5m} \cdot \frac{7 \text{ kg}}{6m}$

$= \frac{3 \cdot 7}{5 \cdot 6} \cdot \frac{\text{kg}}{\text{m}} \cdot \frac{\text{kg}}{\text{m}}$

$= \frac{7 \text{ kg}^2}{10 \text{ m}^2}$

12. 180

13. $\frac{2000 \text{ lb} \cdot (6 \text{ mi/hr})^2}{100 \text{ ft}}$

$= 2000 \text{ lb} \cdot \frac{36 \text{ mi}^2}{\text{hr}^2} \cdot \frac{1}{100 \text{ ft}}$

$= \frac{2000 \cdot 36}{100} \cdot \text{lb} \cdot \frac{\text{mi}^2}{\text{hr}^2} \cdot \frac{1}{\text{ft}}$

$= 720 \frac{\text{lb-mi}^2}{\text{hr}^2\text{-ft}}$

14. $14 \frac{\text{m-kg}}{\text{sec}^2}$

15. $\frac{6 \text{ cm}^2 \cdot 5 \text{ cm/sec}}{2 \text{ sec}^2/\text{cm}^2 \cdot 2 \frac{1}{\text{kg}}}$

$= 6 \text{ cm}^2 \cdot \frac{5 \text{ cm}}{\text{sec}} \cdot \frac{\text{cm}^2}{2 \text{ sec}^2} \cdot \frac{\text{kg}}{2}$

$= \frac{6 \cdot 5}{2 \cdot 2} \cdot \frac{\text{cm}^2 \cdot \text{cm} \cdot \text{cm}^2 \cdot \text{kg}}{\text{sec} \cdot \text{sec}^2}$

$= \frac{15}{2} \frac{\text{cm}^5\text{-kg}}{\text{sec}^3}$

16. 125 ft-lb

17. $72 \text{ in.} = 72 \text{ in.} \cdot \frac{1 \text{ ft}}{12 \text{ in.}}$

$= \frac{72}{12} \cdot \frac{\text{in.}}{\text{in.}} \cdot \text{ft}$

$= 6 \text{ ft}$

18. 1020 min

19. $2 \text{ days} = 2 \text{ days} \cdot \frac{24 \text{ hr}}{1 \text{ day}} \cdot \frac{60 \text{ min}}{1 \text{ hr}} \cdot \frac{60 \text{ sec}}{1 \text{ min}}$

$= 2 \cdot 24 \cdot 60 \cdot 60 \cdot \frac{\text{day}}{\text{day}} \cdot \frac{\text{hr}}{\text{hr}} \cdot \frac{\text{min}}{\text{min}} \cdot \text{sec}$

$= 172,800 \text{ sec}$

20. 0.1 hr

21. $60 \frac{kg}{m} = 60 \frac{kg}{m} \cdot \frac{1000 \ g}{1 \ kg} \cdot \frac{1 \ m}{100 \ cm}$

$\phantom{60 \frac{kg}{m}} = \frac{60 \cdot 1000}{100} \cdot \frac{kg}{kg} \cdot \frac{m}{m} \cdot \frac{g}{cm}$

$\phantom{60 \frac{kg}{m}} = 600 \ \frac{g}{cm}$

22. $30 \frac{mi}{hr}$

23. $216 \ m^2 = 216 \cdot m \cdot m$

$ = 216 \cdot 100 \ cm \cdot 100 \ cm$

$ = 2,160,000 \ cm^2$

24. $0.81 \frac{ton}{yd^3}$

25. $\frac{\$36}{day} = \frac{\$36}{day} \cdot \frac{100 ¢}{\$1} \cdot \frac{1 \ day}{24 \ hr}$

$\phantom{\frac{\$36}{day}} = \frac{36 \cdot 100}{24} \cdot \frac{\$}{\$} \cdot \frac{day}{day} \cdot \frac{¢}{hr}$

$\phantom{\frac{\$36}{day}} = 150 \ \frac{¢}{hr}$

26. 60 man-days

27. $1.73 \frac{mL}{sec} = 1.73 \frac{mL}{sec} \cdot \frac{1 \ L}{1000 \ mL} \cdot \frac{60 \ sec}{1 \ min} \cdot \frac{60 \ min}{1 \ hr}$

$\phantom{1.73 \frac{mL}{sec}} = \frac{1.73 \cdot 60 \cdot 60}{1000} \cdot \frac{mL}{mL} \cdot \frac{sec}{sec} \cdot \frac{min}{min} \cdot \frac{L}{hr}$

$\phantom{1.73 \frac{mL}{sec}} = 6.228 \ \frac{L}{hr}$

28. $180 \frac{cg}{mL}$

29. $186,000 \ \frac{mi}{sec}$

$= 186,000 \ \frac{mi}{sec} \cdot \frac{60 \ sec}{1 \ min} \cdot \frac{60 \ min}{1 \ hr} \cdot \frac{24 \ hr}{1 \ day} \cdot \frac{365 \ days}{1 \ yr}$

$= 186,000 \cdot 60 \cdot 60 \cdot 24 \cdot 365 \ \frac{sec}{sec} \cdot \frac{min}{min} \cdot \frac{hr}{hr} \cdot \frac{day}{day} \cdot \frac{mi}{yr}$

$= 5,865,696,000,000 \ \frac{mi}{yr}$

30. $6,570,000 \ \frac{mi}{yr}$

31. $89.2 \ \frac{ft}{sec} = 89.2 \ \frac{ft}{sec} \cdot \frac{1 \ m}{3.3 \ ft} \cdot \frac{60 \ sec}{1 \ min}$

$\phantom{89.2 \ \frac{ft}{sec}} = \frac{89.2(60)}{3.3} \cdot \frac{ft}{ft} \cdot \frac{sec}{sec} \cdot \frac{m}{min}$

$\phantom{89.2 \ \frac{ft}{sec}} \approx 1621.8 \ \frac{m}{min}$

32. $761.08 \ m^3$

33. $640 \ mi^2 = 640 \ mi^2 \left[\frac{1 \ km}{0.62 \ mi}\right]^2$

$ = \frac{640}{(0.62)^2} \ km^2 \cdot \frac{mi^2}{mi^2}$

$ \approx 1664.93 \ km^2$

34. $208.13 \ \frac{lb}{ft}$

35. $3 \ cm \cdot \frac{5 \ g}{2 \ cm} = \frac{3 \cdot 5}{2} \ g \cdot \frac{cm}{cm} = 7.5 \ g$

A steel rod 3 cm long weighs 7.5 g.

$5 \ m \cdot \frac{100 \ cm}{1 \ m} \cdot \frac{5 \ g}{2 \ cm} = \frac{5 \cdot 100 \cdot 5}{2} \ g \cdot \frac{m}{m} \cdot \frac{cm}{cm} = 1250 \ g$

A steel rod 5 m long weighs 1250 g.

36. 0.08 cm

37. 1 mole of oxygen is 32 grams.

$50 \ moles = 50 \cdot 1 \ mole$

$ = 50 \cdot 32 \ g$

$ = 1600 \ g$

50 moles of oxygen is 1600 grams of oxygen.

38. 888.8 g

39. $303 \ g = 303 \ g \cdot \frac{1 \ mole}{20.2 \ g}$

$ = \frac{303}{20.2} \ moles \cdot \frac{g}{g}$

$ = 15 \ moles$

303 grams of neon is 15 moles of neon.

40. 11.8 moles

41. $E = mc^2$

Substitute 5000 g for m and 2.9979×10^8 m/sec for c.

$E = 5000 \ g(2.9979 \times 10^8 \ m/sec)^2$

$ \approx 44,937 \times 10^{16} \ g \cdot m^2/sec^2$

$ = (4.4937 \times 10^4) \times 10^{16} \ g \cdot m^2/sec^2$

$ = 4.4937 \times 10^{20} \ g \cdot m^2/sec^2$

42. 15.48 g

43. From Exercise 41 we know that the speed of light is 2.9979×10^8 m/sec. We convert m/sec to m/nanosecond.

$2.9979 \times 10^8 \ \frac{m}{sec} \cdot \frac{10^{-9} \ sec}{1 \ nanosecond} =$

$2.9979 \times 10^{-1} \ \frac{m}{nanosecond}$, or $0.29979 \ \frac{m}{nanosecond}$

Light travels 0.29979 m in one nanosecond.

Exercise Set 2.1

1.
$$4x + 12 = 60$$
$$4x + 12 - 12 = 60 - 12$$
$$4x = 48$$
$$\frac{1}{4} \cdot 4x = \frac{1}{4} \cdot 48$$
$$x = 12$$

The solution set is {12}.

2. {24}

3.
$$4 + \frac{1}{2}x = 1$$
$$4 + \frac{1}{2}x - 4 = 1 - 4$$
$$\frac{1}{2}x = -3$$
$$2 \cdot \frac{1}{2}x = 2(-3)$$
$$x = -6$$

The solution set is {-6}.

4. {14}

5.
$$y + 1 = 2y - 7 \quad \text{or} \quad y + 1 = 2y - 7$$
$$1 + 7 = 2y - y \qquad y - 2y = -7 - 1$$
$$8 = y \qquad\qquad -y = -8$$
$$\qquad\qquad\qquad y = 8$$

The solution set is {8}.

6. $\left\{\frac{18}{5}\right\}$

7.
$$5x - 2 + 3x = 2x + 6 - 4x$$
$$8x - 2 = 6 - 2x$$
$$8x + 2x = 6 + 2$$
$$10x = 8$$
$$x = \frac{8}{10}$$
$$x = \frac{4}{5}$$

The solution set is $\left\{\frac{4}{5}\right\}$.

8. {-8}

9.
$$1.9x - 7.8 + 5.3x = 3.0 + 1.8x$$
$$7.2x - 7.8 = 3.0 + 1.8x$$
$$7.2x - 1.8x = 3.0 + 7.8$$
$$5.4x = 10.8$$
$$x = \frac{10.8}{5.4}$$
$$x = 2$$

The solution set is {2}.

10. {-3}

11.
$$7(3x + 6) = 11 - (x + 2)$$
$$21x + 42 = 11 - x - 2$$
$$21x + 42 = 9 - x$$
$$21x + x = 9 - 42$$
$$22x = -33$$
$$x = -\frac{33}{22}$$
$$x = -\frac{3}{2}$$

The solution set is $\left\{-\frac{3}{2}\right\}$.

12. $\left\{-\frac{27}{14}\right\}$

13.
$$2x - (5 + 7x) = 4 - [x - (2x + 3)]$$
$$2x - 5 - 7x = 4 - x + 2x + 3$$
$$-5x - 5 = 7 + x$$
$$-5 - 7 = x + 5x$$
$$-12 = 6x$$
$$-\frac{12}{6} = x$$
$$-2 = x$$

The solution set is {-2}.

14. ∅

15.
$$(2x - 3)(3x - 2) = 0$$
$$2x - 3 = 0 \text{ or } 3x - 2 = 0$$
$$2x = 3 \text{ or } \qquad 3x = 2$$
$$x = \frac{3}{2} \text{ or } \qquad x = \frac{2}{3}$$

The solution set is $\left\{\frac{3}{2}, \frac{2}{3}\right\}$.

16. $\left\{\frac{2}{5}, -\frac{3}{2}\right\}$

17.
$$x(x - 1)(x + 2) = 0$$
$$x = 0 \text{ or } x - 1 = 0 \text{ or } x + 2 = 0$$
$$x = 0 \text{ or } \qquad x = 1 \text{ or } \qquad x = -2$$

The solution set is {0, 1, -2}.

18. {0, -2, 3}

19.
$$3x^2 + x - 2 = 0$$
$$(3x - 2)(x + 1) = 0$$
$$3x - 2 = 0 \text{ or } x + 1 = 0$$
$$x = \frac{2}{3} \text{ or } \qquad x = -1$$

The solution set is $\left\{\frac{2}{3}, -1\right\}$.

20. $\left\{\frac{3}{5}, 1\right\}$

21. $(x - 1)(x + 1) = 5(x - 1)$
    $x^2 - 1 = 5x - 5$
    $x^2 - 5x + 4 = 0$
    $(x - 4)(x - 1) = 0$
    $x - 4 = 0$ or $x - 1 = 0$
    $x = 4$ or $x = 1$

    The solution set is {4, 1}.

22. {8, 3}

23. $x[4(x - 2) - 5(x - 1)] = 2$
    $x(4x - 8 - 5x + 5) = 2$
    $x(-x - 3) = 2$
    $-x^2 - 3x = 2$
    $0 = x^2 + 3x + 2$
    $0 = (x + 2)(x + 1)$
    $x + 2 = 0$ or $x + 1 = 0$
    $x = -2$ or $x = -1$

    The solution set is {-2, -1}.

24. {5, 10}

25. $(3x^2 - 7x - 20)(2x - 5) = 0$
    $(3x + 5)(x - 4)(2x - 5) = 0$
    $3x + 5 = 0$ or $x - 4 = 0$ or $2x - 5 = 0$
    $x = -\frac{5}{3}$ or $x = 4$ or $x = \frac{5}{2}$

    The solution set is $\left\{-\frac{5}{3}, 4, \frac{5}{2}\right\}$.

26. $\left\{-\frac{11}{8}, -\frac{1}{4}, \frac{2}{3}\right\}$

27. $16x^3 = x$
    $16x^3 - x = 0$
    $x(16x^2 - 1) = 0$
    $x(4x + 1)(4x - 1) = 0$
    $x = 0$ or $4x + 1 = 0$ or $4x - 1 = 0$
    $x = 0$ or $x = -\frac{1}{4}$ or $x = \frac{1}{4}$

    The solution set is $\left\{0, -\frac{1}{4}, \frac{1}{4}\right\}$.

28. $\left\{0, -\frac{1}{3}, \frac{1}{3}\right\}$

29. $2x^2 = 6x$
    $2x^2 - 6x = 0$
    $2x(x - 3) = 0$
    $2x = 0$ or $x - 3 = 0$
    $x = 0$ or $x = 3$

    The solution set is {0, 3}.

30. {0, -2}

31. $3y^3 - 5y^2 - 2y = 0$
    $y(3y^2 - 5y - 2) = 0$
    $y(3y + 1)(y - 2) = 0$
    $y = 0$ or $3y + 1 = 0$ or $y - 2 = 0$
    $y = 0$ or $y = -\frac{1}{3}$ or $y = 2$

    The solution set is $\left\{0, -\frac{1}{3}, 2\right\}$

32. $\left\{0, 1, \frac{2}{3}\right\}$

33. $(2x - 3)(3x + 2)(x - 1) = 0$
    $2x - 3 = 0$ or $3x + 2 = 0$ or $x - 1 = 0$
    $x = \frac{3}{2}$ or $x = -\frac{2}{3}$ or $x = 1$

    The solution set is $\left\{\frac{3}{2}, -\frac{2}{3}, 1\right\}$.

34. $\left\{4, -3, \frac{1}{2}\right\}$

35. $(2 - 4y)(y^2 + 3y) = 0$
    $2(1 - 2y)y(y + 3) = 0$
    $1 - 2y = 0$ or $y = 0$ or $y + 3 = 0$
    $y = \frac{1}{2}$ or $y = 0$ or $y = -3$

    The solution set is $\left\{\frac{1}{2}, 0, -3\right\}$.

36. {-3, 3, -6, 6}

37. $x + 4 = 8 + x$
    $-x + x + 4 = -x + 8 + x$
    $4 = 8$

    We get a false equation. There are no solutions, so the solution set is $\emptyset$.

38. $\emptyset$

39.
$$7x^3 + x^2 - 7x - 1 = 0$$
$$x^2(7x + 1) - (7x + 1) = 0$$
$$(x^2 - 1)(7x + 1) = 0$$
$$(x + 1)(x - 1)(7x + 1) = 0$$

$x + 1 = 0$ or $x - 1 = 0$ or $7x + 1 = 0$

$\quad x = -1$ or $\quad x = 1$ or $\quad x = -\frac{1}{7}$

The solution set is $\left\{-1, 1, -\frac{1}{7}\right\}$.

40. $\left\{-\frac{1}{3}, -2, 2\right\}$

41.
$$y^3 + 2y^2 - y - 2 = 0$$
$$y^2(y + 2) - (y + 2) = 0$$
$$(y^2 - 1)(y + 2) = 0$$
$$(y + 1)(y - 1)(y + 2) = 0$$

$y + 1 = 0$ or $y - 1 = 0$ or $y + 2 = 0$

$\quad y = -1$ or $\quad y = 1$ or $\quad y = -2$

The solution set is $\{-1, 1, -2\}$.

42. $\{-1, -5, 5\}$

43.
$$11 + x = x + 11$$
$$-x + 11 + x = -x + x + 11$$
$$11 = 11$$

We get a true equation. The solution set is the set of all real numbers.

44. All real numbers

45.
$x + 6 < 5x - 6$ or	$x + 6 < 5x - 6$
$6 + 6 < 5x - x$	$x - 5x < -6 - 6$
$12 < 4x$	$-4x < -12$
$\frac{12}{4} < x$	$x > \frac{-12}{-4}$
$3 < x$	$x > 3$

The solution set is $\{x \mid x > 3\}$.

46. $\left\{x \mid x > -\frac{4}{5}\right\}$

47.
$$3x - 3 + 2x \geqslant 1 - 7x - 9$$
$$5x - 3 \geqslant -7x - 8$$
$$5x + 7x \geqslant -8 + 3$$
$$12x \geqslant -5$$
$$x \geqslant -\frac{5}{12}$$

The solution set is $\left\{x \mid x \geqslant -\frac{5}{12}\right\}$.

48. $\left\{y \mid y \leqslant -\frac{1}{12}\right\}$

49.
$14 - 5y \leqslant 8y - 8$ or	$14 - 5y \leqslant 8y - 8$
$14 + 8 \leqslant 8y + 5y$	$-5y - 8y \leqslant -8 - 14$
$22 \leqslant 13y$	$-13y \leqslant -22$
$\frac{22}{13} \leqslant y$	$y \geqslant \frac{22}{13}$

The solution set is $\left\{y \mid y \geqslant \frac{22}{13}\right\}$.

50. $\{x \mid x < 5\}$

51.
$$-\frac{3}{4}x \geqslant -\frac{5}{8} + \frac{2}{3}x$$
$$\frac{5}{8} \geqslant \frac{3}{4}x + \frac{2}{3}x$$
$$\frac{5}{8} \geqslant \frac{9}{12}x + \frac{8}{12}x$$
$$\frac{5}{8} \geqslant \frac{17}{12}x$$
$$\frac{12}{17} \cdot \frac{5}{8} \geqslant \frac{12}{17} \cdot \frac{17}{12}x$$
$$\frac{15}{34} \geqslant x$$

The solution set is $\left\{x \mid x \leqslant \frac{15}{34}\right\}$.

52. $\left\{x \mid x \geqslant -\frac{3}{14}\right\}$

53.
$$4x(x - 2) < 2(2x - 1)(x - 3)$$
$$4x(x - 2) < 2(2x^2 - 7x + 3)$$
$$4x^2 - 8x < 4x^2 - 14x + 6$$
$$-8x < -14x + 6$$
$$-8x + 14x < 6$$
$$6x < 6$$
$$x < \frac{6}{6}$$
$$x < 1$$

The solution set is $\{x \mid x < 1\}$.

54. $\{x \mid x > -1\}$

55. $\{x \mid x > 2.5\}$

56. $\{y \mid y \leqslant -7\}$

57.
$$2t^2 = 10$$
$$t^2 = 5$$

The solution set is $\{t \mid t^2 = 5\}$.

58. $\{m \mid m^3 + 3 = m^2 - 2\}$

59. $\sqrt{x - 3}$

   The radicand must be nonnegative. We set
   $x - 3 \geqslant 0$ and solve for x.

   $x - 3 \geqslant 0$

   $\quad x \geqslant 3$

   The meaningful replacements are $\{x \mid x \geqslant 3\}$.

60. $\left\{x \mid x \geqslant \frac{5}{2}\right\}$

61. $\sqrt{3 - 4x}$

   The radicand must be nonnegative. We set
   $3 - 4x \geqslant 0$ and solve for x.

   $3 - 4x \geqslant 0$

   $\quad -4x \geqslant -3$

   $\quad\quad x \leqslant \frac{3}{4}$

   The meaningful replacements are $\left\{x \mid x \leqslant \frac{3}{4}\right\}$.

62. $\{x \mid x \text{ is a real number}\}$

63. $2.905x - 3.214 + 6.789x = 3.012 + 1.805x$

   $\quad 9.694x - 3.214 = 3.012 + 1.805x$

   $\quad 9.694x - 1.805x = 3.012 + 3.214$

   $\quad\quad\quad 7.889x = 6.226$

   $\quad\quad\quad\quad\quad x = \frac{6.226}{7.889}$

   $\quad\quad\quad\quad\quad x \approx 0.7892$

   The solution set is $\{0.7892\}$.

64. $\{-1.3053, 1.9892\}$

65. $\quad 3.12x^2 - 6.715x = 0$

   $\quad x(3.12x - 6.715) = 0$

   $x = 0$ or $3.12x - 6.715 = 0$

   $x = 0$ or $\quad\quad 3.12x = 6.715$

   $x = 0$ or $\quad\quad\quad\quad x \approx 2.1522$

   The solution set is $\{0, 2.1522\}$.

66. $\{0, -1.9492\}$

67. $1.52(6.51x + 7.3) < 11.2 - (7.2x + 13.52)$

   $9.8952x + 11.096 < 11.2 - 7.2x - 13.52$

   $9.8952x + 11.096 < -7.2x - 2.32$

   $\quad 9.8952x + 7.2x < -2.32 - 11.096$

   $\quad\quad\quad 17.0952x < -13.416$

   $\quad\quad\quad\quad\quad x < -\frac{13.416}{17.0952}$

   $\quad\quad\quad\quad\quad x < -0.7848$

   The solution set is $\{x \mid x < -0.7848\}$.

68. $\{y \mid y \geqslant -2.2353\}$

69. $\quad\quad\quad (x + 1)^3 = (x - 1)^3 + 26$

   $x^3 + 3x^2 + 3x + 1 = x^3 - 3x^2 + 3x - 1 + 26$

   $x^3 + 3x^2 + 3x + 1 = x^3 - 3x^2 + 3x + 25$

   $\quad\quad\quad 6x^2 - 24 = 0$

   $\quad\quad\quad 6(x^2 - 4) = 0$

   $\quad 6(x + 2)(x - 2) = 0$

   $x + 2 = 0$ or $x - 2 = 0$

   $\quad x = -2$ or $\quad x = 2$

   The solution set is $\{-2, 2\}$.

70. $\{1\}$

71. $\quad\quad (x^2 - x - 20)(x^2 - 25) = 0$

   $(x - 5)(x + 4)(x + 5)(x - 5) = 0$

   $x - 5 = 0$ or $x + 4 = 0$ or $x + 5 = 0$ or $x - 5 = 0$

   $x = 5$ or $\quad x = -4$ or $\quad x = -5$ or $\quad x = 5$

   The solution set is $\{5, -4, -5\}$.

72. $\{-4, 4, 3, -3, -1\}$

Exercise Set 2.2

1. $3x + 5 = 12$ $3x = 7$

   $\quad 3x = 7$ $x = \frac{7}{3}$

   $\quad\quad x = \frac{7}{3}$

   The solution set is $\left\{\frac{7}{3}\right\}$.

   The solution set is $\left\{\frac{7}{3}\right\}$.

   The equations are equivalent.

2. No

3. $x = 3$ $x^2 = 9$

   The solution set is $\{3\}$. $x = \pm 3$

                 The solution set is $\{-3, 3\}$.

   The equations are not equivalent.

4. Yes

5. $\frac{(x - 3)(x + 9)}{(x - 3)} = x + 9$,   Note: $x \neq 3$

   $\quad\quad x + 9 = x + 9$

   $\quad\quad\quad\quad 9 = 9$

   The solution set is the set of all real numbers
   except 3.

   $x + 9 = x + 9$

   $\quad\quad 9 = 9$

   The solution set is the set of all real numbers.
   Thus, the equations are not equivalent.

6. No

7.
$$\frac{1}{4} + \frac{1}{5} = \frac{1}{t}, \quad \text{LCM is } 20t$$

$$20t\left(\frac{1}{4} + \frac{1}{5}\right) = 20t \cdot \frac{1}{t}$$

$$20t \cdot \frac{1}{4} + 20t \cdot \frac{1}{5} = 20t \cdot \frac{1}{t}$$

$$5t + 4t = 20$$

$$9t = 20$$

$$t = \frac{20}{9}$$

Check:

$$\frac{1}{4} + \frac{1}{5} = \frac{1}{t}$$

$\frac{1}{4} + \frac{1}{5}$	$\frac{1}{\frac{20}{9}}$
$\frac{5}{20} + \frac{4}{20}$	$1 \cdot \frac{9}{20}$
$\frac{9}{20}$	$\frac{9}{20}$ TRUE

The solution set is $\left\{\frac{20}{9}\right\}$.

8. $\{-2\}$

9.
$$\frac{3}{x-8} = \frac{x-5}{x-8}, \quad \text{LCM is } x-8$$

$$(x-8) \cdot \frac{3}{x-8} = (x-8) \cdot \frac{x-5}{x-8}$$

$$3 = x - 5$$

$$8 = x$$

The solution set is $\emptyset$.

Check:

$$\frac{3}{x-8} = \frac{x-5}{x-8}$$

$\frac{3}{8-8}$	$\frac{8-5}{8-8}$
$\frac{3}{0}$	$\frac{3}{0}$

Division by zero is undefined.

10. $\emptyset$

11.
$$\frac{x+2}{4} - \frac{x-1}{5} = 15, \quad \text{LCM is } 20$$

$$20\left(\frac{x+2}{4} - \frac{x-1}{5}\right) = 20 \cdot 15$$

$$5(x+2) - 4(x-1) = 300$$

$$5x + 10 - 4x + 4 = 300$$

$$x + 14 = 300$$

$$x = 286$$

The solution set is $\{286\}$.

12. $\{-1\}$

13.
$$\frac{3x}{x+2} + \frac{6}{x} = \frac{12}{x^2 + 2x}$$

$$\frac{3x}{x+2} + \frac{6}{x} = \frac{12}{x(x+2)}, \quad \text{LCM is } x(x+2)$$

Note that -2 and 0 are not meaningful replacements.

$$x(x+2)\left(\frac{3x}{x+2} + \frac{6}{x}\right) = x(x+2) \cdot \frac{12}{x(x+2)}$$

$$3x \cdot x + 6(x+2) = 12$$

$$3x^2 + 6x + 12 = 12$$

$$3x^2 + 6x = 0$$

$$3x(x+2) = 0$$

$$3x = 0 \quad \text{or} \quad x + 2 = 0$$

$$x = 0 \quad \text{or} \quad x = -2$$

Since we noted at the outset that -2 and 0 are not meaningful replacements, the solution set is $\emptyset$.

14. $\emptyset$

15.
$$\frac{x+2}{2} + \frac{3x+1}{5} = \frac{x-2}{4}, \quad \text{LCM is } 20$$

$$20\left(\frac{x+2}{2} + \frac{3x+1}{5}\right) = 20 \cdot \frac{x-2}{4}$$

$$10(x+2) + 4(3x+1) = 5(x-2)$$

$$10x + 20 + 12x + 4 = 5x - 10$$

$$22x + 24 = 5x - 10$$

$$22x - 5x = -10 - 24$$

$$17x = -34$$

$$x = -2$$

The solution set is $\{-2\}$.

16. $\{2\}$

17.
$$\frac{1}{2} + \frac{2}{x} = \frac{1}{3} + \frac{3}{x}, \quad \text{LCM} = 6x$$

$$6x\left(\frac{1}{2} + \frac{2}{x}\right) = 6x\left(\frac{1}{3} + \frac{3}{x}\right)$$

$$3x + 12 = 2x + 18$$

$$3x - 2x = 18 - 12$$

$$x = 6$$

Check:

$$\frac{1}{2} + \frac{2}{x} = \frac{1}{3} + \frac{3}{x}$$

$\frac{1}{2} + \frac{2}{6}$	$\frac{1}{3} + \frac{3}{6}$
$\frac{1}{2} + \frac{1}{3}$	$\frac{1}{3} + \frac{1}{2}$ TRUE

The solution set is $\{6\}$.

18. $\left\{\frac{11}{30}\right\}$

19.
$$\frac{4}{x^2 - 1} - \frac{2}{x - 1} = \frac{3}{x + 1},$$
$$\text{LCM is } (x + 1)(x - 1)$$
$$(x+1)(x-1)\left[\frac{4}{(x+1)(x-1)} - \frac{2}{x-1}\right] = (x+1)(x-1) \cdot \frac{3}{x+1}$$
$$4 - 2(x + 1) = 3(x - 1)$$
$$4 - 2x - 2 = 3x - 3$$
$$2 - 2x = 3x - 3$$
$$2 + 3 = 3x + 2x$$
$$5 = 5x$$
$$1 = x$$

Check:

$$\frac{\dfrac{4}{x^2 - 1} - \dfrac{2}{x - 1} = \dfrac{3}{x + 1}}{\begin{array}{c|c} \dfrac{4}{1^2 - 1} - \dfrac{2}{1 - 1} & \dfrac{3}{1 + 1} \\ \dfrac{4}{0} - \dfrac{2}{0} & \dfrac{3}{2} \end{array}}$$

Division by zero is undefined.
The solution set is ∅.

20. ∅

21.
$$\frac{490}{x^2 - 49} = \frac{5x}{x - 7} - \frac{35}{x + 7}$$
$$\frac{490}{(x + 7)(x - 7)} = \frac{5x}{x - 7} - \frac{35}{x + 7},$$
$$\text{LCM is } (x + 7)(x - 7)$$

Note that -7 and 7 are not meaningful replacements.
$$(x+7)(x-7)\left[\frac{490}{(x+7)(x-7)}\right] = (x+7)(x-7)\left[\frac{5x}{x-7} - \frac{35}{x+7}\right]$$
$$490 = 5x(x + 7) - 35(x - 7)$$
$$490 = 5x^2 + 35x - 35x + 245$$
$$0 = 5x^2 - 245$$
$$0 = 5(x + 7)(x - 7)$$
$$x + 7 = 0 \quad \text{or} \quad x - 7 = 0$$
$$x = -7 \quad \text{or} \quad x = 7$$

Since we noted at the outset that -7 and 7 are not meaningful replacements, the solution set is ∅.

22. ∅

23.
$$\frac{4}{x} - \frac{4}{x - 6} = \frac{24}{6x - x^2}$$
$$\frac{4}{x} - \frac{4}{x - 6} = \frac{24}{x(6 - x)}$$
$$\frac{4}{x} - \frac{4}{x - 6} = \frac{-1}{-1} \cdot \frac{24}{x(6 - x)}$$
$$\frac{4}{x} - \frac{4}{x - 6} = \frac{-24}{x(x - 6)} \quad \begin{array}{l}[-1\cdot(6 - x) = -6 + x, \text{ or} \\ x - 6]\end{array}$$

Note that 0 and 6 are not meaningful replacements.

$$x(x - 6)\left[\frac{4}{x} - \frac{4}{x - 6}\right] = x(x - 6) \cdot \frac{-24}{x(x - 6)}$$
$$4(x - 6) - 4x = -24$$
$$4x - 24 - 4x = -24$$
$$0 = 0$$

We get an equation that is true for all meaningful replacements. Since we noted at the outset that 0 and 6 are not meaningful replacements, the solution set is the set of all real numbers except 0 and 6.

24. ∅

25.
$$\frac{11 - t^2}{3t^2 - 5t + 2} = \frac{2t + 3}{3t - 2} + \frac{t - 3}{1 - t}$$
$$\frac{11 - t^2}{(3t - 2)(t - 1)} = \frac{2t + 3}{3t - 2} + \frac{t - 3}{1 - t}$$
$$\frac{11 - t^2}{(3t - 2)(t - 1)} = \frac{2t + 3}{3t - 2} + \frac{-1}{-1} \cdot \frac{t - 3}{1 - t}$$
$$\frac{11 - t^2}{(3t - 2)(t - 1)} = \frac{2t + 3}{3t - 2} + \frac{-t + 3}{t - 1},$$
$$\text{LCM is } (3t - 2)(t - 1)$$
$$(3t-2)(t-1)\cdot \frac{11 - t^2}{(3t-2)(t-1)} = (3t-2)(t-1)\left[\frac{2t+3}{3t-2} + \frac{-t+3}{t-1}\right]$$
$$11 - t^2 = (t-1)(2t+3)+(3t-2)(-t+3)$$
$$11 - t^2 = 2t^2+t-3-3t^2+11t-6$$
$$11 - t^2 = -t^2 + 12t - 9$$
$$11 = 12t - 9$$
$$20 = 12t$$
$$\frac{20}{12} = t$$
$$\frac{5}{3} = t$$

The value checks. The solution set is $\left\{\frac{5}{3}\right\}$.

26. {1}

27. $\dfrac{7x}{x - 3} - \dfrac{21}{x} + 11 = \dfrac{63}{x^2 - 3x}$, LCM is $x(x - 3)$

Note that 0 and 3 are not meaningful replacements.
$$x(x - 3)\left[\frac{7x}{x - 3} - \frac{21}{x} + 11\right] = x(x - 3) \cdot \frac{63}{x(x - 3)}$$
$$7x^2 - 21(x - 3) + 11x(x - 3) = 63$$
$$7x^2 - 21x + 63 + 11x^2 - 33x = 63$$
$$18x^2 - 54x = 0$$
$$18x(x - 3) = 0$$
$$18x = 0 \quad \text{or} \quad x - 3 = 0$$
$$x = 0 \quad \text{or} \quad x = 3$$

Since we noted at the outset that 0 and 3 are not meaningful replacements, the solution set is ∅.

28. ∅

29.
$$\frac{2.315}{y} - \frac{12.6}{17.4} = \frac{6.71}{7} + 0.763,$$

$$\text{LCM is } 7(17.4)y$$

$$7(17.4)y\left[\frac{2.315}{y} - \frac{12.6}{17.4}\right] = 7(17.4)y\left[\frac{6.71}{7} + 0.763\right]$$

$$281.967 - 88.2y = 116.754y + 92.9334y$$

$$281.967 - 88.2y = 209.6874y$$

$$281.967 = 297.8874y$$

$$0.94656 \approx y$$

The value checks. The solution set is {0.94656}.

30. {0.0855}

31. $\frac{2x^2}{x - 3} + \frac{4x - 6}{x + 3} = \frac{108}{x^2 - 9}$,   LCM is $(x + 3)(x - 3)$

Note that 3 and -3 are not meaningful replacements.

$$(x+3)(x-3)\left[\frac{2x^2}{x-3} + \frac{4x-6}{x+3}\right] = (x+3)(x-3) \cdot \frac{108}{(x+3)(x-3)}$$

$$2x^2(x+3) + (x-3)(4x-6) = 108$$

$$2x^3+6x^2+4x^2-18x+18 = 108$$

$$2x^3 + 10x^2 - 18x + 18 = 108$$

$$2x^3 + 10x^2 - 18x - 90 = 0$$

$$2(x^3 + 5x^2 - 9x - 45) = 0$$

$$2[x^2(x + 5) - 9(x + 5)] = 0$$

$$2(x^2 - 9)(x + 5) = 0$$

$$2(x + 3)(x - 3)(x + 5) = 0$$

$$x + 3 = 0 \quad \text{or} \quad x - 3 = 0 \quad \text{or} \quad x + 5 = 0$$

$$x = -3 \quad \text{or} \quad x = 3 \quad \text{or} \quad x = -5$$

We noted at the outset that the numbers -3 and 3 are not meaningful replacements. The number -5 checks. The solution set is {-5}.

32. $\left\{\frac{8}{3}\right\}$

33. $\frac{24}{x^2 - 2x + 4} = \frac{3x}{x + 2} + \frac{72}{x^3 + 8}$

$$\text{LCM is } (x + 2)(x^2 - 2x + 4)$$

Note that -2 is not a meaningful replacement.

$$(x+2)(x^2-2x+4) \cdot \frac{24}{x^2-2x+4} =$$

$$(x+2)(x^2-2x+4)\left[\frac{3x}{x+2} + \frac{72}{(x+2)(x^2-2x+4)}\right]$$

$$24(x + 2) = 3x(x^2 - 2x + 4) + 72$$

$$24x + 48 = 3x^3 - 6x^2 + 12x + 72$$

$$0 = 3x^3 - 6x^2 - 12x + 24$$

$$0 = 3(x^3 - 2x^2 - 4x + 8)$$

$$0 = 3[x^2(x-2) - 4(x-2)]$$

$$0 = 3(x^2 - 4)(x - 2)$$

$$0 = 3(x + 2)(x - 2)(x - 2)$$

$$x + 2 = 0 \quad \text{or} \quad x - 2 = 0 \quad \text{or} \quad x - 2 = 0$$

$$x = -2 \quad \text{or} \quad x = 2 \quad \text{or} \quad x = 2$$

We noted at the outset that -2 is not a meaningful replacement. The number 2 checks. The solution set is {2}.

34. {3}

35. $\frac{5}{x - 1} + \frac{9}{x^2 + x + 1} = \frac{15}{x^3 - 1}$

$$\text{LCM is } (x - 1)(x^2 + x + 1)$$

Note that 1 is not a meaningful replacement.

$$(x-1)(x^2+x+1)\left[\frac{5}{x-1} + \frac{9}{x^2+x+1}\right] =$$

$$(x-1)(x^2+x+1) \cdot \frac{15}{(x-1)(x^2+x+1)}$$

$$5(x^2 + x + 1) + 9(x - 1) = 15$$

$$5x^2 + 5x + 5 + 9x - 9 = 15$$

$$5x^2 + 14x - 4 = 15$$

$$5x^2 + 14x - 19 = 0$$

$$(5x + 19)(x - 1) = 0$$

$$5x + 19 = 0 \quad \text{or} \quad x - 1 = 0$$

$$5x = -19 \quad \text{or} \quad x = 1$$

$$x = -\frac{19}{5} \quad \text{or} \quad x = 1$$

The number $-\frac{19}{5}$ checks. We noted at the outset that -1 is not a meaningful replacement. The solution set is $\left\{-\frac{19}{5}\right\}$.

36. $\left\{\frac{23}{7}\right\}$

37. $\frac{7}{x - 9} - \frac{7}{x} = \frac{63}{x^2 - 9x}$,   LCM is $x(x - 9)$

Note that 9 and 0 are not meaningful replacements.

$$x(x - 9)\left[\frac{7}{x - 9} - \frac{7}{x}\right] = x(x - 9) \cdot \frac{63}{x(x - 9)}$$

$$7x - 7(x - 9) = 63$$

$$7x - 7x + 63 = 63$$

$$63 = 63$$

We get an equation that is true for all meaningful replacements. We noted at the outset that 9 and 0 are not meaningful replacements. The solution set is the set of all real numbers except 9 and 0.

38. All real numbers except -13 and -7

39. $\frac{(x - 3)^2}{x - 3} = x - 3$,   LCM is $x - 3$

Note that 3 is not a meaningful replacement.

$$(x - 3) \cdot \frac{(x - 3)^2}{x - 3} = (x - 3)(x - 3)$$

$$(x - 3)^2 = (x - 3)^2$$

We get an equation that is true for all meaningful replacements. We noted at the outset that 3 is not a meaningful replacement. The solution set is the set of all real numbers except 3.

40. All real numbers except 2

<u>41.</u> $\dfrac{x^3 + 8}{x + 2} = x^2 - 2x + 4,$ LCM is $x + 2$

Note that -2 is not a meaningful replacement.

$$(x + 2) \cdot \dfrac{x^3 + 8}{x + 2} = (x + 2)(x^2 - 2x + 4)$$

$$x^3 + 8 = x^3 + 8$$

We get an equation that is true for all meaningful replacements. We noted at the outset that -2 is not a meaningful replacement. The solution set is the set of all real numbers except -2.

<u>42.</u> All real numbers except 2

<u>43.</u> Find the solution set of each equation.

Equation (1): $x^2 - x - 20 = x^2 - 25$

$$-x = -5$$
$$x = 5$$

The solution set is $\{5\}$.

Equation (2): $(x - 5)(x + 4) = (x - 5)(x + 5)$

$$x^2 - x - 20 = x^2 - 25$$

Multiplying, we get equation (1), so the solution set of equation (2) is also $\{5\}$.

Equation (3): $x + 4 = x + 5$

$$4 = 5 \qquad \text{Adding } -x$$

We get a false equation, so the solution set is $\emptyset$.

Equation (4): $4 = 5$

This is a false equation, so the solution set is $\emptyset$.

(1) and (2) have the same solution set, so they are equivalent; (2) and (3) do not have the same solution set, so they are not equivalent; (3) and (4) have the same solution set, so they are equivalent.

<u>44.</u> Yes

<u>45.</u> $x + 4 = 4 + x$

Adding $-x$ to both sides, we get $4 = 4$, a true equation. Thus, the equation is an identity.

<u>46.</u> Yes

<u>47.</u> $\dfrac{x^3 + 8}{x^2 - 4} = \dfrac{x^2 - 2x + 4}{x - 2}$

$$\dfrac{x^3 + 8}{(x + 2)(x - 2)} = \dfrac{x^2 - 2x + 4}{x - 2}$$

$$(x+2)(x-2) \cdot \dfrac{x^3 + 8}{(x+2)(x-2)} = (x+2)(x-2) \cdot \dfrac{x^2-2x+4}{x - 2}$$

$$x^3 + 8 = (x + 2)(x^2 - 2x + 4)$$
$$x^3 + 8 = x^3 + 8$$
$$8 = 8$$

We get an equation that is true for all meaningful replacements of the variables. Thus, the equation is an identity.

<u>48.</u> No

<u>49.</u> $\sqrt{x^2 - 16} = x - 4$

The set of meaningful replacements is $\{x \mid x \leqslant -4 \text{ or } x \geqslant 4\}$. Substitute 5, a meaningful replacement, for x in the equation.

$$\sqrt{5^2 - 16} = 5 - 4$$
$$\sqrt{9} = 1$$
$$3 = 1$$

We get a false equation. Thus, the equation is <u>not</u> true for all meaningful replacements. It is not an identity.

<u>50.</u> $\left\{ -\dfrac{7}{2} \right\}$

Exercise Set 2.3

<u>1.</u> $P = 2\ell + 2w$

$$P - 2\ell = 2w$$

$$\dfrac{P - 2\ell}{2} = w$$

<u>2.</u> $r = \dfrac{C}{2\pi}$

<u>3.</u> $A = \dfrac{1}{2}bh$

$$2A = bh$$

$$\dfrac{2A}{h} = b$$

<u>4.</u> $\pi = \dfrac{r^2}{A}$

<u>5.</u> $d = rt$

$$\dfrac{d}{t} = r$$

<u>6.</u> $a = \dfrac{F}{m}$

<u>7.</u> $E = IR$

$$\dfrac{E}{R} = I$$

<u>8.</u> $m_2 = \dfrac{Fd^2}{km_1}$

<u>9.</u> $\dfrac{P_1 V_1}{T_1} = \dfrac{P_2 V_2}{T_2}$

$$T_1 T_2 \cdot \dfrac{P_1 V_1}{T_1} = T_1 T_2 \cdot \dfrac{P_2 V_2}{T_2}$$

$$T_2 P_1 V_1 = T_1 P_2 V_2$$

$$\dfrac{T_2 P_1 V_1}{P_2 V_2} = T_1$$

<u>10.</u> $V_2 = \dfrac{T_2 P_1 V_1}{T_1 P_2}$

11.
$$S = \frac{H}{m(v_1 - v_2)} \text{ or } \qquad S = \frac{H}{m(v_1 - v_2)}$$

$$m(v_1 - v_2)S = H \qquad\qquad m(v_1 - v_2)S = H$$

$$mSv_1 - mSv_2 = H \qquad\qquad v_1 - v_2 = \frac{H}{mS}$$

$$mSv_1 = H + mSv_2$$

$$v_1 = \frac{H + mSv_2}{mS} \qquad\qquad v_1 = \frac{H}{mS} + v_2$$

12. $v_2 = \dfrac{mSv_1 - H}{mS}$

13.
$$\frac{1}{F} = \frac{1}{m} + \frac{1}{p}$$

$$Fmp \cdot \frac{1}{F} = Fmp\left(\frac{1}{m} + \frac{1}{p}\right)$$

$$mp = Fp + Fm$$

$$mp - Fp = Fm$$

$$p(m - F) = Fm$$

$$p = \frac{Fm}{m - F}$$

14. $F = \dfrac{mp}{p + m}$

15.
$$(x + a)(x - b) = x^2 + 5$$

$$x^2 - bx + ax - ab = x^2 + 5$$

$$ax - bx = 5 + ab$$

$$x(a - b) = 5 + ab$$

$$x = \frac{5 + ab}{a - b}$$

16. $x = \dfrac{c}{2}$

17.
$$10(a + x) = 8(a - x)$$

$$10a + 10x = 8a - 8x$$

$$10x + 8x = 8a - 10a$$

$$18x = -2a$$

$$x = -\frac{2a}{18}$$

$$x = -\frac{a}{9}$$

18. $x = a - 7b$

19. Familiarize: We restate the situation.

The amount invested plus the interest is $702. We let x = the amount originally invested.

Translate:

Invested amount	plus	8% of invested amount	is	$702.
↓	↓	↓	↓	↓
x	+	8%·x	=	702

Carry out:

$$x + 8\%x = 702$$

$$x + 0.08x = 702$$

$$(1 + 0.08)x = 702$$

$$1.08x = 702$$

$$x = 650$$

Check:

$650 + 8% of $650 = $650 + $52 = $702

State:

The amount originally invested was $650.

20. $850

21. Familiarize: We make a drawing.

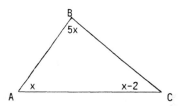

We let x represent the measure of angle A. Then 5x represents angle B, and x − 2 represents angle C. The sum of the angle measures is 180°.

Translate:

Measure of angle A	+	Measure of angle B	+	Measure of angle C	= 180
↓		↓		↓	↓
x	+	5x	+	x − 2	= 180

Carry out:

$$x + 5x + x - 2 = 180$$

$$7x - 2 = 180$$

$$7x = 182$$

$$x = 26$$

If x = 26, then 5x = 5·26, or 130, and x − 2 = 26 − 2, or 24.

Check:

The measure of angle B, 130°, is five times the measure of angle A, 26°. The measure of angle C, 24°, is 2° less than the measure of angle A, 26°. The sum of the angle measures is 26° + 130° + 24°, or 180°.

State:

Angle A measures 26°. Angle B measures 130°, and angle C measures 24°.

22. 40°, 80°, 60°

23. Familiarize: We make a drawing.

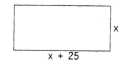

We let x represent the width. Then x + 25 represents the length.

Translate:

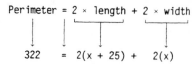

$$322 = 2(x + 25) + 2(x)$$

Carry out:

$322 = 2x + 50 + 2x$

$322 = 4x + 50$

$272 = 4x$

$68 = x$

If x = 68, then x + 25 = 68 + 25, or 93.

Check:

The length is 25 m more than the width: 93 = 68 + 25. The perimeter is 2·93 + 2·68, or 186 + 136, or 322 m.

State:

The length is 93 m; the width is 68 m.

24. 6.5 m, 13 m

25. Familiarize:

We let x represent the score on the fourth test. Then the average of the four scores is

$$\frac{87 + 64 + 78 + x}{4}.$$

Translate:

The average must be 80.

$$\frac{87 + 64 + 78 + x}{4} = 80$$

Carry out:

$87 + 64 + 78 + x = 4·80$

$229 + x = 320$

$x = 91$

Check:

The average of 87, 64, 78, and 91 is $\frac{87 + 64 + 78 + 91}{4}$, or $\frac{320}{4}$, or 80.

State:

The score on the fourth test must be 91%.

26. 83%

27. Familiarize: We make a drawing.

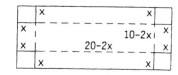

We let x represent the length of a side of the square in each corner. Then the length and width of the resulting base are represented by 20 - 2x and 10 - 2x.

Translate:

The area of the base is 96 cm².

$(20 - 2x)(10 - 2x) = 96$

Carry out:

$200 - 60x + 4x^2 = 96$

$4x^2 - 60x + 104 = 0$

$x^2 - 15x + 26 = 0$

$(x - 13)(x - 2) = 0$

$x - 13 = 0$ or $x - 2 = 0$

$x = 13$ or $x = 2$

Since the length and width of the base cannot be negative, we only consider x = 2.

Check:

If x = 2, then 20 - 2x = 16 and 10 - 2x = 6. The area of the base is 16·6, or 96 cm².

State:

The length of the sides of the squares is 2 cm.

28. 8 cm

29. Familiarize and Translate:

Let x = the former population.

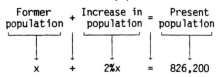

Carry out:

$x + 0.02x = 826,200$

$1.02x = 826,200$

$x = \frac{826,200}{1.02}$

$x = 810,000$

Check:

2%·810,000 = 0.02·810,000 = 16,200. The present population is 810,000 + 16,200, or 826,200.

State:

The former population was 810,000.

30. 720,000

31. Familiarize:

    We first make a drawing. We let r represent the speed of the boat in still water. Then the speed downstream is r + 3, and the speed upstream is r - 3. The time to go downstream is the same as the time to go upstream. We call it t.

    t hours   Downstream
    ────────────────────────────────────
    50 km   r + 3 km/h

    t hours   Upstream
    ◄────────────────────────────────────
    30 km   r - 3 km/h

    We organize the information in a table.

	Distance	Speed	Time
Downstream	50	r + 3	t
Upstream	30	r - 3	t

    Translate:

    Using t = d/r we can get two expressions for t from the table.

    $$t = \frac{50}{r + 3} \quad \text{and} \quad t = \frac{30}{r - 3}$$

    Thus

    $$\frac{50}{r + 3} = \frac{30}{r - 3}$$

    Carry out:

    $$(r + 3)(r - 3) \cdot \frac{50}{r + 3} = (r + 3)(r - 3) \cdot \frac{30}{r - 3}$$

    (Multiplying by the LCM)

    $$50(r - 3) = 30(r + 3)$$
    $$50r - 150 = 30r + 90$$
    $$50r - 30r = 90 + 150$$
    $$20r = 240$$
    $$r = 12$$

    Check:

    If r = 12, the speed downstream is 12 + 3, or 15 km/h, and thus the time downstream is 50/15, or $3.\overline{3}$ hr. If r = 12, the speed upstream is 12 - 3, or 9 km/h, and thus the time upstream is 30/9 or $3.\overline{3}$ hr. The times are equal; the value checks.

    State:

    The speed of the boat in still water is 12 km/h.

32. 4 km/h

33. Familiarize:

    We organize the information in a table. We let r represent the speed of train B. Then the speed of train A is r - 12. The time, t, is the same for each train.

	Distance	Speed	Time
Train A	230	r - 12	t
Train B	290	r	t

    Translate:

    Using t = d/r we get two expressions for t from the table.

    $$t = \frac{230}{r - 12} \quad \text{and} \quad t = \frac{290}{r}$$

    Thus

    $$\frac{230}{r - 12} = \frac{290}{r}$$

    Carry out:

    $$r(r - 12) \cdot \frac{230}{r - 12} = r(r - 12) \cdot \frac{290}{r}$$
    $$230r = 290(r - 12)$$
    $$230r = 290r - 3480$$
    $$3480 = 60r$$
    $$58 = r$$

    Check:

    If r = 58, then the speed of train A is 58 - 12, or 46 mph. Thus the time for train A is 230/46, or 5 hr. If the speed of train B is 58 mph, the time for train B is 290/58, or 5 hr. The times are the same; the value checks.

    State:

    The speed of train A is 46 mph; the speed of train B is 58 mph.

34. Freight train: 66 mph, passenger train: 80 mph

35. Familiarize:

    We first make a drawing.

    Chicago   Cleveland
    ──────────────────   ◄──────────────────
    475 mph   500 mph

    t hours $t - \frac{1}{3}$ hours

    d_1 miles d_2 miles
    ◄──────────────350 miles──────────────►

    We organize the information in a table.

	Distance	Speed	Time
From Chicago	d_1	475	t
From Cleveland	d_2	500	$t - \frac{1}{3}$

Translate:

Using $d = rt$ we get two equations from the table.

$d_1 = 475t$ $\qquad\qquad$ $d_2 = 500\left(t - \frac{1}{3}\right)$

We also know that $d_1 + d_2 = 350$. Thus

$475t + 500\left(t - \frac{1}{3}\right) = 350$

Carry out:

$475t + 500t - \frac{500}{3} = 350$

$975t = \frac{1050}{3} + \frac{500}{3}$

$975t = \frac{1550}{3}$

$2925t = 1550$

$t = \frac{1550}{2925}$

$t = \frac{62}{117}$

Substitute $\frac{62}{117}$ for t and solve for d_2.

$d_2 = 500\left(t - \frac{1}{3}\right)$

$d_2 = 500\left(\frac{62}{117} - \frac{1}{3}\right) = 500 \cdot \frac{23}{117} \approx 98.3$

Check:

If $t = \frac{62}{117}$, then $d_1 = 475 \cdot \frac{62}{117}$, or ≈ 251.7. The sum of the distances is $98.3 + 251.7$, or 350 miles. The value checks.

State:

When the trains meet, they are 98.3 miles from Cleveland.

36. 450 km

37. Familiarize:

A does $\frac{1}{3}$ of the job in 1 hr.

B does $\frac{1}{5}$ of the job in 1 hr.

C does $\frac{1}{7}$ of the job in 1 hr.

Together they do $\frac{1}{3} + \frac{1}{5} + \frac{1}{7} = \frac{71}{105}$ of the job in 1 hr. Together they do $2\left(\frac{1}{3}\right) + 2\left(\frac{1}{5}\right) + 2\left(\frac{1}{7}\right) = \frac{142}{105}$ or $1\frac{37}{105}$ in 2 hr. But $1\frac{37}{105}$ would represent more than 1 job.

We let t = the number of hours required for A, B, and C, working together, to do the job.
Translate:

$t\left(\frac{1}{3}\right) + t\left(\frac{1}{5}\right) + t\left(\frac{1}{7}\right) = 1$, or

$\qquad \frac{t}{3} + \frac{t}{5} + \frac{t}{7} = 1$

Carry out:

$\frac{t}{3} + \frac{t}{5} + \frac{t}{7} = 1$, LCM = $3 \cdot 5 \cdot 7$, or 105

$105\left(\frac{t}{3} + \frac{t}{5} + \frac{t}{7}\right) = 105 \cdot 1$

$35t + 21t + 15t = 105$

$71t = 105$

$t = \frac{105}{71}$

$t = 1\frac{34}{71}$

Check:

In $1\frac{34}{71}$ hr, A will do $\left(\frac{1}{3}\right)\left(\frac{105}{71}\right)$, or $\frac{35}{71}$ of the job. Then B will do $\frac{1}{5}\left(\frac{105}{71}\right)$, or $\frac{21}{71}$ of the job and C will do $\frac{1}{7}\left(\frac{105}{71}\right)$, or $\frac{15}{71}$, or the job. Together they will $\frac{35}{71} + \frac{21}{71} + \frac{15}{71}$, or 1 complete job.

We have another partial check in noting from the familiarization step that the entire job can be done between 1 hr and 2 hr.

State: It will take $1\frac{34}{71}$ hr for A, B, and C to do the job together.

38. $\frac{30}{13}$ hr

39. Familiarize: Let t = the amount of time it takes B to do the job, working alone. In 1 hr A can do $\frac{1}{3.15}$ of the job and B can do $\frac{1}{t}$ of the job. Then in 2.09 hr A can do $2.09\left(\frac{1}{3.15}\right)$ of the job and B can do $2.09\left(\frac{1}{t}\right)$ of the job. If we add these fractional parts we get the entire job, represented by 1.

Translate: We add as described in the previous step and set the result equal to 1.

$2.09\left(\frac{1}{3.15}\right) + 2.09\left(\frac{1}{t}\right) = 1$, or $\frac{2.09}{3.15} + \frac{2.09}{t} = 1$

Carry out: We first multiply by the LCM, 3.15t.

$3.15t\left(\frac{2.09}{3.15} + \frac{2.09}{t}\right) = 3.15t(1)$

$2.09t + 6.5835 = 3.15t$

$6.5835 = 1.06t$

$6.21 \approx t$

Check: In 2.09 hr A will do $\frac{2.09}{3.15}$ of the job and B will do $\frac{2.09}{6.21}$ of the job. Together they will do $\frac{2.09}{3.15} + \frac{2.09}{6.21} \approx 0.66 + 0.34$, or 1 complete job. The answer checks.

State: It would take B about 6.21 hr to do the job, working alone.

40. A: 23.95 hr, B: 51.02 hr

41. a) $A = P(1 + i)^t$
      $A = 1000(1 + 0.0875)^1$
      $= \$1087.50$

    b) $A = P\left(1 + \frac{i}{2}\right)^{2t}$
      $A = 1000\left(1 + \frac{0.0875}{2}\right)^2$
      $= \$1089.41$

    c) $A = P\left(1 + \frac{i}{4}\right)^{4t}$
      $A = 1000\left(1 + \frac{0.0875}{4}\right)^4$
      $= \$1090.41$

    d) $A = P\left(1 + \frac{i}{365}\right)^{365t}$
      $A = 1000\left(1 + \frac{0.0875}{365}\right)^{365}$
      $= \$1091.43$

    e) $A = P\left(1 + \frac{i}{8760}\right)^{8760t}$
      $A = 1000\left(1 + \frac{0.0875}{8760}\right)^{8760}$
      $= \$1091.44$

42. a) $1435.63; b) $1445.04; c) $1449.95;
    d) $1454.94; e) $1454.99

43. Familiarize:
    We let r represent the speed for the slow trip and t the time for the slow trip. Then $r + 4$ and $t - \frac{1}{2}$ represent the speed and time respectively for the fast trip. We organize the information in a table.

	Distance	Speed	Time
Slow trip	144	r	t
Fast trip	144	r + 4	$t - \frac{1}{2}$

    Translate:
    Using $t = d/r$ we get two equations from the table.

    $t = \frac{144}{r}$ and $t - \frac{1}{2} = \frac{144}{r + 4}$

    $\qquad\qquad\qquad$ or $t = \frac{144}{r + 4} + \frac{1}{2}$

    This gives us the following equation:
    $\frac{144}{r} = \frac{144}{r + 4} + \frac{1}{2}$

Carry out:
$2r(r + 4) \cdot \frac{144}{r} = 2r(r + 4)\cdot\left[\frac{144}{r + 4} + \frac{1}{2}\right]$

$\qquad\qquad\qquad$ (Multiplying by the LCM)
$\qquad 288(r + 4) = 288r + r(r + 4)$
$\qquad 288r + 1152 = 288r + r^2 + 4r$
$\qquad\qquad\quad 0 = r^2 + 4r - 1152$
$\qquad\qquad\quad 0 = (r + 36)(r - 32)$

$r + 36 = 0$ or $r - 32 = 0$
$\quad r = -36$ or $\qquad r = 32$

Check:
We only consider $r = 32$ since the speed in this problem cannot be negative. If $r = 32$, then the time for the slow trip is $\frac{144}{32}$, or 4.5 hr. If $r = 32$, then $r + 4 = 36$. Thus, the time for the fast trip is $\frac{144}{36}$, or 4 hr. The time for the fast trip is $\frac{1}{2}$ hr less than the time for the slow trip. The value checks.

State:
The car's speed is 32 mph.

44. 35 mph

45. Familiarize:
    Freeway: $d = rt$
    $\qquad\qquad d = 55\cdot3$, or 165

    City: $t = \frac{d}{r}$
    $\qquad\qquad t = \frac{10}{35}$, or $\frac{2}{7}$

    We organize the information in a table.

	Distance	Speed	Time
Freeway	165	55	3
City	10	35	$\frac{10}{35}$, or $\frac{2}{7}$
Totals	175		$3\frac{2}{7}$ or $\frac{23}{7}$

    Translate:
    Average speed $= \frac{\text{Total distance}}{\text{Total time}}$

    Average speed $= \frac{175}{\frac{23}{7}}$

    Carry out: We simplify.
    Average speed $= \frac{175}{\frac{23}{7}}$

    $\qquad\qquad = 175 \cdot \frac{7}{23}$

    $\qquad\qquad = \frac{1225}{23}$, or $53\frac{6}{23}$

Check: At $53\frac{6}{23}$ mph, the distance traveled in $3\frac{2}{7}$ hr is $\left(53\frac{6}{23}\right)\cdot\left(3\frac{2}{7}\right) = \frac{1225}{23} \cdot \frac{23}{7} = 175$ mi.

This is the total distance the student traveled, so the answer checks.

State: The average speed was $53\frac{6}{23}$ mph.

46. 48 km/h

47. Familiarize:

Making a drawing is helpful.

40 mph	$\frac{d}{40}$ hours		r mph	$\frac{d}{r}$ hours
d miles		.	d miles	

$\longleftarrow$————————2d miles————————$\longrightarrow$

We let d represent the distance of the first half of the trip. Then 2d represents the total distance. We also let r represent the speed for the second half of the trip. The times are represented by $\frac{d}{40}$ and $\frac{d}{r}$. The total time is $\frac{d}{40} + \frac{d}{r}$.

Translate:

Average speed = $\dfrac{\text{Total distance}}{\text{Total time}}$

Substituting we get

$$45 = \frac{2d}{\frac{d}{40} + \frac{d}{r}}.$$

Carry out:

$$45 = \frac{2d}{\frac{d}{40} + \frac{d}{r}}$$

$$45\left(\frac{d}{40} + \frac{d}{r}\right) = 2d$$

$$45\left(\frac{1}{40} + \frac{1}{r}\right)d = 2d$$

$$\frac{45}{40} + \frac{45}{r} = 2$$

$$\frac{45}{r} = 2 - \frac{45}{40}$$

$$\frac{45}{r} = \frac{35}{40}$$

$$40\cdot 45 = r\cdot 35$$

$$1800 = 35r$$

$$\frac{1800}{35} = r$$

$$51\frac{3}{7} = r$$

Check: This value checks.

State: The speed for the second half of the trip would have to be $51\frac{3}{7}$ mph.

49. Familiarize:

It is helpful to make drawings.

At 10:30 the minute hand is 30 units after the 12. The hour hand is $52\frac{1}{2}$ units after the 12. Let x represent the number of units the minute hand moves before the hands are perpendicular for the first time. When the minute hand moves x units, the hour hand moves $\frac{1}{12}$ x units. Then the minute hand is 30 + x units after the 12, and the hour hand is $52\frac{1}{2} + \frac{1}{12}$ x units after the 12. When the hands are perpendicular, they must be 15 units apart.

Translate: We subtract to find the number of units between the hands. This difference must be 15 units.

$$\left(52\frac{1}{2} + \frac{1}{12}x\right) - (30 + x) = 15$$

Carry out:

$$\left(52\frac{1}{2} + \frac{1}{12}x\right) - (30 + x) = 15$$

$$52\frac{1}{2} + \frac{1}{12}x - 30 - x = 15$$

$$-\frac{11}{12}x + 22\frac{1}{2} = 15$$

$$\frac{15}{2} = \frac{11}{12}x$$

$$\frac{12}{11}\cdot\frac{15}{2} = x$$

$$8\frac{2}{11} = \frac{90}{11} = x$$

Check: This answer checks:

State: After 10:30, the hands will first be perpendicular in $8\frac{2}{11}$ minutes, or at $10:38\frac{2}{11}$.

50. $\frac{2}{3}$ hr

51. a) Familiarize: We organize the information in a table. We let d = the distance from Los Angeles at which it takes the same amount of time to return to Los Angeles as it does to go on to Honolulu. We let t = the time for either part of the trip.

	Distance	Speed	Time
Return to L.A.	d	750 – 50, or 700	t
Go on to Honolulu	2574 – d	750 + 50, or 800	t

Translate: Using the formula t = d/r and the rows of the table we get two equations:

$$t = \frac{d}{700} \quad \text{and} \quad t = \frac{2574 - d}{800}$$

Thus,

$$\frac{d}{700} = \frac{2574 - d}{800}.$$

Carry out:

$$\frac{d}{700} = \frac{2574 - d}{800}$$

$$700(800) \cdot \frac{d}{700} = 700(800) \cdot \frac{2574 - d}{800}$$

$$800d = 700(2574 - d)$$

$$800d = 1,801,800 - 700d$$

$$1500d = 1,801,800$$

$$d = 1201.2$$

Check: This answer checks.

State: At a distance of 1201.2 mi from Los Angeles it takes the same amount of time to return to Los Angeles as it does to go on to Honolulu.

b) Since 1187 mi < 1201.2 mi, less time is required to return to Los Angeles.

52. 12 mi

53. Familiarize: Let m = the amount of money in your father's pocket at the outset. After giving your mother half of his money, your father has $\frac{1}{2} \cdot m$, or $\frac{m}{2}$ left. After giving your sister one-fourth of $\frac{m}{2}$, he has $\frac{3}{4} \cdot \frac{m}{2}$, or $\frac{3m}{8}$, left. After giving your brother one-third of $\frac{3m}{8}$, he has $\frac{2}{3} \cdot \frac{3m}{8}$, or $\frac{m}{4}$, left. After giving you one-half of $\frac{m}{4}$, he has $\frac{1}{2} \cdot \frac{m}{4}$, or $\frac{m}{8}$, left. This final amount is $2.

Translate:

Final amount, is $2.

$$\frac{m}{8} = 2$$

Carry out:

$$\frac{m}{8} = 2$$

$$m = 16$$

Check: This value checks.

State: Your father had $16 at the outset.

54. 15

Exercise Set 2.4

1. $\sqrt{-15} = \sqrt{-1 \cdot 15} = \sqrt{-1} \sqrt{15} = i\sqrt{15}$

2. $i\sqrt{17}$

3. $\sqrt{-81} = \sqrt{-1 \cdot 81} = \sqrt{-1} \sqrt{81} = 9i$

4. $5i$

5. $-\sqrt{-12} = -\sqrt{-1 \cdot 4 \cdot 3} = -\sqrt{-1} \sqrt{4} \sqrt{3} = -2i\sqrt{3}$

6. $-2i\sqrt{5}$

7. $\sqrt{-16} + \sqrt{-25} = i\sqrt{16} + i\sqrt{25} = 4i + 5i = 9i$

8. $4i$

9. $\sqrt{-7} - \sqrt{-10} = i\sqrt{7} - i\sqrt{10} = (\sqrt{7} - \sqrt{10})i$

10. $(\sqrt{5} + \sqrt{7})i$

11. $\sqrt{-5} \sqrt{-11} = i\sqrt{5} \cdot i\sqrt{11} = i^2 \sqrt{5} \sqrt{11} = -\sqrt{55}$

12. $-2\sqrt{14}$

13. $-\sqrt{-4} \sqrt{-5} = -(i\sqrt{4} \cdot i\sqrt{5}) = -(i^2 \cdot 2 \cdot \sqrt{5})$
$$= -(-1)2\sqrt{5}$$
$$= 2\sqrt{5}$$

14. $3\sqrt{7}$

15. $\dfrac{-\sqrt{5}}{\sqrt{-2}} = \dfrac{-\sqrt{5}}{i\sqrt{2}} = \dfrac{-\sqrt{5}}{i\sqrt{2}} \cdot \dfrac{i}{i} = \dfrac{-i\sqrt{5}}{i^2\sqrt{2}}$
$$= \dfrac{-i\sqrt{5}}{-1\sqrt{2}}$$
$$= \sqrt{\dfrac{5}{2}}\, i$$

16. $-\sqrt{\dfrac{7}{5}}\, i$

17. $\dfrac{\sqrt{-9}}{-\sqrt{4}} = \dfrac{3i}{-2} = -\dfrac{3}{2}\, i$

18. $\dfrac{5}{4}\, i$

19. $\dfrac{-\sqrt{-36}}{\sqrt{-9}} = \dfrac{-i\sqrt{36}}{i\sqrt{9}} = -\dfrac{6}{3} = -2$

20. $-\dfrac{5}{4}$

21. $i^{18} = (i^2)^9 = (-1)^9 = -1$

22. -1

23. $i^{15} = (i^2)^7 \cdot i = (-1)^7 \cdot i = -1 \cdot i = -i$

24. 1

25. $i^{39} = (i^2)^{19} \cdot i = (-1)^{19} \cdot i = -1 \cdot i = -i$

26. 1

27. $i^{46} = (i^2)^{23} = (-1)^{23} = -1$

28. 1

29. $(2 + 3i) + (4 + 2i) = 2 + 4 + 3i + 2i = 6 + 5i$

30. $11 + i$

31. $(4 + 3i) + (4 - 3i) = 4 + 4 + 3i - 3i = 8$

32. 0

33. $(8 + 11i) - (6 + 7i) = 8 + 11i - 6 - 7i$
$$= 8 - 6 + 11i - 7i$$
$$= 2 + 4i$$

34. $5 - 7i$

35. $2i - (4 + 3i) = 2i - 4 - 3i = -4 - i$

36. $-5 + i$

37. $(1 + 2i)(1 + 3i) = 1 + 3i + 2i + 6i^2$
$$= 1 + 5i - 6 \qquad (i^2 = -1)$$
$$= -5 + 5i$$

38. $13 + i$

39. $(1 + 2i)(1 - 3i) = 1 - 3i + 2i - 6i^2$
$$= 1 - i + 6 \qquad (i^2 = -1)$$
$$= 7 - i$$

40. 13

41. $3i(4 + 2i) = 12i + 6i^2 = 12i - 6$, or $-6 + 12i$

42. $20 + 15i$

43. $(2 + 3i)^2 = 4 + 12i + 9i^2$
$$= 4 + 12i - 9 \qquad (i^2 = -1)$$
$$= -5 + 12i$$

44. $5 - 12i$

45. $4x^2 + 25y^2 = (2x + 5yi)(2x - 5yi)$

Check by multiplying:
$(2x + 5yi)(2x - 5yi) = 4x^2 - 25y^2i^2 = 4x^2 + 25y^2$

46. $(4a + 7bi)(4a - 7bi)$

47.
$$\frac{\begin{array}{r} x^2 - 2x + 5 = 0 \end{array}}{\begin{array}{l} (1 + 2i)^2 - 2(1 + 2i) + 5 \\ 1 + 4i + 4i^2 - 2 - 4i + 5 \\ \qquad\qquad 4i^2 + 4 \\ \qquad\qquad\quad -4 + 4 \\ \qquad\qquad\qquad\quad 0 \end{array}} \Bigg| \begin{array}{l} 0 \\ \\ \\ \\ \\ \end{array}$$

The number $1 + 2i$ is a solution.

48. Yes

49. $4x + 7i = -6 + yi$

We equate the real parts and solve for x.
$$4x = -6$$
$$x = -\frac{3}{2}$$

We equate the imaginary parts and solve for y.
$$7i = yi$$
$$7 = y$$

50. $x = -1$, $y = -\frac{3}{5}$

51. $\dfrac{4 + 3i}{1 - i} = \dfrac{4 + 3i}{1 - i} \cdot \dfrac{1 + i}{1 + i}$
$$= \frac{4 + 7i + 3i^2}{1 - i^2} = \frac{1 + 7i}{2} = \frac{1}{2} + \frac{7}{2}i$$

52. $\dfrac{22}{41} - \dfrac{7}{41}i$

53. $\dfrac{\sqrt{2} + i}{\sqrt{2} - i} = \dfrac{\sqrt{2} + i}{\sqrt{2} - i} \cdot \dfrac{\sqrt{2} + i}{\sqrt{2} + i}$
$$= \frac{2 + 2\sqrt{2}\,i + i^2}{2 - i^2} = \frac{1 + 2\sqrt{2}\,i}{3} = \frac{1}{3} + \frac{2\sqrt{2}}{3}i$$

54. $\dfrac{1}{2} + \dfrac{\sqrt{3}}{2}i$

55. $\dfrac{3 + 2i}{i} = \dfrac{3 + 2i}{i} \cdot \dfrac{-i}{-i}$
$$= \frac{-3i - 2i^2}{-i^2} = \frac{2 - 3i}{1} = 2 - 3i$$

56. $3 - 2i$

57. $\dfrac{1}{2 + i} = \dfrac{1}{2 + i} \cdot \dfrac{2 - i}{2 - i}$
$$= \frac{2i - i^2}{4 - i^2} = \frac{1 + 2i}{5} = \frac{1}{5} + \frac{2}{5}i$$

58. $\dfrac{15}{146} + \dfrac{33}{146}i$

59. $\dfrac{1 - i}{(1 + i)^2} = \dfrac{1 - i}{1 + 2i + i^2} = \dfrac{1 - i}{2i}$
$$= \frac{1 - i}{2i} \cdot \frac{-2i}{-2i} = \frac{-2i + 2i^2}{-4i^2}$$
$$= \frac{-2 - 2i}{4} = -\frac{1}{2} - \frac{1}{2}i$$

60. $-\dfrac{1}{2} + \dfrac{1}{2}i$

61. $\dfrac{3 - 4i}{(2 + i)(3 - 2i)} = \dfrac{3 - 4i}{6 - i - 2i^2}$
$$= \frac{3 - 4i}{8 - i} \cdot \frac{8 + i}{8 + i}$$
$$= \frac{24 - 29i - 4i^2}{64 - i^2}$$
$$= \frac{28 - 29i}{65} = \frac{28}{65} - \frac{29}{65}i$$

62. $\dfrac{719}{3233} + \dfrac{955}{3233}i$

63. $\dfrac{1+i}{1-i} \cdot \dfrac{2-i}{1-i} = \dfrac{2+i-i^2}{1-2i+i^2}$

$\qquad = \dfrac{3+i}{-2i} \cdot \dfrac{2i}{2i}$

$\qquad = \dfrac{6i+2i^2}{-4i^2}$

$\qquad = \dfrac{-2+6i}{4} = -\dfrac{1}{2} + \dfrac{3}{2}i$

64. $-\dfrac{1}{2} - \dfrac{3}{2}i$

65. $\dfrac{3+2i}{1-i} + \dfrac{6+2i}{1-i} = \dfrac{3+6+2i+2i}{1-i}$

$\qquad = \dfrac{9+4i}{1-i} \cdot \dfrac{1+i}{1+i}$

$\qquad = \dfrac{9+13i+4i^2}{1-i^2}$

$\qquad = \dfrac{5+13i}{2} = \dfrac{5}{2} + \dfrac{13}{2}i$

66. $-\dfrac{1}{2} - \dfrac{13}{2}i$

67. The reciprocal of $4+3i$ is $\dfrac{1}{4+3i}$, or

$\dfrac{1}{4+3i} \cdot \dfrac{4-3i}{4-3i} = \dfrac{4-3i}{16-9i^2} = \dfrac{4-3i}{25} = \dfrac{4}{25} - \dfrac{3}{25}i$.

68. $\dfrac{4}{25} + \dfrac{3}{25}i$

69. The reciprocal of $5-2i$ is $\dfrac{1}{5-2i}$, or

$\dfrac{1}{5-2i} \cdot \dfrac{5+2i}{5+2i} = \dfrac{5+2i}{25-4i^2} = \dfrac{5+2i}{29} = \dfrac{5}{29} + \dfrac{2}{29}i$.

70. $\dfrac{2}{29} - \dfrac{5}{29}i$

71. The reciprocal of i is $\dfrac{1}{i}$, or

$\dfrac{1}{i} \cdot \dfrac{-i}{-i} = \dfrac{-i}{-i^2} = \dfrac{-i}{1} = -i$.

72. i

73. The reciprocal of $-4i$ is $\dfrac{1}{-4i}$, or

$\dfrac{1}{-4i} \cdot \dfrac{4i}{4i} = \dfrac{4i}{-16i^2} = \dfrac{4i}{16} = \dfrac{1}{4}i$.

74. $-\dfrac{1}{5}i$

75. $(3+i)x + i = 5i$

$\qquad (3+i)x = 4i$

$\qquad x = \dfrac{4i}{3+i}$

$\qquad x = \dfrac{4i}{3+i} \cdot \dfrac{3-i}{3-i}$

$\qquad x = \dfrac{12i - 4i^2}{9 - i^2} = \dfrac{4+12i}{10}$

$\qquad x = \dfrac{2}{5} + \dfrac{6}{5}i$

76. $\dfrac{12}{5} - \dfrac{1}{5}i$

77. $\qquad 2ix + 5 - 4i = (2+3i)x - 2i$

$\qquad 5 - 4i + 2i = (2+3i)x - 2ix$

$\qquad 5 - 2i = (2+i)x$

$\qquad \dfrac{5-2i}{2+i} = x$

$\qquad \dfrac{2-i}{2-i} \cdot \dfrac{5-2i}{2+i} = x$

$\qquad \dfrac{8-9i}{5} = \dfrac{10-9i+2i^2}{4-i^2} = x$

$\qquad \dfrac{8}{5} - \dfrac{9}{5}i = x$

78. $\dfrac{8}{29} + \dfrac{9}{29}i$

79. $(1+2i)x + 3 - 2i = 4 - 5i + 3ix$

$\qquad (1+2i)x - 3ix = 4 - 5i - 3 + 2i$

$\qquad (1-i)x = 1 - 3i$

$\qquad x = \dfrac{1-3i}{1-i}$

$\qquad x = \dfrac{1-3i}{1-i} \cdot \dfrac{1+i}{1+i}$

$\qquad x = \dfrac{1-2i-3i^2}{1-i^2} = \dfrac{4-2i}{2}$

$\qquad x = 2 - i$

80. $-\dfrac{1}{5} + \dfrac{7}{5}i$

81. $(5+i)x + 1 - 3i = (2-3i)x + 2 - i$

$\qquad (5+i)x - (2-3i)x = 2 - i - 1 + 3i$

$\qquad (3+4i)x = 1 + 2i$

$\qquad x = \dfrac{1+2i}{3+4i}$

$\qquad x = \dfrac{1+2i}{3+4i} \cdot \dfrac{3-4i}{3-4i}$

$\qquad x = \dfrac{3+2i-8i^2}{9-16i^2} = \dfrac{11+2i}{25}$

$\qquad x = \dfrac{11}{25} + \dfrac{2}{25}i$

82. $\dfrac{4}{5} + \dfrac{3}{5}i$

83. For example, $\sqrt{-1}\,\sqrt{-1} = i^2 = -1$
    but $\sqrt{(-1)(-1)} = \sqrt{1} = 1$.

84. For example, $\sqrt{\dfrac{4}{-1}} = \sqrt{-4} = 2i$, but $\dfrac{\sqrt{4}}{\sqrt{-1}} = \dfrac{2}{i} = -2i$

85. Let $z = a + bi$. Then $z \cdot \bar{z} = (a + bi)(a - bi) =$ $a^2 - b^2 i^2 = a^2 + b^2$. Since a and b are real numbers, so is $a^2 + b^2$. Thus $z \cdot \bar{z}$ is real.

86. Let $z = a + bi$. Then $z + \bar{z} = (a + bi) + (a - bi)$ $= 2a$. Since a is a real number, 2a is real. Thus $z + \bar{z}$ is real.

87. Let $z = a + bi$ and $w = c + di$. Then $\overline{z + w} =$ $\overline{(a + bi) + (c + di)} = \overline{(a + c) + (b + d)i} =$ $(a + c) - (b + d)i$. Now $\bar{z} + \bar{w} =$ $\overline{a + bi} + \overline{c + di} = (a - bi) + (c - di) =$ $(a + c) - (b + d)i$, the same result as before. Thus $\overline{z + w} = \bar{z} + \bar{w}$.

88. Let $z = a + bi$ and $w = c + di$. Then $\overline{z \cdot w} =$ $\overline{(a + bi)(c + di)} = \overline{(ac - bd) + (ad + bc)i} =$ $(ac - bd) - (ad + bc)i$. Now $\bar{z} \cdot \bar{w} =$ $\overline{(a + bi)} \cdot \overline{(c + di)} = (a - bi)(c - di) =$ $(ac - bd) - (ad + bc)i$, the same result as before. Thus $\overline{z \cdot w} = \bar{z} \cdot \bar{w}$.

89. By definition of exponents the conjugate of z^n is the conjugate of the product of n factors of z. Using the result of Exercise 88, the conjugate of n factors of z is the product of n factors of $\bar{z}$. Thus $\overline{z^n} = \bar{z}^n$.

90. If z is a real number, then $z = a + 0i = a$ and $\bar{z} = a - 0i = a$. Thus $\bar{z} = z$.

91. $\overline{3z^5 - 4z^2 + 3z - 5}$
    $= \overline{3z^5} - \overline{4z^2} + \overline{3z} - \overline{5}$ By Exercise 87
    $= \bar{3}\,\overline{z^5} - \bar{4}\,\overline{z^2} + \bar{3}\,\bar{z} - \bar{5}$ By Exercise 88
    $= 3\,\overline{z^5} - 4\,\overline{z^2} + 3\,\bar{z} - 5$ By Exercise 90
    $= 3\bar{z}^5 - 4\bar{z}^2 + 3\bar{z} - 5$ By Exercise 89

92. 1

93. Solve $5z - 4\bar{z} = 7 + 8i$ for z.
    Let $z = a + bi$. Then $\bar{z} = a - bi$.
    $5z - 4\bar{z} = 7 + 8i$
    $5(a + bi) - 4(a - bi) = 7 + 8i$
    $5a + 5bi - 4a + 4bi = 7 + 8i$
    $a + 9bi = 7 + 8i$

Equate the real parts.
    $a = 7$
Equate the imaginary parts.
    $9b = 8$
    $b = \dfrac{8}{9}$
Then $z = 7 + \dfrac{8}{9} i$.

94. a

95. Let $z = a + bi$. Then $\bar{z} = a - bi$.
    $\dfrac{1}{2}(\bar{z} - z) = \dfrac{1}{2}[(a - bi) - (a + bi)]$
    $= \dfrac{1}{2}(-2bi) = -bi$

96. $1 - i, -1 + i$

97. $\dfrac{(a + bi)^2}{4} = i$, or $a^2 - b^2 + 2abi = 4i$
    Solving $\begin{matrix} a^2 - b^2 = 0 \\ 2ab = 4 \end{matrix}$, we get $a = \sqrt{2}$, $b = \sqrt{2}$ or $a = -\sqrt{2}$, $b = -\sqrt{2}$. Then $z = \sqrt{2} + i\sqrt{2}$ or $z = -\sqrt{2} - i\sqrt{2}$.

98. $\dfrac{a}{a^2 + b^2} + \dfrac{-b}{a^2 + b^2} i$

99. $\dfrac{w}{z} = \dfrac{c + di}{a + bi} \cdot \dfrac{a - bi}{a - bi} = \dfrac{ac + bd}{a^2 + b^2} + \dfrac{ad - bc}{a^2 + b^2} i$

Exercise Set 2.5

1. $3x^2 = 27$
   $x^2 = 9$ Multiplying by $\dfrac{1}{3}$
   $x = 3$ or $x = -3$ Taking square roots
   The solution set is {3,-3}, or {± 3}.

2. {± 4}

3. $x^2 = -1$
   $x = \sqrt{-1}$ or $x = -\sqrt{-1}$ Taking square roots
   $x = i$ or $x = -i$ Simplifying
   The solution set is {i,-i}, or {± i}.

4. {± 2i}

5. $4x^2 = 20$
   $x^2 = 5$
   $x = \sqrt{5}$ or $x = -\sqrt{5}$
   The solution set is {± $\sqrt{5}$}.

<u>6.</u> $\{\pm\sqrt{7}\}$

<u>7.</u> $10x^2 = 0$
$x^2 = 0$
$x = 0$

The solution set is $\{0\}$.

<u>8.</u> $\{0\}$

<u>9.</u> $2x^2 - 3 = 0$
$2x^2 = 3$
$x^2 = \frac{3}{2}$

$x = \sqrt{\frac{3}{2}}$ or $x = -\sqrt{\frac{3}{2}}$

$x = \sqrt{\frac{3}{2}\cdot\frac{2}{2}}$ or $x = -\sqrt{\frac{3}{2}\cdot\frac{2}{2}}$

$x = \frac{\sqrt{6}}{2}$ or $x = -\frac{\sqrt{6}}{2}$

The solution set is $\left\{\pm\frac{\sqrt{6}}{2}\right\}$.

<u>10.</u> $\left\{\pm\frac{\sqrt{21}}{3}\right\}$

<u>11.</u> $2x^2 + 14 = 0$
$2x^2 = -14$
$x^2 = -7$

$x = \sqrt{-7}$ or $x = -\sqrt{-7}$
$x = i\sqrt{7}$ or $x = -i\sqrt{7}$

The solution set is $\{\pm\ i\sqrt{7}\}$.

<u>12.</u> $\{\pm\ i\sqrt{5}\}$

<u>13.</u> $ax^2 = b$
$x^2 = \frac{b}{a}$

$x = \pm\sqrt{\frac{b}{a}}$

The solution set is $\left\{\pm\sqrt{\frac{b}{a}}\right\}$.

<u>14.</u> $\left\{\pm\sqrt{\frac{k}{\pi}}\right\}$

<u>15.</u> $(x - 7)^2 = 5$
$x - 7 = \pm\sqrt{5}$
$x = 7 \pm\sqrt{5}$

The solution set is $\{7 \pm\sqrt{5}\}$.

<u>16.</u> $\{-3 \pm\sqrt{2}\}$

<u>17.</u> $\frac{4}{9}x^2 - 1 = 0$

$\frac{4}{9}x^2 = 1$

$x^2 = \frac{9}{4}$

$x = \sqrt{\frac{9}{4}}$ or $x = -\sqrt{\frac{9}{4}}$

$x = \frac{3}{2}$ or $x = -\frac{3}{2}$

The solution set is $\left\{\pm\frac{3}{2}\right\}$.

<u>18.</u> $\left\{\pm\frac{5}{4}\right\}$

<u>19.</u> $(x - h)^2 - 1 = a$
$(x - h)^2 = a + 1$
$x - h = \pm\sqrt{a + 1}$
$x = h \pm\sqrt{a + 1}$

The solution set is $\{h \pm\sqrt{a + 1}\}$.

<u>20.</u> $\left\{h \pm\sqrt{\frac{y - k}{a}}\right\}$

<u>21.</u> $x^2 + 6x + 4 = 0$
$x^2 + 6x + 9 - 9 + 4 = 0$ $\left[\frac{1}{2}\cdot 6 = 3,\ 3^2 = 9;\right.$
$\left.\text{thus we add } 9 - 9.\right]$

$(x + 3)^2 - 5 = 0$
$(x + 3)^2 = 5$
$x + 3 = \pm\sqrt{5}$
$x = -3 \pm\sqrt{5}$

The solution set is $\{-3 \pm\sqrt{5}\}$.

<u>22.</u> $\left\{3 \pm\sqrt{13}\right\}$

<u>23.</u> $y^2 + 7y - 30 = 0$
$y^2 + 7y + \frac{49}{4} - \frac{49}{4} - 30 = 0$ $\left[\frac{1}{2}\cdot 7 = \frac{7}{2},\ \left(\frac{7}{2}\right)^2 = \frac{49}{4};\right.$
$\left.\text{thus we add } \frac{49}{4} - \frac{49}{4}\right]$

$\left(y + \frac{7}{2}\right)^2 - \frac{169}{4} = 0$

$\left(y + \frac{7}{2}\right)^2 = \frac{169}{4}$

$y + \frac{7}{2} = \pm\sqrt{\frac{169}{4}} = \pm\frac{13}{2}$

$y = -\frac{7}{2} \pm\frac{13}{2} = \frac{-7 \pm 13}{2}$

$y = \frac{-7 + 13}{2}$ or $y = \frac{-7 - 13}{2}$

$y = 3$ or $y = -10$

The solution set is $\{3, -10\}$.

<u>24.</u> $\{10, -3\}$

<u>25.</u>
$$5x^2 - 4x - 2 = 0$$
$$x^2 - \frac{4}{5}x - \frac{2}{5} = 0 \quad \left(\text{Multiplying by } \frac{1}{5}\right)$$
$$x^2 - \frac{4}{5}x + \frac{4}{25} - \frac{4}{25} - \frac{2}{5} = 0 \quad \left(\text{Adding } \frac{4}{25} - \frac{4}{25}\right)$$
$$\left(x - \frac{2}{5}\right)^2 - \frac{14}{25} = 0$$
$$\left(x - \frac{2}{5}\right)^2 = \frac{14}{25}$$
$$x - \frac{2}{5} = \pm\sqrt{\frac{14}{25}} = \pm\frac{\sqrt{14}}{5}$$
$$x = \frac{2}{5} \pm \frac{\sqrt{14}}{5} = \frac{2 \pm \sqrt{14}}{5}$$

The solution set is $\left\{\dfrac{2 \pm \sqrt{14}}{5}\right\}$.

<u>26.</u> $\left\{\dfrac{7 \pm \sqrt{13}}{12}\right\}$

<u>27.</u>
$$2x^2 + 7x - 15 = 0$$
$$x^2 + \frac{7}{2}x - \frac{15}{2} = 0 \quad \left(\text{Multiplying by } \frac{1}{2}\right)$$
$$x^2 + \frac{7}{2}x + \frac{49}{16} - \frac{49}{16} - \frac{15}{2} = 0 \quad \left(\text{Adding } \frac{49}{16} - \frac{49}{16}\right)$$
$$\left(x + \frac{7}{4}\right)^2 - \frac{169}{16} = 0$$
$$\left(x + \frac{7}{4}\right)^2 = \frac{169}{16}$$
$$x + \frac{7}{4} = \pm\sqrt{\frac{169}{16}} = \pm\frac{13}{4}$$
$$x = -\frac{7}{4} \pm \frac{13}{4} = \frac{-7 \pm 13}{4}$$

$x = \dfrac{-7 + 13}{4}$ or $x = \dfrac{-7 - 13}{4}$

$x = \dfrac{3}{2}$ or $x = -5$

The solution set is $\left\{\dfrac{3}{2}, -5\right\}$.

<u>28.</u> $\left\{\dfrac{5}{3}\right\}$

<u>29.</u>
$$x^2 + 4x = 5$$
$$x^2 + 4x - 5 = 0$$
$$a = 1, \quad b = 4, \quad c = -5$$
$$x = \frac{-b \pm \sqrt{b^2 - 4ac}}{2a}$$
$$x = \frac{-4 \pm \sqrt{4^2 - 4(1)(-5)}}{2(1)} = \frac{-4 \pm \sqrt{16 + 20}}{2}$$
$$x = \frac{-4 \pm \sqrt{36}}{2} = \frac{-4 \pm 6}{2}$$

$x = \dfrac{-4 + 6}{2}$ or $x = \dfrac{-4 - 6}{2}$

$x = 1$ or $x = -5$

The solution set is $\{1, -5\}$.

<u>30.</u> $\{5, -3\}$

<u>31.</u> $2y^2 - 3y - 2 = 0$
$$a = 2, \quad b = -3, \quad c = -2$$
$$y = \frac{-b \pm \sqrt{b^2 - 4ac}}{2a}$$
$$y = \frac{-(-3) \pm \sqrt{(-3)^2 - 4(2)(-2)}}{2(2)} = \frac{3 \pm \sqrt{9 + 16}}{4}$$
$$y = \frac{3 \pm \sqrt{25}}{4} = \frac{3 \pm 5}{4}$$

$y = \dfrac{3 + 5}{4}$ or $y = \dfrac{3 - 5}{4}$

$y = 2$ or $y = -\dfrac{1}{2}$

The solution set is $\left\{2, -\dfrac{1}{2}\right\}$.

<u>32.</u> $\left\{\dfrac{2}{5}, -1\right\}$

<u>33.</u> $3t^2 + 8t + 3 = 0$
$$a = 3, \quad b = 8, \quad c = 3$$
$$t = \frac{-b \pm \sqrt{b^2 - 4ac}}{2a}$$
$$t = \frac{-8 \pm \sqrt{8^2 - 4 \cdot 3 \cdot 3}}{2 \cdot 3} = \frac{-8 \pm \sqrt{64 - 36}}{6}$$
$$t = \frac{-8 \pm \sqrt{28}}{6} = \frac{-8 \pm 2\sqrt{7}}{6} = \frac{2(-4 \pm \sqrt{7})}{6}$$
$$t = \frac{-4 \pm \sqrt{7}}{3}$$

The solution set is $\left\{\dfrac{-4 \pm \sqrt{7}}{3}\right\}$.

<u>34.</u> $\{3 \pm \sqrt{7}\}$

35. $3 + u^2 = 12u$

$u^2 - 12u + 3 = 0$

$a = 1,\quad b = -12,\quad c = 3$

$u = \dfrac{-b \pm \sqrt{b^2 - 4ac}}{2a}$

$u = \dfrac{-(-12) \pm \sqrt{(-12)^2 - 4 \cdot 1 \cdot 3}}{2 \cdot 1} = \dfrac{12 \pm \sqrt{144 - 12}}{2}$

$u = \dfrac{12 \pm \sqrt{132}}{2} = \dfrac{12 \pm 2\sqrt{33}}{2}$

$u = \dfrac{2(6 \pm \sqrt{33})}{2} = 6 \pm \sqrt{33}$

The solution set is $\{6 \pm \sqrt{33}\}$.

36. $\{-2, -4\}$

37. $x^2 - x + 1 = 0$

$a = 1,\qquad b = -1,\qquad c = 1$

$x = \dfrac{-b \pm \sqrt{b^2 - 4ac}}{2a}$

$x = \dfrac{-(-1) \pm \sqrt{(-1)^2 - 4 \cdot 1 \cdot 1}}{2 \cdot 1}$

$x = \dfrac{1 \pm \sqrt{1 - 4}}{2} = \dfrac{1 \pm \sqrt{-3}}{2}$

$x = \dfrac{1 \pm i\sqrt{3}}{2}$

The solution set is $\left\{\dfrac{1 \pm i\sqrt{3}}{2}\right\}$.

38. $\left\{\dfrac{-1 \pm i\sqrt{7}}{2}\right\}$

39. $x^2 + 13 = 4x$

$x^2 - 4x + 13 = 0$

$a = 1,\qquad b = -4,\qquad c = 13$

$x = \dfrac{-b \pm \sqrt{b^2 - 4ac}}{2a}$

$x = \dfrac{-(-4) \pm \sqrt{(-4)^2 - 4 \cdot 1 \cdot 13}}{2 \cdot 1}$

$x = \dfrac{4 \pm \sqrt{16 - 52}}{2} = \dfrac{4 \pm \sqrt{-36}}{2}$

$x = \dfrac{4 \pm 6i}{2} = \dfrac{2(2 \pm 3i)}{2} = 2 \pm 3i$

The solution set is $\{2 \pm 3i\}$.

40. $\left\{\dfrac{-1 \pm 2i}{5}\right\}$

41. $5x^2 = 13x + 17$

$5x^2 - 13x - 17 = 0$ Standard form

$a = 5,\; b = -13,\; c = -17$

$x = \dfrac{-b \pm \sqrt{b^2 - 4ac}}{2a}$

$x = \dfrac{-(-13) \pm \sqrt{(-13)^2 - 4(5)(-17)}}{2 \cdot 5}$

$x = \dfrac{13 \pm \sqrt{169 + 340}}{10}$

$x = \dfrac{13 \pm \sqrt{509}}{10}$

The solution set is $\left\{\dfrac{13 \pm \sqrt{509}}{10}\right\}$.

42. $\left\{\dfrac{5 \pm \sqrt{73}}{6}\right\}$

43. $0.03 + 0.08v = v^2$

$0 = v^2 - 0.08v - 0.03$ Standard form

$a = 1,\; b = -0.08,\; c = -0.03$

$v = \dfrac{-b \pm \sqrt{b^2 - 4ac}}{2a}$

$v = \dfrac{-(-0.08) \pm \sqrt{(-0.08)^2 - 4(1)(-0.03)}}{2 \cdot 1}$

$v = \dfrac{0.08 \pm \sqrt{0.0064 + 0.12}}{2} = \dfrac{0.08 \pm \sqrt{0.1264}}{2}$

$v = \dfrac{0.08 \pm \sqrt{0.0016 \cdot 79}}{2} = \dfrac{0.08 \pm 0.04\sqrt{79}}{2}$

$v = \dfrac{2(0.04 \pm 0.02\sqrt{79})}{2} = 0.04 \pm 0.02\sqrt{79}$

The solution set is $\{0.04 \pm 0.02\sqrt{79}\}$.

44. $\left\{\dfrac{1 \pm \sqrt{17}}{3}\right\}$

45. $\dfrac{1}{x} + \dfrac{1}{x + 3} = 7$, LCM is $x(x + 3)$

$x(x + 3)\left[\dfrac{1}{x} + \dfrac{1}{x + 3}\right] = x(x + 3) \cdot 7$

$x + 3 + x = 7x^2 + 21x$

$2x + 3 = 7x^2 + 21x$

$0 = 7x^2 + 19x - 3$ Standard form

$a = 7,\; b = 19,\; c = -3$

$x = \dfrac{-b \pm \sqrt{b^2 - 4ac}}{2a}$

$x = \dfrac{-19 \pm \sqrt{19^2 - 4(7)(-3)}}{2 \cdot 7}$

$x = \dfrac{-19 \pm \sqrt{361 + 84}}{14} = \dfrac{-19 \pm \sqrt{445}}{14}$

Both values check.

The solution set is $\left\{\dfrac{-19 \pm \sqrt{445}}{14}\right\}$.

46. $\left\{\pm\dfrac{2\sqrt{30}}{5}\right\}$

47.
$$1 + \frac{x+5}{(x+1)^2} - \frac{2}{x+1} = 0, \text{ LCM is } (x+1)^2$$

$$(x+1)^2\left[1 + \frac{x+5}{(x+1)^2} - \frac{2}{x+1}\right] = (x+1)^2 \cdot 0$$

$$(x+1)^2 + x + 5 - 2(x+1) = 0$$
$$x^2 + 2x + 1 + x + 5 - 2x - 2 = 0$$
$$x^2 + x + 4 = 0$$

$a = 1, b = 1, c = 4$

$$x = \frac{-b \pm \sqrt{b^2 - 4ac}}{2a}$$

$$x = \frac{-1 \pm \sqrt{1^2 - 4\cdot1\cdot4}}{2\cdot1} = \frac{-1 \pm \sqrt{1-16}}{2}$$

$$x = \frac{-1 \pm \sqrt{-15}}{2} = \frac{-1 \pm \sqrt{-1\cdot15}}{2}$$

$$x = \frac{-1 \pm i\sqrt{15}}{2}$$

Both values check.

The solution set is $\left\{\dfrac{-1 \pm i\sqrt{15}}{2}\right\}$.

48. $\left\{\dfrac{-1 \pm i\sqrt{43}}{2}\right\}$

49. $x^2 - 6x + 9 = 0$

$a = 1, \quad b = -6, \quad c = 9$

We compute the discriminant.
$$b^2 - 4ac = (-6)^2 - 4\cdot1\cdot9$$
$$= 36 - 36$$
$$= 0$$

Since $b^2 - 4ac = 0$, there is just one solution, and it is a real number.

50. One real solution

51. $x^2 + 7 = 0$

$a = 1, \quad b = 0, \quad c = 7$

We compute the discriminant.
$$b^2 - 4ac = 0^2 - 4\cdot1\cdot7$$
$$= -28$$

Since $b^2 - 4ac < 0$, there are two nonreal solutions.

52. Two nonreal solutions

53. $x^2 - 2 = 0$

$a = 1, \quad b = 0, \quad c = -2$

We compute the discriminant.
$$b^2 - 4ac = 0^2 - 4\cdot1\cdot(-2)$$
$$= 8$$

Since $b^2 - 4ac > 0$, there are two real solutions.

54. Two real solutions

55. $4x^2 - 12x + 9 = 0$

$a = 4, \quad b = -12, \quad c = 9$

We compute the discriminant.
$$b^2 - 4ac = (-12)^2 - 4\cdot4\cdot9$$
$$= 144 - 144$$
$$= 0$$

Since $b^2 - 4ac = 0$, there is just one solution, and it is a real number.

56. Two real solutions

57. $x^2 - 2x + 4 = 0$

$a = 1, \quad b = -2, \quad c = 4$

We compute the discriminant.
$$b^2 - 4ac = (-2)^2 - 4\cdot1\cdot4$$
$$= 4 - 16$$
$$= -12$$

Since $b^2 - 4ac < 0$, ther are two nonreal solutions.

58. Two nonreal solutions

59. $9t^2 - 3t = 0$

$a = 9, \quad b = -3, \quad c = 0$

We compute the discriminant.
$$b^2 - 4ac = (-3)^2 - 4\cdot9\cdot0$$
$$= 9 - 0$$
$$= 9$$

Since $b^2 - 4ac > 0$, there are two real solutions.

60. Two real solutions

61. $y^2 = \dfrac{1}{2}y + \dfrac{3}{5}$

$$y^2 - \frac{1}{2}y - \frac{3}{5} = 0 \qquad \text{(Standard form)}$$

$a = 1, \quad b = -\dfrac{1}{2}, \quad c = -\dfrac{3}{5}$

We compute the discriminant.
$$b^2 - 4ac = \left(-\frac{1}{2}\right)^2 - 4\cdot1\cdot\left(-\frac{3}{5}\right)$$
$$= \frac{1}{4} + \frac{12}{5}$$
$$= \frac{53}{20}$$

Since $b^2 - 4ac > 0$, there are two real solutions.

62. Two real solutions

63. $4x^2 - 4\sqrt{3}\,x + 3 = 0$

    $a = 4, \qquad b = -4\sqrt{3}, \qquad c = 3$

    We compute the discriminant.

    $b^2 - 4ac = (-4\sqrt{3})^2 - 4\cdot4\cdot3$

    $\qquad\qquad = 48 - 48$

    $\qquad\qquad = 0$

    Since $b^2 - 4ac = 0$, there is just one solution, and it is a real number.

64. Two real solutions

65. The solutions are -11 and 9.

    $\qquad\qquad x = -11 \text{ or} \qquad x = 9$

    $x + 11 = 0 \quad\text{ or } x - 9 = 0$

    $(x + 11)(x - 9) = 0$

    $\qquad x^2 + 2x - 99 = 0$

66. $x^2 - 16 = 0$

67. The solutions are both 7.

    $\qquad\qquad x = 7 \text{ or} \qquad x = 7$

    $x - 7 = 0 \text{ or } x - 7 = 0$

    $(x - 7)(x - 7) = 0$

    $\qquad x^2 - 14x + 49 = 0$

68. $x^2 + \dfrac{4}{3}x + \dfrac{4}{9} = 0$, or $9x^2 + 12x + 4 = 0$

69. The solutions are $-\dfrac{2}{5}$ and $\dfrac{6}{5}$.

    $\qquad x = -\dfrac{2}{5} \text{ or} \qquad x = \dfrac{6}{5}$

    $x + \dfrac{2}{5} = 0 \quad\text{ or } x - \dfrac{6}{5} = 0$

    $\left[x + \dfrac{2}{5}\right]\left[x - \dfrac{6}{5}\right] = 0$

    $\qquad x^2 - \dfrac{4}{5}x - \dfrac{12}{25} = 0$

    or

    $25x^2 - 20x - 12 = 0$

70. $x^2 + \dfrac{3}{4}x + \dfrac{1}{8} = 0$, or $8x^2 + 6x + 1 = 0$

71. The solutions are $\dfrac{c}{2}$ and $\dfrac{d}{2}$.

    $\qquad x = \dfrac{c}{2} \text{ or} \qquad x = \dfrac{d}{2}$

    $x - \dfrac{c}{2} = 0 \text{ or } x - \dfrac{d}{2} = 0$

    $\left[x - \dfrac{c}{2}\right]\left[x - \dfrac{d}{2}\right] = 0$

    $x^2 - \left[\dfrac{c + d}{2}\right]x + \dfrac{cd}{4} = 0$

    or $\quad 4x^2 - 2(c + d)x + cd = 0$

72. $12x^2 - (4k + 3m)x + km = 0$

73. The solutions are $\sqrt{2}$ and $3\sqrt{2}$.

    $\qquad x = \sqrt{2} \text{ or} \qquad x = 3\sqrt{2}$

    $x - \sqrt{2} = 0 \quad\text{ or } x - 3\sqrt{2} = 0$

    $(x - \sqrt{2})(x - 3\sqrt{2}) = 0$

    $\qquad x^2 - 4\sqrt{2}\,x + 6 = 0$

74. $x^2 - \sqrt{3}\,x - 6 = 0$

75. The solutions are $3i$ and $-3i$.

    $\qquad x = 3i \text{ or} \qquad x = -3i$

    $x - 3i = 0 \quad\text{ or } x + 3i = 0$

    $(x - 3i)(x + 3i) = 0$

    $\qquad x^2 - 9i^2 = 0$

    $\qquad\qquad x^2 + 9 = 0$

76. $x^2 + 16 = 0$

77. $x^2 - 0.75x - 0.5 = 0$

    $a = 1, \quad b = -0.75, \quad c = -0.5$

    $x = \dfrac{-(-0.75) \pm \sqrt{(-0.75)^2 - 4(1)(-0.5)}}{2(1)}$

    $x = \dfrac{0.75 \pm \sqrt{0.5625 + 2}}{2} = \dfrac{0.75 \pm \sqrt{2.5625}}{2}$

    $x = \dfrac{0.75 + \sqrt{2.5625}}{2} \text{ or } x = \dfrac{0.75 - \sqrt{2.5625}}{2}$

    $x \approx 1.1754 \qquad\qquad \text{or } x \approx -0.4254$

    The solution set is $\{1.1754, -0.4254\}$.

78. $\{1.8693, -0.3252\}$

79. $\qquad\qquad x + \dfrac{1}{x} = \dfrac{13}{6}, \quad$ LCM is $6x$

    $\qquad 6x\left[x + \dfrac{1}{x}\right] = 6x \cdot \dfrac{13}{6}$

    $\qquad\qquad 6x^2 + 6 = 13x$

    $\qquad 6x^2 - 13x + 6 = 0$

    $(3x - 2)(2x - 3) = 0$

    $3x - 2 = 0 \text{ or } 2x - 3 = 0$

    $\qquad x = \dfrac{2}{3} \text{ or} \qquad x = \dfrac{3}{2}$

    Both values check.

    The solution set is $\left\{\dfrac{2}{3}, \dfrac{3}{2}\right\}$.

80. $\left\{\dfrac{3}{2}, 6\right\}$

81. $x^2 + x - \sqrt{2} = 0$

$a = 1, \quad b = 1, \quad c = -\sqrt{2}$

$x = \dfrac{-1 \pm \sqrt{1^2 - 4(1)(-\sqrt{2})}}{2 \cdot 1} = \dfrac{-1 \pm \sqrt{1 + 4\sqrt{2}}}{2}$

The solution set is $\left\{\dfrac{-1 \pm \sqrt{1 + 4\sqrt{2}}}{2}\right\}$.

82. $\left\{\dfrac{-\sqrt{5} \pm \sqrt{5 + 4\sqrt{3}}}{2}\right\}$

83. $(2t - 3)^2 + 17t = 15$

    $4t^2 - 12t + 9 + 17t = 15$

       $4t^2 + 5t - 6 = 0$

       $(4t - 3)(t + 2) = 0$

$4t - 3 = 0 \text{ or } t + 2 = 0$

    $t = \dfrac{3}{4}$ or $t = -2$

The solution set is $\left\{\dfrac{3}{4}, -2\right\}$.

84. $\{2, -3\}$

85. $(x + 3)(x - 2) = 2(x + 11)$

      $x^2 + x - 6 = 2x + 22$

      $x^2 - x - 28 = 0$

$a = 1, \quad b = -1, \quad c = -28$

$x = \dfrac{-(-1) \pm \sqrt{(-1)^2 - 4(1)(-28)}}{2 \cdot 1}$

$x = \dfrac{1 \pm \sqrt{1 + 112}}{2} = \dfrac{1 \pm \sqrt{113}}{2}$

The solution set is $\left\{\dfrac{1 \pm \sqrt{113}}{2}\right\}$.

86. $\{-4, 3\}$

87. $2x^2 + (x - 4)^2 = 5x(x - 4) + 24$

    $2x^2 + x^2 - 8x + 16 = 5x^2 - 20x + 24$

            $0 = 2x^2 - 12x + 8$

            $0 = x^2 - 6x + 4$

$a = 1, \quad b = -6, \quad c = 4$

$x = \dfrac{-(-6) \pm \sqrt{(-6)^2 - 4 \cdot 1 \cdot 4}}{2 \cdot 1} = \dfrac{6 \pm \sqrt{36 - 16}}{2}$

$x = \dfrac{6 \pm \sqrt{20}}{2} = \dfrac{6 \pm 2\sqrt{5}}{2}$

$x = \dfrac{2(3 \pm \sqrt{5})}{2} = 3 \pm \sqrt{5}$

The solution set is $\{3 \pm \sqrt{5}\}$.

88. $\{-12, 4\}$

89. a) $\dfrac{-b + \sqrt{b^2 - 4ac}}{2a} + \dfrac{-b - \sqrt{b^2 - 4ac}}{2a} = \dfrac{-2b}{2a} = -\dfrac{b}{a}$

   b) $\dfrac{-b + \sqrt{b^2 - 4ac}}{2a} \cdot \dfrac{-b - \sqrt{b^2 - 4ac}}{2a}$

       $= \dfrac{b^2 - (b^2 - 4ac)}{4a^2}$

       $= \dfrac{4ac}{4a^2}$

       $= \dfrac{c}{a}$

90. a) $k = 2;$ b) $\dfrac{11}{2}$

91. a) $kx^2 - 2x + k = 0$

    $k(-3)^2 - 2(-3) + k = 0$ (Substituting -3 for x)

       $9k + 6 + k = 0$

          $10k = -6$

            $k = -\dfrac{3}{5}$

   b) $-\dfrac{3}{5}x^2 - 2x - \dfrac{3}{5} = 0$ $\left(\text{Substituting } -\dfrac{3}{5} \text{ for } k\right)$

      $3x^2 + 10x + 3 = 0$ (Multiplying by -5)

   $(3x + 1)(x + 3) = 0$

   $3x + 1 = 0$ or $x + 3 = 0$

       $3x = -1$ or $x = -3$

        $x = -\dfrac{1}{3}$ or $x = -3$

The other solution is $-\dfrac{1}{3}$.

92. a) $k = 2;$ b) $1 - i$

93. a) $x^2 - (6 + 3i)x + k = 0$

    $3^2 - (6 + 3i) \cdot 3 + k = 0$ (Substituting 3 for x)

      $9 - 18 - 9i + k = 0$

             $k = 9 + 9i$

   b) $x^2 - (6 + 3i)x + 9 + 9i = 0$

    $x = \dfrac{-[-(6+3i)] \pm \sqrt{[-(6+3i)]^2 - 4(1)(9+9i)}}{2 \cdot 1}$

    $x = \dfrac{6 + 3i \pm \sqrt{36 + 36i - 9 - 36 - 36i}}{2}$

    $x = \dfrac{6 + 3i \pm \sqrt{-9}}{2} = \dfrac{6 + 3i \pm 3i}{2}$

    $x = \dfrac{6 + 3i + 3i}{2}$ or $x = \dfrac{6 + 3i - 3i}{2}$

    $x = \dfrac{6 + 6i}{2}$ or $x = \dfrac{6}{2}$

    $x = 3 + 3i$ or $x = 3$

The other solution is $3 + 3i$.

94. -1

95. Write $ax^2 + bx + c = 0$ as $x^2 - \left(-\dfrac{b}{a}\right)x + \dfrac{c}{a} = 0$.

From Exercise 89 we know that $-b/a$ is the sum of the solutions and c/a is the product of the solutions. Then we have $x^2 - \sqrt{3}x + 8 = 0$.

96. a) $x^2 - 2x + \dfrac{1}{4} = 0$, or $4x^2 - 8x + 1 = 0$;

b) $x^2 - \dfrac{g^2 - h^2}{hg}x - 1 = 0$, or

$hgx^2 - (g^2 - h^2)x - hg = 0$;

c) $x^2 - 4x + 29 = 0$

97. $3x^2 - hx + 4k = 0$

$a = 3$, $b = -h$, $c = 4k$

From Exercise 89 we know that the sum of the solutions is $-b/a$, or $-(-h)/3 = h/3$ and the product of the solutions is c/a, or $4k/3$. Then we have

$\dfrac{h}{3} = -12$ and $\dfrac{4k}{3} = 20$

$h = -36$ and $4k = 60$

$h = -36$ and $k = 15$.

98. 1

99. Let $1 \cdot x^2 + 2kx - 5 = 0$ have solutions r and s.

Then $r + s = \dfrac{-2k}{1}$, or $-2k$ and $rs = -5/1$, or -5.
Adding $2rs = -10$ and $r^2 + s^2 = 26$ we obtain
$r^2 + 2rs + s^2 = 16$ so that $r + s = \pm 4$. Then
$r + s = -2k = \pm 4$, so $k = -2$ or $k = 2$, or
$|k| = 2$.

100. $ax^2 + bx + c = 0$ has solution set

$S_1 = \left\{\dfrac{-b + \sqrt{b^2 - 4ac}}{2a}, \dfrac{-b - \sqrt{b^2 - 4ac}}{2a}\right\}$.

$cy^2 + by + a = 0$ has $y = \dfrac{-b \pm \sqrt{b^2 - 4ac}}{2c}$.

Then $\dfrac{1}{y} = \dfrac{2c}{-b \pm \sqrt{b^2 - 4ac}} \cdot \dfrac{-b \mp \sqrt{b^2 - 4ac}}{-b \mp \sqrt{b^2 - 4ac}}$

$= \dfrac{-b \mp \sqrt{b^2 - 4ac}}{2a}$, so that the

reciprocals of the solutions of the second equation form the set

$S_2 = \left\{\dfrac{-b - \sqrt{b^2 - 4ac}}{2a}, \dfrac{-b + \sqrt{b^2 - 4ac}}{2a}\right\}$ which is

the same as S_1.

Exercise Set 2.6

1. $F = \dfrac{kM_1M_2}{d^2}$

$Fd^2 = kM_1M_2$

$d^2 = \dfrac{kM_1M_2}{F}$

$d = \sqrt{\dfrac{kM_1M_2}{F}}$ (Taking the positive square root)

2. $c = \sqrt{\dfrac{E}{m}}$

3. $S = \dfrac{1}{2}at^2$

$2S = at^2$

$\dfrac{2S}{a} = t^2$

$\sqrt{\dfrac{2S}{a}} = t$ (Taking the positive square root)

4. $r = \sqrt{\dfrac{S}{4\pi}}$, or $r = \dfrac{1}{2}\sqrt{\dfrac{S}{\pi}}$

5. $s = -16t^2 + v_0t$

$0 = -16t^2 + v_0t - s$

$a = -16$, $b = v_0$, $c = -s$

$t = \dfrac{-v_0 \pm \sqrt{v_0{}^2 - 4(-16)(-s)}}{2(-16)}$

$t = \dfrac{-v_0 \pm \sqrt{v_0{}^2 - 64s}}{-32}$

or

$t = \dfrac{v_0 \pm \sqrt{v_0{}^2 - 64s}}{32}$

6. $r = \dfrac{-3\pi h + \sqrt{9\pi^2h^2 + 8\pi A}}{4\pi}$

7. $d = \dfrac{n^2 - 3n}{2}$

$2d = n^2 - 3n$

$0 = n^2 - 3n - 2d$

$a = 1$, $b = -3$, $c = -2d$

$n = \dfrac{-(-3) \pm \sqrt{(-3)^2 - 4(1)(-2d)}}{2 \cdot 1}$

$n = \dfrac{3 \pm \sqrt{9 + 8d}}{2}$, or just

$n = \dfrac{3 + \sqrt{9 + 8d}}{2}$ since the negative square root would result in a negative solution.

8. $t = \dfrac{\pi \pm \sqrt{\pi^2 - 12k\sqrt{2}}}{2\sqrt{2}}$

9. $A = P(1 + i)^2$

$\frac{A}{P} = (1 + i)^2$

$\pm \sqrt{\frac{A}{P}} = 1 + i$

$-1 \pm \sqrt{\frac{A}{P}} = i$, or just

$-1 + \sqrt{\frac{A}{P}} = i$ since the negative square root would result in a negative solution.

10. $i = 2\left[-1 + \sqrt{\frac{A}{P}}\right]$

11. $A = P(1 + i)^t$

$7290 = 6250(1 + i)^2$ (Substituting)

$\frac{7290}{6250} = (1 + i)^2$

$\pm \sqrt{\frac{729}{625}} = 1 + i$

$\pm \frac{27}{25} = 1 + i$

$-1 \pm \frac{27}{25} = i$

$-1 + \frac{27}{25} = i$ or $-1 - \frac{27}{25} = i$

$\frac{2}{25} = i$ or $-\frac{52}{25} = i$

Since the interest rate cannot be negative, $i = \frac{2}{25} = 0.08 = 8\%$.

12. 4%

13. $A = P(1 + i)^t$

$4410 = 4000(1 + i)^2$ (Substituting)

$\frac{4410}{4000} = (1 + i)^2$

$\pm \sqrt{\frac{4410}{4000}} = 1 + i$

$-1 \pm 1.05 = i$

$-1 + 1.05 = i$ or $-1 - 1.05 = i$

$0.05 = i$ or $-2.05 = i$

Since the interest rate cannot be negative, $i = 0.05 = 5\%$.

14. 10%

15. $d = \frac{n^2 - 3n}{2}$

$27 = \frac{n^2 - 3n}{2}$ (Substituting 27 for d)

$54 = n^2 - 3n$

$0 = n^2 - 3n - 54$

$0 = (n - 9)(n + 6)$

$n - 9 = 0$ or $n + 6 = 0$

$n = 9$ or $n = -6$

The number of sides must be positive. Thus the number of sides is 9.

16. 11

17. Familiarize: We make a drawing and label it with the known and unknown data.

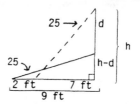

Translate: We use the Pythagorean theorem. From the taller triangle we get

$h^2 + 7^2 = 25^2$.   (1)

From the other triangle we get

$(h - d)^2 + 9^2 = 25^2$.   (2)

Carry out: We solve equation (1) for h and get h = 24. Then we substitute 24 for h in equation (2).

$(24 - d)^2 + 9^2 = 25^2$, or

$d^2 - 48d + 32 = 0$

Using the quadratic formula, we get

$d = \frac{48 \pm \sqrt{2176}}{2} = \frac{48 \pm 8\sqrt{34}}{2} = 24 \pm 4\sqrt{34}$.

Check and State: The length $24 + 4\sqrt{34}$ is not a solution since it exceeds the original length. The number $24 - 4\sqrt{34} \approx 0.7$ checks and is the solution. Therefore, the top of the ladder moves down the wall about 0.7 ft when the bottom is moved 2 ft.

18. 1.2 ft

19. Familiarize and Translate:

Using the Pythagorean theorem we determine the length of the other leg.

$10^2 = 6^2 + a^2$

$100 = 36 + a^2$

$64 = a^2$

$8 = a$

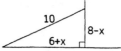

We let x represent the amount the ladder is pulled away and pulled down. The legs then become 6 + x and 8 - x. Then we again use the Pythagorean theorem.

$$10^2 = (6 + x)^2 + (8 - x)^2$$

Carry out:

$$100 = 36 + 12x + x^2 + 64 - 16x + x^2$$
$$100 = 2x^2 - 4x + 100$$
$$0 = 2x^2 - 4x$$
$$0 = x^2 - 2x$$
$$0 = x(x - 2)$$

$$x = 0 \text{ or } x - 2 = 0$$
$$x = 0 \text{ or } \quad x = 2$$

We only consider x = 2 since x ≠ 0 if the ladder is pulled away.

Check and State:

When x = 2, the legs of the triangle become 6 + 2, or 8, and 8 - 2, or 6. The Pythagorean relationship still holds.

$$8^2 + 6^2 = 10^2$$
$$64 + 36 = 100$$

The ladder is pulled away 2 ft and pulled down 2 ft.

20. 7 ft

21. Familiarize:

We make a drawing. We let h represent the height and h + 3 represent the base.

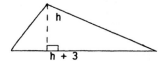

Translate:

$$A = \frac{1}{2} \cdot \text{base} \cdot \text{height} \quad \text{(Area of triangle)}$$
$$18 = \frac{1}{2}(h + 3)h \quad \text{(Substituting)}$$

Carry out:
$$36 = h^2 + 3h$$
$$0 = h^2 + 3h - 36$$
$$a = 1, \quad b = 3, \quad c = -36$$
$$h = \frac{-3 \pm \sqrt{3^2 - 4(1)(-36)}}{2 \cdot 1}$$
$$h = \frac{-3 \pm \sqrt{9 + 144}}{2} = \frac{-3 \pm \sqrt{153}}{2}$$
$$h = \frac{-3 + \sqrt{153}}{2} \quad \text{(h cannot be negative)}$$
$$h \approx 4.7$$

Check and State:

When h = 4.7, h + 3 = 7.7 and the area of the triangle is $\frac{1}{2}(4.7)(7.7) \approx 18.1$.

The height is 4.7 cm.

22. $90\sqrt{2} \approx 127.3$ ft

23. Familiarize: We make a drawing.

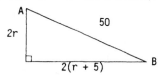

We let r represent the speed of train A. Then r + 5 represents the speed of train B. The distances they travel in 2 hours are 2r and 2(r + 5). After 2 hours they are 50 miles apart.

Translate:

Using the Pythagorean theorem we have an equation.

$$(2r)^2 + [2(r + 5)]^2 = 50^2$$

Carry out:
$$4r^2 + 4(r + 5)^2 = 2500$$
$$r^2 + (r + 5)^2 = 625$$
$$r^2 + r^2 + 10r + 25 = 625$$
$$2r^2 + 10r - 600 = 0$$
$$r^2 + 5r - 300 = 0$$
$$(r + 20)(r - 15) = 0$$

$$r + 20 = 0 \quad \text{or } r - 15 = 0$$
$$r = -20 \text{ or } \quad r = 15$$

Check and State:

Since speed in this problem must be positive, we only check 15. If r = 15, then r + 5 = 20. In 2 hours train A will travel 2·15, or 30 miles, and train B will travel 2·20, or 40 miles.

We now calculate how far apart trains A and B will be after 2 hours.

$$30^2 + 40^2 = d^2$$
$$900 + 1600 = d^2$$
$$2500 = d^2$$
$$50 = d$$

The value checks.

The speed of train A is 15 mph; the speed of train B is 20 mph.

24. A: 24 km/h, B: 10 km/h

25. a) $s = 4.9t^2 + v_0 t$

     $75 = 4.9t^2 + 0 \cdot t$ (Substituting 75 for
                                    s and 0 for v_0)

     $75 = 4.9t^2$

     $\dfrac{75}{4.9} = t^2$

     $\sqrt{\dfrac{75}{4.9}} = t$

     $3.9 \approx t$

     It takes 3.9 sec to reach the ground.

  b) $s = 4.9t^2 + v_0 t$

     $75 = 4.9t^2 + 30t$ (Substituting 75 for s
                               and 30 for v_0)

     $0 = 4.9t^2 + 30t - 75$

     $a = 4.9, \quad b = 30, \quad c = -75$

     $t = \dfrac{-30 \pm \sqrt{30^2 - 4(4.9)(-75)}}{2(4.9)}$

     $t = \dfrac{-30 \pm \sqrt{900 + 1470}}{9.8}$

     $t = \dfrac{-30 + \sqrt{2370}}{9.8}$ (t must be positive)

     $t \approx 1.9$

     It takes 1.9 sec to reach the ground.

  c) $s = 4.9t^2 + v_0 t$

     $s = 4.9(2)^2 + 30(2)$ (Substituting 2 for t
                                  and 30 for v_0)

     $s = 19.6 + 60$

     $s = 79.6$

     The object will fall 79.6 m.

26. a) 10.1 sec; b) 7.5 sec; c) 272.5 m

27. Familiarize and Translate:
     It helps to make a drawing.

     We let s represent the length of the side. Then
     s + 1.341 represents the diagonal. Using the
     Pythagorean theorem we have an equation.

     $$s^2 + s^2 = (s + 1.341)^2$$

Carry out:

$$2s^2 = s^2 + 2.682s + 1.798281$$

$s^2 - 2.682s - 1.798281 = 0$

$s = \dfrac{-(-2.682) \pm \sqrt{(-2.682)^2 - 4(1)(-1.798281)}}{2 \cdot 1}$

$s = \dfrac{2.682 \pm \sqrt{7.193124 + 7.193124}}{2}$

$s = \dfrac{2.682 + \sqrt{14.386248}}{2}$

$s \approx 3.2$

Check and State:

The value checks. The length of the side is
3.2 cm.

28. 8.01 cm, 2.22 cm

29. Familiarize and Translate: From the drawing in
     the text we see that the area of the original
     garden is $80 \cdot 60$ and the area of the new garden
     is $(80 - 2x)(60 - 2x)$, where x is the width of
     the sidewalk. We know the area of the new garden
     is 2/3 of the original area, so we can write an
     equation.

     $(80 - 2x)(60 - 2x) = \dfrac{2}{3} \cdot 80 \cdot 60$, or

     $x^2 - 70x + 400 = 0$

     Carry out: Using the quadratic formula, we get

     $x = \dfrac{70 \pm \sqrt{3300}}{2} = \dfrac{70 \pm 10\sqrt{33}}{2} = 35 \pm 5\sqrt{33}.$

     Check and State: The number $35 + 5\sqrt{33}$ is not a
     solution since twice that number exceeds both
     the original length and width of the garden.
     The number $35 - 5\sqrt{33} \approx 6.3$ checks and is the
     solution. Therefore, the sidewalk is about 6.3
     ft wide.

30. $\sqrt{304} \approx 17.4$ mph

31. Familiarize: We make a drawing, letting x
     represent the length of a side of the square
     cut from each corner.

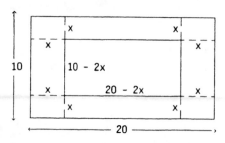

     The dimensions of the resulting base are 10 - 2x
     and 20 - 2x.

     Translate: The area of the resulting base is
     90 cm², so we can write an equation.

     $(20 - 2x)(10 - 2x) = 90$, or

     $2x^2 - 30x + 55 = 0$

Carry out: Using the quadratic formula, we get

$$x = \frac{30 \pm \sqrt{460}}{4} = \frac{30 \pm 2\sqrt{115}}{4} = \frac{15 \pm \sqrt{115}}{2}.$$

Check and State: The number $\frac{15 + \sqrt{115}}{2}$ is not a solution since twice that number exceeds the width of the piece of tin. The number $\frac{15 - \sqrt{115}}{2} \approx 2.1$ checks and is the solution. Therefore, the length of the sides of the squares is about 2.1 cm.

32. $15 - \sqrt{51} \approx 7.9$ cm

33. Familiarize: We can use the formula d = rt in the form $t = \frac{d}{r}$. Let x represent the speed on the first part of the trip. Then x - 10 represents the speed on the second part of the trip. The time for the first part of the trip is $\frac{50}{x}$, and the time for the second part is $\frac{80}{x - 10}$.

Translate: We know that the total time for the trip is 2 hr, so we can write an equation.

$$\frac{50}{x} + \frac{80}{x - 10} = 2, \text{ or}$$
$$x^2 - 75x + 250 = 0 \quad \text{Clearing fractions and simplifying}$$

Carry out: Using the quadratic formula, we get

$$x = \frac{75 \pm \sqrt{4625}}{2} = \frac{75 \pm 5\sqrt{185}}{2}.$$

Check and State: The number $x = \frac{75 - 5\sqrt{185}}{2}$ is not a solution since x - 10 < 0. The number $x = \frac{75 + 5\sqrt{185}}{2} \approx 71.5$ mph checks and is the solution. This is the speed on the first part of the trip. The speed on the second part is approximately 71.5 - 10, or 61.5 mph.

34. 20 mph

35. Familiarize: We will use the formula $t = \frac{d}{r}$ as we did in Exercise 33. Let x represent the speed of the boat in still water. Then the time to travel upstream is $\frac{12}{x - 3}$ and the time to travel downstream is $\frac{12}{x + 3}$.

Translate: We know the total time for the round trip is 2 hr, so we can write an equation.

$$\frac{12}{x - 3} + \frac{12}{x + 3} = 2, \text{ or}$$
$$x^2 - 12x - 9 = 0 \quad \text{Clearing fractions and simplifying}$$

Carry out: Using the quadratic formula, we get

$$x = \frac{12 \pm \sqrt{180}}{2} = \frac{12 \pm 6\sqrt{5}}{2} = 6 \pm 3\sqrt{5}$$

Check and State: The number $6 - 3\sqrt{5}$ is not a solution, because the speed cannot be negative. The number $6 + 3\sqrt{5} \approx 12.7$ checks and is the solution. The speed of the boat in still water is $6 + 3\sqrt{5} \approx 12.7$ mph.

36. $\frac{3 + \sqrt{89}}{2} \approx 6.2$ hr

37. Familiarize and Translate:

We let x represent the number of students in the group at the outset. Then x - 3 represents the number of students at the last minute.

$$\text{Total cost} = (\text{Cost per person}) \cdot (\text{Number of persons})$$
$$\frac{\text{Total cost}}{\text{Number of persons}} = \text{Cost per person}$$

Using this formula we get the following:

$$\begin{array}{l} \text{Cost per person} \\ \text{at the outset} \end{array} = \frac{140}{x}$$

$$\begin{array}{l} \text{Cost per person at} \\ \text{the last minute} \end{array} = \frac{140}{x - 3}$$

The cost at the last minute was $15 more than the cost at the outset. This gives us an equation.

$$\frac{140}{x - 3} = 15 + \frac{140}{x}$$

Carry out:

$$x(x - 3) \cdot \frac{140}{x - 3} = x(x - 3)\left[15 + \frac{140}{x}\right]$$
$$\qquad \text{(Multiplying by the LCM)}$$
$$140x = 15x(x - 3) + 140(x - 3)$$
$$140x = 15x^2 - 45x + 140x - 420$$
$$0 = 15x^2 - 45x - 420$$
$$0 = x^2 - 3x - 28$$
$$0 = (x - 7)(x + 4)$$

$$x - 7 = 0 \text{ or } x + 4 = 0$$
$$x = 7 \text{ or } \quad x = -4$$

Check and State:

Since the number of people cannot be negative, we only check x = 7. When x = 7, the cost at the outset is 140/7, or $20. When x = 7, the cost at the last minute is 140/(7 - 3), or 140/4, or $35 and $35 is $15 more than $20. The value checks.

Thus, 7 students were in the group at the outset.

38. 12

39. Familiarize and Translate:

Let x represent the number of shares bought. Then $\frac{720}{x}$ represents the cost of each share. When the cost per share is reduced by \$15, the new cost is $\frac{720}{x}$ - 15. When the number of shares purchased is increased by 4, the new number of shares is x + 4. The total cost remains the same.

$$720 = \left(\frac{720}{x} - 15\right) \cdot (x + 4)$$

Carry out:

$$720 = 720 + \frac{2880}{x} - 15x - 60$$

$$60 = \frac{2880}{x} - 15x$$

$$60x = 2880 - 15x^2$$

$$15x^2 + 60x - 2880 = 0$$

$$x^2 + 4x - 192 = 0$$

$$(x + 16)(x - 12) = 0$$

$$x + 16 = 0 \quad \text{or} \quad x - 12 = 0$$
$$x = -16 \quad \text{or} \quad x = 12$$

Check and State:

The number of shares must be positive. We only check x = 12. The cost per share when x = 12 is 720/12, or \$60. The cost per share when the number is increased by 4 is 720/16, or \$45, and \$45 is \$15 less than \$60. The value checks.

Thus, 12 shares of stock were bought.

40. \$8

41. $kx^2 + (3 - 2k)x - 6 = 0$

$$a = k, \quad b = 3 - 2k, \quad c = -6$$

$$x = \frac{-(3 - 2k) \pm \sqrt{(3 - 2k)^2 - 4(k)(-6)}}{2k}$$

$$= \frac{-3 + 2k \pm \sqrt{4k^2 + 12k + 9}}{2k}$$

$$= \frac{-3 + 2k \pm \sqrt{(2k + 3)^2}}{2k}$$

$$= \frac{-3 + 2k \pm (2k + 3)}{2k}$$

$$x = \frac{-3 + 2k + 2k + 3}{2k} \quad \text{or} \quad x = \frac{-3 + 2k - 2k - 3}{2k}$$

$$x = \frac{4k}{2k} \quad \text{or} \quad x = \frac{-6}{2k}$$

$$x = 2 \quad \text{or} \quad x = -\frac{3}{k}$$

The solution set is $\left\{2, -\frac{3}{k}\right\}$.

42. $\left\{\frac{1}{1 - k}, 1\right\}$

43. $$(m + n)^2x^2 + (m + n)x = 2$$

$$(m + n)^2x^2 + (m + n)x - 2 = 0$$

$$a = (m + n)^2, \quad b = m + n, \quad c = -2$$

$$x = \frac{-(m + n) \pm \sqrt{(m + n)^2 - 4(m + n)^2(-2)}}{2(m + n)^2}$$

$$= \frac{-(m + n) \pm \sqrt{9(m + n)^2}}{2(m + n)^2}$$

$$= \frac{-(m + n) \pm 3(m + n)}{2(m + n)^2}$$

$$x = \frac{2(m + n)}{2(m + n)^2} \quad \text{or} \quad x = \frac{-4(m + n)}{2(m + n)^2}$$

$$x = \frac{1}{m + n} \quad \text{or} \quad x = \frac{-2}{m + n}$$

The solution set is $\left\{\frac{1}{m + n}, -\frac{2}{m + n}\right\}$.

44. a) $\{-y, 4y\}$; b) $\left\{-x, \frac{x}{4}\right\}$

45. $$A = P(1 + i)^t$$

$$13,704 = 9826(1 + i)^3 \quad \text{(Substituting)}$$

$$\frac{13,704}{9826} = (1 + i)^3$$

$$\sqrt[3]{\frac{13,704}{9826}} = 1 + i$$

$$-1 + \sqrt[3]{\frac{13,704}{9826}} = i$$

$$0.117 \approx i$$

The interest rate is 11.7%.

46. $\frac{27}{4}\sqrt{3} \approx 11.7$

47. Familiarize:
We first make a drawing.

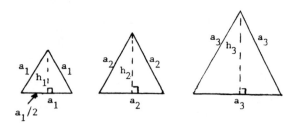

We find the height of the first triangle.

$$\left(\frac{a_1}{2}\right)^2 + (h_1)^2 = a_1^2$$

$$(h_1)^2 = a_1^2 - \frac{a_1^2}{4} = \frac{3a_1^2}{4}$$

$$h_1 = \frac{a_1\sqrt{3}}{2}$$

Next we find the area of the first triangle.

$$A_1 = \frac{1}{2} \cdot a_1 \cdot \frac{a_1\sqrt{3}}{2} = \frac{a_1^2\sqrt{3}}{4}$$

Similarly we find the areas of the other two triangles:

$$A_2 = \frac{a_2{}^2\sqrt{3}}{4} \quad \text{and} \quad A_3 = \frac{a_3{}^2\sqrt{3}}{4}.$$

Translate:

The sum of the areas of the first two triangles equals the area of the third triangle.

$$\frac{a_1{}^2\sqrt{3}}{4} + \frac{a_2{}^2\sqrt{3}}{4} = \frac{a_3{}^2\sqrt{3}}{4}$$

Carry out:

$$\frac{\sqrt{3}}{4}(a_1{}^2 + a_2{}^2) = \frac{\sqrt{3}}{4} \cdot a_3{}^2$$

$$a_1{}^2 + a_2{}^2 = a_3{}^2$$

$$\sqrt{a_1{}^2 + a_2{}^2} = a_3 \quad \text{Taking the positive square root}$$

Check and State: The answer checks.

Thus the length of a side of the third triangle is $\sqrt{a_1{}^2 + a_2{}^2}$.

<u>48.</u> The triangles have the same area (300 ft²).

<u>49.</u>

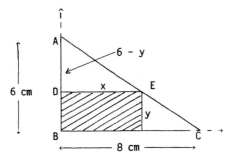

Familiarize:

Triangle ABC is similar to triangle ADE.

Translate:

The area of the rectangle is 12 cm², so we have

$$xy = 12, \quad \text{or} \quad y = \frac{12}{x}.$$

Using similar triangles, we have

$$\frac{6 - y}{x} = \frac{6}{8}, \quad \text{or} \quad 48 - 8y = 6x.$$

Carry out:

$$48 - 8y = 6x$$

$$48 - 8\left[\frac{12}{x}\right] = 6x \quad \left[\text{Substituting } \frac{12}{x} \text{ for } y\right]$$

$$48 - \frac{96}{x} = 6x, \quad \text{LCM is } x$$

$$48x - 96 = 6x^2 \quad \text{(Multiplying by } x)$$

$$0 = 6x^2 - 48x + 96$$

$$0 = 6(x - 4)^2$$

$$0 = (x - 4)^2 \quad \left[\text{Multiplying by } \frac{1}{6}\right]$$

$$0 = x - 4$$

$$4 = x$$

When $x = 4$, $y = \frac{12}{4} = 3$.

Check and State: The answer checks.

The dimensions of the rectangle are 4 cm by 3 cm.

<u>50.</u> $\frac{5\sqrt{7}}{4} \approx 3.3$ sec

Exercise Set 2.7

<u>1.</u> $\sqrt{3x - 4} = 1$ Check:

$$(\sqrt{3x - 4})^2 = 1^2$$

$$3x - 4 = 1$$

$$3x = 5$$

$$x = \frac{5}{3}$$

$$\begin{array}{c|c} \sqrt{3x - 4} = 1 & \\ \hline \sqrt{3 \cdot \frac{5}{3} - 4} & 1 \\ \sqrt{5 - 4} & \\ \sqrt{1} & \\ 1 & \text{TRUE} \end{array}$$

The solution set is $\left\{\frac{5}{3}\right\}$.

<u>2.</u> {-63}

<u>3.</u> $\sqrt[4]{x^2 - 1} = 1$ Check:

$$(\sqrt[4]{x^2 - 1})^4 = 1^4$$

$$x^2 - 1 = 1$$

$$x^2 = 2$$

$$x = \pm\sqrt{2}$$

$$\begin{array}{c|c} \sqrt[4]{x^2 - 1} = 1 & \\ \hline \sqrt[4]{(\pm\sqrt{2})^2 - 1} & 1 \\ \sqrt[4]{2 - 1} & \\ \sqrt[4]{1} & \\ 1 & \text{TRUE} \end{array}$$

The solution set is $\{\pm\sqrt{2}\}$.

<u>4.</u> {168}

<u>5.</u> $\sqrt{y - 1} + 4 = 0$

$$\sqrt{y - 1} = -4$$

Note:

The principal square root is never negative. Thus, there is no solution.

If we do not observe the above fact, we can continue and reach the same answer.

$$(\sqrt{y - 1})^2 = (-4)^2$$ Check:

$$y - 1 = 16$$

$$y = 17$$

$$\begin{array}{c|c} \sqrt{y - 1} + 4 = 0 & \\ \hline \sqrt{17 - 1} + 4 & 0 \\ \sqrt{16} + 4 & \\ 4 + 4 & \\ 8 & \text{FALSE} \end{array}$$

Since 17 does not check, there is no solution. The solution set is ∅.

6. $\emptyset$

7. $\sqrt{x - 3} + \sqrt{x + 2} = 5$

$$\sqrt{x + 2} = 5 - \sqrt{x - 3}$$

$$(\sqrt{x + 2})^2 = (5 - \sqrt{x - 3})^2$$

$$x + 2 = 25 - 10\sqrt{x - 3} + (x - 3)$$

$$x + 2 = 22 - 10\sqrt{x - 3} + x$$

$$10\sqrt{x - 3} = 20$$

$$\sqrt{x - 3} = 2$$

$$(\sqrt{x - 3})^2 = 2^2$$

$$x - 3 = 4$$

$$x = 7$$

Check: $\dfrac{\sqrt{x - 3} + \sqrt{x + 2} = 5}{}$

$\sqrt{7 - 3} + \sqrt{7 + 2}$ | 5

$\sqrt{4} + \sqrt{9}$

$2 + 3$

5 | TRUE

The solution set is {7}.

8. {9}

9. $\sqrt{3x - 5} + \sqrt{2x + 3} + 1 = 0$

$$\sqrt{3x - 5} + \sqrt{2x + 3} = -1$$

Note:

The principal square root is never negative. Thus the sum of two principal square roots cannot equal -1. There is no solution. The solution set is $\emptyset$.

10. {2}

11. $\sqrt[3]{6x + 9} + 8 = 5$

$$\sqrt[3]{6x + 9} = -3$$

$$(\sqrt[3]{6x + 9})^3 = (-3)^3$$

$$6x + 9 = -27$$

$$6x = -36$$

$$x = -6$$

Check:

$\dfrac{\sqrt[3]{6x + 9} + 8 = 5}{}$

$\sqrt[3]{6(-6) + 9} + 8$ | 5

$\sqrt[3]{-27} + 8$

$-3 + 8$

5 | TRUE

The solution set is {-6}.

12. $\left\{\dfrac{28}{3}\right\}$

13. $\sqrt{6x + 7} = x + 2$

$$(\sqrt{6x + 7})^2 = (x + 2)^2$$

$$6x + 7 = x^2 + 4x + 4$$

$$0 = x^2 - 2x - 3$$

$$0 = (x - 3)(x + 1)$$

$$x - 3 = 0 \text{ or } x + 1 = 0$$

$$x = 3 \text{ or } \quad x = -1$$

Both values check. The solution set is {3, -1}.

14. $\left\{\dfrac{1}{3}, -1\right\}$

15. $\sqrt{20 - x} = \sqrt{9 - x} + 3$

$$(\sqrt{20 - x})^2 = (\sqrt{9 - x} + 3)^2$$

$$20 - x = (9 - x) + 6\sqrt{9 - x} + 9$$

$$20 - x = 18 - x + 6\sqrt{9 - x}$$

$$2 = 6\sqrt{9 - x}$$

$$1 = 3\sqrt{9 - x}$$

$$1^2 = (3\sqrt{9 - x})^2$$

$$1 = 9(9 - x)$$

$$1 = 81 - 9x$$

$$9x = 80$$

$$x = \frac{80}{9}$$

The value checks. The solution set is $\left\{\dfrac{80}{9}\right\}$.

16. {-1}

17. $\sqrt{x} - \sqrt{3x - 3} = 1$

$$\sqrt{x} = \sqrt{3x - 3} + 1$$

$$(\sqrt{x})^2 = (\sqrt{3x - 3} + 1)^2$$

$$x = (3x - 3) + 2\sqrt{3x - 3} + 1$$

$$2 - 2x = 2\sqrt{3x - 3}$$

$$1 - x = \sqrt{3x - 3}$$

$$(1 - x)^2 = (\sqrt{3x - 3})^2$$

$$1 - 2x + x^2 = 3x - 3$$

$$x^2 - 5x + 4 = 0$$

$$(x - 4)(x - 1) = 0$$

$$x = 4 \text{ or } x = 1$$

The number 4 does not check, but 1 does. The solution set is {1}.

18. {0, 4}

19. $\sqrt{2y - 5} - \sqrt{y - 3} = 1$

$\qquad \sqrt{2y - 5} = \sqrt{y - 3} + 1$

$\qquad (\sqrt{2y - 5})^2 = (\sqrt{y - 3} + 1)^2$

$\qquad 2y - 5 = (y - 3) + 2\sqrt{y - 3} + 1$

$\qquad y - 3 = 2\sqrt{y - 3}$

$\qquad (y - 3)^2 = (2\sqrt{y - 3})^2$

$\qquad y^2 - 6y + 9 = 4(y - 3)$

$\qquad y^2 - 6y + 9 = 4y - 12$

$\qquad y^2 - 10y + 21 = 0$

$\qquad (y - 7)(y - 3) = 0$

$y = 7 \quad$ or $\quad y = 3$

Both numbers check. The solution set is {7, 3}.

20. $\{6 + 2\sqrt{3}\}$

21. $\qquad \sqrt{7.35x + 8.051} = 0.345x + 0.067$

$\qquad (\sqrt{7.35x + 8.051})^2 = (0.345x + 0.067)^2$

$\qquad 7.35x + 8.051 = 0.119025x^2 + 0.04623x +$
$\qquad\qquad\qquad\qquad\qquad\qquad\qquad 0.004489$

$\qquad 0 = 0.119025x^2 - 7.30377x -$
$\qquad\qquad\qquad\qquad\qquad\qquad 8.046511$

$x = \dfrac{-(-7.30377) \pm \sqrt{(-7.30377)^2 - 4(0.119025)(-8.046511)}}{2(0.119025)}$

$x \approx 62.4459 \quad$ or $\quad x \approx -1.0826$

Since −1.0826 does not check and 62.4459 does check, the solution set is {62.4459}.

22. {0.1444}

23. $\qquad x^{1/3} = -2$

$\qquad (x^{1/3})^3 = (-2)^3 \qquad\qquad (x^{1/3} = \sqrt[3]{x})$

$\qquad x = -8$

The value checks. The solution set is {−8}.

24. {32}

25. $\qquad t^{1/4} = 3$

$\qquad (t^{1/4})^4 = 3^4 \qquad\qquad (t^{1/4} = \sqrt[4]{t})$

$\qquad t = 81$

The value checks. The solution set is {81}.

26. Ø

27. $\qquad 8 = \dfrac{1}{\sqrt{x}}$

$\qquad 8\sqrt{x} = 1$

$\qquad \sqrt{x} = \dfrac{1}{8}$

$\qquad (\sqrt{x})^2 = \left[\dfrac{1}{8}\right]^2$

$\qquad x = \dfrac{1}{64}$

The value checks. The solution set is $\left\{\dfrac{1}{64}\right\}$.

28. $\left\{\dfrac{1}{9}\right\}$

29. $\qquad \sqrt[3]{m} = -5$

$\qquad (\sqrt[3]{m})^3 = (-5)^3$

$\qquad m = -125$

The value checks. The solution set is {−125}.

30. Ø

31. For L: $\qquad T = 2\pi\sqrt{\dfrac{L}{g}} \qquad$ For g: $\qquad T = 2\pi\sqrt{\dfrac{L}{g}}$

$\qquad\qquad \dfrac{T}{2\pi} = \sqrt{\dfrac{L}{g}} \qquad\qquad\qquad\qquad \dfrac{T}{2\pi} = \sqrt{\dfrac{L}{g}}$

$\qquad\qquad \left[\dfrac{T}{2\pi}\right]^2 = \left[\sqrt{\dfrac{L}{g}}\right]^2 \qquad\qquad \left[\dfrac{T}{2\pi}\right]^2 = \left[\sqrt{\dfrac{L}{g}}\right]^2$

$\qquad\qquad \dfrac{T^2}{4\pi^2} = \dfrac{L}{g} \qquad\qquad\qquad\qquad \dfrac{T^2}{4\pi^2} = \dfrac{L}{g}$

$\qquad\qquad \dfrac{gT^2}{4\pi^2} = L \qquad\qquad\qquad\qquad gT^2 = 4\pi^2 L$

$\qquad\qquad\qquad\qquad\qquad\qquad\qquad\qquad\qquad g = \dfrac{4\pi^2 L}{T^2}$

32. $c = \sqrt{H^2 - d^2}$

33. $V = 1.2\sqrt{h}$

$\qquad V = 1.2\sqrt{30{,}000} \qquad$ (Substituting)

$\qquad \approx 208$

You can see about 208 miles to the horizon.

34. 10 mi

35. $\qquad V = 1.2\sqrt{h}$

$\qquad 144 = 1.2\sqrt{h} \qquad$ (Substituting)

$\qquad \dfrac{144}{1.2} = \sqrt{h}$

$\qquad 120 = \sqrt{h}$

$\qquad 14{,}400 = h$

The airplane is 14,400 ft high.

36. 84 ft

37. $(x - 5)^{2/3} = 2$

$\sqrt[3]{(x - 5)^2} = 2$

$(\sqrt[3]{(x - 5)^2})^3 = 2^3$

$(x - 5)^2 = 2^3$

$x^2 - 10x + 25 = 8$

$x^2 - 10x + 17 = 0$

$a = 1, \quad b = -10, \quad c = 17$

$x = \dfrac{-(-10) \pm \sqrt{(-10)^2 - 4\cdot 1\cdot 17}}{2\cdot 1}$

$= \dfrac{10 \pm \sqrt{100 - 68}}{2} = \dfrac{10 \pm \sqrt{32}}{2}$

$= \dfrac{10 \pm 4\sqrt{2}}{2} = 5 \pm 2\sqrt{2}$

Both values check. The solution set is
$\{5 \pm 2\sqrt{2}\}$.

38. $\{3 \pm 2\sqrt{2}\}$.

39. $\dfrac{x + \sqrt{x + 1}}{x - \sqrt{x + 1}} = \dfrac{5}{11}$

$11(x + \sqrt{x + 1}) = 5(x - \sqrt{x + 1})$

$11x + 11\sqrt{x + 1} = 5x - 5\sqrt{x + 1}$

$6x = -16\sqrt{x + 1}$

$(6x)^2 = (-16\sqrt{x + 1})^2$

$36x^2 = 256(x + 1)$

$36x^2 - 256x - 256 = 0$

$9x^2 - 64x - 64 = 0$

$(9x + 8)(x - 8) = 0$

$9x + 8 = 0 \quad \text{or} \quad x - 8 = 0$

$x = -\dfrac{8}{9} \text{ or } \qquad x = 8$

Only $-\dfrac{8}{9}$ checks. The solution set is $\left\{-\dfrac{8}{9}\right\}$.

40. Ø

41. $\sqrt{x + 2} - \sqrt{x - 2} = \sqrt{2x}$

$(\sqrt{x + 2} - \sqrt{x - 2})^2 = (\sqrt{2x})^2$

$x + 2 - 2\sqrt{(x + 2)(x - 2)} + x - 2 = 2x$

$-2\sqrt{(x + 2)(x - 2)} = 0$

$\sqrt{(x + 2)(x - 2)} = 0$

$(\sqrt{(x + 2)(x - 2)})^2 = 0^2$

$(x + 2)(x - 2) = 0$

$x + 2 = 0 \text{ or } x - 2 = 0$

$x = -2 \text{ or } \qquad x = 2$

Only 2 checks. The solution set is $\{2\}$.

42. $\{1\}$

43. $\sqrt[4]{x + 2} = \sqrt{3x + 1}$

$(\sqrt[4]{x + 2})^4 = (\sqrt{3x + 1})^4$

$x + 2 = (3x + 1)^2$

$x + 2 = 9x^2 + 6x + 1$

$0 = 9x^2 + 5x - 1$

$a = 9, \quad b = 5, \quad c = -1$

$x = \dfrac{-5 \pm \sqrt{5^2 - 4(9)(-1)}}{2\cdot 9} = \dfrac{-5 \pm \sqrt{25 + 36}}{18}$

$= \dfrac{-5 \pm \sqrt{61}}{18}$

Only $\dfrac{-5 + \sqrt{61}}{18}$ checks. The solution set
is $\left\{\dfrac{-5 + \sqrt{61}}{18}\right\}$.

44. $\left\{\dfrac{5}{4}\right\}$

45. $\dfrac{14}{3 + \sqrt{7 + x}} - \dfrac{\sqrt{7 + x}}{2} = 0$, LCM is $2(3 + \sqrt{7 + x})$

$2(3 + \sqrt{7 + x})\left[\dfrac{14}{3 + \sqrt{7 + x}} - \dfrac{\sqrt{7 + x}}{2}\right] =$

$2(3 + \sqrt{7 + x})\cdot 0$

$28 - 3\sqrt{7 + x} - (7 + x) = 0$

$21 - x = 3\sqrt{7 + x}$

$(21 - x)^2 = (3\sqrt{7 + x})^2$

$441 - 42x + x^2 = 9(7 + x)$

$441 - 42x + x^2 = 63 + 9x$

$x^2 - 51x + 378 = 0$

$(x - 9)(x - 42) = 0$

$x = 9 \quad \text{or} \quad x = 42$

Only 9 checks. The solution set is $\{9\}$.

46. $\{6\}$

47. $\sqrt{3x + 1} - \sqrt{2x} = \dfrac{5}{\sqrt{3x + 1}}$,

LCM is $\sqrt{3x + 1}$

$\sqrt{3x + 1}(\sqrt{3x + 1} - \sqrt{2x}) = \sqrt{3x + 1} \cdot \dfrac{5}{\sqrt{3x + 1}}$

$(3x + 1) - \sqrt{6x^2 + 2x} = 5$

$3x - 4 = \sqrt{6x^2 + 2x}$

$(3x - 4)^2 = (\sqrt{6x^2 + 2x})^2$

$9x^2 - 24x + 16 = 6x^2 + 2x$

$3x^2 - 26x + 16 = 0$

$(3x - 2)(x - 8) = 0$

$x = \dfrac{2}{3} \quad \text{or} \quad x = 8$

Only 8 checks. The solution set is $\{8\}$.

48. $\left\{\frac{25}{13}\right\}$

49. $\sqrt{15 + \sqrt{2x + 80}} = 5$

$\left(\sqrt{15 + \sqrt{2x + 80}}\right)^2 = 5^2$

$15 + \sqrt{2x + 80} = 25$

$\sqrt{2x + 80} = 10$

$(\sqrt{2x + 80})^2 = 10^2$

$2x + 80 = 100$

$2x = 20$

$x = 10$

This number checks. The solution set is {10}.

50.   {-1}

Exercise Set 2.8

1. $x - 10\sqrt{x} + 9 = 0$

We substitute u for $\sqrt{x}$.

$u^2 - 10u + 9 = 0$

$(u - 9)(u - 1) = 0$

$u - 9 = 0$ or $u - 1 = 0$

$u = 9$ or $u = 1$

Now we substitute $\sqrt{x}$ for u and solve for x.

$\sqrt{x} = 9$ or $\sqrt{x} = 1$

$x = 81$ or $x = 1$

Check:

For 81:   For 1:

$x - 10\sqrt{x} + 9 = 0$		$x - 10\sqrt{x} + 9 = 0$	
$81 - 10\sqrt{81} + 9$	0	$1 - 10\sqrt{1} + 9$	0
$81 - 10 \cdot 9 + 9$		$1 - 10 + 9$	
$81 - 90 + 9$			0
0			

The solutions are 81 and 1. The solution set is {81, 1}.

2. $\left\{\frac{1}{4}, 16\right\}$

3. $x^4 - 10x^2 + 25 = 0$

We substitute u for x^2.

$u^2 - 10u + 25 = 0$

$(u - 5)(u - 5) = 0$

$u - 5 = 0$ or $u - 5 = 0$

$u = 5$ or $u = 5$

Now we substitute x^2 for u and solve for x.

$x^2 = 5$ or $x^2 = 5$

$x = \pm\sqrt{5}$ or $x = \pm\sqrt{5}$

The solutions are $\pm\sqrt{5}$. The solution set is $\{\pm\sqrt{5}\}$.

4. $\{\pm 1, \pm\sqrt{2}\}$

5. $t^{2/3} + t^{1/3} - 6 = 0$

We substitute u for $t^{1/3}$.

$u^2 + u - 6 = 0$

$(u + 3)(u - 2) = 0$

$u + 3 = 0$ or $u - 2 = 0$

$u = -3$ or $u = 2$

We now substitute $t^{1/3}$ for u and solve for t.

$t^{1/3} = -3$ or $t^{1/3} = 2$

$t = (-3)^3$ or $t = 2^3$

$t = -27$ or $t = 8$

The solutions are -27 and 8. The solution set is {-27, 8}.

6.   {64, -8}

7. $z^{1/2} = z^{1/4} + 2$

$z^{1/2} - z^{1/4} - 2 = 0$

We substitute u for $z^{1/4}$.

$u^2 - u - 2 = 0$

$(u - 2)(u + 1) = 0$

$u - 2 = 0$ or $u + 1 = 0$

$u = 2$ or $u = -1$

Next we substitute $z^{1/4}$ for u and solve for z.

$z^{1/4} = 2$ or $z^{1/4} = -1$

$z = 2^4$ or $z = (-1)^4$

$z = 16$ or $z = 1$

The number 16 checks, but 1 does not. The solution set is {16}.

8.   {729}

9. $(x^2 - 6x)^2 - 2(x^2 - 6x) - 35 = 0$

We substitute u for $x^2 - 6x$.
$$u^2 - 2u - 35 = 0$$
$$(u - 7)(u + 5) = 0$$

$u - 7 = 0$ or $u + 5 = 0$
  $u = 7$ or $u = -5$

Next we substitute $x^2 - 6x$ for u and solve for x.
   $x^2 - 6x = 7$ or $x^2 - 6x = -5$
$x^2 - 6x - 7 = 0$ or $x^2 - 6x + 5 = 0$
$(x - 7)(x + 1) = 0$ or $(x - 5)(x - 1) = 0$

$x - 7 = 0$ or $x + 1 = 0$ or $x - 5 = 0$ or $x - 1 = 0$
  $x = 7$ or $x = -1$ or $x = 5$ or $x = 1$

The solutions are 7, -1, 5, and 1. The solution set is {7, -1, 5, 1}.

10. {1}

11. $(y^2 - 5y)^2 + (y^2 - 5y) - 12 = 0$

We substitute u for $y^2 - 5y$.
$$u^2 + u - 12 = 0$$
$$(u + 4)(u - 3) = 0$$

$u + 4 = 0$ or $u - 3 = 0$
  $u = -4$ or $u = 3$

Next we substitute $y^2 - 5y$ for u and solve for y.
   $y^2 - 5y = -4$ or $y^2 - 5y = 3$
$y^2 - 5y + 4 = 0$ or $y^2 - 5y - 3 = 0$

$(y - 4)(y - 1) = 0$ or $y = \dfrac{-(-5) \pm \sqrt{(-5)^2 - 4(1)(-3)}}{2 \cdot 1}$

$y - 4 = 0$ or $y - 1 = 0$ or $y = \dfrac{5 \pm \sqrt{25 + 12}}{2}$

  $y = 4$ or $y = 1$ or $y = \dfrac{5 \pm \sqrt{37}}{2}$

The solutions are 4, 1, and $\dfrac{5 \pm \sqrt{37}}{2}$. The

solution set is $\left\{ 4,\ 1,\ \dfrac{5 \pm \sqrt{37}}{2} \right\}$.

12. $\left\{ \dfrac{1}{2},\ -1,\ -\dfrac{3}{2},\ 1 \right\}$

13. $w^4 - 4w^2 - 2 = 0$

First we substitute u for w^2.
  $u^2 - 4u - 2 = 0$

$u = \dfrac{-(-4) \pm \sqrt{(-4)^2 - 4(1)(-2)}}{2 \cdot 1}$

$u = \dfrac{4 \pm \sqrt{16 + 8}}{2} = \dfrac{4 \pm \sqrt{24}}{2}$

$u = \dfrac{4 \pm 2\sqrt{6}}{2} = 2 \pm \sqrt{6}$

Next we substitute w^2 for u and solve for w.

$w^2 = 2 + \sqrt{6}$ or $w^2 = 2 - \sqrt{6}$

$w = \pm \sqrt{2 + \sqrt{6}}$ or $w = \pm \sqrt{2 - \sqrt{6}}$

The solutions are $\pm \sqrt{2 + \sqrt{6}}$ and $\pm \sqrt{2 - \sqrt{6}}$.

The solution set is $\left\{ \pm \sqrt{2 + \sqrt{6}},\ \pm \sqrt{2 - \sqrt{6}} \right\}$.

14. $\left\{ \pm \sqrt{\dfrac{5 + \sqrt{5}}{2}},\ \pm \sqrt{\dfrac{5 - \sqrt{5}}{2}} \right\}$

15. $x^{-2} - x^{-1} - 6 = 0$

We substitute u for x^{-1}.
   $u^2 - u - 6 = 0$
$(u - 3)(u + 2) = 0$

$u - 3 = 0$ or $u + 2 = 0$
    $u = 3$ or $u = -2$

Then we substitute x^{-1} for u and solve for x.
$x^{-1} = 3$ or $x^{-1} = -2$

 $\dfrac{1}{x} = 3$ or $\dfrac{1}{x} = -2$

 $x = \dfrac{1}{3}$ or $x = -\dfrac{1}{2}$

The solutions are $\dfrac{1}{3}$ and $-\dfrac{1}{2}$. The solution set is $\left\{ \dfrac{1}{3},\ -\dfrac{1}{2} \right\}$.

16. $\left\{ \dfrac{4}{5},\ -1 \right\}$

17. $2x^{-2} + x^{-1} = 1$
       $2x^{-2} + x^{-1} - 1 = 0$

We substitute u for x^{-1}.
    $2u^2 + u - 1 = 0$
 $(2u - 1)(u + 1) = 0$

$2u - 1 = 0$ or $u + 1 = 0$
    $u = \dfrac{1}{2}$ or $u = -1$

Next we substitute x^{-1} for u and solve for x.
$x^{-1} = \dfrac{1}{2}$ or $x^{-1} = -1$

 $\dfrac{1}{x} = \dfrac{1}{2}$ or $\dfrac{1}{x} = -1$

 $x = 2$ or $x = -1$

The solutions are 2 and -1. The solution set is {2, -1}.

18. $\left\{ -\dfrac{1}{10},\ 1 \right\}$

19. $x^4 - 24x^2 - 25 = 0$

We substitute u for x^2.

$u^2 - 24u - 25 = 0$

$(u - 25)(u + 1) = 0$

$u - 25 = 0$ or $u + 1 = 0$

$u = 25$ or $u = -1$

Then we substitute x^2 for u and solve for x.

$x^2 = 25$ or $x^2 = -1$

$x = \pm 5$ or $x = \pm i$

The solutions are ± 5 and $\pm i$. The solution set is $\{\pm 5, \pm i\}$.

20. $\{\pm 3, \pm 2i\}$

21. $\left(\dfrac{x^2 - 2}{x}\right)^2 - 7\left(\dfrac{x^2 - 2}{x}\right) - 18 = 0$

We first substitute u for $\dfrac{x^2 - 2}{x}$.

$u^2 - 7u - 18 = 0$

$(u - 9)(u + 2) = 0$

$u - 9 = 0$ or $u + 2 = 0$

$u = 9$ or $u = -2$

Then we substitute $\dfrac{x^2 - 2}{x}$ for u and solve for x.

$\dfrac{x^2 - 2}{x} = 9$ or $\dfrac{x^2 - 2}{x} = -2$

$x^2 - 2 = 9x$ or $x^2 - 2 = -2x$

$x^2 - 9x - 2 = 0$ or $x^2 + 2x - 2 = 0$

$x = \dfrac{-(-9) \pm \sqrt{(-9)^2 - 4(1)(-2)}}{2 \cdot 1}$ or $x = \dfrac{-2 \pm \sqrt{2^2 - 4(1)(-2)}}{2 \cdot 1}$

$x = \dfrac{9 \pm \sqrt{89}}{2}$ or $x = \dfrac{-2 \pm 2\sqrt{3}}{2}$

                             or $x = -1 \pm \sqrt{3}$

The solutions are $\dfrac{9 \pm \sqrt{89}}{2}$ and $-1 \pm \sqrt{3}$. The solution set is $\left\{\dfrac{9 \pm \sqrt{89}}{2}, -1 \pm \sqrt{3}\right\}$.

22. $\left\{\dfrac{5 \pm \sqrt{21}}{2}, \dfrac{3 \pm \sqrt{5}}{2}\right\}$

23. $\dfrac{x}{x - 1} - 6\sqrt{\dfrac{x}{x - 1}} - 40 = 0$

We substitute u for $\sqrt{\dfrac{x}{x - 1}}$.

$u^2 - 6u - 40 = 0$

$(u - 10)(u + 4) = 0$

$u - 10 = 0$ or $u + 4 = 0$

$u = 10$ or $u = -4$

We then substitute $\sqrt{\dfrac{x}{x - 1}}$ for u and solve for x.

$\sqrt{\dfrac{x}{x - 1}} = 10$ or $\sqrt{\dfrac{x}{x - 1}} = -4$

$\dfrac{x}{x - 1} = 100$ No solution

$x = 100x - 100$

$100 = 99x$

$\dfrac{100}{99} = x$

The solution is $\dfrac{100}{99}$. The solution set is $\left\{\dfrac{100}{99}\right\}$.

24. $\left\{\dfrac{1}{98}\right\}$

25. $5\left(\dfrac{x + 2}{x - 2}\right)^2 = 3\left(\dfrac{x + 2}{x - 2}\right) + 2$

$5\left(\dfrac{x + 2}{x - 2}\right)^2 - 3\left(\dfrac{x + 2}{x - 2}\right) - 2 = 0$

Substitute u for $\dfrac{x + 2}{x - 2}$.

$5u^2 - 3u - 2 = 0$

$(5u + 2)(u - 1) = 0$

$5u + 2 = 0$ or $u - 1 = 0$

$u = -\dfrac{2}{5}$ or $u = 1$

Then substitute $\dfrac{x + 2}{x - 2}$ for u and solve for x.

$\dfrac{x + 2}{x - 2} = -\dfrac{2}{5}$ or $\dfrac{x + 2}{x - 2} = 1$

$5(x + 2) = -2(x - 2)$ or $x + 2 = x - 2$

$5x + 10 = -2x + 4$ or $2 = -2$

$7x = -6$ No solution

$x = -\dfrac{6}{7}$

The solution is $-\dfrac{6}{7}$. The solution set is $\left\{-\dfrac{6}{7}\right\}$.

26. $\left\{-\dfrac{5}{2}, -5\right\}$

27. $\left(\dfrac{x^2 - 1}{x}\right)^2 - \left(\dfrac{x^2 - 1}{x}\right) - 2 = 0$

Substitute u for $\dfrac{x^2 - 1}{x}$.

$u^2 - u - 2 = 0$

$(u - 2)(u + 1) = 0$

$u - 2 = 0$ or $u + 1 = 0$

$u = 2$ or $u = -1$

Then substitute $\dfrac{x^2 - 1}{x}$ for u and solve for x.

$\dfrac{x^2 - 1}{x} = 2$ or $\dfrac{x^2 - 1}{x} = -1$

$x^2 - 1 = 2x$ or $x^2 - 1 = -x$

$x^2 - 2x - 1 = 0$ or $x^2 + x - 1 = 0$

$x = \dfrac{-(-2) \pm \sqrt{(-2)^2 - 4(1)(-1)}}{2 \cdot 1}$ or $x = \dfrac{-1 \pm \sqrt{1^2 - 4(1)(-1)}}{2 \cdot 1}$

$$x = \frac{2 \pm \sqrt{8}}{2}$$

$$x = \frac{-1 \pm \sqrt{5}}{2}$$

$$x = \frac{2 \pm 2\sqrt{2}}{2}$$

$$x = 1 \pm \sqrt{2}$$

The solutions are $1 \pm \sqrt{2}$ and $\frac{-1 \pm \sqrt{5}}{2}$. The solution set is $\left\{1 \pm \sqrt{2}, \frac{-1 \pm \sqrt{5}}{2}\right\}$.

28. $\left\{\frac{56}{5}, 0\right\}$

29. Familiarize and Translate:

We let t_1 represent the time it takes for the object to fall to the ground and t_2 represent the time it takes the sound to get back. The total amount of time, 3 sec, is the sum of t_1 and t_2. This gives us an equation.

$$t_1 + t_2 = 3 \qquad (1)$$

We use the formula $s = 16t^2 + v_0 t$ to find an expression for t_1. Since the stone is dropped, v_0 is 0.

$$s = 16t_1{}^2$$

$$\frac{s}{16} = t_1{}^2$$

$$\sqrt{\frac{s}{16}} = t_1$$

$$\frac{\sqrt{s}}{4} = t_1 \qquad (2)$$

We use $d = rt$ (here we use $s = rt$) to find an expression for t_2.

$$s = 1100 \cdot t_2 \qquad \text{(Substituting 1100 for r)}$$

$$\frac{s}{1100} = t_2 \qquad (3)$$

We substitute (2) and (3) in (1) and obtain

$$\frac{\sqrt{s}}{4} + \frac{s}{1100} = 3$$

Carry out:

$$275\sqrt{s} + s = 3300 \qquad \text{(Multiplying by 1100)}$$

$$s + 275\sqrt{s} - 3300 = 0$$

Substitute u for $\sqrt{s}$.

$$u^2 + 275u - 3300 = 0$$

Using the quadratic formula, we get $u \approx 11.52$. Then $u = \sqrt{s} \approx 11.52$ and $s \approx 132.7$.

Check and State:
The value checks. The cliff is 132.7 ft high.

30. 229.9 ft

31. Familiarize: We will use the formula $A = P(1 + i)^t$. At the beginning of the third year the $2000 investment has grown to $2000(1 + i)^2$ and the $1200 investment has grown to $1200(1 + i)$.

Translate: We know the total in both accounts is $3573.80, so we can write an equation.

$$2000(1 + i)^2 + 1200(1 + i) = 3573.80, \text{ or}$$

$$2000(1 + i)^2 + 1200(1 + i) - 3573.80 = 0$$

Carry out: This equation is reducible to quadratic with $u = 1 + i$. Substituting, we get

$$2000u^2 + 1200u - 3573.80 = 0.$$

Using the quadratic formula and taking the positive solution, we get $u = 1.07$. Substituting $1 + i$ for u, we get

$$1 + i = 1.07$$
$$i = 0.07.$$

Check: We check by substituting 0.07 for i back through the related equations.

State: The interest rate is 0.07, or 7%.

32. 9%

33.
$$6.75x = \sqrt{35x} + 5.36$$

$$6.75x - \sqrt{35}\sqrt{x} - 5.36 = 0 \qquad (\sqrt{35x} = \sqrt{35}\sqrt{x})$$

Substitute u for $\sqrt{x}$.

$$6.75u^2 - \sqrt{35}u - 5.36 = 0$$

$$u = \frac{-(-\sqrt{35}) \pm \sqrt{(-\sqrt{35})^2 - 4(6.75)(-5.36)}}{2(6.75)}$$

$$= \frac{\sqrt{35} \pm \sqrt{35 + 144.72}}{13.5}$$

$$= \frac{\sqrt{35} \pm \sqrt{179.72}}{13.5}$$

$$u \approx 1.4313 \text{ or } u \approx -0.5548$$

Substitute $\sqrt{x}$ for u and solve for x.

$$\sqrt{x} \approx 1.4313 \text{ or } \sqrt{x} \approx -0.5548$$
$$x \approx 2.05 \qquad \text{No solution}$$

The value checks. The solution set is {2.05}.

34. $\{\pm 1.99, \pm 0.90i\}$

35. $9x^{3/2} - 8 = x^3$

$$0 = x^3 - 9x^{3/2} + 8$$

Substitute u for $x^{3/2}$.

$$0 = u^2 - 9u + 8$$
$$0 = (u - 8)(u - 1)$$

$$u - 8 = 0 \text{ or } u - 1 = 0$$
$$u = 8 \text{ or } \quad u = 1$$

Substitute $x^{3/2}$ for u and solve for x.

$x^{3/2} = 8$ or $x^{3/2} = 1$

$x^{1/2} = 2$ or $x^{1/2} = 1$ (Taking cube roots)

  x = 4 or   x = 1

Both values check. The solution set is {4, 1}.

36. $\left\{-\frac{3}{2},\ -1\right\}$

37. $\sqrt{x-3} - \sqrt[4]{x-3} = 2$

Substitute u for $\sqrt[4]{x-3}$.

    $u^2 - u - 2 = 0$

  $(u - 2)(u + 1) = 0$

u - 2 = 0 or u + 1 = 0

    u = 2 or   u = -1

Substitute $\sqrt[4]{x-3}$ for u and solve for x.

$\sqrt[4]{x-3} = 2$ or $\sqrt[4]{x-3} = -1$

  x - 3 = 16   No solution

        x = 19

The value checks. The solution set is {19}.

38. {9}

39. $x^6 - 28x^3 + 27 = 0$

Substitute u for x^3.

    $u^2 - 28u + 27 = 0$

  $(u - 27)(u - 1) = 0$

u = 27 or u = 1

Substitute x^3 for u and solve for x.

        $x^3 = 27$ or $x^3 = 1$

        $x^3 - 27 = 0$ or $x^3 - 1 = 0$

$(x-3)(x^2 + 3x + 9) = 0$ or $(x-1)(x^2 + x + 1) = 0$

Using the principle of zero products and, where necessary, the quadratic formula, we find that the solution set is

$\left\{3,\ \frac{-3 \pm 3i\sqrt{3}}{2},\ 1,\ \frac{-1 \pm i\sqrt{3}}{2}\right\}.$

40. $\left\{-2,\ 1 \pm i\sqrt{3},\ 1,\ \frac{-1 \pm i\sqrt{3}}{2}\right\}$

41. $(x^2 - 5x - 2)^2 - 5(x^2 - 5x - 2) + 4 = 0$

Substitute u for $x^2 - 5x - 2$.

    $u^2 - 5u + 4 = 0$

  $(u - 4)(u - 1) = 0$

u = 4 or u = 1

Substitute $x^2 - 5x - 2$ for u and solve for x.

  $x^2 - 5x - 2 = 4$ or $x^2 - 5x - 2 = 1$

  $x^2 - 5x - 6 = 0$ or $x^2 - 5x - 3 = 0$

$(x + 1)(x - 6) = 0$ or $x = \dfrac{-(-5) \pm \sqrt{(-5)^2 - 4(1)(-3)}}{2 \cdot 1}$

x + 1 = 0 or x - 6 = 0 or $x = \dfrac{5 \pm \sqrt{25 + 12}}{2}$

    x = -1 or   x = 6 or $x = \dfrac{5 \pm \sqrt{37}}{2}$

The solution set is $\left\{-1,\ 6,\ \dfrac{5 \pm \sqrt{37}}{2}\right\}.$

42. $\left\{\dfrac{4}{9},\ \dfrac{9}{4}\right\}$

43. $\left(y + \dfrac{2}{y}\right)^2 + 3y + \dfrac{6}{y} = 4$

$\left(y + \dfrac{2}{y}\right)^2 + 3\left(y + \dfrac{2}{y}\right) - 4 = 0$

Substitute u for $y + \dfrac{2}{y}$.

    $u^2 + 3u - 4 = 0$

  $(u + 4)(u - 1) = 0$

u = -4 or u = 1

Substitute $y + \dfrac{2}{y}$ for u and solve for y.

    $y + \dfrac{2}{y} = -4$ or $y + \dfrac{2}{y} = 1$

    $y^2 + 2 = -4y$ or $y^2 + 2 = y$

$y^2 + 4y + 2 = 0$ or $y^2 - y + 2 = 0$

$y = \dfrac{-4 \pm \sqrt{4^2 - 4 \cdot 1 \cdot 2}}{2 \cdot 1}$ or $y = \dfrac{-(-1) \pm \sqrt{(-1)^2 - 4 \cdot 1 \cdot 2}}{2 \cdot 1}$

$y = \dfrac{-4 \pm \sqrt{8}}{2}$ or $y = \dfrac{1 \pm \sqrt{-7}}{2}$

$y = \dfrac{-4 \pm 2\sqrt{2}}{2}$ or $y = \dfrac{1 \pm i\sqrt{7}}{2}$

$y = -2 \pm \sqrt{2}$ or $y = \dfrac{1 \pm i\sqrt{7}}{2}$

The solution set is $\left\{-2 \pm \sqrt{2},\ \dfrac{1 \pm i\sqrt{7}}{2}\right\}.$

44. $\left\{\dfrac{-3 \pm \sqrt{39 + 2\sqrt{33}}}{2}\right\}$

45.
$$\frac{2x + 1}{x} = 3 + 7\sqrt{\frac{2x + 1}{x}}$$

$$\frac{2x + 1}{x} - 7\sqrt{\frac{2x + 1}{x}} - 3 = 0$$

Substitute u for $\sqrt{\frac{2x + 1}{x}}$.

$$u^2 - 7u - 3 = 0$$

$$u = \frac{-(-7) \pm \sqrt{(-7)^2 - 4(1)(-3)}}{2 \cdot 1}$$

$$u = \frac{7 \pm \sqrt{61}}{2}$$

Substitute $\sqrt{\frac{2x + 1}{x}}$ for u and solve for x.

$$\sqrt{\frac{2x + 1}{x}} = \frac{7 + \sqrt{61}}{2} \quad \text{or} \quad \sqrt{\frac{2x + 1}{x}} = \frac{7 - \sqrt{61}}{2}$$

$$\left(\sqrt{\frac{2x + 1}{x}}\right)^2 = \left(\frac{7 + \sqrt{61}}{2}\right)^2 \qquad \text{No solution since}$$
$$\frac{7 - \sqrt{61}}{2} < 0$$

$$\frac{2x + 1}{x} = \frac{49 + 14\sqrt{61} + 61}{4}$$

$$8x + 4 = 110x + 14x\sqrt{61}$$

$$4 = 102x + 14x\sqrt{61}$$

$$4 = (102 + 14\sqrt{61})x$$

$$\frac{4}{102 + 14\sqrt{61}} = x$$

Rationalizing the denominator, we get

$x = \frac{-51 + 7\sqrt{61}}{194}$. The solution set is

$\left\{\frac{-51 + 7\sqrt{61}}{194}\right\}$.

46. 8%

Exercise Set 2.9

1. $y = kx$
$0.6 = k(0.4)$ (Substituting)
$\frac{0.6}{0.4} = k$
$\frac{3}{2} = k$

The equation of variation is $y = \frac{3}{2}x$.

2. $y = \frac{125}{32}x$

3. $y = \frac{k}{x}$
$125 = \frac{k}{32}$ (Substituting)
$4000 = k$
The equation of variation is $y = \frac{4000}{x}$.

4. $y = \frac{0.32}{x}$

5. $y = kx$
$8.6 = k(1.6)$ (Substituting)
$\frac{8.6}{1.6} = k$
$5.375 = k$

The equation of variation is $y = 5.375x$.

6. $y = \frac{20.52}{x}$

7. $y = \frac{k}{x^2}$
$0.15 = \frac{k}{(0.1)^2}$ (Substituting)
$0.15 = \frac{k}{0.01}$
$0.01(0.15) = k$
$0.0015 = k$
The equation of variation is $y = \frac{0.0015}{x^2}$.

8. $y = 0.8xz$

9. $y = k \cdot \frac{xz}{w}$
$\frac{3}{2} = k \cdot \frac{2 \cdot 3}{4}$ (Substituting)
$\frac{3}{2} = k \cdot \frac{6}{4}$
$\frac{4}{6} \cdot \frac{3}{2} = k$
$1 = k$

The equation of variation is $y = \frac{xz}{w}$.

10. $y = 0.3xz^2$

11. $y = k \cdot \frac{xz}{w^2}$
$\frac{12}{5} = k \cdot \frac{16 \cdot 3}{5^2}$ (Substituting)
$\frac{12}{5} = k \cdot \frac{48}{25}$
$\frac{25}{48} \cdot \frac{12}{5} = k$
$\frac{5}{4} = k$

The equation of variation is $y = \frac{5}{4} \cdot \frac{xz}{w^2}$.

12. $y = \frac{1}{5} \cdot \frac{xz}{wp}$

13. $y = kx$

We double x.

$y = k(2x)$

$y = 2 \cdot kx$

Thus, y is doubled.

14. y is multiplied by $\frac{1}{3}$

15. $y = \frac{k}{x^2}$

We multiply x by n.

$y = \frac{k}{(nx)^2}$

$y = \frac{k}{n^2 x^2}$

$y = \frac{1}{n^2} \cdot \frac{k}{x^2}$

Thus, y is multiplied by $\frac{1}{n^2}$.

16. y is multiplied by n^2

17. $A = kN$

$42,600 = k \cdot 60,000$ (Substituting)

$0.71 = k$

The equation of variation is A = 0.71N.

$A = 0.71N$

$A = 0.71(750,000)$ (Substituting)

$A = 532,500$

Thus, 532,500 tons will enter the atmosphere.

18. 220 in^3

19. $L = \frac{kwh^2}{\ell}$

The width and height are doubled, and the length is halved.

$L = \frac{k \cdot 2w \cdot (2h)^2}{\frac{\ell}{2}}$

$L = k \cdot 2w \cdot 4h^2 \cdot \frac{2}{\ell}$

$L = 16 \cdot \frac{kwh^2}{\ell}$

Thus, L is multiplied by 16.

20. 256

21. $d = k\sqrt{h}$

$28.97 = k\sqrt{19.5}$ (Substituting)

$\frac{28.97}{\sqrt{19.5}} = k$

$d = \frac{28.97}{\sqrt{19.5}} \sqrt{h}$ (Equation of variation)

$54.32 = \frac{28.97}{\sqrt{19.5}} \sqrt{h}$ (Substituting)

$\frac{54.32\sqrt{19.5}}{28.97} = \sqrt{h}$

$\left(\frac{54.32\sqrt{19.5}}{28.97}\right)^2 = (\sqrt{h})^2$

$68.6 \approx h$

One must be about 68.6 m above sea level.

22. 1.26 ohms

23. $A = ks^2$

$168.54 = k(5.3)^2$ (Substituting)

$168.54 = 28.09k$

$6 = k$

$A = 6s^2$ (Equation of variation)

$A = 6(10.2)^2$ (Substituting)

$A = 624.24$

The area is 624.24 m^2.

24. 22.5 W/m^2

25. $A = \frac{kR}{I}$

$2.92 = \frac{k \cdot 85}{262}$ (Substituting)

$2.92 \cdot \frac{262}{85} = k$

$9 \approx k$

$A = \frac{9R}{I}$ (Equation of variation)

$2.92 = \frac{9R}{300}$ (Substituting)

$\frac{2.92(300)}{9} = k$

$97 \approx k$

He would have given up 97 runs.

26. 96 cm^3

27. If p varies directly as q,

then p = kq.

Thus, $q = \frac{1}{k} p$,

so q varies directly as p.

28. $u = \dfrac{k}{v}$ gives $v = \dfrac{k}{u}$; and $u = \dfrac{k}{v}$ gives $\dfrac{1}{u} = \left(\dfrac{1}{k}\right)v$

or $\dfrac{1}{u} = (k_2)v$, where $k_2 = \dfrac{1}{k}$.

29. $$A = kd^2$$
$$\pi r^2 = kd^2$$
$$\pi\left(\dfrac{d}{2}\right)^2 = kd^2$$
$$\dfrac{\pi}{4} \cdot d^2 = kd^2$$
$$\dfrac{\pi}{4} = k$$

30. t varies directly as $\sqrt{P}$

31. a) $$N = \dfrac{kP_1P_2}{d^2}$$
$$11{,}153 = \dfrac{k(741{,}952)(364{,}040)}{(174)^2}$$
$$\dfrac{(11{,}153)(174)^2}{(741{,}952)(364{,}040)} = k$$
$$0.001 \approx k$$

The equation of variation is $N = \dfrac{0.001P_1P_2}{d^2}$.

b) $N = \dfrac{0.001(741{,}952)(1{,}027{,}974)}{(446)^2}$

$N \approx 3834$

The average number of daily phone calls is about 3834.

c) $4270 = \dfrac{0.001(741{,}952)(7{,}322{,}564)}{d^2}$

$d^2 = \dfrac{0.001(741{,}952)(7{,}322{,}564)}{4270}$

$d \approx 1128$

The distance is about 1128 km.

d) Division by 0 is undefined.

Exercise Set 3.1

1. The origin has coordinates (0,0). The x-coordinate of each ordered pair tells us how far to move to the left or right of the y-axis. The y-coordinate tells us how far to move up or down from the x-axis.

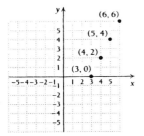

2.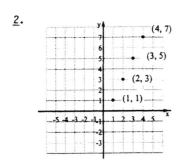

3.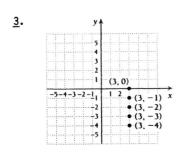

4.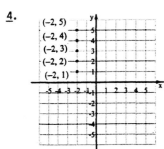

5.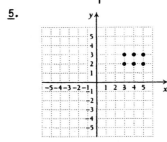

6.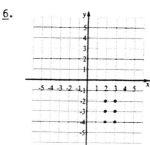

7.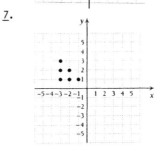

8.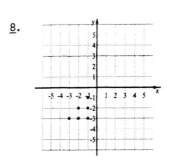

9.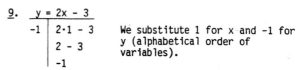

$y = 2x - 3$	
-1	$2 \cdot 1 - 3$
	$2 - 3$
	-1

We substitute 1 for x and -1 for y (alphabetical order of variables).

The equation $-1 = -1$ is true, so $(1,-1)$ is a solution.

$y = 2x - 3$	
3	$2 \cdot 0 - 3$
	$0 - 3$
	-3

The equation $3 = -3$ is false, so $(0,3)$ is not a solution.

10. Yes, yes

11.
$3s + t = 4$	
$3 \cdot 3 + 4$	4
$9 + 4$	
13	

We substitute 3 for s and 4 for t (alphabetical order of variables).

The equation $13 = 4$ is false, so $(3,4)$ is not a solution.

$$\frac{3s + t = 4}{\begin{array}{c|c} 3(-3) + 5 & 4 \\ -9 + 5 & \\ -4 & \end{array}}$$

The equation -4 = 4 is false, so (-3,5) is not a solution.

12. No, no

13.
$$\frac{2a + 5b = 3}{\begin{array}{c|c} 2 \cdot 0 + 5 \cdot \frac{3}{5} & 3 \\ 0 + 3 & \\ 3 & \end{array}}$$

$\left(0, \frac{3}{5}\right)$ is a solution.

$$\frac{2a + 5b = 3}{\begin{array}{c|c} 2\left(-\frac{1}{2}\right) + 5\left(-\frac{4}{5}\right) & 3 \\ -1 - 4 & \\ -5 & \end{array}}$$

$\left(-\frac{1}{2}, -\frac{4}{5}\right)$ is not a solution.

14. Yes, yes

15.
$$\frac{r^2 - s^2 = 3}{\begin{array}{c|c} 2^2 - (-1)^2 & 3 \\ 4 - 1 & \\ 3 & \end{array}}$$

(2, -1) is a solution.

$$\frac{r^2 - s^2 = 3}{\begin{array}{c|c} (-0.75)^2 - (2.75)^2 & 3 \\ 0.5625 - 7.5625 & \\ -7 & \end{array}}$$

(-0.75, 2.75) is not a solution.

16. No, yes

17. $y = x$

We select numbers for x and find the corresponding values for y.

x	y	(x,y)
-4	-4	(-4,-4)
-2	-2	(-2,-2)
0	0	(0,0)
1	1	(1,1)
3	3	(3,3)

Plot these points and draw the line.

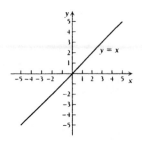

The graph shows the relation {(x,y)|y = x}.

18.

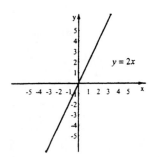

19. $y = -2x$

We select numbers for x and find the corresponding values for y.

x	y	(x,y)
-2	4	(-2,4)
0	0	(0,0)
1	-2	(1,-2)
3	-6	(3,-6)

Plot these points and draw the line.

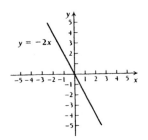

The graph shows the relation {(x,y)|y = -2x}.

20.

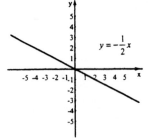

21. y = x + 3

   We select numbers for x and find the corresponding
   values for y.

x	y	(x,y)
0	3	(0,3)
2	5	(2,5)
-3	0	(-3,0)
-5	-2	(-5,-2)

   Plot these solutions and draw the line.

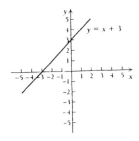

   The graph shows the relation {(x,y)|y = x + 3}.

22.

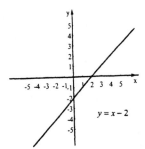

23. y = 3x - 2

   We select numbers for x and find the corresponding
   values for y.

x	y	(x,y)
2	4	(2,4)
0	-2	(0,-2)
-1	-5	(-1,-5)

   Plot these solutions and draw the line.

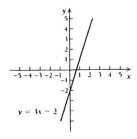

   The graph shows the relation {(x,y)|y = 3x - 2}.

24.

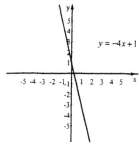

25. y = x²

   We select numbers for x and find the corresponding
   values for y.

x	y	(x,y)
-2	4	(-2,4)
-1	1	(-1,1)
0	0	(0,0)
1	1	(1,1)
2	4	(2,4)

   Plot these ordered pairs and connect the points
   with a smooth curve.

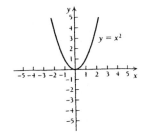

   The graph shows the relation {(x,y)|y = x²}.

26.

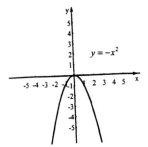

27. y = x² + 2

   We select numbers for x and find the corresponding
   values for y.

x	y
-2	6
-1	3
0	2
1	3
2	6

   We plot these ordered pairs and connect the points
   with a smooth curve.

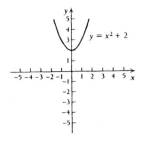

The graph shows the relation $\{(x,y) | y = x^2 + 2\}$.

28.

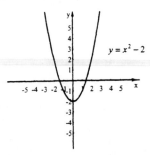

29. $x = y^2 + 2$

Here we select numbers for y and find the corresponding values for x.

x	y
6	-2
3	-1
2	0
3	1
6	2

We plot these points and connect them with a smooth curve.

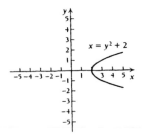

The graph shows the relation $\{(x,y) | x = y^2 + 2\}$.

30.

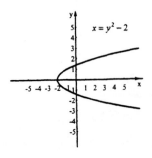

31. $y = |x + 1|$

We find numbers that satisfy the equation.

x	y
-5	4
-3	2
-1	0
2	3
4	5

We plot these points and connect them.

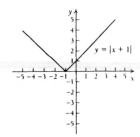

The graph shows the relation $\{(x,y) | y = |x + 1|\}$.

32.

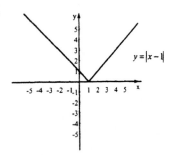

33. $x = |y + 1|$

Here we choose numbers for y and find the corresponding values for x.

x	y
4	-5
1	-2
0	-1
1	0
4	3
5	4

We plot these points and connect them.

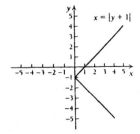

The graph shows the relation $\{(x,y) | x = |y + 1|\}$.

72

34.

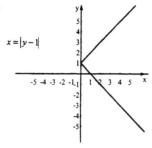

$x = |y - 1|$

35. xy = 10

We find numbers that satisfy the equation. We then plot these points and connect them. Note that neither x nor y can be 0.

x	y
-6	$-\frac{5}{3}$
-5	-2
-3	$-\frac{10}{3}$
-2	-5
-1	-10

x	y
1	10
2	5
3	$\frac{10}{3}$
5	2
6	$\frac{5}{3}$

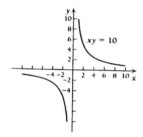

$xy = 10$

The graph shows the relation {(x, y)|xy = 10}.

36.

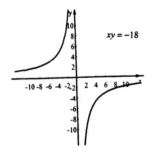

$xy = -18$

37. $y = \frac{1}{x}$

We find numbers that satisfy the equation. We then plot these points and connect them. Note that neither x nor y can be 0.

x	y
-5	$-\frac{1}{5}$
-3	$-\frac{1}{3}$
-2	$-\frac{1}{2}$
-1	-1
$-\frac{1}{2}$	-2
$-\frac{1}{4}$	-4

x	y
$\frac{1}{4}$	4
$\frac{1}{2}$	2
1	1
2	$\frac{1}{2}$
3	$\frac{1}{3}$
5	$\frac{1}{5}$

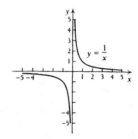

$y = \frac{1}{x}$

The graph shows the relation $\left\{(x, y)\,\middle|\,y = \frac{1}{x}\right\}$.

38.

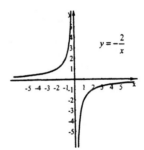

$y = -\frac{2}{x}$

39. $y = \frac{1}{x^2}$

We find numbers that satisfy the equation. We plot these points and connect them. Note that neither x nor y can be 0.

x	y
-3	$\frac{1}{9}$
-2	$\frac{1}{4}$
-1	1
$-\frac{1}{2}$	4
$-\frac{1}{4}$	16

x	y
3	$\frac{1}{9}$
2	$\frac{1}{4}$
1	1
$\frac{1}{2}$	4
$\frac{1}{4}$	16

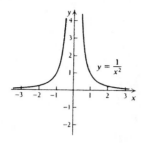

The graph shows the relation $\left\{(x, y)\middle| y = \frac{1}{x^2}\right\}$.

40.

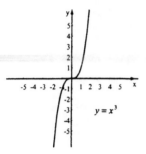

41. $y = \sqrt{x}$

The meaningful replacements for x are x ≥ 0. We choose meaningful replacements for x and find corresponding values of y.

x	y
0	0
2	1.414
4	2
6	2.449
9	3

We plot these points and connect them.

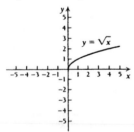

The graph shows the relation $\{(x, y)\mid y = \sqrt{x}\}$.

42.

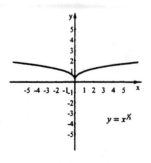

43. $y = 8 - x^2$

We choose numbers for x and find the corresponding values of y.

x	y
-3	-1
-2	4
-1	7
0	8
1	7
2	4
3	-1

We plot these ordered pairs and connect the points with a smooth curve.

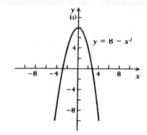

The graph shows the relation $\{(x, y)\mid y = 8 - x^2\}$.

44.

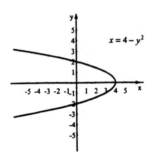

45. $y = x^2 + 1$

x	y
0	1
1	2
-1	2
2	5
-2	5

$y = (-x)^2 + 1$

x	y
0	1
1	2
-1	2
2	5
-2	5

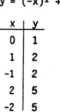

The graphs are the same. The equations are equivalent.

46. $y = x^2 - 2$ $y = 2 - x^2$

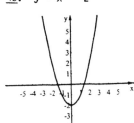   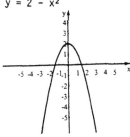

The graphs are not the same. The equations are not equivalent.

47. To determine the domain, think of drawing vertical lines to the x-axis from the left and right extremities of the relation. (In this case the extreme points lie on the x-axis.) The domain consists of the numbers where the lines cross the x-axis and all numbers in between.

To determine the range, think of drawing horizontal lines to the y-axis from the upper and lower extremities of the relation. (In this case the extreme points lie on the y-axis.) The range consists of the numbers where the lines cross the y-axis and all numbers in between.

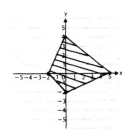

Range = $\{y \mid -2 \leqslant y \leqslant 4\}$

Domain = $\{x \mid -2 \leqslant x \leqslant 5\}$

48. Domain: $\{x \mid -5 \leqslant x \leqslant 5\}$; range: $\{y \mid -2 \leqslant y \leqslant 5\}$

49. a)

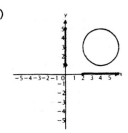

b) Domain: $\{x \mid 2 \leqslant x \leqslant 6\}$

c) Range: $\{y \mid 1 \leqslant y \leqslant 5\}$

50. a)

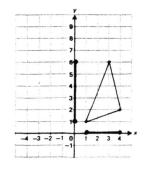

b) Domain: $\{x \mid 1 \leqslant x \leqslant 4\}$

c) Range: $\{y \mid 1 \leqslant y \leqslant 6\}$

51. $y = -x^2$

Domain: The set of real numbers

Range: $\{y \mid y \leqslant 0\}$

52. Domain: The set of real numbers

Range: The set of real numbers

53. $x = |y + 1|$

Domain: $\{x \mid x \geqslant 0\}$

Range: The set of real numbers

54. Domain: $\{x \mid x \neq 0\}$

Range: $\{y \mid y \neq 0\}$

55. $y = \sqrt{x}$

Domain: $\{x \mid x \geqslant 0\}$

Range: $\{y \mid y \geqslant 0\}$

56. Domain: The set of real numbers

Range: The set of real numbers

57. $y = 8 - x^2$

Domain: The set of real numbers

Range: $\{y \mid y \leqslant 8\}$

58. Domain: $\{x \mid x \leqslant 4\}$

Range: The set of real numbers

59.

x	y
-1	0
$-\frac{1}{2}$	$\pm \frac{1}{2}$
0	± 1
$\frac{1}{2}$	$\pm \frac{1}{2}$
1	0

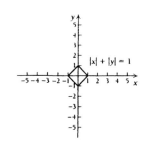

$|x| + |y| = 1$

60.

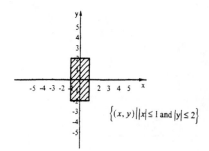

$\{(x, y)\,|\,|x| \le 1 \text{ and } |y| \le 2\}$

64. $y = x^{\frac{2}{3}}$

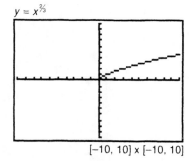

$[-10, 10] \times [-10, 10]$

61. $y = |x| + x$

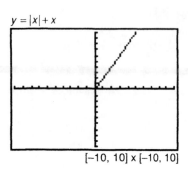

$[-10, 10] \times [-10, 10]$

65. $y = x^2|x|$

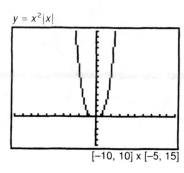

$[-10, 10] \times [-5, 15]$

62. $y = x|x|$

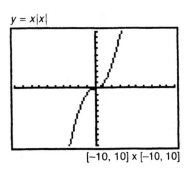

$[-10, 10] \times [-10, 10]$

66. $y = |x^3|$

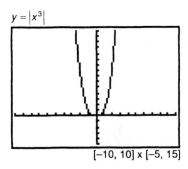

$[-10, 10] \times [-5, 15]$

63. $y = |x^2 - 4|$

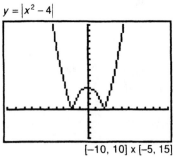

$[-10, 10] \times [-5, 15]$

67. $y = \frac{1}{3}x^3 - x + \frac{2}{3}$

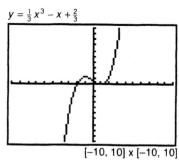

$[-10, 10] \times [-10, 10]$

68. $y = -\frac{1}{3}x^3 + \frac{1}{2}x^2 + 2x - 1$

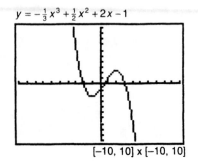

$[-10, 10] \times [-10, 10]$

69. $y = 1 + \sqrt{2 - x}$

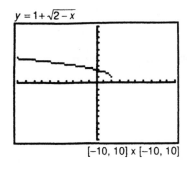

[−10, 10] x [−10, 10]

70. $y = 1 + \dfrac{2}{x + 1}$

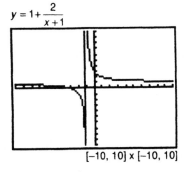

[−10, 10] x [−10, 10]

Exercise Set 3.2

1.   We substitute into the distance formula.

$d = \sqrt{(x_2 - x_1)^2 + (y_2 - y_1)^2}$

$d = \sqrt{[1 - (-3)]^2 + [1 - (-2)]^2}$

$d = \sqrt{4^2 + 3^2} = \sqrt{16 + 9}$

$d = \sqrt{25} = 5$

2. $3\sqrt{5} \approx 6.708$

3.   We substitute into the distance formula.

$d = \sqrt{(x_2 - x_1)^2 + (y_2 - y_1)^2}$

$d = \sqrt{(3 - 0)^2 + [-4 - (-7)]^2}$

$d = \sqrt{3^2 + 3^2} = \sqrt{9 + 9}$

$d = \sqrt{18} = \sqrt{9 \cdot 2}$

$d = 3\sqrt{2} \approx 4.243$

4. $4\sqrt{2} \approx 5.657$

5. $d = \sqrt{(x_2 - x_1)^2 + (y_2 - y_1)^2}$ (Distance formula)

$d = \sqrt{(2a - a)^2 + [5 - (-3)]^2}$ (Substituting)

$d = \sqrt{a^2 + 8^2}$

$d = \sqrt{a^2 + 64}$

6. $\sqrt{64 + k^2}$

7. $d = \sqrt{(x_2 - x_1)^2 + (y_2 - y_1)^2}$ (Distance formula)

$d = \sqrt{(a - 0)^2 + (b - 0)^2}$ (Substituting)

$d = \sqrt{a^2 + b^2}$

8. $\sqrt{5} \approx 2.236$

9. $d = \sqrt{(x_2 - x_1)^2 + (y_2 - y_1)^2}$ (Distance formula)

$d = \sqrt{(-\sqrt{a} - \sqrt{a})^2 + (\sqrt{b} - \sqrt{b})^2}$ (Substituting)

$d = \sqrt{(-2\sqrt{a})^2 + 0^2} = \sqrt{4a}$

$d = 2\sqrt{a}$

10. $2\sqrt{f^2 + c^2}$

11. $d = \sqrt{(x_2 - x_1)^2 + (y_2 - y_1)^2}$ (Distance formula)

$d = \sqrt{(18.9431 - 7.3482)^2 + [-17.9054 - (-3.0991)]^2}$ (Substituting)

$d = \sqrt{(11.5949)^2 + (-14.8063)^2}$

$d \approx \sqrt{134.4417 + 219.2265}$

$d \approx \sqrt{353.6682} \approx 18.806$

12. $\sqrt{163{,}598.3346} \approx 404.473$

13.   First we find the squares of the distances between the points:

Between (9, 6) and (−1, 2):

$d_1^2 = (-1 - 9)^2 + (2 - 6)^2 = 100 + 16 = 116$

Between (−1, 2) and (1, −3):

$d_2^2 = [1 - (-1)]^2 + (-3 - 2)^2 = 4 + 25 = 29$

Between (9, 6) and (1, −3):

$d_3^2 = (1 - 9)^2 + (-3 - 6)^2 = 64 + 81 = 145$

Since $116 + 29 = 145$, it follows from the Pythagorean theorem that the points are the vertices of a right triangle.

14.   Yes

15. $\left(\dfrac{x_1 + x_2}{2}, \dfrac{y_1 + y_2}{2} \right)$ (Midpoint formula)

$\left(\dfrac{-4 + 3}{2}, \dfrac{7 + (-9)}{2} \right)$ (Substituting)

$\left(\dfrac{-1}{2}, \dfrac{-2}{2} \right)$

The midpoint is $\left(-\dfrac{1}{2}, -1 \right)$.

16.   (5, −1)

17. $\left[\dfrac{x_1 + x_2}{2}, \dfrac{y_1 + y_2}{2}\right]$ (Midpoint formula)

$\left[\dfrac{a + a}{2}, \dfrac{b + (-b)}{2}\right]$ (Substituting)

$\left[\dfrac{2a}{2}, \dfrac{0}{2}\right]$

The midpoint is (a, 0).

18. (0, d)

19. $\left[\dfrac{x_1 + x_2}{2}, \dfrac{y_1 + y_2}{2}\right]$ (Midpoint formula)

$\left[\dfrac{-3.895 + 2.998}{2}, \dfrac{8.1212 + (-8.6677)}{2}\right]$

     (Substituting)

$\left[\dfrac{-0.897}{2}, \dfrac{-0.5465}{2}\right]$

The midpoint is (-0.4485, -0.27325).

20. (4.652, 6.95205)

21. We write standard form.

$(x - 0)^2 + (y - 0)^2 = 6^2$

The center is (0, 0), and the radius is 6.

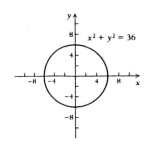

22. Center: (0, 0); radius: 5

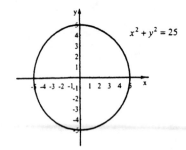

23. We write standard form.

$(x - 3)^2 + (y - 0)^2 = (\sqrt{3})^2$

The center is (3, 0), and the radius is $\sqrt{3}$.

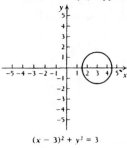

24. Center: (4, 1); radius: $\sqrt{2}$

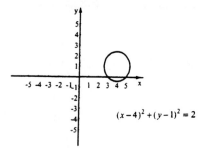

25. We write standard form.

$[x - (-1)^2] + [y - (-3)]^2 = 2^2$

The center is (-1, -3), and the radius is 2.

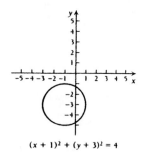

26. Center: (2, -3); radius: 1

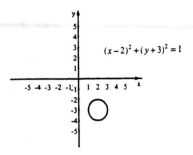

27. $(x - 8)^2 + (y + 3)^2 = 40$

$(x - 8)^2 + [y - (-3)]^2 = (\sqrt{40})^2 = (2\sqrt{10})^2$

Center: (8, -3),   Radius: $2\sqrt{10}$

28. Center: (-5, 1),   Radius: $5\sqrt{3}$

29. $(x - 3)^2 + y^2 = \frac{1}{25}$

$(x - 3)^2 + (y - 0)^2 = \left(\frac{1}{5}\right)^2$

Center: (3, 0),   Radius: $\frac{1}{5}$

30. Center: (0, 1),   Radius: $\frac{1}{2}$

31. $x^2 + y^2 + 8x - 6y - 15 = 0$

We complete the square twice to get standard form:

$(x^2 + 8x \quad\quad) + (y^2 - 6y \quad\quad) = 15$

We take half the coefficient of the x-term and square it, obtaining 16. We add 16 - 16 in the first parentheses. Similarly, we add 9 - 9 in the second parentheses.

$(x^2 + 8x + 16 - 16) + (y^2 - 6y + 9 - 9) = 15$

Next we rearrange and factor:

$(x^2 + 8x + 16) + (y^2 - 6y + 9) - 16 - 9 = 15$

$(x + 4)^2 + (y - 3)^2 = 40$

$[x - (-4)]^2 + (y - 3)^2 = (2\sqrt{10})^2$

Center: (-4, 3),   Radius: $2\sqrt{10}$

32. Center: $\left(-\frac{25}{2}, -5\right)$,   Radius: $\frac{\sqrt{677}}{2}$

33. $x^2 + y^2 + 6x = 0$

$(x^2 + 6x \quad\quad) + y^2 = 0$

$(x^2 + 6x + 9 - 9) + y^2 = 0$ Completing the square once

$(x^2 + 6x + 9) + y^2 - 9 = 0$

$(x + 3)^2 + y^2 = 9$

$[x - (-3)]^2 + (y - 0)^2 = 3^2$

Center: (-3, 0),   Radius: 3

34. Center: (2, 0),   Radius: 2

35. $x^2 + y^2 + 8x = 84$

$(x^2 + 8x \quad\quad) + y^2 = 84$

$(x^2 + 8x + 16 - 16) + y^2 = 84$ Completing the square once

$(x^2 + 8x + 16) + y^2 - 16 = 84$

$(x + 4)^2 + y^2 = 100$

$[x - (-4)]^2 + (y - 0)^2 = 10^2$

Center: (-4, 0),   Radius: 10

36. Center: (0, 5),   Radius: 10

37. $x^2 + y^2 + 21x + 33y + 17 = 0$

$(x^2 + 21x \quad\quad) + (y^2 + 33y \quad\quad) = -17$

$\left[x^2 + 21x + \frac{441}{4} - \frac{441}{4}\right] + \left[y^2 + 33y + \frac{1089}{4} - \frac{1089}{4}\right] = -17$

Completing the square twice

$\left[x^2 + 21x + \frac{441}{4}\right] + \left[y^2 + 33y + \frac{1089}{4}\right] - \frac{441}{4} - \frac{1089}{4} = -17$

$\left[x + \frac{21}{2}\right]^2 + \left[y + \frac{33}{2}\right]^2 = \frac{1462}{4}$

$\left[x - \left(-\frac{21}{2}\right)\right]^2 + \left[y - \left(-\frac{33}{2}\right)\right]^2 = \left(\frac{\sqrt{1462}}{2}\right)^2$

Center: $\left(-\frac{21}{2}, -\frac{33}{2}\right)$,   Radius: $\frac{\sqrt{1462}}{2}$

38. Center: $\left(\frac{7}{2}, -\frac{3}{2}\right)$,   Radius: $\frac{7\sqrt{2}}{2}$

39. $x^2 + y^2 + 8.246x - 6.348y - 74.35 = 0$

$(x^2 + 8.246x \quad\quad) +$

$\quad (y^2 - 6.348y \quad\quad) = 74.35$

$(x^2 + 8.246x + 16.999 - 16.999) +$

$\quad (y^2 - 6.348y + 10.074 - 10.074) = 74.35$

$(x^2 + 8.246x + 16.999) + (y^2 - 6.348y + 10.074) -$

$\quad 16.999 - 10.074 = 74.35$

$(x + 4.123)^2 + (y - 3.174)^2 = 101.423$

$[x - (-4.123)]^2 + (y - 3.174)^2 = (10.071)^2$

Center: (-4.123, 3.174),   Radius: 10.071

40. Center: (-12.537, -5.002), Radius: 13.044

41. $9x^2 + 9y^2 = 1$

$x^2 + y^2 = \frac{1}{9}$ Multiplying by $\frac{1}{9}$

$(x - 0)^2 + (y - 0)^2 = \left(\frac{1}{3}\right)^2$

Center: (0, 0),   Radius: $\frac{1}{3}$

42. Center: (0, 0),   Radius: $\frac{1}{4}$

43. Since the center is (0, 0), we have

$(x - 0)^2 + (y - 0)^2 = r^2$ or $x^2 + y^2 = r^2$

The circle passes through (-3, 4). We find r^2 by substituting -3 for x and 4 for y.

$(-3)^2 + 4^2 = r^2$

$9 + 16 = r^2$

$25 = r^2$

Then $x^2 + y^2 = 25$ is an equation of the circle.

44. $(x - 3)^2 + (y + 2)^2 = 64.$

45. Since the center is (-4, 1), we have

$[x - (-4)]^2 + (y - 1)^2 = r^2$, or

$(x + 4)^2 + (y - 1)^2 = r^2$.

The circle passes through (-2, 5). We find r^2 by substituting -2 for x and 5 for y.

$(-2 + 4)^2 + (5 - 1)^2 = r^2$

$4 + 16 = r^2$

$20 = r^2$

Then $(x + 4)^2 + (y - 1)^2 = 20$ is an equation of the circle.

46. $(x + 3)^2 + (y + 3)^2 = 54.4$

47.

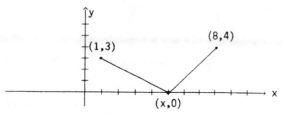

Between (x, 0) and (1, 3):

$d = \sqrt{(1 - x)^2 + (3 - 0)^2}$

$= \sqrt{1 - 2x + x^2 + 9} = \sqrt{x^2 - 2x + 10}$

Between (x, 0) and (8, 4):

$d = \sqrt{(8 - x)^2 + (4 - 0)^2}$

$= \sqrt{64 - 16x + x^2 + 16} = \sqrt{x^2 - 16x + 80}$

Thus,

$\sqrt{x^2 - 2x + 10} = \sqrt{x^2 - 16x + 80}$

$x^2 - 2x + 10 = x^2 - 16x + 80$

$14x = 70$

$x = 5$

The point on the x-axis equidistant from the points (1, 3) and (8, 4) is (5, 0).

48. (0, 4)

49. If the circle with center (2, 4) is tangent to the x-axis, the radius is 4. We substitute 2 for h, 4 for k, and 4 for r.

$(x - h)^2 + (y - k)^2 = r^2$

$(x - 2)^2 + (y - 4)^2 = 4^2 = 16$

50. $(x + 3)^2 + (y + 2)^2 = 9$

51.

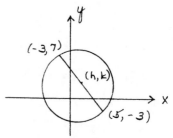

The center of the circle is the midpoint of the line segment with endpoints (5, -3) and (-3, 7):

$\left(\frac{5 + (-3)}{2}, \frac{-3 + 7}{2}\right)$, or (1, 2)

r^2 is the square of the distance from (1, 2) to (5, -3) (or to (-3, 7)):

$r^2 = (5 - 1)^2 + (-3 - 2)^2 = 16 + 25 = 41$

The equation of the circle is

$(x - 1)^2 + (y - 2)^2 = 41$.

52. $\left(x - \frac{5}{2}\right)^2 + \left(y - \frac{5}{2}\right)^2 = \frac{145}{2}$

53. $C = 2\pi r$

$10\pi = 2\pi r$

$5 = r$

Then $[x - (-8)]^2 + (y - 5)^2 = 5^2$, or

$(x + 8)^2 + (y - 5)^2 = 25$.

54. $(x + 3)^2 + (y + 8)^2 = 36$

55.

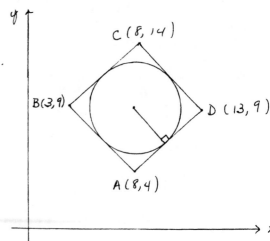

The center is the midpoint of AC or BD. We use AC:

$\left(\frac{8 + 8}{2}, \frac{4 + 14}{2}\right)$, or (8, 9)

The radius is the distance from the center to the midpoint of a side of the square. We will use side AD. The midpoint of AD is

$\left(\frac{8 + 13}{2}, \frac{4 + 9}{2}\right)$, or $\left(\frac{21}{2}, \frac{13}{2}\right)$. Then

$$r = \sqrt{\left(8 - \frac{21}{2}\right)^2 + \left(9 - \frac{13}{2}\right)^2} = \sqrt{\frac{25}{2}} = \frac{5\sqrt{2}}{2}.$$

56. $a_1 \approx 2.68$ ft, $a_2 \approx 37.32$ ft

57.
$$\begin{array}{c|c} x^2 + y^2 = 1 \\ \hline (-1)^2 + 0^2 & 1 \\ 1 & \end{array}$$

$(-1, 0)$ lies on the unit circle.

58. Yes

59.
$$\begin{array}{c|c} x^2 + y^2 = 1 \\ \hline (0.838670568)^2 + (-0.544639035)^2 & 1 \\ 0.703368321 + 0.296631678 & \\ 1 & \end{array}$$

$(0.838670568, -0.544639035)$ lies on the unit circle.

60. No

61.
$$\begin{array}{c|c} x^2 + y^2 = 1 \\ \hline \left(\frac{\sqrt{2}}{2}\right)^2 + \left(\frac{\sqrt{2}}{2}\right)^2 & 1 \\ \frac{2}{4} + \frac{2}{4} & \\ 1 & \end{array}$$

$\left(\frac{\sqrt{2}}{2}, \frac{\sqrt{2}}{2}\right)$ lies on the unit circle.

62. No

63.
$$\begin{array}{c|c} x^2 + y^2 = 1 \\ \hline 0^2 + 2^2 & 1 \\ 4 & \end{array}$$

$(0, 2)$ does not lie on the unit circle.

64. Yes

65. See the answer section in the text.

66. For two points on a vertical line, (x, y_1) and (x, y_2), $d = \sqrt{(x - x)^2 + (y_2 - y_1)^2} = |y_2 - y_1|$. For two points on a horizontal line, (x_1, y) and (x_2, y), $d = \sqrt{(x_2 - x_1)^2 + (y - y)^2} = |x_2 - x_1|$.

67. The result of combining the graphs is a circle with center $(0, 0)$ and radius 2.

68. Graph $y = \sqrt{9 - x^2}$ and $y = -\sqrt{9 - x^2}$ on the same set of axes.

Exercise Set 3.3

1. This is not a function. There are two members of the range, Mets and Giants, that correspond to New York. There are other instances that also show that this is not a function, but only one case is necessary.

2. Yes

3. This is a function. Each member of the domain corresponds to exactly one member of the range.

4. No

5. This is a function, because no two ordered pairs have the same first coordinate and different second coordinates.

6. No

7. This is not a function, because the ordered pairs $(-4,-4)$ and $(-4,4)$ have the same first coordinate and different second coordinates.

8. No

9. This is a function, because no two ordered pairs have the same first coordinate and different second coordinates.

10. No

11. $f(x) = 5x^2 + 4x$

a) $f(0) = 5 \cdot 0^2 + 4 \cdot 0 = 0 + 0 = 0$

b) $f(-1) = 5(-1)^2 + 4(-1) = 5 - 4 = 1$

c) $f(3) = 5 \cdot 3^2 + 4 \cdot 3 = 45 + 12 = 57$

d) $f(t) = 5t^2 + 4t$

e) $f(t - 1) = 5(t - 1)^2 + 4(t - 1)$
$$= 5(t^2 - 2t + 1) + 4(t - 1)$$
$$= 5t^2 - 10t + 5 + 4t - 4$$
$$= 5t^2 - 6t + 1$$

f) $f(a + h) = 5(a + h)^2 + 4(a + h)$
$$= 5(a^2 + 2ah + h^2) + 4(a + h)$$
$$= 5a^2 + 10ah + 5h^2 + 4a + 4h$$

$f(a) = 5a^2 + 4a$

$$\frac{f(a + h) - f(a)}{h} = \frac{(5a^2+10ah+5h^2+4a+4h)-(5a^2+4a)}{h}$$
$$= \frac{10ah + 5h^2 + 4h}{h}$$
$$= \frac{h(10a + 5h + 4)}{h}$$
$$= 10a + 5h + 4$$

12. a) 1; b) 6; c) 22; d) $3t^2 - 2t + 1$;
e) $3a^2 + 6ah + 3h^2 - 2a - 2h + 1$;
f) $6a + 3h - 2$

13. $f(x) = 2|x| + 3x$

 a) $f(1) = 2|1| + 3\cdot1 = 2 + 3 = 5$

 b) $f(-2) = 2|-2| + 3(-2) = 4 - 6 = -2$

 c) $f(-4) = 2|-4| + 3(-4) = 8 - 12 = -4$

 d) $f(2y) = 2|2y| + 3(2y) = 4|y| + 6y$

 e) $f(a + h) = 2|a + h| + 3(a + h)$

   $= 2|a + h| + 3a + 3h$

 f) $\dfrac{f(a + h) - f(a)}{h}$

   $= \dfrac{(2|a + h| + 3a + 3h) - (2|a| + 3a)}{h}$

   $= \dfrac{2|a + h| - 2|a| + 3h}{h}$

14. a) -1; b) -4; c) -56; d) $27y^3 - 6y$;

 e) $4 + 10h + 6h^2 + h^3$; f) $10 + 6h + h^2$

15. $f(x) = \dfrac{x}{2 - x}$

 a) $f(1) = \dfrac{1}{2 - 1} = \dfrac{1}{1} = 1$

 b) $f(2) = \dfrac{1}{2 - 2} = \dfrac{1}{0}$; division by 0 is not defined, so $f(2)$ does not exist.

 c) $f(-3) = \dfrac{-3}{2 - (-3)} = \dfrac{-3}{5}$, or $-\dfrac{3}{5}$

 d) $f(-16) = \dfrac{-16}{2 - (-16)} = \dfrac{-16}{18} = -\dfrac{8}{9}$

 e) $f(x + h) = \dfrac{x + h}{2 - (x + h)} = \dfrac{x + h}{2 - x - h}$

 f) $\dfrac{f(x + h) - f(x)}{h} = \dfrac{\dfrac{x + h}{2 - x - h} - \dfrac{x}{2 - x}}{h}$

   $= \dfrac{\dfrac{x+h}{2-x-h} - \dfrac{x}{2-x}}{h} \cdot \dfrac{(2-x-h)(2-x)}{(2-x-h)(2-x)}$

   $= \dfrac{(x+h)(2-x) - x(2-x-h)}{h(2-x-h)(2-x)}$

   $= \dfrac{2x-x^2+2h-hx-2x+x^2+xh}{h(2-x-h)(2-x)}$

   $= \dfrac{2h}{h(2 - x - h)(2 - x)}$

   $= \dfrac{2}{(2 - x - h)(2 - x)}$

16. a) $\dfrac{1}{8}$; b) 0; c) does not exist; d) $\dfrac{81}{53}$;

 e) $\dfrac{x + h - 4}{x + h + 3}$; f) $\dfrac{7}{(x + h + 3)(x + 3)}$

17. $f(x) = \dfrac{x^2 - x - 2}{2x^2 - 5x - 3}$

 a) $f(0) = \dfrac{0^2 - 0 - 2}{2\cdot0^2 - 5\cdot0 - 3} = \dfrac{-2}{-3} = \dfrac{2}{3}$

 b) $f(4) = \dfrac{4^2 - 4 - 2}{2\cdot4^2 - 5\cdot4 - 3} = \dfrac{10}{9}$

 c) $f(-1) = \dfrac{(-1)^2 - (-1) - 2}{2(-1)^2 - 5(-1) - 3} = \dfrac{0}{4} = 0$

 d) $f(3) = \dfrac{3^2 - 3 - 2}{2\cdot3^2 - 5\cdot3 - 3} = \dfrac{4}{0}$

   Division by 0 is not defined, so $f(3)$ does not exist.

 e) $f(2 - h) = \dfrac{(2 - h)^2 - (2 - h) - 2}{2(2 - h)^2 - 5(2 - h) - 3}$

   $= \dfrac{4 - 4h + h^2 - 2 + h - 2}{2(4 - 4h + h^2) - 10 + 5h - 3}$

   $= \dfrac{-3h + h^2}{8 - 8h + 2h^2 - 10 + 5h - 3}$

   $= \dfrac{-3h + h^2}{-5 - 3h + 2h^2}$, or $\dfrac{h^2 - 3h}{2h^2 - 3h - 5}$

 f) $f(a + b) = \dfrac{(a + b)^2 - (a + b) - 2}{2(a + b)^2 - 5(a + b) - 3}$

   $= \dfrac{a^2 + 2ab + b^2 - a - b - 2}{2(a^2 + 2ab + b^2) - 5a - 5b - 3}$

   $= \dfrac{a^2 + 2ab + b^2 - a - b - 2}{2a^2 + 4ab + 2b^2 - 5a - 5b - 3}$

18. a) $\dfrac{\sqrt{26}}{5}$; b) $\dfrac{\sqrt{2}}{3}$; c) does not exist as a real number; d) does not exist as a real number;

 e) $\sqrt{\dfrac{5 + 3h}{11 + 2h}}$; f) $\sqrt{\dfrac{3a - 3b + 4}{2a - 2b + 5}}$

19. $f(x) = x^2$

 $f(a + h) = (a + h)^2 = a^2 + 2ah + h^2$

 $f(a) = a^2$

 $\dfrac{f(a + h) - f(a)}{h} = \dfrac{(a^2 + 2ah + h^2) - (a^2)}{h}$

   $= \dfrac{2ah + h^2}{h} = \dfrac{h(2a + h)}{h}$

   $= 2a + h$

20. $3a^2 + 3ah + h^2$

21. $f(x) = x + \sqrt{x^2 - 1}$

 $f(0) = 0 + \sqrt{0^2 - 1} = 0 + \sqrt{-1}$

   Since $\sqrt{-1}$ is not a real number, $f(0)$ does not exist as a real number.

 $f(2) = 2 + \sqrt{2^2 - 1} = 2 + \sqrt{4 - 1} = 2 + \sqrt{3}$

 $f(10) = 10 + \sqrt{10^2 - 1} = 10 + \sqrt{100 - 1} =$

   $10 + \sqrt{99} = 10 + \sqrt{9\cdot11} = 10 + 3\sqrt{11}$

22. 0; does not exist as a real number; $\dfrac{1}{\sqrt{3}}$, or $\dfrac{\sqrt{3}}{3}$

23. $f(x) = 7x + 4$

 There are no restrictions on the numbers we can substitute into this formula. Thus, the domain is the entire set of real numbers.

24. All real numbers

25. $f(x) = 4 - \dfrac{2}{x}$

   Division by 0 is undefined. Thus, $x \neq 0$. The domain is all real numbers except 0, or $\{x \mid x \neq 0\}$.

26. $\{x \mid x \neq 0\}$

27. $f(x) = \sqrt{7x + 4}$

   Since this formula is meaningful only if the radicand is nonnegative, we want the replacements for x which make the following inequality true.

   $7x + 4 \geqslant 0$

   $\quad 7x \geqslant -4$

   $\quad\ \ x \geqslant -\dfrac{4}{7}$

   The domain is $\left\{ x \mid x \geqslant -\dfrac{4}{7} \right\}$.

28. $\{x \mid x \geqslant 3\}$

29. $f(x) = \dfrac{1}{x^2 - 4}$

   This formula is meaningful as long as a replacement for x does not make the denominator 0. To find those replacements which do make the denominator 0, we solve $x^2 - 4 = 0$.

   $x^2 - 4 = 0$

   $(x + 2)(x - 2) = 0$

   $x = -2 \text{ or } x = 2$

   Thus the domain consists of all real numbers except -2 and 2. This set can be named $\{x \mid x \neq -2 \text{ and } x \neq 2\}$.

30. $\{x \mid x \neq -3 \text{ and } x \neq 3\}$

31. Since $(-1,2)$ is in the function, $f(-1) = 2$.
   Since $(7,9)$ is in the function, $f(7) = 9$.
   Since $(5,-6)$ is in the function, $f(5) = -6$.
   Since $(-3,4)$ is in the function, $f(-3) = 4$.

   The domain is $\{-1,-3,5,7\}$.
   The range is $\{2,4,-6,9\}$.

32. Domain: $\{2,-4,6,8\}$; range: $\{-3,5,-7\}$

33. $f(z) = z^2 - 4z + i$

   $f(3 + i) = (3 + i)^2 - 4(3 + i) + i$

   $\qquad\quad = 9 + 6i + i^2 - 12 - 4i + i$

   $\qquad\quad = i^2 + 3i - 3$

   $\qquad\quad = -1 + 3i - 3$

   $\qquad\quad = -4 + 3i$

34. $8 - 11i$

35. $f(x) = \dfrac{1}{x} \qquad\qquad f(x + h) = \dfrac{1}{x + h}$

   $\dfrac{f(x + h) - f(x)}{h} = \dfrac{\dfrac{1}{x + h} - \dfrac{1}{x}}{h}$

   $\qquad\qquad = \dfrac{\dfrac{1}{x + h} \cdot \dfrac{x}{x} - \dfrac{1}{x} \cdot \dfrac{x + h}{x + h}}{h}$

   $\qquad\qquad = \dfrac{\dfrac{x - (x + h)}{x(x + h)}}{h}$

   $\qquad\qquad = \dfrac{-h}{x(x + h)} \cdot \dfrac{1}{h}$

   $\qquad\qquad = \dfrac{-1}{x(x + h)}$

36. $\dfrac{-2x - h}{x^2(x + h)^2}$

37. $f(x) = \sqrt{x} \qquad\qquad f(x + h) = \sqrt{x + h}$

   $\dfrac{f(x + h) - f(x)}{h} = \dfrac{\sqrt{x + h} - \sqrt{x}}{h}$

   $\qquad\quad = \dfrac{\sqrt{x + h} - \sqrt{x}}{h} \cdot \dfrac{\sqrt{x + h} + \sqrt{x}}{\sqrt{x + h} + \sqrt{x}}$

   $\qquad\quad = \dfrac{(x + h) - x}{h(\sqrt{x + h} + \sqrt{x})}$

   $\qquad\quad = \dfrac{h}{h(\sqrt{x + h}) + \sqrt{x})}$

   $\qquad\quad = \dfrac{1}{\sqrt{x + h} + \sqrt{x}}$

38. No

39. $\left\{ x \mid x \neq -\dfrac{3}{4} \text{ and } x \neq 2 \right\}$

40. $\{x \mid x \neq 0 \text{ and } x \neq -2 \text{ and } x \neq 2\}$

41. $\{x \mid x \neq 0 \text{ and } x \neq -2 \text{ and } x \neq 1\}$

42. All real numbers

43. $\{x \mid x \neq 2 \text{ and } x \neq -1 \text{ and } x \geqslant -3\}$

44. $\left\{ x \mid x \neq \dfrac{5}{2} \text{ and } x \geqslant 0 \right\}$

45. All real numbers

46. $\{x \mid x > 0\}$

Exercise Set 3.4

1. Graph: 8x - 3y = 24

First find two of its points. Here the easiest points are the intercepts.

y-intercept: Set x = 0 and find y.

$$8x - 3y = 24$$
$$8 \cdot 0 - 3y = 24$$
$$-3y = 24$$
$$y = -8$$

The y-intercept is (0, -8).

x-intercept: Set y = 0 and find x.

$$8x - 3y = 24$$
$$8x - 3 \cdot 0 = 24$$
$$8x = 24$$
$$x = 3$$

The x-intercept is (3, 0).

Plot these two points and draw a line through them.

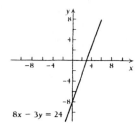

A third point should be used as a check.

Set x = 2 and find y.

$$8x - 3y = 24$$
$$8 \cdot 2 - 3y = 24$$
$$16 - 3y = 24$$
$$-3y = 8$$
$$y = -\frac{8}{3}$$

The ordered pair $\left(2, -\frac{8}{3}\right)$ is also a point on the line.

2. 5x - 10y = 50

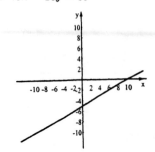

3. Graph: 3x + 12 = 4y

First find two of its points. Here the easiest points to find are the intercepts.

y-intercept: Set x = 0 and find y.

$$3x + 12 = 4y$$
$$3 \cdot 0 + 12 = 4y$$
$$12 = 4y$$
$$3 = y$$

The y-intercept is (0, 3).

x-intercept: Set y = 0 and find x.

$$3x + 12 = 4y$$
$$3x + 12 = 4 \cdot 0$$
$$3x + 12 = 0$$
$$3x = -12$$
$$x = -4$$

The x-intercept is (-4, 0).

Plot these points and draw a line through them. A third point should be used as a check.

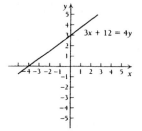

4. 4x - 20 = 5y

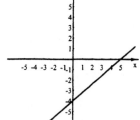

5. Graph: y = -2

The equation y = -2 says that the second coordinate of every ordered pair of the graph is -2. Below is a table of a few ordered pairs that are solutions of y = -2. (It might help to think of y = -2 as $0 \cdot x + y = -2$.)

x	y	
-3	-2	(x can be any number,
0	-2	but y must be -2)
1	-2	
4	-2	

Plot these points and draw the line through them.

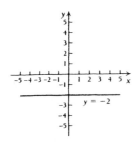

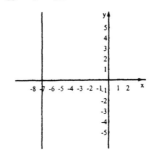

__8.__ 19 = 5 - 2x

__6.__ 2y - 3 = 9

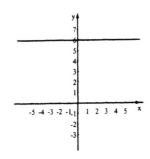

__9.__ Find the slope of the line containing (6, 2) and (-2, 1).

$$m = \frac{\text{change in } y}{\text{change in } x} = \frac{2 - 1}{6 - (-2)} = \frac{1}{8}$$

$$\text{or} = \frac{1 - 2}{-2 - 6} = \frac{-1}{-8} = \frac{1}{8}$$

Note that it does not matter in which order we choose the points, so long as we take the differences in the same order. We get the same slope either way.

__10.__ $\frac{3}{2}$

__7.__ Graph: 5x + 2 = 17

          5x = 15

           x = 3

The equation 5x + 2 = 17 is equivalent to x = 3. The equation x = 3 says that the first coordinate of every ordered pair of the graph is 3. Below is a table of a few ordered pairs that are solutions of x = 3. (It might help to think of x = 3 as x + 0·y = 3.)

x	y	
3	0	(y can be any number,
3	-2	but x must be 3)
3	4	
3	-1	

Plot these point and draw a line through them.

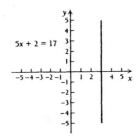

__11.__ Find the slope of the line containing (2, -4) and (4, -3). Let (4, -3) be (x_1, y_1) and (2, -4) be (x_2, y_2).

$$m = \frac{y_2 - y_1}{x_2 - x_1} = \frac{-4 - (-3)}{2 - 4} = \frac{-1}{-2} = \frac{1}{2}$$

__12.__ $-\frac{11}{10}$

__13.__ Find the slope of the line containing $(\pi, 5)$ and $(\pi, 4)$. Let $(\pi, 4)$ be (x_1, y_1) and $(\pi, 5)$ be (x_2, y_2).

$$m = \frac{y_2 - y_1}{x_2 - x_1} = \frac{5 - 4}{\pi - \pi} = \frac{1}{0}$$

The slope is not defined, because we cannot divide by 0.

__14.__ 0

__15.__ $m = \frac{920.58}{13,740} = 0.067$

The road grade is 6.7%.

We can think of an equation giving the height y as a function of the horizontal distance x as an equation of variation where y varies directly as x and the variation constant is 0.067. The equation is y = 0.067x.

__16.__ 4%; y = 0.04x

17. We express the slope of the treadmill, 8%, as a decimal quantity. Let h = the height of the vertical end of the treadmill.

$$0.08 = \frac{h}{5}$$

$$0.4 = h$$

The end of the treadmill is 0.4 ft high.

18. 30 ft

19. Find an equation of the line through (3, 2) with m = 4.

$(y - y_1) = m(x - x_1)$ (Point-slope equation)

$(y - 2) = 4(x - 3)$ (Substituting)

$y - 2 = 4x - 12$

$y = 4x - 10$

20. $y = -2x + 15$

21. Find an equation of the line with y-intercept -5 and m = 2.

$y = mx + b$ (Slope-intercept equation)

$y = 2x + (-5)$ (Substituting)

$y = 2x - 5$

22. $y = \frac{1}{4} x + \pi$

23. Find an equation of the line through (-4, 7) with $m = -\frac{2}{3}$.

$(y - y_1) = m(x - x_1)$ (Point-slope equation)

$(y - 7) = -\frac{2}{3} [x - (-4)]$ (Substituting)

$y - 7 = -\frac{2}{3} (x + 4)$

$y - 7 = -\frac{2}{3} x - \frac{8}{3}$

$y = -\frac{2}{3} x + \frac{13}{3}$

24. $y = \frac{3}{4} x - \frac{11}{4}$

25. Find an equation of the line through (5, -8) with m = 0.

$(y - y_1) = m(x - x_1)$ (Point-slope equation)

$[y - (-8)] = 0(x - 5)$

$y + 8 = 0$

$y = -8$

26. x = 5

27. Find an equation of the line containing (1, 4) and (5, 6). Let $(x_1, y_1) = (1, 4)$ and $(x_2, y_2) = (5, 6)$.

$y - y_1 = \frac{y_2 - y_1}{x_2 - x_1} (x - x_1)$ (Two-point equation)

$y - 4 = \frac{6 - 4}{5 - 1} (x - 1)$ (Substituting)

$y - 4 = \frac{2}{4} (x - 1)$

$y - 4 = \frac{1}{2} x - \frac{1}{2}$

$y = \frac{1}{2} x + \frac{7}{2}$

28. $y = \frac{3}{4} x + \frac{3}{2}$

29. Find an equation of the line containing (-2, 5) and (-4, -7). Let $(x_1, y_1) = (-2, 5)$ and $(x_2, y_2) = (-4, -7)$.

$y - y_1 = \frac{y_2 - y_1}{x_2 - x_1} (x - x_1)$ (Two-point equation)

$y - 5 = \frac{-7 - 5}{-4 - (-2)} [x - (-2)]$ (Substituting)

$y - 5 = \frac{-12}{-2} (x + 2)$

$y - 5 = 6x + 12$

$y = 6x + 17$

30. $y = -\frac{54}{7} x + \frac{152}{35}$

31. Find an equation of the line containing (3, 6) and (-2, 6). Let (x_1, y_1) be (3, 6) and (x_2, y_2) be (-2, 6).

$y - y_1 = \frac{y_2 - y_1}{x_2 - x_1} (x - x_1)$ (Two-point equation)

$y - 6 = \frac{6 - 6}{-2 - 3} (x - 3)$ (Substituting)

$y - 6 = \frac{0}{-5} (x - 3)$

$y - 6 = 0$

$y = 6$

32. $x = -\frac{3}{8}$

33. $y = 2x + 3$

$y = mx + b$ (Slope-intercept equation)

The slope is 2, and the y-intercept is 3.

34. m = -1, b = 6

35. $2y = -6x + 10$

$y = -3x + 5$

$y = mx + b$ (Slope-intercept equation)

The slope is -3, and the y-intercept is 5.

36. m = 4, b = -3

37. 3x - 4y = 12

$$-4y = -3x + 12$$

$$y = \frac{3}{4} x - 3$$

$$\underset{\uparrow}{y} = \underset{\uparrow}{mx} + \underset{}{b} \qquad \text{(Slope-intercept equation)}$$

The slope is $\frac{3}{4}$, and the y-intercept is -3.

38. $m = -\frac{5}{2}$, $b = -\frac{7}{2}$

39. 3y + 10 = 0

$$3y = -10$$

$$y = -\frac{10}{3}$$

or

$$y = 0 \cdot x + \left(-\frac{10}{3}\right)$$

$$\underset{\uparrow}{y} = \underset{\uparrow}{mx} + \underset{}{b} \qquad \text{(Slope-intercept equation)}$$

The slope is 0, and the y-intercept is $-\frac{10}{3}$.

40. m = 0, b = 7

41. Graph $y = -\frac{3}{2} x$.

First we plot the y-intercept (0, 0). We think of the slope as $\frac{-3}{2}$. Starting at the y-intercept and using the slope, we find another point by moving 3 units down and 2 units right. We get to a new point (2, -3). By thinking of the slope as $\frac{3}{-2}$ we can start again at the y-intercept and find another point by moving 3 units up and 2 units left. We get to another point on the line, (-2, 3). We draw the line through these points.

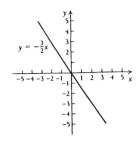

42.

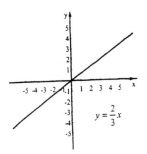

43. Graph $y = -\frac{5}{2} x - 2$.

First we plot the y-intercept (0, -2). We think of the slope as $\frac{-5}{2}$. Starting at the y-intercept and using the slope, we find another point by moving 5 units down and 2 units right. We get to a new point, (2, -7). By thinking of the slope as $\frac{5}{-2}$ we can start again at the y-intercept and find another point by moving 5 units up and 2 units left. We get to another point on the line, (-2, 3). We draw the line through these points.

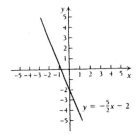

44.

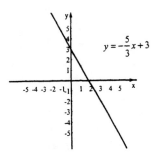

<u>45.</u> Graph $y = \frac{1}{2}x + 1$.

First we plot the y-intercept (0, 1). We consider the slope $\frac{1}{2}$. Starting at the y-intercept and using the slope, we find another point by moving 1 unit up and 2 units right. We get to a new point, (2, 2). By thinking of the slope as $\frac{-1}{-2}$ we can start again at the y-intercept and find another point by moving 1 unit down and 2 units left. We get to another point on the line, (-2, 0). We draw the line through these points.

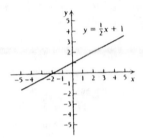

<u>46.</u>

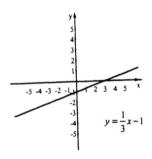

<u>47.</u> Graph $y = \frac{4}{3} - \frac{1}{3}x$, or $y = -\frac{1}{3}x + \frac{4}{3}$.

First we plot the y-intercept $\left(0, \frac{4}{3}\right)$. We think of the slope as $\frac{-1}{3}$. Starting at the y-intercept and using the slope, we find another point by moving 1 unit down and 3 units right. We get to a new point, $\left(3, \frac{1}{3}\right)$. By thinking of the slope as $\frac{1}{-3}$ we can start again at the y-intercept and find another pont by moving 1 unit up and 3 units left. We get to another point on the line, $\left(-3, \frac{7}{3}\right)$. We draw the line through these points.

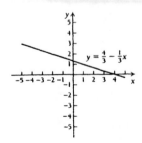

<u>48.</u>

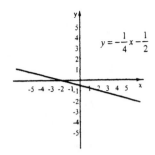

<u>49.</u> $y - y_1 = m(x - x_1)$ (Point-slope equation)

$y - (-2.563) = 3.516(x - 3.014)$ (Substituting)

$y + 2.563 = 3.516x - 10.597224$

$y = 3.516x - 13.1602$

<u>50.</u> $y = 1.2222x + 1.0949$

<u>51.</u> T(d) = 10d + 20

a)   T(5 km) = 10·5 + 20 = 50 + 20 = 70°C

T(20 km) = 10·20 + 20 = 200 + 20 = 220°C

T(1000 km) = 10·1000 + 20 = 10,000 + 20 = 10,020°C

b) This is a linear function. We plot the points (5, 70), (20, 220), and (1000, 10,020) and draw the graph.

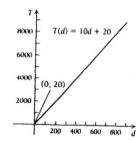

c) The depth must be nonnegative, so $d \geqslant 0$. If we go more than 5600 km into the earth, we begin to emerge on the other side. Thus $d \leqslant 5600$.

The domain of the function is $\{d \mid 0 \leqslant d \leqslant 5600\}$.

<u>52.</u> a) -1.18 mm, 5.97 mm, 10.26 mm

b)

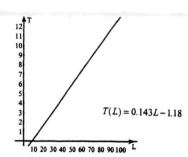

c) {L|L ≥ 8.25}

53. V(t) = $5200 - $512.50t

a) V(0) = $5200 - $512.50(0) = $5200 - $0 =
        $5200

V(1) = $5200 - $512.50(1) = $5200 - $512.50 =
        $4687.50

V(2) = $5200 - $512.50(2) = $5200 - $1025 =
        $4175

V(3) = $5200 - $512.50(3) = $5200 - $1537.50 =
        $3662.50

V(8) = $5200 - $512.80(8) = $5200 - $4100 =
        $1100

b) This is a linear function. We plot the
    points (0, $5200) and (8, $1100) and draw
    the graph.

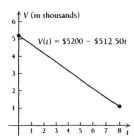

54. a) 207.25 cm; b) 206.3 cm; c) For each
     function, x > 0. The upper bound
     of the domains depends on the maximum
     height of men and women in the population
     from which the bone came.

55. Solve each equation for y.

$2x - 5y = -3$ $2x + 5y = 4$
$-5y = -2x - 3$ $5y = -2x + 4$
$y = \frac{2}{5}x + \frac{3}{5}$ $y = -\frac{2}{5}x + \frac{4}{5}$

The slopes are $\frac{2}{5}$ and $-\frac{2}{5}$. The slopes are not
equal. The product of the slopes is not -1.
Thus, the lines are neither parallel nor
perpendicular.

56. Parallel

57. Solve each equation for y.

$y = 4x - 5$ $4y = 8 - x$
                      $y = -\frac{1}{4}x + 2$

The slopes are 4 and $-\frac{1}{4}$. Their product is -1,
so the lines are perpendicular.

58. Perpendicular

59. We first solve for y.
$3x - y = 7$
$-y = -3x + 7$
$y = 3x - 7$

The slope of the given line is 3. The line
parallel to the given line will have slope 3.
Since the line contains the point (0, 3), we
know that the y-intercept is 3. We use the
slope-intercept equation.
$y = mx + b$
$y = 3x + 3$
The perpendicular line has slope $-\frac{1}{3}$ and
y-intercept 3. We use the slope-intercept
equation again.
$y = mx + b$
$y = -\frac{1}{3}x + 3$

60. Parallel line: $y = -2x - 13$
     Perpendicular line: $y = \frac{1}{2}x - 3$

61. We first solve for y.
$5x - 2y = 4$
$-2y = -5x + 4$
$y = \frac{5}{2}x - 2$

The slope of the given line is $\frac{5}{2}$. The line
parallel to the given line will have slope $\frac{5}{2}$.
We use the point-slope equation to write an
equation with slope $\frac{5}{2}$ and containing the point
(-3, -5).
$y - y_1 = m(x - x_1)$
$y - (-5) = \frac{5}{2}[x - (-3)]$
$y + 5 = \frac{5}{2}x + \frac{15}{2}$
$y = \frac{5}{2}x + \frac{5}{2}$

The slope of the perpendicular line is $-\frac{2}{5}$.

$y - y_1 = m(x - x_1)$ (Point-slope equation)
$y - (-5) = -\frac{2}{5}[x - (-3)]$ [Substituting -3 for x_1, -5 for y_1, and $-\frac{2}{5}$ for m]

$y + 5 = -\frac{2}{5}(x + 3)$
$y + 5 = -\frac{2}{5}x - \frac{6}{5}$
$y = -\frac{2}{5}x - \frac{31}{5}$

62. Parallel line: $y = -\frac{3}{4}x + \frac{1}{4}$
     Perpendicular line: $y = \frac{4}{3}x - 6$

89

63. x = -1 is the equation of a vertical line. The line parallel to the given line is a vertical line containing the point (3, -3), or x = 3.

The line perpendicular to the given line is a horizontal line containing the point (3, -3), or y = -3.

64. Parallel line: y = -5

Perpendicular line: x = 4

65.
$$f(x) = mx + b \qquad \text{(Linear function)}$$
$$f(3x) = 3f(x) \qquad \text{(Given)}$$
$$m(3x) + b = 3(mx + b)$$
$$3mx + b = 3mx + 3b$$
$$b = 3b$$
$$0 = 2b$$
$$0 = b$$

Thus, if f(3x) = 3f(x), then
$$f(x) = mx + 0$$
$$f(x) = mx$$

66. f(x) = x + b

67.
$$f(x) = mx + b \qquad \text{(Linear function)}$$
$$f(c + d) = m(c + d) + b = mc + md + b$$
$$f(c) + f(d) = (mc + b) + (md + b) = mc + md + 2b$$

mc + md + b ≠ mc + md + 2b, for example, when m = 0, b = 1

Thus, f(c + d) = f(c) + f(d) is false.

68. False

69.
$$f(x) = mx + b \qquad \text{(Linear function)}$$
$$f(kx) = m(kx) + b = mkx + b$$
$$kf(x) = k(mx + b) = mkx + kb$$

mkx + b ≠ mkx + kb, for example, when m = 0, b = 1, k = 2

Thus, f(kx) = kf(x) is false.

70. False

71. If A(9, 4), B(-1, 2), and C(4, 3) are on the same line, then the slope of the segment $\overline{AB}$ must be the same as the slope of the segment $\overline{BC}$. (Note that B is a point of each segment.)

Slope of $\overline{AB}$ = $\dfrac{4 - 2}{9 - (-1)}$ = $\dfrac{2}{10}$ = $\dfrac{1}{5}$

Slope of $\overline{BC}$ = $\dfrac{2 - 3}{-1 - 4}$ = $\dfrac{-1}{-5}$ = $\dfrac{1}{5}$

Since the slopes are the same, and B is on both lines, A, B, and C are on the same line.

72. No

73.
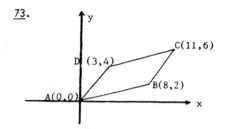

Slope of $\overline{AB}$ = $\dfrac{2 - 0}{8 - 0}$ = $\dfrac{2}{8}$ = $\dfrac{1}{4}$

Slope of $\overline{CD}$ = $\dfrac{6 - 4}{11 - 3}$ = $\dfrac{2}{8}$ = $\dfrac{1}{4}$

Slope of $\overline{BC}$ = $\dfrac{6 - 2}{11 - 8}$ = $\dfrac{4}{3}$

Slope of $\overline{DA}$ = $\dfrac{4 - 0}{3 - 0}$ = $\dfrac{4}{3}$

$\overline{AB}$ and $\overline{CD}$ have the same slope.
$\overline{BC}$ and $\overline{DA}$ have the same slope.

74. The product of the slopes is -1; that is, each is the negative reciprocal of the other.

75. Consider a line containing the ordered pairs (0, 32) and (100, 212).

The slope of the line is $\dfrac{212 - 32}{100 - 0}$, or $\dfrac{9}{5}$.

$$F(C) = mC + b \qquad \text{(Linear function)}$$
$$F(C) = \dfrac{9}{5}C + b \qquad \left[\text{Substituting } \tfrac{9}{5} \text{ for } m\right]$$

We now determine b.

$$32 = \dfrac{9}{5} \cdot 0 + b \qquad \text{(Substituting 0 for C and 32 for F)}$$
$$32 = b$$

The linear function is
$$F(C) = \dfrac{9}{5}C + 32.$$

76. $C = \dfrac{5}{9}(F - 32)$

77.
$$P = mQ + b, \quad m \ne 0 \qquad \text{(Linear function)}$$

We solve for Q.
$$P - b = mQ$$
$$\dfrac{P - b}{m} = Q$$
$$\dfrac{P}{m} - \dfrac{b}{m} = Q \qquad \text{(Linear function)}$$

78. By definition, y = kx. Thus, y is a linear function of x.

79. The slope of the line containing (-1, 4) and (2, -3) is $\frac{4 - (-3)}{-1 - 2}$, or $-\frac{7}{3}$. The slope of the line parallel to this line and containing (4, -2) is also $-\frac{7}{3}$.

$y - y_1 = m(x - x_1)$ (Point-slope equation)

$y - (-2) = -\frac{7}{3}(x - 4)$ $\left[\begin{array}{l}\text{Substituting 4 for } x_1\text{, -2} \\ \text{for } y_1\text{, and } -\frac{7}{3} \text{ for m}\end{array}\right]$

$y + 2 = -\frac{7}{3}x + \frac{28}{3}$

$y = -\frac{7}{3}x + \frac{22}{3}$

The slope of the line perpendicular to the given line is $\frac{3}{7}$.

$y - y_1 = m(x - x_1)$ (Point-slope equation)

$y - (-2) = \frac{3}{7}(x - 4)$ (Substituting)

$y + 2 = \frac{3}{7}x - \frac{12}{7}$

$y = \frac{3}{7}x - \frac{26}{7}$

80. Parallel line: $y = \frac{2}{5}x + \frac{17}{5}$

Perpendicular line: $y = -\frac{5}{2}x + \frac{1}{2}$

81. The slope of the line containing (-3, k) and (4, 8) is

$\frac{8 - k}{4 - (-3)} = \frac{8 - k}{7}$.

The slope of the line containing (6, 4) and (2, -5) is

$\frac{-5 - 4}{2 - 6} = \frac{-9}{-4} = \frac{9}{4}$.

The lines are parallel when their slopes are equal:

$\frac{8 - k}{7} = \frac{9}{4}$

$32 - 4k = 63$

$-4k = 31$

$k = -\frac{31}{4}$

82. $\frac{100}{9}$

83. The slope of the line segment with endpoints (-1, 3) and (-6, 7) is $\frac{3 - 7}{-1 - (-6)} = -\frac{4}{5}$.

The midpoint of the given line segement is $\left(\frac{-1 + (-6)}{2}, \frac{3 + 7}{2}\right)$, or $\left(-\frac{7}{2}, 5\right)$.

The perpendicular bisector of the given line segment has slope $\frac{5}{4}$ and contains the point $\left(-\frac{7}{2}, 5\right)$.

$y - y_1 = m(x - x_1)$

$y - 5 = \frac{5}{4}\left[x - \left(-\frac{7}{2}\right)\right]$

$y - 5 = \frac{5}{4}x + \frac{35}{8}$

$y = \frac{5}{4}x + \frac{75}{8}$

84. $y = -\frac{2}{3}x - 8$

85. See the answer section in the text.

86. $x^2 + y^2 = a^2$ gives $b^2 + c^2 = a^2$ or $c^2 = -(b^2 - a^2)$. Also

slope of $\overline{AB}$ · slope of $\overline{BC} = \frac{c - 0}{b - (-a)} \cdot \frac{c - 0}{b - a}$

$= \frac{c^2}{b^2 - a^2} = \frac{-(b^2 - a^2)}{b^2 - a^2}$

$= -1$.

Then $\overline{AB}$ is perpendicular to $\overline{BC}$, so ∠ ABC is a right angle.

Exercise Set 3.5

1. $f(x) = |x| + 2$

We compute some function values.

$f(-3) = |-3| + 2 = 3 + 2 = 5$

$f(-1) = |-1| + 2 = 1 + 2 = 3$

$f(0) = |0| + 2 = 0 + 2 = 2$

$f(2) = |2| + 2 = 2 + 2 = 4$

$f(4) = |4| + 2 = 4 + 2 = 6$

x	f(x)	(x, f(x))
-3	5	(-3,5)
-1	3	(-1,3)
0	2	(0,2)
2	4	(2,4)
4	6	(4,6)

We plot these points, look for a pattern, and sketch the graph.

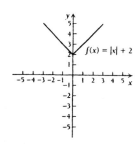

<u>2.</u>

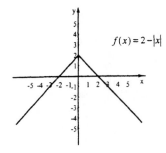

$f(x) = 2 - |x|$

<u>3.</u> $g(x) = 4 - x^2$

We compute some function values.
$g(-3) = 4 - (-3)^2 = 4 - 9 = -5$
$g(-2) = 4 - (-2)^2 = 4 - 4 = 0$
$g(0) = 4 - 0^2 = 4 - 0 = 4$
$g(1) = 4 - 1^2 = 4 - 1 = 3$
$g(2) = 4 - 2^2 = 4 - 4 = 0$

x	g(x)	(x, g(x))
-3	-5	(-3,-5)
-2	0	(-2,0)
0	4	(0,4)
1	3	(1,3)
2	0	(2,0)

We plot these points, look for a pattern, and sketch the graph.

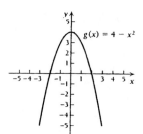

$g(x) = 4 - x^2$

<u>4.</u>

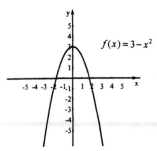

$f(x) = 3 - x^2$

<u>5.</u> $f(x) = \dfrac{2}{x}$

We compute some function values. Note that x cannot be 0.

$f(-4) = \dfrac{2}{-4} = -\dfrac{1}{2}$ $f\left(\dfrac{1}{2}\right) = \dfrac{2}{\frac{1}{2}} = 4$

$f(-2) = \dfrac{2}{-2} = -1$ $f(2) = \dfrac{2}{2} = 1$

$f\left(-\dfrac{1}{2}\right) = \dfrac{2}{-\frac{1}{2}} = -4$ $f(4) = \dfrac{2}{4} = \dfrac{1}{2}$

x	f(x)	(x, f(x))
-4	$-\dfrac{1}{2}$	$\left(-4, -\dfrac{1}{2}\right)$
-2	-1	(-2,-1)
$-\dfrac{1}{2}$	-4	$\left(-\dfrac{1}{2}, -4\right)$
$\dfrac{1}{2}$	4	$\left(\dfrac{1}{2}, 4\right)$
2	1	(2,1)
4	$\dfrac{1}{2}$	$\left(4, \dfrac{1}{2}\right)$

We plot these points, look for a pattern, and sketch the graph.

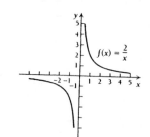

$f(x) = \dfrac{2}{x}$

<u>6.</u>

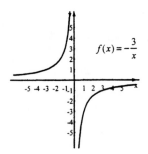

$f(x) = -\dfrac{3}{x}$

<u>7.</u> $f(x) = x^2 - 3$

We compute some function values.
$f(-3) = (-3)^2 - 3 = 9 - 3 = 6$
$f(-1) = (-1)^2 - 3 = 1 - 3 = -2$
$f(0) = 0^2 - 3 = 0 - 3 = -3$
$f(2) = 2^2 - 3 = 4 - 3 = 1$
$f(3) = 3^2 - 3 = 9 - 3 = 6$
We plot these points, look for a pattern, and sketch the graph.

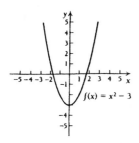

$f(x) = x^2 - 3$

8.

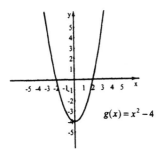

$g(x) = x^2 - 4$

9. g(x) = 3

The function value is 3 for every value of x. For example,

g(-5) = 3,   g(1) = 3,

g(-3) = 3,   g(2) = 3,

g(0) = 3,   g(4) = 3.

The graph is a horizontal line 3 units above the x-axis. We sketch the graph.

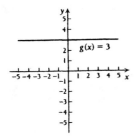

$g(x) = 3$

10.

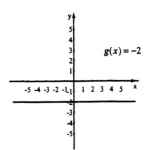

$g(x) = -2$

11. f(x) = x³ - 3x

We compute some function values.

f(-3) = (-3)³ - 3(-3) = -27 + 9 = -18

f(-2) = (-2)³ - 3(-2) = -8 + 6 = -2

f(-1) = (-1)³ - 3(-1) = -1 + 3 = 2

f(0) = 0³ - 3·0 = 0 - 0 = 0

f(1) = 1³ - 3·1 = 1 - 3 = -2

f(2) = 2³ - 3·2 = 8 - 6 = 2

f(3) = 3³ - 3·3 = 27 - 9 = 18

We plot these points, look for a pattern, and sketch the graph.

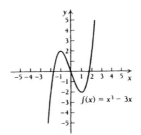

$f(x) = x^3 - 3x$

12.

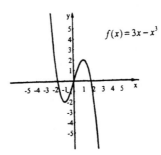

$f(x) = 3x - x^3$

13. f(x) = |x| + x

We compute some function values.

f(-4) = |-4| + (-4) = 4 - 4 = 0

f(-2) = |-2| + (-2) = 2 - 2 = 0

f(0) = |0| + 0 = 0 + 0 = 0

f(1) = |1| + 1 = 1 + 1 = 2

f(3) = |3| + 3 = 3 + 3 = 6

We plot these points, look for a pattern, and sketch the graph.

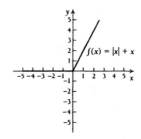

$f(x) = |x| + x$

14.

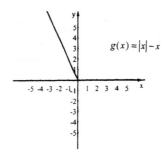

$g(x) = |x| - x$

15. $f(x) = \sqrt{x + 3}$

We compute some function values. Note that the meaningful replacements for x are x ≥ -3.

$f(-3) = \sqrt{-3 + 3} = \sqrt{0} = 0$

$f(-2) = \sqrt{-2 + 3} = \sqrt{1} = 1$

$f(1) = \sqrt{1 + 3} = \sqrt{4} = 2$

$f(6) = \sqrt{6 + 3} = \sqrt{9} = 3$

We plot these points, look for a pattern, and sketch the graph.

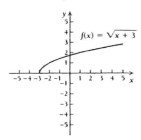

$f(x) = \sqrt{x + 3}$

16.

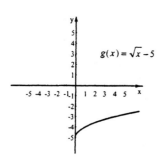

$g(x) = \sqrt{x} - 5$

17.   This is not the graph of a function, because it fails the vertical line test. We can find a vertical line which intersects the graph at more than one point.

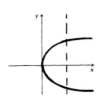

18.   Yes

19.   This is the graph of a function. There is no vertical line which intersects the graph at more than one point.

20.   Yes

21.   This is not the graph of a function. We can find a vertical line which intersects the graph in more than one point.

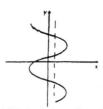

22.   Yes

23.   This is the graph of a function. There is no vertical line which intersects the graph in more than one point.

24.   Yes

25.   The endpoints 0 and 5 are not included in the interval, so we use parentheses. Interval notation is (0,5).

26.   [-1,2]

27.   The endpoint -9 is included in the interval, so we use a bracket before the -9. The endpoint -4 is not included, so we use a parenthesis after the -4. Interval notation is [-9,4).

28.   (-9,-5]

29.   Both endpoints are included in the interval, so we use brackets. Interval notation is [x, x + h].

30.   (x, x + h]

31.   The endpoint p is not included in the interval, so we use a parenthesis before the p. The interval is of unlimited extent in the positive direction, so we use the infinity symbol ∞. Interval notation is (p,∞).

32.   (-∞,q]

33.   This is a closed interval, so we use brackets. Interval notation is [-3,3].

34.   (-4,4)

35.   This is a half-open interval. We use a bracket on the left and a parenthesis on the right. Interval notation is [-14,-11).

36.   (6,20]

37. This interval is of unlimited extent in the negative direction, and the endpoint -4 is included. Interval notation is $(-\infty, -4]$.

38. $(-5, \infty)$

39. This interval is of unlimited extent in the negative direction, and the endpoint 3.8 is not included. Interval notation is $(-\infty, 3.8)$.

40. $[\sqrt{3}, \infty)$

41. a) Increasing,   b) Neither
    c) Decreasing,   d) Neither

42. a) Neither,   b) Decreasing
    c) Increasing,   d) Neither

43. Graph $f(x) = \begin{cases} 1 & \text{for } x < 0 \\ -1 & \text{for } x \geqslant 0. \end{cases}$

We graph $f(x) = 1$ for inputs less than 0. Note that $f(x) = 1$ <u>only</u> for numbers less than 0. We use an open circle at the point (0, 1).

We graph $f(x) = -1$ for inputs greater than or equal to 0. Note that $f(x) = -1$ <u>only</u> for numbers greater than or equal to 0. We use a solid circle at the point (0, -1).

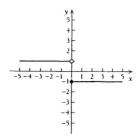

44.

$f(x) = \begin{cases} 2 & \text{for } x \text{ an integer} \\ -2 & \text{for } x \text{ not an integer} \end{cases}$

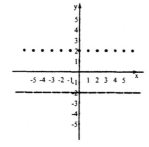

45. $f(x) = \begin{cases} 3 & \text{for } x \leqslant -3 \\ |x| & \text{for } -3 < x \leqslant 3 \\ -3 & \text{for } x > 3 \end{cases}$

We graph $f(x) = 3$ for inputs less than or equal to -3. Note that $f(x) = 3$ <u>only</u> for numbers less than or equal to -3.

We graph $f(x) = |x|$ for inputs greater than -3 and less than or equal to 3. Note that $f(x) = |x|$ <u>only</u> on the interval (-3, 3]. We use a solid circle at the point (3, 3).

We graph $f(x) = -3$ for inputs greater than 3. Note that $f(x) = -3$ <u>only</u> for numbers greater than 3. We use an open circle at the point (3, -3).

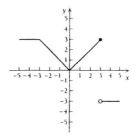

46.

$f(x) = \begin{cases} -2x - 6 & \text{for } x \leq -2 \\ 2 - x^2 & \text{for } -2 < x < 2 \\ 2x - 6 & \text{for } x \geq 2 \end{cases}$

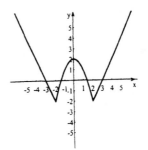

47. $f(x) = \begin{cases} \dfrac{x^2 - 1}{x - 1}, & \text{for } x \neq 1 \\ -2, & \text{for } x = 1 \end{cases}$

When $x \neq 1$, the denominator of $\dfrac{x^2 - 1}{x - 1}$ is nonzero, so we can simplify.

$$\frac{x^2 - 1}{x - 1} = \frac{(x + 1)(x - 1)}{x - 1} = x + 1$$

Thus, $f(x) = x + 1$ for $x \neq 1$. The graph of this part of the function consists of a line with a hole at the point (1,2). At $x = 1$, $f(1) = -2$, so the point (1,-2) is plotted below (1,2).

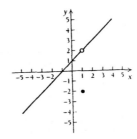

48.
$$f(x) = \begin{cases} \dfrac{x^2 - 9}{x + 3} & \text{for } x \neq -3 \\ 4 & \text{for } x = -3 \end{cases}$$

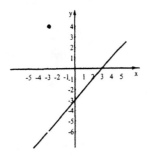

49.
$$p(x) = \begin{cases} 29\text{¢} & \text{if } 0 < x \leqslant 1 \\ 52\text{¢} & \text{if } 1 < x \leqslant 2 \\ 75\text{¢} & \text{if } 2 < x \leqslant 3 \\ 98\text{¢} & \text{if } 3 < x \leqslant 4 \\ \$1.21 & \text{if } 4 < x \leqslant 5 \\ \$1.44 & \text{if } 5 < x \leqslant 6 \\ \$1.67 & \text{if } 6 < x \leqslant 7 \\ \$1.90 & \text{if } 7 < x \leqslant 8 \\ \$2.13 & \text{if } 8 < x \leqslant 9 \\ \$2.36 & \text{if } 9 < x \leqslant 10 \\ \$2.59 & \text{if } 10 < x \leqslant 11 \\ \$2.82 & \text{if } 11 < x \leqslant 12 \end{cases}$$

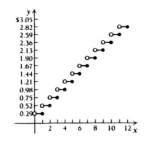

50.
$$f(x) = \begin{cases} 3 + x & \text{for } x \leq 0 \\ \sqrt{x} & \text{for } 0 < x < 4 \\ x^2 - 4x - 1 & \text{for } x \geq 4 \end{cases}$$

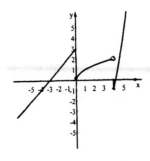

51. f(x) = INT(x - 2)

We can define this function by a piecewise function with an infinite number of statements.

$$f(x) = \text{INT}(x - 2) = \begin{cases} \vdots \\ -4 & \text{if } -2 \leqslant x < -1 \\ -3 & \text{if } -1 \leqslant x < 0 \\ -2 & \text{if } 0 \leqslant x < 1 \\ -1 & \text{if } 1 \leqslant x < 2 \\ 0 & \text{if } 2 \leqslant x < 3 \\ 1 & \text{if } 3 \leqslant x < 4 \\ \vdots \end{cases}$$

We graph the function.

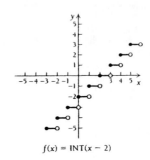

$f(x) = \text{INT}(x - 2)$

52.

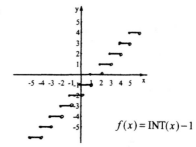

$f(x) = \text{INT}(x) - 1$

53. f(x) = INT(x) + 2

We can define the function by a piecewise function with an infinite number of statements.

$$f(x) = \text{INT}(x) + 2 = \begin{cases} \vdots \\ -1 & \text{if } -3 \leqslant x < -2 \\ 0 & \text{if } -2 \leqslant x < -1 \\ 1 & \text{if } -1 \leqslant x < 0 \\ 2 & \text{if } 0 \leqslant x < 1 \\ \vdots \end{cases}$$

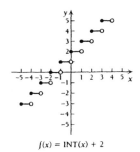

$f(x) = \text{INT}(x) + 2$

54.

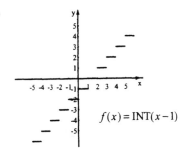

$f(x) = \text{INT}(x-1)$

55. $P(y) = 0.1522y - 298.592$

a) $P(1993) = 0.1522(1993) - 298.592 =$
   $303.3346 - 298.592 \approx 4.74$

   The average price of a ticket in 1993 will be $4.74.

   $P(1995) = 0.1522(1995) - 298.592 =$
   $303.639 - 298.592 \approx 5.05$

   The average price of a ticket in 1995 will be $5.05.

   $P(2000) = 0.1522(2000) - 298.592 =$
   $304.4 - 298.592 \approx 5.81$

   The average price of a ticket in 2000 will be $5.81.

b) $8.00 = 0.1522y - 298.592$ [Substituting 8.00 for $P(y)$]

   $306.592 = 0.1522y$

   $2014 \approx y$

   The average price of a ticket will be $8.00 in 2014.

56. a) 30;   b) 90

57. $E(T) = 1000(100 - T) + 580(100 - T)^2$

a) $E(99.5) = 1000(100 - 99.5) + 580(100 - 99.5)^2$
   $= 1000(0.5) + 580(0.5)^2$
   $= 500 + 580(0.25) = 500 + 145$
   $= 645$ m above sea level

b) $E(100) = 1000(100 - 100) + 580(100 - 100)^2$
   $= 1000 \cdot 0 + 580(0)^2 = 0 + 0$
   $= 0$ m above sea level, or at sea level

58. 0.4 acres, 20.4 acres, 50.6 acres, 125.5 acres, 416.9 acres, 1033.6 acres

59. a) $L = W + 4$, so $W = L - 4$.
   Area = length × width, so
   $A(L) = L(L - 4)$, or $L^2 - 4L$.

b) $L = W + 4$, and area = length × width, so
   $A(W) = (W + 4)W$, or $W^2 + 4W$.

60. a) $A(h) = \frac{1}{2}h(2h - 3)$;

b) $A(b) = \frac{1}{2}b\left[\frac{b + 3}{2}\right]$

61. First express y in terms of x.
   $2x + 2y = 34$
   $2y = 34 - 2x$
   $y = \frac{1}{2}(34 - 2x)$
   $y = 17 - x$
   The area is given by $A = xy$, so
   $A(x) = x(17 - x)$, or $17x - x^2$.

62. $A(x) = x(24 - x)$, or $24x - x^2$.

63. The diameter of the circle is 2·8, or 16 ft. We can use the Pythagorean theorem to express y in terms of x.
   $x^2 + y^2 = 16^2$
   $x^2 + y^2 = 256$
   $y^2 = 256 - x^2$
   $y = \sqrt{256 - x^2}$ Taking the positive square root
   The area is given by $A = xy$, so
   $A(x) = x\sqrt{256 - x^2}$.

64. $A(x) = x\sqrt{400 - x^2}$

65. We make a drawing.

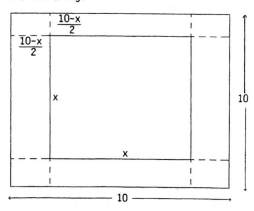

The volume is given by V = length × width × height, so we have

$$V(x) = x \cdot x \cdot \frac{10 - x}{2}$$

$$V(x) = \frac{10x^2 - x^3}{2}$$

$$V(x) = 5x^2 - \frac{x^3}{2}.$$

66. $A(x) = x(30 - x)$, or $30x - x^2$

67. First we use the volume to express y in terms of x.

$$x \cdot x \cdot y = 108$$

$$y = \frac{108}{x^2}$$

Then we write an expression for the surface area.

$$SA(x) = x^2 + 4 \cdot x \cdot \frac{108}{x^2}$$

$$SA(x) = x^2 + \frac{432}{x}$$

68. $C(x) = 2.5x^2 + \frac{3200}{x}$

69. We will use similar triangles, expressing all distances in feet. $\left[6 \text{ in.} = \frac{1}{2} \text{ ft}, s \text{ in.} = \frac{s}{12} \text{ ft}, \text{ and } d \text{ yd} = 3d \text{ ft}\right]$ We have

$$\frac{3d}{7} = \frac{\frac{1}{2}}{\frac{s}{12}}$$

$$\frac{s}{12} \cdot 3d = 7 \cdot \frac{1}{2}$$

$$\frac{sd}{4} = \frac{7}{2}$$

$$d = \frac{4}{s} \cdot \frac{7}{2}, \text{ so}$$

$$d(s) = \frac{14}{s}.$$

70. $h(d) = \sqrt{d^2 - 13,690,000}$

71. a) The volume of the tank is the sum of the volume of a sphere with radius r and a right circular cylinder with radius r and height 6 ft.

$$V(r) = \frac{4}{3}\pi r^3 + 6\pi r^2$$

b) The surface area of the tank is the sum of the surface area of a sphere with radius r and the lateral surface area of a right circular cylinder with radius r and height 6 ft.

$$S(r) = 4\pi r^2 + 12\pi r$$

72. $A(a) = \frac{a^2 \sqrt{3}}{4}$

73. The distance from A to S is $4 - x$.
Using the Pythagorean theorem, we find that the distance from S to C is $\sqrt{1 + x^2}$.
Then $C(x) = 3000(4 - x) + 5000\sqrt{1 + x^2}$.

74. a) $h(r) = \frac{30 - 5r}{3}$;

b) $V(r) = \pi r^2 \left[\frac{30 - 5r}{3}\right]$;

c) $V(h) = \pi \left[\frac{30 - 3h}{5}\right]^2 h$

75. $f(x) = \frac{|x|}{x}$

We compute some function values, plot these points, look for a pattern, and sketch the graph. Note that x cannot be zero.

x	f(x)
-3	-1
-2	-1
-1	-1
1	1
2	1
3	1

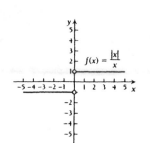

76.

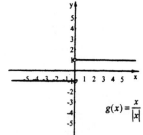

77. $INT(x) = 4$ for $4 \le x < 5$, so the possible inputs are $\{x \mid 4 \le x < 5\}$.

78. $\{x \mid -3 \le x < -2\}$

79. a) $INT\left[\frac{547}{3}\right] = INT\left[182\frac{1}{3}\right] = 182$

The bowling average is 182.

b) $INT\left[\frac{4621}{27}\right] = INT\left[171\frac{4}{27}\right] = 171$

The bowling average is 171.

80. a) 0, 1, 3, 6; b) Domain: The set of all real numbers, range: $\{0,1,2,3,4,5,6,7,8,9\}$;
c) 22

81.

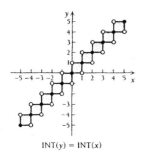

INT(y) = INT(x)

This is not the graph of a function, because it fails the vertical line test.

82.

$$f(x) = \begin{cases} 4x^2 - 5x + 3, & x \le -1 \\ 4.4x^3, & -1 < x < 4 \\ |{-}3x^2 - 4x|, & x \ge 4 \end{cases}$$ DOT MODE

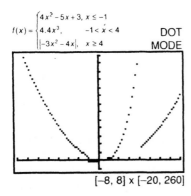

[−8, 8] x [−20, 260]

83.

$$f(x) = \begin{cases} x|x - x^2|, & x < 0 \\ \dfrac{x^3}{1+x}, & x \ge 0 \end{cases}$$ DOT MODE

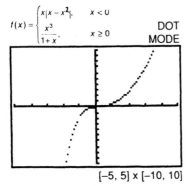

[−5, 5] x [−10, 10]

84. About 71.6 lb

Exercise Set 3.6

1. Test for symmetry with respect to the x-axis.

$3y = x^2 + 4$ (Original equation)
$3(-y) = x^2 + 4$ (Replacing y by -y)
$-3y = x^2 + 4$ (Simplifying)

Since the resulting equation is not equivalent to the original, the graph is not symmetric with respect to the x-axis.

Test for symmetry with respect to the y-axis.

$3y = x^2 + 4$ (Original equation)
$3y = (-x)^2 + 4$ (Replacing x by -x)
$3y = x^2 + 4$ (Simplifying)

Since the resulting equation is equivalent to the original, the graph is symmetric with respect to the y-axis.

Test for symmetry with respect to the origin.

$3y = x^2 + 4$ (Original equation)
$3(-y) = (-x)^2 + 4$ (Replacing x by -x and y by -y)
$-3y = x^2 + 4$ (Simplifying)

Since the resulting equation is not equivalent to the original, the graph is not symmetric with respect to the origin.

2. x-axis: no, y-axis: yes, origin: no

3. Test for symmetry with respect to the x-axis.

$y^3 = 2x^2$ (Original equation)
$(-y)^3 = 2x^2$ (Replacing y by -y)
$-y^3 = 2x^2$ (Simplifying)

Since the resulting equation is not equivalent to the original, the graph is not symmetric with respect to the x-axis.

Test for symmetry with respect to the y-axis.

$y^3 = 2x^2$ (Original equation)
$y^3 = 2(-x)^2$ (Replacing x by -x)
$y^3 = 2x^2$ (Simplifying)

Since the resulting equation is equivalent to the original, the graph is symmetric with respect to the y-axis.

Test for symmetry with respect to the origin.

$y^3 = 2x^2$ (Original equation)
$(-y)^3 = 2(-x)^2$ (Replacing x by -x and y by -y)
$-y^3 = 2x^2$ (Simplifying)

Since the resulting equation is not equivalent to the original, the graph is not symmetric with respect to the origin.

4. x-axis: no, y-axis: yes, origin: no

5. Test for symmetry with respect to the x-axis.

$2x^4 + 3 = y^2$ (Original equation)
$2x^4 + 3 = (-y)^2$ (Replacing y by -y)
$2x^4 + 3 = y^2$ (Simplifying)

Since the resulting equation is equivalent to the original, the graph is symmetric with respect to the x-axis.

Test for symmetry with respect to the y-axis.

$2x^4 + 3 = y^2$ (Original equation)
$2(-x)^4 + 3 = y^2$ (Replacing x by -x)
$2x^4 + 3 = y^2$ (Simplifying)

Since the resulting equation is equivalent to the original, the graph is symmetric with respect to the y-axis.

Test for symmetry with respect to the origin.

$$2x^4 + 3 = y^2 \qquad \text{(Original equation)}$$
$$2(-x)^4 + 3 = (-y)^2 \qquad \text{(Replacing x by -x and y by -y)}$$
$$2x^4 + 3 = y^2 \qquad \text{(Simplifying)}$$

Since the resulting equation is equivalent to the original, the graph is symmetric with respect to the origin.

6. All yes

7. Test for symmetry with respect to the x-axis.

$$2y^2 = 5x^2 + 12 \qquad \text{(Original equation)}$$
$$2(-y)^2 = 5x^2 + 12 \qquad \text{(Replacing y by -y)}$$
$$2y^2 = 5x^2 + 12 \qquad \text{(Simplifying)}$$

Since the resulting equation is equivalent to the original, the graph is symmetric with respect to the x-axis.

Test for symmetry with respect to the y-axis.

$$2y^2 = 5x^2 + 12 \qquad \text{(Original equation)}$$
$$2y^2 = 5(-x)^2 + 12 \qquad \text{(Replacing x by -x)}$$
$$2y^2 = 5x^2 + 12 \qquad \text{(Simplifying)}$$

Since the resulting equation is equivalent to the original, the graph is symmetric with respect to the y-axis.

Test for symmetry with respect to the origin.

$$2y^2 = 5x^2 + 12 \qquad \text{(Original equation)}$$
$$2(-y)^2 = 5(-x)^2 + 12 \qquad \text{(Replacing x by -x and y by -y)}$$
$$2y^2 = 5x^2 + 12 \qquad \text{(Simplifying)}$$

Since the resulting equation is equivalent to the original, the graph is symmetric with respect to the origin.

8. All yes

9. Test for symmetry with respect to the x-axis.

$$2x - 5 = 3y \qquad \text{(Original equation)}$$
$$2x - 5 = 3(-y) \qquad \text{(Replacing y by -y)}$$
$$2x - 5 = -3y \qquad \text{(Simplifying)}$$

Since the resulting equation is not equivalent to the original, the graph is not symmetric with respect to the x-axis.

Test for symmetry with respect to the y-axis.

$$2x - 5 = 3y \qquad \text{(Original equation)}$$
$$2(-x) - 5 = 3y \qquad \text{(Replacing x by -x)}$$
$$-2x - 5 = 3y \qquad \text{(Simplifying)}$$

Since the resulting equation is not equivalent to the original, the graph is not symmetric with respect to the y-axis.

Test for symmetry with respect to the origin.

$$2x - 5 = 3y \qquad \text{(Original equation)}$$
$$2(-x) - 5 = 3(-y) \qquad \text{(Replacing x by -x and y by -y)}$$
$$-2x - 5 = -3y \qquad \text{(Simplifying)}$$
or
$$2x + 5 = 3y \qquad \text{(Multiplying by -1)}$$

Since the resulting equation is not equivalent to the original, the graph is not symmetric with respect to the origin.

10. All no

11. Test for symmetry with respect to the a-axis.

$$3b^3 = 4a^3 + 2 \qquad \text{(Original equation)}$$
$$3(-b)^3 = 4a^3 + 2 \qquad \text{(Replacing b by -b)}$$
$$-3b^3 = 4a^3 + 2 \qquad \text{(Simplifying)}$$

Since the resulting equation is not equivalent to the original, the graph is not symmetric with respect to the a-axis.

Test for symmetry with respect to the b-axis.

$$3b^3 = 4a^3 + 2 \qquad \text{(Original equation)}$$
$$3b^3 = 4(-a)^3 + 2 \qquad \text{(Replacing a by -a)}$$
$$3b^3 = -4a^3 + 2 \qquad \text{(Simplifying)}$$

Since the resulting equation is not equivalent to the original, the graph is not symmetric with respect to the b-axis.

Test for symmetry with respect to the origin.

$$3b^3 = 4a^3 + 2 \qquad \text{(Original equation)}$$
$$3(-b)^3 = 4(-a)^3 + 2 \qquad \text{(Replacing a by -a and b by -b)}$$
$$-3b^3 = -4a^3 + 2 \qquad \text{(Simplifying)}$$
or
$$3b^3 = 4a^3 - 2 \qquad \text{(Multiplying by -1)}$$

Since the resulting equation is not equivalent to the original, the graph is not symmetric with respect to the origin.

12. All no

13.
$$3x^2 - 2y^2 = 3 \qquad \text{(Original equation)}$$
$$3(-x)^2 - 2(-y)^2 = 3 \qquad \text{(Replacing x by -x and y by -y)}$$
$$3x^2 - 2y^2 = 3 \qquad \text{(Simplifying)}$$

Since the resulting equation is equivalent to the original equation, the graph is symmetric with respect to the origin.

14. Yes

15.
$$5x - 5y = 0 \qquad \text{(Original equation)}$$
$$5(-x) - 5(-y) = 0 \qquad \text{(Replacing x by -x and y by -y)}$$
$$-5x + 5y = 0 \qquad \text{(Simplifying)}$$
$$5x - 5y = 0 \qquad \text{(Multiplying by -1)}$$

Since the resulting equation is equivalent to the original equation, the graph is symmetric with respect to the origin.

16. Yes

17.
$$3x + 3y = 0 \quad \text{(Original equation)}$$
$$3(-x) + 3(-y) = 0 \quad \text{(Replacing x by -x and y by -y)}$$
$$-3x - 3y = 0 \quad \text{(Simplifying)}$$
$$3x + 3y = 0 \quad \text{(Multiplying by -1)}$$

Since the resulting equation is equivalent to the original equation, the graph <u>is symmetric</u> with respect to the origin.

18. Yes

19.
$$3x = \frac{5}{y} \quad \text{(Original equation)}$$
$$3(-x) = \frac{5}{-y} \quad \text{(Replacing x by -x and y by -y)}$$
$$-3x = -\frac{5}{y} \quad \text{(Simplifying)}$$
$$3x = \frac{5}{y} \quad \text{(Multiplying by -1)}$$

Since the resulting equation is equivalent to the original, the graph <u>is symmetric</u> with respect to the origin.

20. Yes

21.
$$y = |2x| \quad \text{(Original equation)}$$
$$-y = |2(-x)| \quad \text{(Replacing x by -x and y by -y)}$$
$$-y = |-2x|$$
$$-y = |2x|$$

Since the resulting equation is not equivalent to the original, the graph <u>is not symmetric</u> with respect to the origin.

22. No

23.
$$3a^2 + 4a = 2b \quad \text{(Original equation)}$$
$$3(-a)^2 + 4(-a) = 2(-b) \quad \text{(Replacing a by -a and b by -b)}$$
$$3a^2 - 4a = -2b$$
or
$$-3a^2 + 4a = 2b$$

Since the resulting equation is not equivalent to the original, the graph <u>is not symmetric</u> with respect to the origin.

24. No

25.
$$3x = 4y \quad \text{(Original equation)}$$
$$3(-x) = 4(-y) \quad \text{(Replacing x by -x and y by -y)}$$
$$-3x = -4y \quad \text{(Simplifying)}$$
$$3x = 4y \quad \text{(Multiplying by -1)}$$

Since the resulting equation is equivalent to the original, the graph <u>is symmetric</u> with respect to the origin.

26. Yes

27.
$$xy = 12 \quad \text{(Original equation)}$$
$$(-x)(-y) = 12 \quad \text{(Replacing x by -x and y by -y)}$$
$$xy = 12 \quad \text{(Simplifying)}$$

Since the resulting equation is equivalent to the original, the graph <u>is symmetric</u> with respect to the origin.

28. Yes

29. a) The graph is symmetric with respect to the y-axis. Thus the function is <u>even</u>.

Reflect the graph across the origin. Are the graphs the same? No.

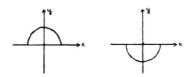

The graph is not symmetric with respect to the origin. Thus the function is not odd.

b) The graph is symmetric with respect to the y-axis. Thus the function is <u>even</u>.

Reflect the graph across the origin. Are the graphs the same? No.

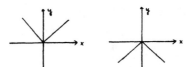

The graph is not symmetric with respect to the origin. Thus the function is not odd.

c) Reflect the graph across the y-axis. Are the graphs the same? No.

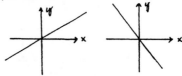

The graph is not symmetric with respect to the y-axis. Thus the function is not even.

Reflect the graph across the origin. Are the graphs the same? Yes.

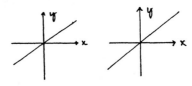

The graph is symmetric with respect to the origin. Thus the function is <u>odd</u>.

d) Reflect the graph across the y-axis. Are the graphs the same? No.

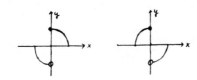

The graph is not symmetric with respect to the y-axis. Thus the function is not even.

Reflect the graph across the origin. Are the graphs the same? No.

The graph is not symmetric with respect to the origin. Thus the function is not odd.

Therefore the function is <u>neither</u> even nor odd.

30. a) Odd; b) even; c) neither; d) neither

31. $f(x) = 2x^2 + 4x$

$f(-x) = 2(-x)^2 + 4(-x) = 2x^2 - 4x$

$-f(x) = -(2x^2 + 4x) = -2x^2 - 4x$

Since $f(x)$ and $f(-x)$ are not the same for all x in the domain, <u>f is not an even function.</u>

Since $f(-x)$ and $-f(x)$ are not the same for all x in the domain, f <u>is not an odd function.</u>

Thus, $f(x) = 2x^2 + 4x$ is <u>neither</u> even nor odd.

32. Odd

33. $f(x) = 3x^4 - 4x^2$

$f(-x) = 3(-x)^4 - 4(-x)^2 = 3x^4 - 4x^2$

$-f(x) = -(3x^4 - 4x^2) = -3x^4 + 4x^2$

Since $f(x)$ and $f(-x)$ are the same for all x in the domain, <u>f is an even function.</u>

Since $f(-x)$ and $-f(x)$ are not the same for all x in the domain, <u>f is not an odd function.</u>

34. Even

35. $f(x) = 7x^3 + 4x - 2$

$f(-x) = 7(-x)^3 + 4(-x) - 2 = -7x^3 - 4x - 2$

$-f(x) = -(7x^3 + 4x - 2) = -7x^3 - 4x + 2$

Since $f(x)$ and $f(-x)$ are not the same for all x in the domain, <u>f is not an even function.</u>

Since $f(-x)$ and $-f(x)$ are not the same for all x in the domain, <u>f is not an odd function.</u>

Thus, $f(x) = 7x^3 + 4x - 2$ is <u>neither</u> even nor odd.

36. Odd

37. $f(x) = |3x|$

$f(-x) = |3(-x)| = |-3x| = |3x|$

$-f(x) = -|3x|$

Since $f(x)$ and $f(-x)$ are the same for all x in the domain, <u>f is an even function.</u>

Since $f(-x)$ and $-f(x)$ are not the same for all x in the domain, <u>f is not an odd function.</u>

38. Even

39. $f(x) = x^{17}$

$f(-x) = (-x)^{17} = -x^{17}$

$-f(x) = -(x^{17}) = -x^{17}$

Since $f(x)$ and $f(-x)$ are not the same for all x in the domain, <u>f is not an even function.</u>

Since $f(-x)$ and $-f(x)$ are the same for all x in the domain, <u>f is an odd function.</u>

40. Odd

41. $f(x) = x - |x|$

$f(-x) = (-x) - |(-x)| = -x - |x|$

$-f(x) = -(x - |x|) = -x + |x|$

Since $f(x)$ and $f(-x)$ are not the same for all x in the domain, <u>f is not an even function.</u>

Since $f(-x)$ and $-f(x)$ are not the same for all x in the domain, <u>f is not an odd function.</u>

Therefore, $f(x) = x - |x|$ is <u>neither</u> even nor odd.

42. Neither

43. $f(x) = \sqrt[3]{x}$

$f(-x) = \sqrt[3]{-x} = -\sqrt[3]{x}$

$-f(x) = -(\sqrt[3]{x}) = -\sqrt[3]{x}$

Since $f(x)$ and $f(-x)$ are not the same for all x in the domain, <u>f is not an even function.</u>

Since $f(-x)$ and $-f(x)$ are the same for all x in the domain, <u>f is an odd function.</u>

44. Even

45. $f(x) = 0$

$f(-x) = 0$

$-f(x) = -(0) = 0$

Since $f(x)$ and $f(-x)$ are the same for all x in the domain, <u>f is an even function.</u>

Since $f(-x)$ and $-f(x)$ are the same for all x in the domain, <u>f is an odd function.</u>

46. Odd

47. The vertices of the original polygon are
(0, 4), (4, 4), (-2, -2), and (1, -2).

When the graph is reflected across the x-axis
the vertices will be
(0, -4), (4, -4), (-2, 2), and (1, 2).

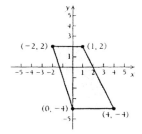

48.

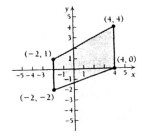

49. The vertices of the original polygon are
(0, 4), (4, 4), (1, -2), and (-2, -2).

When the graph is reflected across the
line y = x, the vertices will be
(4, 0), (4, 4), (-2, 1), and (-2, -2).

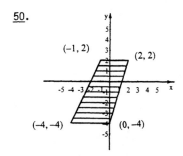

50.

51. See the answer section in the text.

52. Neither

53. Odd

54. Even

55. Neither

56. Odd

57. Even

Exercise Set 3.7

1. f(x) = x - 3, g(x) = x + 4

   a) (f + g)(x) = f(x) + g(x) = (x - 3) + (x + 4) =
   2x + 1

   (f - g)(x) = f(x) - g(x) = (x - 3) - (x + 4) =
   -7

   fg(x) = f(x)g(x) = (x - 3)(x + 4) =
   x² + x - 12

   ff(x) = [f(x)]² = (x - 3)² = x² - 6x + 9

   (f/g)(x) = $\frac{f(x)}{g(x)}$ = $\frac{x - 3}{x + 4}$

   (g/f)(x) = $\frac{g(x)}{f(x)}$ = $\frac{x + 4}{x - 3}$

   f ∘ g(x) = f(g(x)) = f(x + 4) = (x + 4) - 3 =
   x + 1

   g ∘ f(x) = g(f(x)) = g(x - 3) = (x - 3) + 4 =
   x + 1

   b) The domain of f is the set of all real
   numbers. The domain of g is the set of all
   real numbers. Then the domain of f + g,
   f - g, fg, and ff is the set of all real
   numbers. Since g(-4) = 0, the domain of
   f/g is {x|x is a real number and x ≠ -4}.
   Since f(3) = 0, the domain of g/f is
   {x|x is a real number and x ≠ 3}. The
   domain of f ∘ g and of g ∘ f is the set of
   all real numbers.

2. a) (f + g)(x) = x² + 2x + 4,
   (f - g)(x) = x² - 2x - 6,
   fg(x) = 2x³ + 5x² - 2x - 5,
   ff(x) = x⁴ - 2x² + 1,
   (f/g)(x) = $\frac{x² - 1}{2x + 5}$, (g/f)(x) = $\frac{2x + 5}{x² - 1}$,
   f ∘ g(x) = 4x² + 20x + 24,
   g ∘ f(x) = 2x² + 3

   b) The domain of f + g, f - g, fg and ff is the
   set of all real numbers. The domain of
   f/g is $\left\{x \middle| x \text{ is a real number and } x ≠ -\frac{5}{2}\right\}$.
   The domain of g/f is {x|x is a real number and
   x ≠ 1 and x ≠ -1}. The domain of f ∘ g and of
   g ∘ f is the set of all real numbers.

3. $f(x) = x^3$, $g(x) = 2x^2 + 9x - 3$

    a) $(f + g)(x) = f(x) + g(x) = x^3 + 2x^2 + 9x - 3$

       $(f - g)(x) = f(x) - g(x) = x^3 - 2x^2 - 9x + 3$

       $fg(x) = f(x)g(x) = x^3(2x^2 + 9x - 3) =$
            $2x^5 + 9x^4 - 3x^3$

       $ff(x) = [f(x)]^2 = (x^3)^2 = x^6$

       $(f/g)(x) = \dfrac{f(x)}{g(x)} = \dfrac{x^3}{2x^2 + 9x - 3}$

       $(g/f)(x) = \dfrac{2x^2 + 9x - 3}{x^3}$

       $f \circ g(x) = f(g(x)) = f(2x^2 + 9x - 3) =$
            $(2x^2 + 9x - 3)^3$

       $g \circ f(x) = g(f(x)) = g(x^3) = 2x^6 + 9x^3 - 3$

    b) The domain of f is the set of all real numbers. The domain of g is the set of all real numbers. Then the domain of $f + g$, $f - g$, fg, and ff is the set of all real numbers. Since $g(x) = 0$ when $x = \dfrac{-9 \pm \sqrt{105}}{4}$, the domain of f/g is $\left\{x \mid x \text{ is a real number} \text{ and } x \ne \dfrac{-9 \pm \sqrt{105}}{4}\right\}$. Since $f(0) = 0$, the domain of g/f is $\{x \mid x \text{ is a real number and } x \ne 0\}$. The domain of $f \circ g$ and of $g \circ f$ is the set of all real numbers.

4. a) $(f + g)(x) = x^2 + \sqrt{x}$, $(f - g)(x) = x^2 - \sqrt{x}$, $fg(x) = x^2\sqrt{x}$, $ff(x) = x^4$, $(f/g)(x) = x^2/\sqrt{x} = x\sqrt{x}$, $(g/f)(x) = \dfrac{\sqrt{x}}{x^2}$, $f \circ g(x) = (\sqrt{x})^2 = x$, $g \circ f(x) = \sqrt{x^2} = |x|$

    b) The domain of $f + g$, $f - g$, and fg is $\{x \mid x \text{ is a real number and } x \geqslant 0\}$. The domain of ff is the set of all real numbers. The domain of f/g and of g/f is $\{x \mid x \text{ is a real number and } x > 0\}$. The domain of $f \circ g$ is $\{x \mid x \text{ is a real number and } x \geqslant 0\}$, and the domain of $g \circ f$ is the set of all real numbers.

In Exercises 5 - 20, $f(x) = x^2 - 4$ and $g(x) = 2x + 5$.

5. $(f - g)(3) = f(3) - g(3) = (3^2 - 4) - (2 \cdot 3 + 5) =$
         $5 - 11 = -6$

6. 0

7. $(f - g)(x) = f(x) - g(x) = (x^2 - 4) - (2x + 5) =$
         $x^2 - 2x - 9$

8. $x^2 + 2x + 1$

9. $fg(3) = f(3)g(3) = (3^2 - 4)(2 \cdot 3 + 5) = 5 \cdot 11 = 55$

10. -1

    $(g/f)(-2) = \dfrac{g(-2)}{f(-2)} = \dfrac{2(-2) + 5}{(-2)^2 - 4} = \dfrac{-4 + 5}{4 - 4} = \dfrac{1}{0}$, which does not exist

12. Does not exist

13. $fg(x) = f(x)g(x) = (x^2 - 4)(2x + 5) =$
         $2x^3 + 5x^2 - 8x - 20$

14. $\dfrac{x^2 - 4}{2x + 5}$

15. $(g/f)(x) = \dfrac{g(x)}{f(x)} = \dfrac{2x + 5}{x^2 - 4}$

16. $x^4 - 8x^2 + 16$

17. $f \circ g(x) = f(g(x)) = f(2x + 5) = (2x + 5)^2 - 4 =$
         $4x^2 + 20x + 25 - 4 = 4x^2 + 20x + 21$

18. $x^4 - 8x^2 + 12$

19. $g \circ g(x) = g(g(x)) = g(2x + 5) = 2(2x + 5) + 5 =$
         $4x + 10 + 5 = 4x + 15$

20. $2x^2 - 3$

21. $R(x) = 60x - 0.4x^2$, $C(x) = 3x + 13$

    a) $P(x) = R(x) - C(x)$
         $= (60x - 0.4x^2) - (3x + 13)$
         $= -0.4x^2 + 57x - 13$

    b) $R(20) = 60(20) - 0.4(20)^2$
         $= 1200 - 0.4(400) = 1200 - 160$
         $= 1040$

       $C(20) = 3(20) + 13 = 60 + 13$
         $= 73$

       $P(20) = R(20) - C(20)$
         $= 1040 - 73$
         $= 967$

22. a) $P(x) = -0.001x^2 + 13.8x - 60$

    b) $R(100) = 1500$, $C(100) = 190$, $P(100) = 1310$

23. $f \circ g(x) = f(g(x)) = f\left(\dfrac{5}{4}x\right) = \dfrac{4}{5} \cdot \dfrac{5}{4}x = x$

    $g \circ f(x) = g(f(x)) = g\left(\dfrac{4}{5}x\right) = \dfrac{5}{4} \cdot \dfrac{4}{5}x = x$

24. $f \circ g(x) = x$
    $g \circ f(x) = x$

25. $f \circ g(x) = f(g(x)) = f\left(\dfrac{x + 7}{3}\right) = 3\left(\dfrac{x + 7}{3}\right) - 7 =$
         $x + 7 - 7 = x$

    $g \circ f(x) = g(f(x)) = g(3x - 7) = \dfrac{(3x - 7) + 7}{3} =$
         $\dfrac{3x}{3} = x$

26. $f \circ g(x) = x$
    $g \circ f(x) = x$

27. $f \circ g(x) = f(g(x)) = f(\sqrt[3]{x + 1}) = (\sqrt[3]{x + 1})^3 - 1 =$
     $x + 1 - 1 = x$

     $g \circ f(x) = g(f(x)) = g(x^3 - 1) = \sqrt[3]{(x^3 - 1) + 1} =$
     $\sqrt[3]{x^3} = x$

28. $f \circ g(x) = x$
     $g \circ f(x) = x$

29. $f \circ g(x) = f(g(x)) = f(x^2 - 5) = \sqrt{x^2 - 5 + 5} =$
     $\sqrt{x^2} = |x|$

     $g \circ f(x) = g(f(x)) = g(\sqrt{x + 5}) = (\sqrt{x + 5})^2 - 5 =$
     $x + 5 - 5 = x$

30. $f \circ g(x) = x$
     $g \circ f(x) = |x|$

31. $f \circ g(x) = f(g(x)) = f\left[\dfrac{1}{1 + x}\right] = \dfrac{1 - \left[\dfrac{1}{1 + x}\right]}{\dfrac{1}{1 + x}} =$

     $\dfrac{\dfrac{1 + x - 1}{1 + x}}{\dfrac{1}{1 + x}} = \dfrac{x}{1 + x} \cdot \dfrac{1 + x}{1} = x$

     $g \circ f(x) = g(f(x)) = g\left[\dfrac{1 - x}{x}\right] = \dfrac{1}{1 + \left[\dfrac{1 - x}{x}\right]} =$

     $\dfrac{1}{\dfrac{x + 1 - x}{x}} = \dfrac{1}{\dfrac{1}{x}} = 1 \cdot \dfrac{x}{1} = x$

32. $f \circ g(x) = \dfrac{-8x^2 - 2x + 6}{17x^2 - 22x + 10}$

     $g \circ f(x) = \dfrac{-x^2 - 7}{3x^2 - 7}$

33. $f \circ g(x) = f(g(x)) = f(12) = -6$

     $g \circ f(x) = g(f(x)) = g(-6) = 12$

34. $f \circ g(x) = 20$, $g \circ f(x) = 5$

35. $h(x) = (4 - 3x)^5$

     This is 4 - 3x to the 5th power. The most
     obvious answer is $f(x) = x^5$ and $g(x) = 4 - 3x$.

36. $f(x) = \sqrt[3]{x}$, $g(x) = x^2 - 8$

37. $h(x) = \dfrac{1}{(x - 1)^4}$

     This is 1 divided by (x - 1) to the 4th power.
     One obvious answer is $f(x) = \dfrac{1}{x^4}$ and $g(x) =$
     $(x - 1)$.

38. $f(x) = \dfrac{1}{\sqrt{x}}$, $g(x) = 3x + 7$

39. $f(x) = \dfrac{x - 1}{x + 1}$, $g(x) = x^3$

40. $f(x) = |x|$, $g(x) = 9x^2 - 4$

41. $f(x) = x^6$, $g(x) = \dfrac{2 + x^3}{2 - x^3}$

42. $f(x) = x^4$, $g(x) = \sqrt{x} - 3$

43. $f(x) = \sqrt{x}$, $g(x) = \dfrac{x - 5}{x + 2}$

44. $f(x) = \sqrt{1 + x}$, $g(x) = \sqrt{1 + x}$

45. $f(x) = x^5 + x^4 + x^3 - x^2 + 4x$, $g(x) = x + 3$

46. $f(x) = 4x^{2/3} + 5$, $g(x) = x - 1$

47. a) Recall that distance = rate × time. Then
         $a(t) = 250t$.

     b) The plane's distance from the control tower
         is the sum of 300 ft and the distance a the
         plane has traveled down the runway. Thus,
         $P(a) = 300 + a$.

     c) $(P \circ a)(t) = P(a(t)) = 300 + 250t$
         This function gives the plane's distance
         from the control tower in terms of the time t
         the plane travels.

48. a) $r(t) = 3t$

     b) $A(r) = \pi r^2$

     c) $(A \circ r)(t) = 9\pi t^2$
     This function gives the area of the ripple in
     terms of the time, t.

49. First we graph both functions on the same axes.

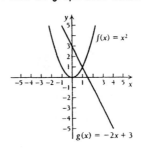

     Then we graph $y = (f + g)(x)$ by adding second
     coordinates. The following table shows some
     examples.

First coordinate	Second coordinates		
x	f(x)	g(x)	(f + g)(x)
-1	1	5	6
0	0	3	3
1	1	1	2
2	4	-1	3
3	9	-3	6

We sketch the graph of y = (f + g)(x).

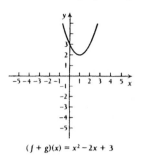

$$(f + g)(x) = x^2 - 2x + 3$$

First coordinate	Second coordinates		
x	f(x)	g(x)	(f + g)(x)
0	0	1	1
0.5	0.707	0.75	1.457
1	1	0	1
1.5	1.225	-1.25	-0.025
2	1.414	-3	-1.586
3	1.732	-8	-6.268

We sketch the graph of y = (f + g)(x).

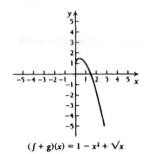

$$(f + g)(x) = 1 - x^2 + \sqrt{x}$$

<u>50.</u>

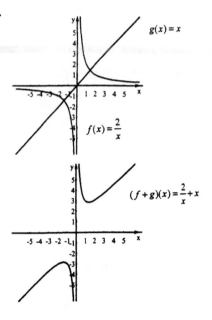

<u>52.</u>

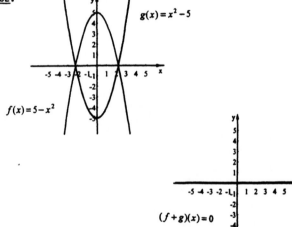

$$(f + g)(x) = 0$$

<u>51.</u>   First we graph both functions on the same axes.

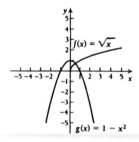

Then we graph y = (f + g)(x) by adding second coordinates. The following table shows some examples. Note that we can only add second coordinates for first coordinates x such that x ⩾ 0.

<u>53.</u> f ∘ g(x) = f(g(x)) = f(2x - 1) = 3(2x - 1) + b = 6x - 3 + b

g ∘ f(x) = g(f(x)) = g(3x + b) = 2(3x + b) - 1 = 6x + 2b - 1

Now f ∘ g(x) = g ∘ f(x) means
      6x - 3 + b = 6x + 2b - 1, or
            -2 = b

<u>54.</u> m = $\frac{1}{3}$, b = $\frac{4}{3}$

55. $f \circ f(x) = f(f(x)) = \dfrac{1}{1 - \dfrac{1}{1-x}} = \dfrac{1}{\dfrac{1-x-1}{1-x}} =$

$\dfrac{1}{\dfrac{-x}{1-x}} = \dfrac{1-x}{-x}$, or $\dfrac{x-1}{x}$

$f \circ f \circ f(x) = f[f(f(x))] = \dfrac{1}{1 - \dfrac{x-1}{x}} =$

$\dfrac{1}{\dfrac{x-(x-1)}{x}} = \dfrac{1}{\dfrac{1}{x}} = x$

56. Let $f(x)$ and $g(x)$ be even functions. Then by definition $f(x) = f(-x)$ and $g(x) = g(-x)$. Now $f \circ g(x) = f(g(x)) = f(g(-x)) = f \circ g(-x)$ and $g \circ f(x) = g(f(x)) = g(f(-x)) = g \circ f(-x)$. Thus the composition of two even functions is even.

57. See the answer section in the text.

58. Let $f(x)$ and $g(x)$ be odd functions. By definition $f(-x) = -f(x)$, or $f(x) = -f(-x)$, and $g(-x) = -g(x)$, or $g(x) = -g(-x)$. Then $fg(x) = f(x) \cdot g(x) = [-f(-x)] \cdot [-g(-x)] = f(-x) \cdot g(-x) = fg(-x)$, so fg is even.

59. See the answer section in the text.

60. Let $f(x)$ be an even function, and let $g(x)$ be an odd function. By definition $f(x) = f(-x)$ and $g(-x) = -g(x)$, or $g(x) = -g(-x)$. Then $fg(x) = f(x) \cdot g(x) = f(-x) \cdot [-g(-x)] = -f(-x) \cdot g(-x) = -fg(-x)$, and fg is odd.

61. See the answer section in the text.

62. $E(-x) = \dfrac{f(-x) + f(-(-x))}{2} = \dfrac{f(-x) + f(x)}{2} = E(x)$.

Thus $E(x)$ is even.

63. See the answer section in the text.

64. $E(x) + 0(x) = \dfrac{f(x) + f(-x)}{2} + \dfrac{f(x) - f(-x)}{2} = \dfrac{2f(x)}{2} = f(x)$

65. $f(x) = |x| + \text{INT}(x)$

This function can also be expressed as a piecewise function with an infinite number of statements:

$$f(x) = |x| + \text{INT}(x) = \begin{cases} \vdots \\ -x - 3 & \text{if } -3 \leqslant x < -2 \\ -x - 2 & \text{if } -2 \leqslant x < -1 \\ -x - 1 & \text{if } -1 \leqslant x < 0 \\ x & \text{if } 0 \leqslant x < 1 \\ x + 1 & \text{if } 1 \leqslant x < 2 \\ x + 2 & \text{if } 2 \leqslant x < 3 \\ \vdots \end{cases}$$

We sketch a graph of the function.

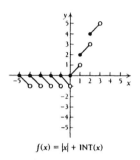

$f(x) = |x| + \text{INT}(x)$

Exercise Set 3.8

---

1. $f(x) = |x| - 3$

By Theorem 5, the graph of $f(x) = |x| - 3$ is a shift of the graph of $f(x) = |x|$ downward 3 units.

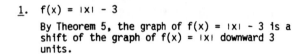

| ----- | $f(x) = |x|$ |
| ——— | $f(x) = |x| - 3$ |

2. $f(x) = 2 + |x|$

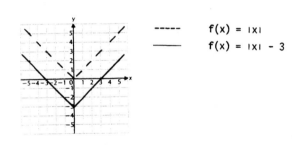

<u>3.</u> f(x) = |x - 1|

The equation f(x) = |x - 1| is obtained from the equation f(x) = |x| is obtained by replacing x by x - 1. By Theorem 6, the graph of f(x) = |x - 1| is a shift of the graph of f(x) = |x| one unit to the right.

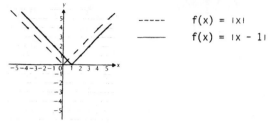

```
-----        f(x) = |x|
_____        f(x) = |x - 1|
```

<u>4.</u> f(x) = |x + 2|

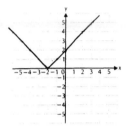

<u>5.</u> f(x) = -4|x|

By Theorem 7, the graph of f(x) = -4|x| can be obtained from the graph of f(x) = |x| by stretching vertically and reflecting across the x-axis. The y-coordinate of each ordered-pair solution of f(x) = |x| is multiplied by -4.

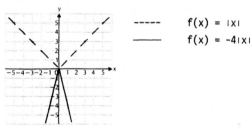

```
-----        f(x) = |x|
_____        f(x) = -4|x|
```

<u>6.</u> f(x) = 3|x|

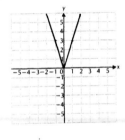

<u>7.</u> f(x) = $\frac{1}{3}$|x|

By Theorem 7, the graph of f(x) = $\frac{1}{3}$|x| can be obtained from the graph of f(x) = |x| by shrinking vertically. The y-coordinate of each ordered-pair solution of f(x) = |x| is multiplied by $\frac{1}{3}$.

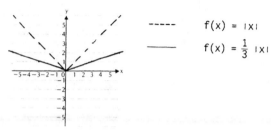

```
-----        f(x) = |x|
_____        f(x) = $\frac{1}{3}$ |x|
```

<u>8.</u> f(x) = - $\frac{1}{4}$ |x|

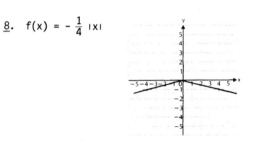

<u>9.</u> f(x) = |2x|, or f(x) = 2|x|

By Theorem 7, the graph of f(x) = |2x| can be obtained from the graph of f(x) = |x| by stretching vertically. The y-coordinate of each ordered-pair solution of f(x) = |x| is multiplied by 2. (Equivalently, we could use Theorem 4 and shrink the graph of f(x) = |x| horizontally. The result is the same in either case.)

```
-----        f(x) = |x|
_____        f(x) = |2x|
```

<u>10.</u> f(x) = $\left|\frac{x}{3}\right|$

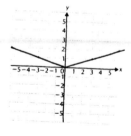

11. $f(x) = |x - 2| + 3$

We translate the graph of $f(x) = |x|$ two units to the right (Theorem 6) and upward 3 units (Theorem 5).

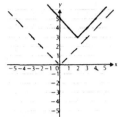

----- $f(x) = |x|$

—— $f(x) = |x - 2| + 3$

12. $f(x) = 2|x + 1| - 3$

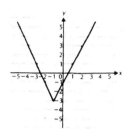

13. $f(x) = -3|x - 2|$

We translate the graph of $f(x) = |x|$ to the right 2 units (Theorem 6), stretch it vertically, and reflect it across the x-axis (Theorem 7).

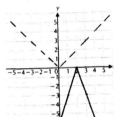

----- $f(x) = |x|$

—— $f(x) = -3|x - 2|$

14. $f(x) = \frac{1}{3}|x + 2| + 1$

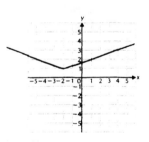

15. $y = 2 + f(x)$

We translate the graph of $y = f(x)$ upward 2 units (Theorem 5).

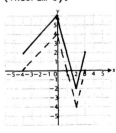

----- $y = f(x)$

—— $y = 2 + f(x)$

16. $y = f(x) - 1$

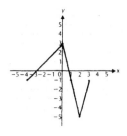

17. $y = f(x - 1)$

We translate the graph of $y = f(x)$ one unit to the right (Theorem 6).

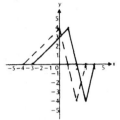

----- $y = f(x)$

—— $y = f(x - 1)$

18. $y = f(x + 2)$

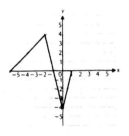

19. $y = -2f(x)$

We stretch the graph of $y = f(x)$ vertically and reflect it across the x-axis (Theorem 7). The y-coordinate of each ordered-pair solution of $y = f(x)$ is multiplied by -2.

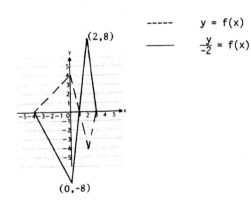

----- $y = f(x)$

—— $\frac{y}{-2} = f(x)$

20. $y = 3f(x)$

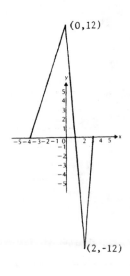

(0,12)

(2,-12)

21. $y = \frac{1}{3}f(x)$

We shrink the graph of $y = f(x)$ vertically (Theorem 7). The y-coordinate of each ordered-pair solution of $y = f(x)$ is multiplied by $\frac{1}{3}$.

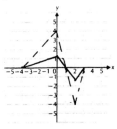

----- $y = f(x)$

——— $y = \frac{1}{3}f(x)$

22. $y = -\frac{1}{2}f(x)$

23. $y = f(2x)$

We shrink the graph of $y = f(x)$ horizontally (Theorem 8). The x-coordinate of each ordered-pair solution of $y = f(x)$ is divided by 2.

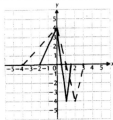

----- $y = f(x)$

——— $y = f(2x)$

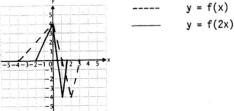

24. $y = f(3x)$

25. $y = f(-2x)$

We shrink the graph of $y = f(x)$ horizontally and reflect it across the y-axis (Theorem 8). The x-coordinate of each ordered-pair solution of $y = f(x)$ is divided by -2.

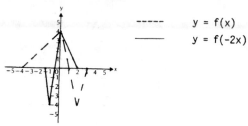

----- $y = f(x)$

——— $y = f(-2x)$

26. $y = f(-3x)$

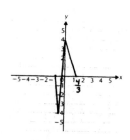

27. $y = f\left(\frac{x}{-2}\right)$, or $y = f\left(-\frac{1}{2}x\right)$

We stretch the graph of $y = f(x)$ horizontally and reflect it across the y-axis (Theorem 8). The x-coordinate of each ordered-pair solution of $y = f(x)$ is divided by $-\frac{1}{2}$.

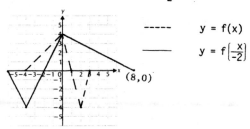

----- $y = f(x)$

——— $y = f\left(\frac{x}{-2}\right)$

(8,0)

28. $y = f\left(\frac{1}{3}x\right)$

(-12,0)   (9,0)

29. y = f(x - 2) + 3

We translate the graph of y = f(x) two units to the right (Theorem 6) and upward 3 units (Theorem 5).

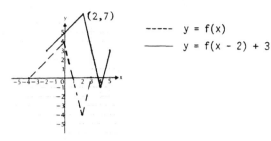

    ----- y = f(x)
    ——— y = f(x - 2) + 3

30. y = -3f(x - 2)

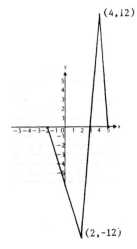

31. y = 2·f(x + 1) - 2

We translate the graph of y = f(x) one unit to the left to obtain the graph of y = f(x + 1). (Theorem 6)

Then we stretch the graph of y = f(x + 1) vertically, multiplying each y-coordinate by 2, to obtain the graph of y = 2·f(x + 1). (Theorem 7) Finally we translate the graph of y = 2·f(x + 1) downward 2 units to obtain the graph of y = 2·f(x + 1) - 2. (Theorem 5)

    ----- y = f(x)
    ——— y = 2·f(x + 1) - 2

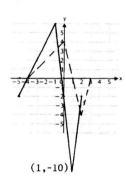

32. $y = \frac{1}{2}f(x + 2) - 1$

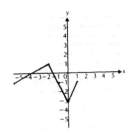

33. $y = -\frac{1}{2}f(x - 3) + 2$

We translate the graph of y = f(x) three units to the right to obtain the graph of y = f(x - 3). (Theorem 6)

Then we shrink the graph of y = f(x - 3) vertically and reflect it across the x-axis, multiplying each y-coordinate by $-\frac{1}{2}$, to obtain the graph of $y = -\frac{1}{2}f(x - 3)$. (Theorem 7)

Finally we translate the graph of $y = -\frac{1}{2}f(x - 3)$ upward 2 units to obtain the graph of $y = -\frac{1}{2}f(x - 3) + 2$. (Theorem 5)

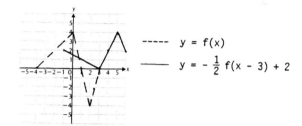

    ----- y = f(x)
    ——— $y = -\frac{1}{2} f(x - 3) + 2$

34. y = -3f(x - 1) + 4

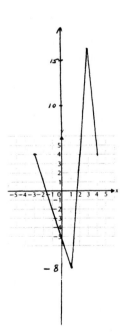

35. $y = -2f(x + 1) - 1$

We translate the graph of $y = f(x)$ one unit to the left to obtain the graph of $y = f(x + 1)$. (Theorem 6)

Then we stretch the graph of $y = f(x + 1)$ vertically and reflect it across the x-axis, multiplying each y-coordinate by -2, to obtain the graph of $y = -2f(x + 1)$. (Theorem 7) Finally we translate the graph of $y = -2f(x + 1)$ downward 1 unit to obtain the graph of $y = -2f(x + 1) - 1$. (Theorem 5)

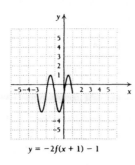

$y = -2f(x + 1) - 1$

36.

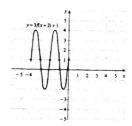

37. $y = \frac{5}{2}f(x - 3) - 2$

We translate the graph of $y = f(x)$ three units to the right to obtain the graph of $y = f(x - 3)$. (Theorem 6)

Then we stretch the graph of $y = f(x - 3)$ vertically, multiplying each y-coordinate by $\frac{5}{2}$, to obtain the graph of $y = \frac{5}{2} f(x - 3)$.

(Theorem 7)

Finally we translate the graph of $y = \frac{5}{2}f(x - 3)$ downward 2 units to obtain the graph of $y = \frac{5}{2}f(x - 3) - 2$. (Theorem 5)

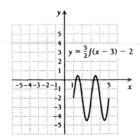

38. $y = -\frac{2}{3}f(x - 4) + 3$

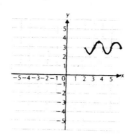

39. $y = -\sqrt{2}\, f(x + 1.8)$

The graph of $y = f(x)$ is translated 1.8 units to the left to obtain the graph of $y = f(x + 1.8)$. Then the graph of $y = f(x + 1.8)$ is stretched vertically and reflected across the x-axis, multiplying each y-coordinate by $-\sqrt{2}$.

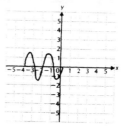

40. $y = \frac{\sqrt{3}}{2} \cdot f(x - 2.5) - 5.3$

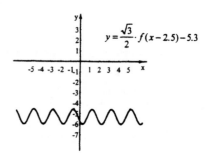

41. The original graph is shifted upward two units. A formula for the transformed function is $g(x) = f(x) + 2$, or $g(x) = x^3 - 3x^2 + 2$.

42. $h(x) = \frac{1}{2}(x^3 - 3x^2)$

43. The original graph is shifted to the left one unit. A formula for the transformed function is $k(x) = f(x + 1)$, or $k(x) = (x + 1)^3 - 3(x + 1)^2$.

44. $t(x) = (x - 2)^3 - 3(x - 2)^2 + 1$

45. `---------` (1) y = f(x) is the given function

    `— — —` (2) y = 3f(2x) is the result of shrinking (1) horizontally by a factor of $\frac{1}{2}$ and stretching vertically by a factor of 3

    `————` (3) $y = 3f\left[2\left(x + \frac{1}{4}\right)\right]$, the required graph, is a shift of (2) to the left $\frac{1}{4}$ unit.

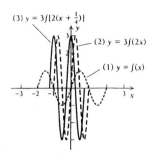

46.

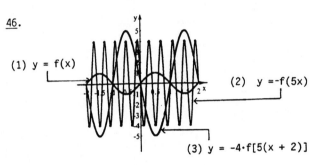

(1) y = f(x)

(2) y = -f(5x)

(3) y = -4·f[5(x + 2)]

47.

$f(x) = x^2|x|$

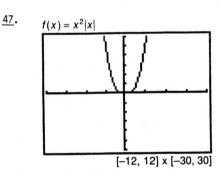

[−12, 12] x [−30, 30]

$f(x) = (4.7x)^2|4.7x|$

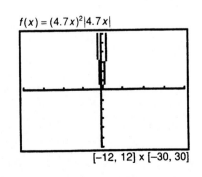

[−12, 12] x [−30, 30]

$f(x) = -2.5x^2|x|$

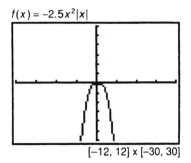

[−12, 12] x [−30, 30]

$f(x) = x^2|x| - 11.5$

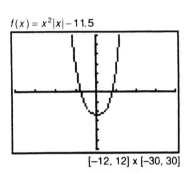

[−12, 12] x [−30, 30]

$f(x) = (x + 9.9)^2|x + 9.9|$

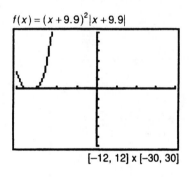

[−12, 12] x [−30, 30]

48. $f(x) = \dfrac{1 - 4x^3}{x^2}$

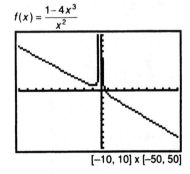

[−10, 10] x [−50, 50]

$f(x) = \dfrac{1-4x^3}{x^2} + 34.1$

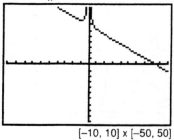

[−10, 10] x [−50, 50]

$f(x) = \dfrac{1-4x^3}{3x^2}$

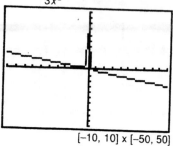

[−10, 10] x [−50, 50]

$f(x) = \dfrac{1-4(0.35x)^3}{(0.35x)^2}$

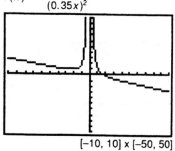

[−10, 10] x [−50, 50]

$f(x) = \dfrac{1-4(x-8.1)^3}{(x-8.1)^2}$

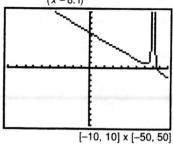

[−10, 10] x [−50, 50]

Exercise Set 4.1

<u>1</u>. $f(x) = x^2$

$f(x) = (x - 0)^2 + 0$ [In the form
$f(x) = a(x - h)^2 + k$]

a) The vertex is (0, 0).

b) The line of symmetry goes through the vertex;
it is the line x = 0.

c) Since the coefficient of x^2 is positive
(1 > 0), the parabola opens upward. Thus we
have a <u>minimum</u> value which is 0.

<u>2</u>. a) (0, 0),   b) x = 0,   c) 0 is a maximum

<u>3</u>. $f(x) = -2(x - 9)^2$

$f(x) = -2(x - 9)^2 + 0$ [In the form
$f(x) = a(x - h)^2 + k$]

a) The vertex is (9, 0).

b) The line of symmetry goes through the vertex;
it is the line x = 9.

c) Since the coefficient of x^2 is negative
(-2 < 0), the parabola opens downward. Thus
we have a <u>maximum</u> value which is 0.

<u>4</u>. a) (7, 0),   b) x = 7,   c) 0 is a minimum

<u>5</u>. $f(x) = 2(x - 1)^2 - 4$

$f(x) = 2(x - 1)^2 + (-4)$ [In the form
$f(x) = a(x - h)^2 + k$]

a) The vertex is (1, -4)

b) The line of symmetry goes through the vertex;
it is the line x = 1.

c) Since the coefficient of x^2 is positive
(2 > 0), the parabola opens upward. Thus we
have a <u>minimum</u> value which is -4.

<u>6</u>. a) (-4, -3),   b) x = -4,   c) -3 is a maximum

<u>7</u>. a) $f(x) = -x^2 + 2x + 3$

$= -(x^2 - 2x) + 3$

$= -(x^2 - 2x + 1 - 1) + 3$

(Adding 1 - 1 inside parentheses)

$= -(x^2 - 2x + 1) + 1 + 3$

$= -(x - 1)^2 + 4$ [In the form
$f(x) = a(x - h)^2 + k$]

b) The vertex is (1, 4).

c) Since the coefficient of x^2 is negative
(-1 < 0), the parabola opens downward. Thus
we have a <u>maximum</u> value which is 4.

<u>8</u>. a) $f(x) = -(x - 4)^2 + 9$

b) (4, 9)

c) 9 is a maximum

<u>9</u>. a) $f(x) = x^2 + 3x$

$= x^2 + 3x + \frac{9}{4} - \frac{9}{4}$ $\left[\text{Adding } \frac{9}{4} - \frac{9}{4}\right]$

$= \left[x + \frac{3}{2}\right]^2 - \frac{9}{4}$

$= \left[x - \left[-\frac{3}{2}\right]\right]^2 + \left[-\frac{9}{4}\right]$

[In the form $f(x) = a(x - h)^2 + k$]

b) The vertex is $\left[-\frac{3}{2}, -\frac{9}{4}\right]$.

c) Since the coefficient of x^2 is positive
(1 > 0), the parabola opens upward. Thus we
have a <u>minimum</u> value which is $-\frac{9}{4}$.

<u>10</u>. a) $f(x) = \left[x - \frac{9}{2}\right]^2 + \left[-\frac{81}{4}\right]$

b) $\left[\frac{9}{2}, -\frac{81}{4}\right]$

c) $-\frac{81}{4}$ is a minimum

<u>11</u>. a) $f(x) = -\frac{3}{4} x^2 + 6x$

$= -\frac{3}{4} (x^2 - 8x)$

$= -\frac{3}{4} (x^2 - 8x + 16 - 16)$

(Adding 16 - 16 inside
parentheses)

$= -\frac{3}{4} (x^2 - 8x + 16) + 12$

$= -\frac{3}{4} (x - 4)^2 + 12$

[In the form $f(x) = a(x - h)^2 + k$]

b) The vertex is (4, 12).

c) Since the coefficient of x^2 is negative
$\left[-\frac{3}{4} < 0\right]$, the parabola opens downward. Thus
we have a <u>maximum</u> value which is 12.

<u>12</u>. a) $f(x) = \frac{3}{2}[x - (-1)]^2 + \left[-\frac{3}{2}\right]$

b) $\left[-1, -\frac{3}{2}\right]$

c) $-\frac{3}{2}$ is a minimum

13. a) $f(x) = 3x^2 + x - 4$

$$= 3\left(x^2 + \tfrac{1}{3} x\right) - 4$$

$$= 3\left(x^2 + \tfrac{1}{3} x + \tfrac{1}{36} - \tfrac{1}{36}\right) - 4$$

$\left(\text{Adding } \tfrac{1}{36} - \tfrac{1}{36} \text{ inside parentheses}\right)$

$$= 3\left(x^2 + \tfrac{1}{3} x + \tfrac{1}{36}\right) - \tfrac{1}{12} - 4$$

$$= 3\left(x + \tfrac{1}{6}\right)^2 - \tfrac{49}{12}$$

$$= 3\left[x - \left(-\tfrac{1}{6}\right)\right]^2 + \left(-\tfrac{49}{12}\right)$$

[In the form $f(x) = a(x - h)^2 + k$]

b) The vertex is $\left(-\tfrac{1}{6}, -\tfrac{49}{12}\right)$.

c) Since the coefficient of x^2 is positive ($3 > 0$), the parabola opens upward. Thus we have a <u>minimum</u> value which is $-\tfrac{49}{12}$.

14. a) $f(x) = -2\left(x - \tfrac{1}{4}\right)^2 + \left(-\tfrac{7}{8}\right)$

b) $\left(\tfrac{1}{4}, -\tfrac{7}{8}\right)$

c) $-\tfrac{7}{8}$ is a maximum

15. $f(x) = -5(x + 2)^2$

$f(x) = -5[x - (-2)]^2 + 0$

The value 0 is a maximum since the coefficient -5 is negative.

16. Minimum: 0

17. $f(x) = 8(x - 1)^2 + 5$

The value 5 is a minimum since the coefficient 8 is positive.

18. Maximum: -11

19. $f(x) = -4x^2 + x - 13$

$a = -4, b = 1, c = -13$

Since we want only the maximum or minimum value of the function, we find the second coordinate of the vertex:

$$-\frac{b^2 - 4ac}{4a} = -\frac{1^2 - 4(-4)(-13)}{4(-4)}$$

$$= -\frac{207}{16}$$

The value $-\tfrac{207}{16}$ is a maximum since the coefficient of x^2 is negative.

20. Minimum: 3.89625

21. $f(x) = \tfrac{1}{2}x^2 - \tfrac{2}{5}x - \tfrac{67}{100}$

$a = \tfrac{1}{2}, b = -\tfrac{2}{5}, c = -\tfrac{67}{100}$

The second coordinate of the vertex is

$$-\frac{b^2 - 4ac}{4a} = -\frac{\left(-\tfrac{2}{5}\right)^2 - 4\left(\tfrac{1}{2}\right)\left(-\tfrac{67}{100}\right)}{4\left(\tfrac{1}{2}\right)}$$

$$= -\tfrac{3}{4}.$$

The value $-\tfrac{3}{4}$ is a minimum since the coefficient of x^2 is positive.

22. Maximum: -16.0106

23. $q(x) = -\$120,000x^2 + \$430,000x - \$240,000$

$a = -\$120,000, b = \$430,000, c = -\$240,000$

The second coordinate of the vertex is

$$-\frac{b^2 - 4ac}{4a} =$$

$$-\frac{(\$430,000)^2 - 4(-\$120,000)(-\$240,000)}{4(-\$120,000)} =$$

$$\$\frac{435,625}{3}$$

The value $\$\frac{435,625}{3}$ is a maximum since the coefficient of x^2 is negative.

24. Minimum: $-\dfrac{\pi^2 - 8\sqrt{10}}{8\sqrt{2}} \approx 1.3637$

25. $f(x) = -x^2 + 2x + 3$

$$= -(x^2 - 2x) + 3$$

$$= -(x^2 - 2x + 1 - 1) + 3 \quad (\text{Adding } 1 - 1)$$

$$= -(x^2 - 2x + 1) + 1 + 3$$

$$= -(x - 1)^2 + 4$$

It is not necessary to rewrite the function in the form $f(x) = a(x - h)^2 + k$, but it can be helpful. Since the coefficient of x^2 is negative, the parabola opens downward. The vertex is $(1, 4)$. The line of symmetry is $x = 1$. We calculate several input-output pairs and plot these points.

x	f(x)
1	4
0	3
2	3
-1	0
3	0

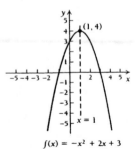

$f(x) = -x^2 + 2x + 3$
$= -(x - 1)^2 + 4$

26.

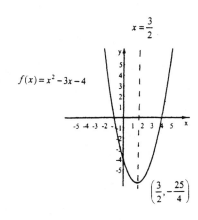

$f(x) = x^2 - 3x - 4$

$x = \frac{3}{2}$

$\left(\frac{3}{2}, -\frac{25}{4}\right)$

27. $f(x) = x^2 - 8x + 19$

$= (x^2 - 8x + 16 - 16) + 19$ (Adding 16 - 16)

$= (x^2 - 8x + 16) - 16 + 19$

$= (x - 4)^2 + 3$

Since the coefficient of x^2 is positive, the parabola opens upward. The vertex is (4, 3). The line of symmetry is x = 4. We calculate several input-output pairs and plot these points.

x	f(x)
4	3
3	4
5	4
2	7
6	7

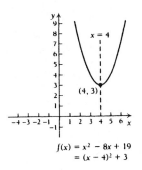

$f(x) = x^2 - 8x + 19$
$= (x - 4)^2 + 3$

28.

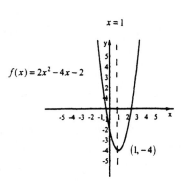

$x = -4$

$f(x) = -x^2 - 8x - 17$

$(-4, -1)$

29. $f(x) = -\frac{1}{2}x^2 - 3x + \frac{1}{2}$

$= -\frac{1}{2}(x^2 + 6x) + \frac{1}{2}$

$= -\frac{1}{2}(x^2 + 6x + 9 - 9) + \frac{1}{2}$ (Adding 9 - 9)

$= -\frac{1}{2}(x^2 + 6x + 9) + \frac{9}{2} + \frac{1}{2}$

$= -\frac{1}{2}(x + 3)^2 + 5$

$= -\frac{1}{2}[x - (-3)]^2 + 5$

Since the coefficient of x^2 is negative, the parabola opens downward. The vertex is (-3, 5). The line of symmetry is x = -3. We calculate several input-output pairs and plot these points.

x	f(x)
-3	5
-4	$\frac{9}{2}$
-2	$\frac{9}{2}$
-5	3
-1	3

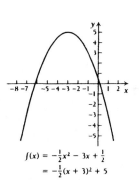

$f(x) = -\frac{1}{2}x^2 - 3x + \frac{1}{2}$
$= -\frac{1}{2}(x + 3)^2 + 5$

30.

$x = 1$

$f(x) = 2x^2 - 4x - 2$

$(1, -4)$

31. $f(x) = 3x^2 - 24x + 50$

        $= 3(x^2 - 8x) + 50$

        $= 3(x^2 - 8x + 16 - 16) + 50$ (Adding 16 - 16)

        $= 3(x^2 - 8x + 16) - 48 + 50$

        $= 3(x - 4)^2 + 2$

Since the coefficient of x^2 is positive, the parabola opens upward. The vertex is (4, 2). The line of symmetry is $x = 4$. We calculate several input-output pairs and plot these points.

x	f(x)
4	2
3	5
5	5
2	14
6	14

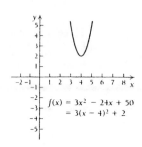

$f(x) = 3x^2 - 24x + 50$
$= 3(x - 4)^2 + 2$

32.

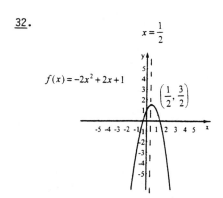

$x = \frac{1}{2}$

$f(x) = -2x^2 + 2x + 1$

$\left(\frac{1}{2}, \frac{3}{2}\right)$

33. $f(x) = -x^2 + 2x + 3$

We solve the equation $-x^2 + 2x + 3 = 0$.

    $x^2 - 2x - 3 = 0$ (Multiplying by -1)

$(x + 1)(x - 3) = 0$

$x + 1 = 0$ or $x - 3 = 0$

     $x = -1$ or $x = 3$

Thus the x-intercepts are (-1, 0) and (3, 0).

34. (-1, 0), (4, 0)

35. $f(x) = x^2 - 8x + 5$

We solve the equation $x^2 - 8x + 5 = 0$.

$a = 1,$ $b = -8,$ $c = 5$

$x = \dfrac{-(-8) \pm \sqrt{(-8)^2 - 4 \cdot 1 \cdot 5}}{2 \cdot 1} = \dfrac{8 \pm \sqrt{64 - 20}}{2}$

$= \dfrac{8 \pm \sqrt{44}}{2} = \dfrac{8 \pm 2\sqrt{11}}{2} = 4 \pm \sqrt{11}$

Thus the x-intercepts are $(4 - \sqrt{11}, 0)$ and $(4 + \sqrt{11}, 0)$.

36. None

37. $f(x) = -5x^2 + 6x - 5$

We solve the equation

  $-5x^2 + 6x - 5 = 0.$

   $5x^2 - 6x + 5 = 0$

$a = 5,$ $b = -6,$ $c = 5$

$b^2 - 4ac$ (the discriminant) $= -64 < 0$

Thus, there are no x-intercepts.

38. $\left(\dfrac{-1 - \sqrt{41}}{4}, 0\right),$ $\left(\dfrac{-1 + \sqrt{41}}{4}, 0\right)$

39. Familiarize and Translate: Let x and y represent the numbers, and let Q = the product. We have $x + y = 50$, so $y = 50 - x$. Then Q is given by

     Q = xy

     $Q(x) = x(50 - x) = 50x - x^2.$ (Substituting 50 - x for y)

Carry out: We want to maximize this function. We find the first coordinate of the vertex of the function:

$$-\frac{b}{2a} = -\frac{50}{2(-1)} = 25$$

Since the leading coefficient of $Q(x) = 50x - x^2$, -1, is negative, we know that the function has a maximum value. Note that $y = 50 - 25$, or 25, when $x = 25$.

Check: We go over our work. We could also examine the graph of the function and estimate the maximum value.

State: The two numbers that have the maximum product are 25 and 25.

40. 8, -8

41. Familiarize and Translate: We first draw a picture. Let h = the height and b = the base.

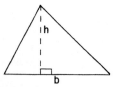

Since the sum of the base and height is 20 cm, we have

b + h = 20 and b = 20 - h.

Then the area is given by

$$A = \frac{1}{2} bh$$

$$A(h) = \frac{1}{2} (20 - h)h = 10h - \frac{1}{2} h^2.$$

Carry out: We find the first coordinate of the vertex:

$$-\frac{b}{2a} = -\frac{10}{2\left(-\frac{1}{2}\right)} = 10$$

Since the leading coefficient of the function, $-\frac{1}{2}$, is negative, we know that the function has a maximum value. The maximum value occurs when h = 10. Note that b = 20 - 10, or 10, when h = 10.

Check: We go over our work. We could also examine the graph of the function and estimate the maximum value.

State: The area is a maximum when the base is 10 cm and the height is 10 cm.

<u>42.</u> Base: 69 yd, height: 69 yd

<u>43.</u> Familiarize and Translate: Using the drawing in the text, we see that the area is given by A(x) = (120 - 2x)x, or A(x) = 120x - 2x².

Carry out: We find the vertex of the function. The first coordinate of the vertex is

$$-\frac{b}{2a} = -\frac{120}{2(-2)} = 30.$$

We find the second coordinate by substituting:

$$A(30) = 120(30) - 2(30)^2 = 3600 - 1800 = 1800$$

The vertex is (30, 1800). Since the leading coefficient of A(x), -2, is negative, we know that the function has a maximum value. The maximum value is 1800 when x = 30. Note that 120 - 2x = 120 - 2(30) = 60 when x = 30.

Check: We go over our work. We could also examine the graph of the function and estimate the maximum value.

State: Dimensions of 30 yd by 60 yd will maximize the area. The maximum area is 1800 yd².

<u>44.</u> 4800 yd²

<u>45.</u> Familiarize and Translate: We make a drawing. Let ℓ = the length and w = the width of the room.

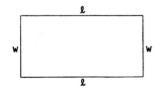

Since the perimeter is 54 ft, we have

2ℓ + 2w = 54 and ℓ = 27 - w.

Then the area is given by

$$A = \ell w$$

$$A(w) = (27 - w)w = 27w - w^2.$$

Carry out: We find the vertex of the function. The first coordinate of the vertex is given by

$$-\frac{b}{2a} = -\frac{27}{2(-1)} = 13.5.$$

We find the second coordinate by substituting:

A(13.5) = 27(13.5) - (13.5)² = 364.5 - 182.25 = 182.25

The vertex is (13.5, 182.25). Since the leading coefficient of A(w), -1, is negative, we know the function has a maximum value. The maximum value is 182.25 when w = 13.5. Note that ℓ = 27 - 13.5 = 13.5 when w = 13.5.

Check: We go over our work. We could also examine the graph of the function and estimate the maximum value.

State: The dimensions are 13.5 ft by 13.5 ft. The area is 182.25 ft².

<u>46.</u> 72.25 ft²

<u>47.</u> P(x) = R(x) - C(x)

P(x) = (50x - 0.5x²) - (4x + 10)

P(x) = -0.5x² + 46x - 10

We find the vertex of P(x). The first coordinate is

$$-\frac{b}{2a} = -\frac{46}{2(-0.5)} = 46.$$

We substitute to find the second coordinate:

P(46) = -0.5(46)² + 46(46) - 10

= -1058 + 2116 - 10

= 1048

The vertex is (46, 1048). Since the leading coefficient of P(x), -0.5, is negative, we know that the function has a maximum value. Thus, the maximum profit is $1048, and it occurs when 46 units are produced and sold.

<u>48.</u> 40 units, $797

<u>49.</u> P(x) = R(x) - C(x)

P(x) = 2x - (0.01x² + 0.6x + 30)

P(x) = -0.01x² + 1.4x - 30

We find the vertex of P(x). The first coordinate is

$$-\frac{b}{2a} = -\frac{1.4}{2(-0.01)} = 70.$$

We find the second coordinate by substituting:

P(70) = -0.01(70)² + 1.4(70) - 30

= -49 + 98 - 30

= 19

The vertex is (70, 19). Since the leading coefficient of P(x), -0.01, is negative, we know that the function has a maximum value. Thus, the maximum profit is $19, and it occurs when 70 units are produced and sold.

50. 1900 units, $3550

51. $s(t) = -4.9t^2 + v_0t + h$

$s(t) = -4.9t^2 + 147t + 560$ (Substituting)

a) We find the vertex of $s(t)$. The first coordinate is

$$-\frac{b}{2a} = -\frac{147}{2(-4.9)} = 15.$$

We find the second coordinate by substituting:

$s(15) = -4.9(15)^2 + 147(15) + 560 = 1662.5$

The vertex is (15, 1662.5). Since the leading coefficient of $s(t)$, -4.9, is negative, we know the function has a maximum value. Thus, the rocket has a maximum height of 1662.5 m, and it is attained 15 sec after the end of the burn.

b) We set $s(t) = 0$ and solve for t. We use the quadratic formula.

$-4.9t^2 + 147t + 560 = 0$

$a = -4.9$, $b = 147$, $c = 560$

$$t = \frac{-b \pm \sqrt{b^2 - 4ac}}{2a}$$

$$t = \frac{-147 - \sqrt{147^2 - 4(-4.9)(560)}}{2(-4.9)}$$ Taking the negative square root ($t > 0$)

$t \approx 33.420$

The rocket will reach the ground about 33.420 sec after the end of the burn.

52. The dimensions of the rectangular part of the window are $\frac{24}{x + 4}$ ft by $\frac{24}{x + 4}$ ft.

53. $f(x) = |x^2 - 1|$

First graph $f(x) = x^2 - 1$.

x	f(x)
0	-1
-1	0
1	0
2	3
-2	3

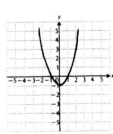

Then reflect the negative values across the x-axis. The point (0, -1) is now (0, 1).

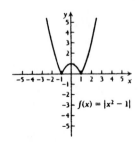

54.

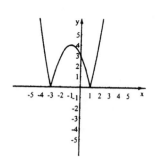

55. $f(x) = 2.31x^2 - 3.135x - 5.89$

$$-\frac{b^2 - 4ac}{4a} = -\frac{(-3.135)^2 - 4(2.31)(-5.89)}{4(2.31)} \approx$$
-6.9537

The value of -6.9537 is a minimum since the coefficient of x^2 is positive.

56. Maximum: 7.0141

57. $f(x) = ax^2 + 3x - 8$

The first coordinate of the vertex, $-\frac{b}{2a}$, is -2:

$$-\frac{3}{2a} = -2$$

$$\frac{3}{4} = a$$

58. $\pm 4\sqrt{47}$

59. $f(x) = -0.2x^2 - 3x + c$

The second coordinate of the vertex, $-\frac{b^2 - 4ac}{4a}$, is -225:

$$-\frac{(-3)^2 - 4(-0.2)(c)}{4(-0.2)} = -225$$

$$\frac{9 + 0.8c}{0.8} = -225$$

$$9 + 0.8c = -180$$

$$0.8c = -189$$

$$c = -236.25, \text{ or } -\frac{945}{4}$$

60. $f(x) = \frac{6}{49}x^2 - \frac{48}{49}x - \frac{149}{49}$

61. $f(x) = qx^2 - 2x + q$

The first coordinate of the vertex is

$$-\frac{b}{2a} = -\frac{-2}{2q} = \frac{1}{q}.$$

We substitute to find the second coordinate of the vertex:

$$f\left(\frac{1}{q}\right) = q\left(\frac{1}{q^2}\right) - 2\left(\frac{1}{q}\right) + q$$

$$= \frac{1}{q} - \frac{2}{q} + q$$

$$= \frac{-1 + q^2}{q}, \text{ or } \frac{q^2 - 1}{q}$$

The vertex is $\left(\frac{1}{q}, \frac{q^2 - 1}{q}\right)$.

62. $x = \dfrac{p}{6}$

63. $f(x) = x^2 - 6x + c$

The second coordinate of the vertex, $-\dfrac{b^2 - 4ac}{4a}$, is 0.

$$-\frac{(-6)^2 - 4 \cdot 1 \cdot c}{4 \cdot 1} = 0$$

$$36 - 4c = 0$$

$$36 = 4c$$

$$9 = c$$

64. 0

65. Familiarize: Let x = the number of \$1 increases in the ticket price. When the ticket price is $6 + x$, then the average attendance will be $70,000 - 10,000x$.

Translate:

Revenue =

$\begin{bmatrix} \text{Price per} \\ \text{ticket} \end{bmatrix} \cdot \begin{bmatrix} \text{Average} \\ \text{attendance} \end{bmatrix} + \$1.50 \cdot \begin{bmatrix} \text{Average} \\ \text{attendance} \end{bmatrix}$

$R(x) =$
$(6 + x) \cdot (70,000 - 10,000x) + 1.5(70,000 - 10,000x)$

$R(x) =$
$420,000 + 10,000x - 10,000x^2 + 105,000 - 15,000x$

$R(x) = -10,000x^2 - 5000x + 525,000$

Carry out: We find the value of x for which $R(x)$ is a maximum. The first coordinate of the vertex is

$$-\frac{b}{2a} = -\frac{-5000}{2(-10,000)} = -0.25.$$

Since the coefficient of x^2 is negative, $R(x)$ has a maximum value. It occurs when $x = -0.25$. When $x = -0.25$, the price per ticket is $6 + x =$ \$6 + (-\$0.25), or \$5.75. The number of people that will attend is $70,000 - 10,000(-0.25)$, or $72,500$.

Check: We go over our work. We could also examine the graph of the function and estimate the first coordinate of the vertex.

State: \$5.75 per ticket should be charged to maximize revenue. At that price, 72,500 people will attend.

66. \$26

67. Familiarize: Let x = the number of additional trees that are planted per acre. Then the total number of trees per acre will be $20 + x$, and the yield per tree will be $30 - x$ bushels.

Translate:

Yield $= \begin{bmatrix} \text{Number of trees} \\ \text{per acre} \end{bmatrix} \cdot \begin{bmatrix} \text{Yield} \\ \text{per tree} \end{bmatrix}$

$Y(x) = \quad (20 + x) \quad \cdot \quad (30 - x)$

$Y(x) = \quad 600 + 10x - x^2$

Carry out: We find the value of x for which $Y(x)$ is a maximum. The first coordinate of the vertex of $Y(x)$ is

$$-\frac{b}{2a} = -\frac{10}{2(-1)} = 5.$$

Since the coefficient of x^2 is negative, the function has a maximum value. It occurs when $x = 5$. When $x = 5$, the number of trees per acre is $20 + 5$, or 25 trees.

Check: We go over our work. We could also examine the graph of the function and estimate the first coordinate of the vertex.

State: In order to get the highest yield, 25 trees per acre should be planted.

68. \$6.50

69. Familiarize: First we find the radius r of a circle with circumference x:

$$2\pi r = x$$

$$r = \frac{x}{2\pi}$$

Then we find the length s of a side of a square with perimeter $24 - x$:

$$4s = 24 - x$$

$$s = \frac{24 - x}{4}$$

Translate:

S = area of circle + area of square

$S = \qquad \pi r^2 \qquad + \qquad s^2$

$S(x) = \quad \pi\left[\dfrac{x}{2\pi}\right]^2 \quad + \quad \left[\dfrac{24 - x}{4}\right]^2$

$S(x) = \dfrac{x^2}{4\pi} + \left[\dfrac{24 - x}{4}\right]^2$

$S(x) = \left[\dfrac{1}{4\pi} + \dfrac{1}{16}\right]x^2 - 3x + 36$

Carry out: We find the value of x for which $S(x)$ is a minimum. The first coordinate of the vertex of $S(x)$ is

$$-\frac{b}{2a} = -\frac{-3}{2\left[\dfrac{1}{4\pi} + \dfrac{1}{16}\right]} = \frac{3}{\dfrac{1}{2\pi} + \dfrac{1}{8}} = \frac{24\pi}{4 + \pi}.$$

Since the coefficient of x^2 is positive, the function has a minimum value. It occurs when $x = \dfrac{24\pi}{4 + \pi}$. Thus one piece will be $\dfrac{24\pi}{4 + \pi}$ in., and then the other piece will be $24 - \dfrac{24\pi}{4 + \pi}$, or $\dfrac{96}{4 + \pi}$ in.

Check: We go over our work.

State: One piece of string should be $\dfrac{24\pi}{4 + \pi}$ in. and then the other piece would be $\dfrac{96}{4 + \pi}$ in.

70. 6 cm by $\dfrac{9}{2}$ cm, 27 cm²

71. Vertex: (0.9, -10.0)

    x-intercepts: (-0.8, 0), (2.6, 0)

72. Vertex: (-9.1, 50.0)

    x-intercepts: none

73. a) 12.9 sec

    b) 920.6 m

    c) 26.6 sec

Exercise Set 4.2

---

1. The intersection of two sets consists of those element <u>common</u> to the sets.

    {3, 4, 5, 8, 10} ∩ {1, 2, 3, 4, 5, 6, 7}

    = {3, 4, 5}

    Only the elements 3, 4, and 5 are in <u>both</u> sets.

2. {1, 2, 3, 4, 5, 6, 7, 8, 10}

3. The union of two sets consists of the elements that are in <u>one or both</u> of the sets.

    {0, 2, 4, 6, 8} ∪ {4, 6, 9} = {0, 2, 4, 6, 8, 9}

    Note that the elements 4 and 6 are in both sets, but each is listed only once in the union.

4. {4, 6}

5. {a, b, c} ∩ {c, d} = {c}

    Since the element c is the only element which is in <u>both</u> sets, the intersection only contains c.

6. {a, b, c, d}

7. Graph: {x⎮7 ⩽ x} ∪ {x⎮x < 9}

    Graph the solution sets separately, and then find the union.

    Graph: {x⎮7 ⩽ x}   (7 ⩽ x means x ⩾ 7)

    The solid circle at 7 indicates that 7 is in the solution set.

    Graph: {x⎮x < 9}

    The open circle at 9 indicates that 9 is not in the solution set.

    Now graph the union:

    {x⎮7 ⩽ x} ∪ {x⎮x < 9} = The set of all real numbers, or (-∞, ∞).

8.

9. Graph: $\{x|-\frac{1}{2} \leqslant x\} \cap \{x|x < \frac{1}{2}\}$

    Graph the solution sets separately, and then find the intersection.

    Graph: $\{x|-\frac{1}{2} \leqslant x\}$ $\left[-\frac{1}{2} \leqslant x \text{ means } x \geqslant -\frac{1}{2}\right]$

    The solid circle indicates that $-\frac{1}{2}$ is in the solution set.

    Graph: $\{x|x < \frac{1}{2}\}$

    The open circle at $\frac{1}{2}$ indicates that $\frac{1}{2}$ is not in the solution set.

    Now graph the intersection:

    $\{x|-\frac{1}{2} \leqslant x\} \cap \{x|x < \frac{1}{2}\} = \{x|-\frac{1}{2} \leqslant x < \frac{1}{2}\}$, or

    $\left[-\frac{1}{2}, \frac{1}{2}\right)$

10.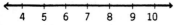

11. Graph: {x⎮x < -π} ∪ {x⎮x > π}

    Graph the solution sets separately, and then find the union.

    Graph: {x⎮x < -π}

    The open circle at -π indicates that -π is not in the solution set.

    Graph: {x⎮x > π}

    The open circle at π indicates that π is not in the solution set.

    Now graph the union:

    {x⎮x < -π} ∪ {x⎮x > π} = {x⎮x < -π or x > π}, or

    (-∞, -π) ∪ (π, ∞)

12.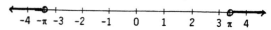

13. Graph: {x|x < -7} ∪ {x|x = -7}

    Graph the solution sets separately, and then find the union.

    Graph: {x|x < -7}

    The open circle at -7 indicates that -7 is not in the solution set.

    Graph: {x|x = -7}

    The solid circle at -7 indicates that -7 is in the solution set.

    Now graph the union:

    {x|x < -7} ∪ {x|x = -7} = {x|x ⩽ -7}, or (-∞, -7]

14.

15. Graph: {x|x ⩾ 5} ∩ {x|x ⩽ -3}

    Graph the solution sets separately, and then find the intersection.

    Graph: {x|x ⩾ 5}

    The solid circle at 5 indicates that 5 is in the solution set.

    Graph: {x|x ⩽ -3}

    The solid circle at -3 indicates that -3 is in the solution set.

    There are no elements in the intersection.

    {x|x ⩾ 5} ∩ {x|x ⩽ -3} = ∅

16.

17. -2 ⩽ x + 1 < 4

    -3 ⩽ x < 3   (Adding -1)

    The solution set is {x|-3 ⩽ x < 3}, or [-3, 3).

18. {x|-5 < x ⩽ 3}, or (-5, 3]

19. 5 ⩽ x - 3 ⩽ 7

    8 ⩽ x ⩽ 10   (Adding 3)

    The solution set is {x|8 ⩽ x ⩽ 10}, or [8, 10].

20. {x|3 < x < 11}, or (3, 11)

21. -3 ⩽ x + 4 ⩽ -3

    -7 ⩽ x ⩽ -7

    The solution set is {-7}.

22. ∅

23. -2 < 2x + 1 < 5

    -3 < 2x < 4   (Adding -1)

    $-\frac{3}{2}$ < x < 2 $\left[\text{Multiplying by } \frac{1}{2}\right]$

    The solution set is $\left\{x \middle| -\frac{3}{2} < x < 2\right\}$, or $\left[-\frac{3}{2}, 2\right]$.

24. $\left\{x \middle| -\frac{4}{5} \leqslant x \leqslant \frac{2}{5}\right\}$, or $\left[-\frac{4}{5}, \frac{2}{5}\right]$

25. -4 ⩽ 6 - 2x < 4

    -10 ⩽ -2x < -2   (Adding -6)

    5 ⩾ x > 1 $\left[\text{Multiplying by } -\frac{1}{2}\right]$

    or 1 < x ⩽ 5

    The solution set is {x|1 < x ⩽ 5}, or (1, 5].

26. {x|-1 ⩽ x < 2}, or [-1, 2)

27. $-5 < \frac{1}{2}(3x + 1) \leqslant 7$

    -10 < 3x + 1 ⩽ 14   (Multiplying by 2)

    -11 < 3x ⩽ 13   (Adding -1)

    $-\frac{11}{3} < x \leqslant \frac{13}{3}$ $\left[\text{Multiplying by } \frac{1}{3}\right]$

    The solution set is $\left\{x \middle| -\frac{11}{3} < x \leqslant \frac{13}{3}\right\}$, or $\left[-\frac{11}{3}, \frac{13}{3}\right]$.

28. $\left\{x \middle| \frac{7}{4} < x \leqslant \frac{13}{6}\right\}$, or $\left[\frac{7}{4}, \frac{13}{6}\right]$

29. 3x ⩽ -6 or x - 1 > 0

    x ⩽ -2 or   x > 1

    The solution set is {x|x ⩽ -2 or x > 1}, or (-∞, -2] ∪ (1, ∞).

30. The set of all real numbers, or (-∞, ∞)

31. 2x + 3 ⩽ -4 or 2x + 3 ⩾ 4

    2x ⩽ -7 or   2x ⩾ 1

    $x \leqslant -\frac{7}{2}$ or $x \geqslant \frac{1}{2}$

    The solution set is $\left\{x \middle| x \leqslant -\frac{7}{2} \text{ or } x \geqslant \frac{1}{2}\right\}$, or $\left[-\infty, -\frac{7}{2}\right] \cup \left[\frac{1}{2}, \infty\right)$.

32. $\left\{x \middle| x < -\frac{4}{3} \text{ or } x > 2\right\}$, or $\left[-\infty, -\frac{4}{3}\right] \cup (2, \infty)$

33. $2x - 20 < -0.8$ or $2x - 20 > 0.8$

$\qquad 2x < 19.2$ or $\qquad 2x > 20.8$

$\qquad x < 9.6$ or $\qquad x > 10.4$

The solution set is $\{x|x < 9.6$ or $x > 10.4\}$, or $(-\infty, 9.6) \cup (10.4, \infty)$.

34. $\{x|x \leqslant -3$ or $x \geqslant -\frac{7}{5}\}$, or

$(-\infty, -3] \cup [-\frac{7}{5}, \infty)$

35. $x + 14 \leqslant -\frac{1}{4}$ or $x + 14 \geqslant \frac{1}{4}$

$\qquad x \leqslant -\frac{57}{4}$ or $\qquad x \geqslant -\frac{55}{4}$

The solution set is $\{x|x \leqslant -\frac{57}{4}$ or $x \geqslant -\frac{55}{4}\}$, or

$(-\infty, -\frac{57}{4}] \cup [-\frac{55}{4}, \infty)$.

36. $\{x|x < \frac{17}{2}$ or $x > \frac{19}{2}\}$, or

$(-\infty, \frac{17}{2}] \cup [\frac{19}{2}, \infty)$

37. We first make a drawing.

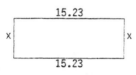

Let x represent the width. Then $2x + 2(15.23)$, or $2x + 30.46$, represents the perimeter. The perimeter is greater than 40.23 cm and less than 137.8 cm. We now have the following inequality.

$40.23 < 2x + 30.46 < 137.8$

$9.77 < 2x < 107.34$ $\qquad$ (Adding -30.46)

$4.885 < x < 53.67$ $\qquad$ [Multiplying by $\frac{1}{2}$]

Widths greater than 4.885 cm and less than 53.67 cm will give a perimeter greater than 40.23 cm and less than 137.8 cm. We can express this as $\{x|4.885$ cm $< x < 53.67$ cm$\}$.

38. $\{b|0$ m $< b \leqslant 40.72$ m$\}$

39. Let x represent the score on the fourth test. The average of the four scores is $\frac{83 + 87 + 93 + x}{4}$. This average must be greater than or equal to 90. We solve the following inequality.

$\frac{83 + 87 + 93 + x}{4} \geqslant 90$

$83 + 87 + 93 + x \geqslant 360$

$263 + x \geqslant 360$

$x \geqslant 97$

The score on the fourth test must be greater than or equal to 97. It will of course be less than or equal to 100%. We can express this as $\{x|97\% \leqslant x \leqslant 100\%\}$. Yes, an A is possible.

40. $\{x|132\% \leqslant x \leqslant 172\%\}$; no

41. We substitute $\frac{5}{9}$ (F - 32) for C in the inequality.

$1083° \leqslant \frac{5}{9}$ (F - 32) $\leqslant 2580°$

$1949.4° \leqslant$ F - 32 $\leqslant 4644°$ $\quad$ [Multiplying by $\frac{9}{5}$]

$1981.4° \leqslant$ F $\leqslant 4676°$

42. $\{d|33$ ft $\leqslant d \leqslant 297$ ft$\}$

43. $2 < 0.027d + 0.32 < 4$

$2000 < 27d + 320 < 4000$ $\quad$ (Clearing decimals)

$1680 < 27d < 3680$

$\frac{1680}{27} < d < \frac{3680}{27}$

$62\frac{2}{9} < d < 136\frac{8}{27}$

The solution set is $\{d|62\frac{2}{9}$ mi $< d < 136\frac{8}{27}$ mi$\}$.

44. $\{t|2\frac{14}{41}$ yr $< t < 6\frac{10}{41}$ yr$\}$

45. $x \leqslant 3x - 2 \leqslant 2 - x$

$x \leqslant 3x - 2$ and $3x - 2 \leqslant 2 - x$

$-2x \leqslant -2$ $\qquad$ and $\qquad 4x \leqslant 4$

$x \geqslant 1$ $\qquad$ and $\qquad x \leqslant 1$

The word "and" corresponds to set intersection.

The solution set is

$\{x|x \geqslant 1\} \cap \{x|x \leqslant 1\}$, or $\{1\}$.

46. $\{x|-\frac{1}{4} < x \leqslant \frac{5}{9}\}$, or $(-\frac{1}{4}, \frac{5}{9}]$

47. $(x + 1)^2 > x(x - 3)$

$x^2 + 2x + 1 > x^2 - 3x$

$2x + 1 > -3x$

$5x > -1$

$x > -\frac{1}{5}$

The solution set is $\{x|x > -\frac{1}{5}\}$, or $(-\frac{1}{5}, \infty)$.

48. $\{x|x > \frac{13}{5}\}$, or $(\frac{13}{5}, \infty)$

49. $(x + 1)^2 \leqslant (x + 2)^2 \leqslant (x + 3)^2$

$x^2 + 2x + 1 \leqslant x^2 + 4x + 4 \leqslant x^2 + 6x + 9$

$2x + 1 \leqslant 4x + 4 \leqslant 6x + 9$

$2x + 1 \leqslant 4x + 4$ and $4x + 4 \leqslant 6x + 9$

$-2x \leqslant 3$ and $-2x \leqslant 5$

$x \geqslant -\frac{3}{2}$ and $x \geqslant -\frac{5}{2}$

The word "and" corresponds to set intersection.

The solution set is

$\{x | x > -\frac{3}{2}\} \cap \{x | x \geqslant -\frac{5}{2}\}$, or $\{x | x \geqslant -\frac{3}{2}\}$, or

$[-\frac{3}{2}, \infty)$.

50. $\{x | -1 < x \leqslant 1\}$, or $(-1, 1]$

51. $f(x) = \dfrac{\sqrt{x + 2}}{\sqrt{x - 2}}$

The radicand in the denominator must be greater than 0. Thus, $x - 2 > 0$, or $x > 2$. Also the radicand in the numerator must be greater than or equal to 0. Thus, $x + 2 \geqslant 0$, or $x \geqslant -2$. The domain of $f(x)$ is the intersection of the two solution sets.

$\{x | x > 2\} \cap \{x | x \geqslant -2\} = \{x | x > 2\}$.

The domain of $f(x)$ is $\{x | x > 2\}$, or $(2, \infty)$.

52. $\{x | -5 < x \leqslant 3\}$, or $(-5, 3]$

53. $\{x | -5.2 < x \leqslant 9\}$, or $(-5.2, 9]$

54. $\{x | 0.5 \leqslant x < 4.1\}$, or $[0.5, 4.1)$

55. $\{x | x < -0.2 \text{ or } x \geqslant 6.2\}$, or $(-\infty, -0.2) \cup [6.2, \infty)$

56. $\{x | x \leqslant -2.2 \text{ or } x > -1.2\}$, or $(-\infty, 2.2] \cup (-1.2, \infty)$

Exercise Set 4.3

1. $|x| = 7$

To solve we look for all numbers x whose distance from 0 is 7. There are two of them, so there are two solutions, -7 and 7. The graph is as follows.

$\{-7, 7\}$

2. $\{-4.5, 4.5\}$

3. $|x| < 7$

To solve we look for all numbers x whose distance from 0 is less than 7. These are the numbers between -7 and 7. The solution set and its graph are as follows.

$\{x | -7 < x < 7\}$, or $(-7, 7)$

4. $\{x | -4.5 \leqslant x \leqslant 4.5\}$, or $[-4.5, 4.5]$

5. $|x| \geqslant 4.5$

To solve we look for all numbers x whose distance from 0 is greater than or equal to 4.5. The solution set and its graph are as follows.

$\{x | x \leqslant -4.5 \text{ or } x \geqslant 4.5\}$, or $(-\infty, -4.5] \cup [4.5, \infty)$

6. $\{x | x < -7 \text{ or } x > 7\}$, or $(-\infty, -7) \cup (7, \infty)$

7. $|x - 1| = 4$

Method 1:

The solutions are those numbers x whose distance from 1 is 4. Thus to find the solutions graphically we locate 1. Then we locate those numbers 4 units to the left and 4 units to the right.

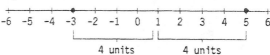

The solution set is $\{-3, 5\}$.

Method 2:

$|x - 1| = 4$

$x - 1 = -4$ or $x - 1 = 4$ (Property A1)

$x = -3$ or $x = 5$ (Adding 1)

The solution set is $\{-3, 5\}$.

8. $\{2, 12\}$

9. $|x + 8| < 9$, or $|x - (-8)| < 9$

Method 1:

The solutions are those numbers x whose distance from -8 is less than 9. Thus to find the solutions graphically we locate -8. Then we locate those numbers that are less than 9 units to the left and less than 9 units to the right.

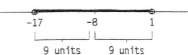

The solution set is $\{x | -17 < x < 1\}$, or $(-17, 1)$.

Method 2:

$|x + 8| < 9$

$-9 < x + 8 < 9$ (Property A2)

$-17 < x < 1$ (Adding -8)

The solution set is {x|-17 < x < 1}, or (-17, 1).

10. {x|-16 ≤ x ≤ 4}, or [-16, 4]

11. $|x + 8| ⩾ 9$, or $|x - (-8)| ⩾ 9$

Method 1:

The solutions are those numbers x whose distance from -8 is greater than or equal to 9. Thus to find the solutions graphically we locate -8. Then we locate those numbers that are greater than or equal to 9 units to the left and greater than or equal to 9 units to the right.

The solution set is {x|x ≤ -17 or x ⩾ 1}, or (-∞, -17] ∪ [1, ∞).

Method 2:

$|x + 8| ⩾ 9$

$x + 8 ≤ -9$ or $x + 8 ⩾ 9$ (Property A3)

$x ≤ -17$ or $x ⩾ 1$ (Adding -8)

The solution set is {x|x ≤ -17 or x ⩾ 1}, or (-∞, -17] ∪ [1, ∞).

12. {x|x < -16 or x > 4}, or (-∞, -16) ∪ (4, ∞)

13. $\left| x - \frac{1}{4} \right| < \frac{1}{2}$

Method 1:

The solutions are those numbers x whose distance from $\frac{1}{4}$ is less than $\frac{1}{2}$. Thus to find the solutions graphically we locate $\frac{1}{4}$. Then we locate those numbers that are less than $\frac{1}{2}$ unit to the left and less than $\frac{1}{2}$ unit to the right.

The solution set is $\left\{ x \left| -\frac{1}{4} < x < \frac{3}{4} \right. \right\}$, or $\left(-\frac{1}{4}, \frac{3}{4} \right)$.

Method 2:

$\left| x - \frac{1}{4} \right| < \frac{1}{2}$

$-\frac{1}{2} < x - \frac{1}{4} < \frac{1}{2}$ (Property A2)

$-\frac{1}{4} < x < \frac{3}{4}$ $\left[\text{Adding } \frac{1}{4} \right]$

The solution set is $\left\{ x \left| -\frac{1}{4} < x < \frac{3}{4} \right. \right\}$, or $\left(-\frac{1}{4}, \frac{3}{4} \right)$.

14. {x|0.3 ≤ x ≤ 0.7}, or [0.3, 0.7]

15. $|3x| = 1$

$3x = -1$ or $3x = 1$ (Property A1)

$x = -\frac{1}{3}$ or $x = \frac{1}{3}$ $\left[\text{Multiplying by } \frac{1}{3} \right]$

The solution set is $\left\{ -\frac{1}{3}, \frac{1}{3} \right\}$.

16. $\left\{ -\frac{4}{5}, \frac{4}{5} \right\}$

17. $|3x + 2| = 1$

$3x + 2 = -1$ or $3x + 2 = 1$ (Property A1)

$3x = -3$ or $3x = -1$ (Adding -2)

$x = -1$ or $x = -\frac{1}{3}$ $\left[\text{Multiplying by } \frac{1}{3} \right]$

The solution set is $\left\{ -1, -\frac{1}{3} \right\}$.

18. $\left\{ -\frac{4}{7}, \frac{12}{7} \right\}$

19. $|3x| < 1$

$-1 < 3x < 1$ (Property A2)

$-\frac{1}{3} < x < \frac{1}{3}$ $\left[\text{Multiplying by } \frac{1}{3} \right]$

The solution set is $\left\{ x \left| -\frac{1}{3} < x < \frac{1}{3} \right. \right\}$, or $\left(-\frac{1}{3}, \frac{1}{3} \right)$.

20. $\left\{ x \left| -\frac{4}{5} ≤ x ≤ \frac{4}{5} \right. \right\}$, or $\left[-\frac{4}{5}, \frac{4}{5} \right]$

21. $|2x + 3| ≤ 9$

$-9 ≤ 2x + 3 ≤ 9$ (Property A2)

$-12 ≤ 2x ≤ 6$ (Adding -3)

$-6 ≤ x ≤ 3$ $\left[\text{Multiplying by } \frac{1}{2} \right]$

The solution set is {x|-6 ≤ x ≤ 3}, or [-6, 3].

22. {x|-8 < x < 5}, or (-8, 5)

23. $|x - 5| > 0.1$

$x - 5 < -0.1$ or $x - 5 > 0.1$ (Property A3)

$x < 4.9$ or $x > 5.1$ (Adding 5)

The solution set is {x|x < 4.9 or x > 5.1}, or (-∞, 4.9) ∪ (5.1, ∞).

24. {x|x ≤ 6.6 or x ⩾ 7.4}, or (-∞, 6.6] ∪ [7.4, ∞)

25. $\left| x + \frac{2}{3} \right| ≤ \frac{5}{3}$

$-\frac{5}{3} ≤ x + \frac{2}{3} ≤ \frac{5}{3}$ (Property A2)

$-\frac{7}{3} ≤ x ≤ \frac{3}{3}$ $\left[\text{Adding } -\frac{2}{3} \right]$

The solution set is $\left\{ x \left| -\frac{7}{3} ≤ x ≤ 1 \right. \right\}$, or $\left[-\frac{7}{3}, 1 \right]$.

26. $\left\{x\middle| -1 < x < -\frac{1}{2}\right\}$, or $\left[-1, -\frac{1}{2}\right]$

27. $|6 - 4x| \leqslant 8$

$\quad -8 \leqslant 6 - 4x \leqslant 8$ (Property A2)

$\quad -14 \leqslant -4x \leqslant 2$ (Adding -6)

$\quad \frac{14}{4} \geqslant x \geqslant -\frac{2}{4}$ $\left[\text{Multiplying by} -\frac{1}{4}\right]$

$\quad \frac{7}{2} \geqslant x \geqslant -\frac{1}{2}$

The solution set is $\left\{x\middle| -\frac{1}{2} \leqslant x \leqslant \frac{7}{2}\right\}$, or $\left[-\frac{1}{2}, \frac{7}{2}\right]$.

28. $\left\{x\middle| x > \frac{15}{2} \text{ or } x < -\frac{5}{2}\right\}$, or $\left[-\infty, -\frac{5}{2}\right] \cup \left[\frac{15}{2}, \infty\right]$

29. $\left|\frac{2x + 1}{3}\right| > 5$

$\quad \frac{2x + 1}{3} < -5 \text{ or } \frac{2x + 1}{3} > 5$ (Property A3)

$\quad 2x + 1 < -15 \text{ or } 2x + 1 > 15$ (Multiplying by 3)

$\quad\quad 2x < -16 \text{ or }\quad 2x > 14$ (Adding -1)

$\quad\quad\quad x < -8 \text{ or }\quad\quad x > 7$ $\left[\text{Multiplying by} \frac{1}{2}\right]$

The solution set is $\{x | x < -8 \text{ or } x > 7\}$, or $(-\infty, -8) \cup (7, \infty)$.

30. $\left\{x\middle| -\frac{22}{3} \leqslant x \leqslant 6\right\}$, or $\left[-\frac{22}{3}, 6\right]$

31. $\left|\frac{13}{4} + 2x\right| > \frac{1}{4}$

$\quad \frac{13}{4} + 2x < -\frac{1}{4} \text{ or } \frac{13}{4} + 2x > \frac{1}{4}$ (Property A3)

$\quad\quad 2x < -\frac{14}{4} \text{ or }\quad 2x > -\frac{12}{4}$

$\quad\quad 2x < -\frac{7}{2} \text{ or }\quad 2x > -3$

$\quad\quad\quad x < -\frac{7}{4} \text{ or }\quad\quad x > -\frac{3}{2}$

The solution set is $\left\{x\middle| x < -\frac{7}{4} \text{ or } x > -\frac{3}{2}\right\}$, or $\left[-\infty, -\frac{7}{4}\right] \cup \left[-\frac{3}{2}, \infty\right]$.

32. $\left\{x\middle| -\frac{2}{3} < x < \frac{1}{9}\right\}$, or $\left[-\frac{2}{3}, \frac{1}{9}\right]$

33. $\left|\frac{3 - 4x}{2}\right| \leqslant \frac{3}{4}$

$\quad -\frac{3}{4} \leqslant \frac{3 - 4x}{2} \leqslant \frac{3}{4}$ (Property A2)

$\quad -3 \leqslant 6 - 8x \leqslant 3$ (Multiplying by 4)

$\quad -9 \leqslant -8x \leqslant -3$ (Adding -6)

$\quad \frac{9}{8} \geqslant x \geqslant \frac{3}{8}$ $\left[\text{Multiplying by} -\frac{1}{8}\right]$

The solution set is $\left\{x\middle| \frac{3}{8} \leqslant x \leqslant \frac{9}{8}\right\}$, or $\left[\frac{3}{8}, \frac{9}{8}\right]$.

34. $\left\{x\middle| x \leqslant -\frac{3}{4} \text{ or } x \geqslant \frac{7}{4}\right\}$, or $\left[-\infty, -\frac{3}{4}\right] \cup \left[\frac{7}{4}, \infty\right]$

35. $|x| = -3$

Since $|x| \geqslant 0$ for all x, there is no x such that $|x|$ would be negative. There is no solution. The solution set is $\emptyset$.

36. $\emptyset$

37. $|2x - 4| < -5$

Since $|2x - 4| \geqslant 0$ for all x, there is no x such that $|2x - 4|$ would be less than -5. There is no solution. The solution set is $\emptyset$.

38. $\emptyset$

39. $|x + 17.217| > 5.0012$

$\quad x + 17.217 < -5.0012 \text{ or } x + 17.217 > 5.0012$

(Property A3)

$\quad\quad x < -22.2182 \text{ or } x > -12.2158$

(Adding -17.217)

The solution set is $\{x | x < -22.2182 \text{ or } x > -12.2158\}$, or $(-\infty, -22.2182) \cup (-12.2158, \infty)$.

40. $\{x | 1.9234 < x < 2.1256\}$, or $(1.9234, 2.1256)$

41. $|-2.1437x + 7.8814| \geqslant 9.1132$

$\quad -2.1437x + 7.8814 \leqslant -9.1132$

$\quad\quad -2.1437x \leqslant -16.9946$

$\quad\quad\quad x \geqslant 7.9277$

_or_

$\quad -2.1437x + 7.8814 \geqslant 9.1132$

$\quad\quad -2.1437x \geqslant 1.2318$

$\quad\quad\quad x \leqslant -0.5746$

The solution set is $\{x | x \leqslant -0.5746 \text{ or } x \geqslant 7.9277\}$, or $(-\infty, -0.5746] \cup [7.9277, \infty)$.

42. $\{x | 0.9841 \leqslant x \leqslant 4.9808\}$, or $[0.9841, 4.9808]$

43. $|2x - 8| = |x + 3|$

$\quad 2x - 8 = x + 3 \text{ or } 2x - 8 = -(x + 3)$

$\quad\quad x = 11 \quad\quad \text{ or } 2x - 8 = -x - 3$

$\quad\quad x = 11 \quad\quad \text{ or }\quad\quad 3x = 5$

$\quad\quad x = 11 \quad\quad \text{ or }\quad\quad x = \frac{5}{3}$

The solution set is $\left\{\frac{5}{3}, 11\right\}$.

44. $\left\{-\frac{11}{2}, \frac{3}{4}\right\}$

45. $\left|\dfrac{2x+3}{6}\right| = \left|\dfrac{4-5x}{8}\right|$

$\dfrac{2x+3}{6} = \dfrac{4-5x}{8}$ or $\dfrac{2x+3}{6} = -\left(\dfrac{4-5x}{8}\right)$

$4(2x+3) = 3(4-5x)$ or $\dfrac{2x+3}{6} = \dfrac{5x-4}{8}$

$8x + 12 = 12 - 15x$ or $4(2x+3) = 3(5x-4)$

$23x = 0$ or $8x + 12 = 15x - 12$

$x = 0$ or $24 = 7x$

$x = 0$ or $\dfrac{24}{7} = x$

The solution set is $\left\{0, \dfrac{24}{7}\right\}$.

46. $\left\{-\dfrac{5}{86}, \dfrac{85}{14}\right\}$

47. $|x - 2| = x - 2$

By the definition of absolute value,

$|x - 2| = x - 2$ for $x - 2 \geqslant 0$, or $x \geqslant 2$.

The solution set is $\{x | x \geqslant 2\}$, or $[2, \infty)$.

48. $\left\{\dfrac{4}{5}, 2\right\}$

49. $|7x - 2| = x + 5$

$7x - 2 = x + 5$ or $7x - 2 = -(x + 5)$

$6x = 7$ or $7x - 2 = -x - 5$

$x = \dfrac{7}{6}$ or $8x = -3$

$x = \dfrac{7}{6}$ or $x = -\dfrac{3}{8}$

The solution set is $\left\{-\dfrac{3}{8}, \dfrac{7}{6}\right\}$.

50. $\left\{\dfrac{5}{2}, \dfrac{11}{4}\right\}$

51. $|3x - 4| \leqslant 2x + 1$

$-(2x + 1) \leqslant 3x - 4 \leqslant 2x + 1$

$-2x - 1 \leqslant 3x - 4 \leqslant 2x + 1$

$-1 \leqslant 5x - 4 \leqslant 4x + 1$ (Adding 2x)

We write the conjunction, solve each inequality separately, and then find the intersection of the solution sets.

$-1 \leqslant 5x - 4$ and $5x - 4 \leqslant 4x + 1$

$3 \leqslant 5x$ and $x \leqslant 5$

$\dfrac{3}{5} \leqslant x$ and $x \leqslant 5$

The solution set is $\left[\dfrac{3}{5}, \infty\right) \cap (-\infty, 5] = \left[\dfrac{3}{5}, 5\right]$, or $\left\{x \left| \dfrac{3}{5} \leqslant x \leqslant 5\right.\right\}$.

52. $\left\{x \left| -6 \leqslant x \leqslant -\dfrac{1}{2}\right.\right\}$, or $\left[-6, -\dfrac{1}{2}\right]$

53. $|3x - 4| \geqslant 2x + 1$

$3x - 4 \leqslant -(2x + 1)$ or $3x - 4 \geqslant 2x + 1$

$3x - 4 \leqslant -2x - 1$ or $x \geqslant 5$

$5x \leqslant 3$ or $x \geqslant 5$

$x \leqslant \dfrac{3}{5}$ or $x \geqslant 5$

The solution set is $\left\{x \left| x \leqslant \dfrac{3}{5} \text{ or } x \geqslant 5\right.\right\}$, or $\left(-\infty, \dfrac{3}{5}\right] \cup [5, \infty)$.

54. $\left\{x \left| x \leqslant -6 \text{ or } x \geqslant -\dfrac{1}{2}\right.\right\}$, or $\left(-\infty, -6\right] \cup \left[-\dfrac{1}{2}, \infty\right)$

55. $|4x - 5| = |x| + 1$

If $x \geqslant 0$, then $|x| = x$ and we solve:

$4x - 5 = x + 1$ or $4x - 5 = -(x + 1)$

$3x = 6$ or $4x - 5 = -x - 1$

$x = 2$ or $5x = 4$

$x = 2$ or $x = \dfrac{4}{5}$

If $x < 0$, then $|x| = -x$ and we solve:

$4x - 5 = -x + 1$ or $4x - 5 = -(-x + 1)$

$5x = 6$ or $4x - 5 = x - 1$

$x = \dfrac{6}{5}$ or $3x = 4$

$x = \dfrac{6}{5}$ or $x = \dfrac{4}{3}$

Since $\dfrac{6}{5} > 0$ and $\dfrac{4}{3} > 0$, this case yields no solution.

The solution set is $\left\{\dfrac{4}{5}, 2\right\}$.

56. $\{5, -11\}$

57. $\big||x| - 1\big| = 3$

$|x| - 1 = -3$ or $|x| - 1 = 3$ (Property A1)

$|x| = -2$ or $|x| = 4$

The solution set of $|x| = -2$ is $\emptyset$.

The solution set of $|x| = 4$ is $\{-4, 4\}$.

$\emptyset \cup \{-4, 4\} = \{-4, 4\}$

The solution set of $\big||x| - 1\big| = 3$ is $\{-4, 4\}$.

58. The set of all real numbers, or $(-\infty, \infty)$

59. $|x + 2| \leqslant |x - 5|$

Divide the set of reals into three intervals:

$x \geqslant 5$

$-2 \leqslant x < 5$

$x < -2$

Find the solution set of $|x + 2| \leqslant |x - 5|$ for each interval. Then take the union of the three solution sets.

If $x \geqslant 5$, then $|x + 2| = x + 2$ and $|x - 5| = x - 5$.

Solve: $x \geqslant 5$ and $x + 2 \leqslant x - 5$

$x \geqslant 5$ and $2 \leqslant -5$ (False)

The solution set for this interval is $\emptyset$.

If $-2 \leqslant x < 5$, $|x + 2| = x + 2$ and $|x - 5| = -(x - 5)$.

Solve: $-2 \leqslant x < 5$ and $x + 2 \leqslant -(x - 5)$

$-2 \leqslant x < 5$ and $x + 2 \leqslant -x + 5$

$-2 \leqslant x < 5$ and $2x \leqslant 3$

$-2 \leqslant x < 5$ and $x \leqslant \frac{3}{2}$

The solution set for this interval is $\left\{x \mid -2 \leqslant x \leqslant \frac{3}{2}\right\}$.

If $x < -2$, then $|x + 2| = -(x + 2)$ and $|x - 5| = -(x - 5)$.

Solve: $x < -2$ and $-(x + 2) \leqslant -(x - 5)$

$x < -2$ and $-x - 2 \leqslant -x + 5$

$x < -2$ and $-2 \leqslant 5$

The solution set for this interval is $\{x \mid x < -2\}$.

The <u>union</u> of the above three solution sets is $\left\{x \mid x \leqslant \frac{3}{2}\right\}$, or $\left(-\infty, \frac{3}{2}\right]$. This set is the solution of $|x + 2| \leqslant |x - 5|$.

60. $\left\{x \mid x < \frac{1}{2}\right\}$, or $\left(-\infty, \frac{1}{2}\right]$

61. $|x| + |x - 1| < 10$

If $x < 0$, then $|x| = -x$ and $|x - 1| = -(x - 1)$.

Solve: $x < 0$ and $-x + [-(x - 1)] < 10$

$x < 0$ and $-x - x + 1 < 10$

$x < 0$ and $-2x + 1 < 10$

$x < 0$ and $-2x < 9$

$x < 0$ and $x > -\frac{9}{2}$

The solution set for this interval is $\left\{x \mid -\frac{9}{2} < x < 0\right\}$.

If $x \geqslant 0$, then $|x| = x$ and we have:

$x \geqslant 0$ and $x + |x - 1| < 10$

$x \geqslant 0$ and $|x - 1| < 10 - x$

$x \geqslant 0$ and $-(10 - x) < x - 1 < 10 - x$

$x \geqslant 0$ and $x - 10 < x - 1$ and $x - 1 < 10 - x$

$x \geqslant 0$ and $-10 < -1$ and $x < \frac{11}{2}$

The solution set for this interval is $\left\{x \mid 0 \leqslant x < \frac{11}{2}\right\}$.

The <u>union</u> of the two solution sets above is $\left\{x \mid -\frac{9}{2} < x < \frac{11}{2}\right\}$, or $\left(-\frac{9}{2}, \frac{11}{2}\right)$. This set is the solution set of $|x| + |x - 1| < 10$.

62. The set of all real numbers, or $(-\infty, \infty)$

63. $|x - 3| + |2x + 5| > 6$

Divide the set of real numbers into three intervals:

$x \geqslant 3$

$-\frac{5}{2} \leqslant x < 3$

$x < -\frac{5}{2}$

Find the solution set of $|x - 3| + |2x + 5| > 6$ for each interval. Then take the union of the three solution sets.

If $x \geqslant 3$, then $|x - 3| = x - 3$ and $|2x + 5| = 2x + 5$.

Solve: $x \geqslant 3$ and $x - 3 + 2x + 5 > 6$

$x \geqslant 3$ and $3x > 4$

$x \geqslant 3$ and $x > \frac{4}{3}$

The solution set for this interval is $[3, \infty)$.

If $-\frac{5}{2} \leqslant x < 3$, then $|x - 3| = -(x - 3)$ and $|2x + 5| = 2x + 5$.

Solve: $-\frac{5}{2} \leqslant x < 3$ and $-(x - 3) + 2x + 5 > 6$

$-\frac{5}{2} \leqslant x < 3$ and $-x + 3 + 2x + 5 > 6$

$-\frac{5}{2} \leqslant x < 3$ and $x > -2$

The solution set for this interval is $(-2, 3)$.

If $x < -\frac{5}{2}$, then $|x - 3| = -(x - 3)$ and $|2x + 5| = -(2x + 5)$.

Solve: $x < -\frac{5}{2}$ and $-(x - 3) + [-(2x + 5)] > 6$

$x < -\frac{5}{2}$ and $-x + 3 - 2x - 5 > 6$

$x < -\frac{5}{2}$ and $-3x > 8$

$x < -\frac{5}{2}$ and $x < -\frac{8}{3}$

The solution set for this interval is $\left(-\infty, -\frac{8}{3}\right)$.

The union of the above solution sets is

$\left[-\infty, -\frac{8}{3}\right] \cup (-2, \infty)$. This is the solution set of $|x - 3| + |2x + 5| > 6$.

64. Ø

65. $|x - 3| + |2x + 5| + |3x - 1| = 12$ (1)

If $x < -\frac{5}{2}$, we have $-(x - 3) - (2x + 5) - (3x - 1) = 12$, and $x = -\frac{13}{6}$. Since $-\frac{13}{6} < -\frac{5}{2}$, this case yields no solution.

If $x \leqslant 0$, we have
$-(x - 3) + |2x + 5| - (3x - 1) = 12$ or
$|2x + 5| = 4x + 8$; then $x = -\frac{3}{2}$. We now notice that x must be less than, say, 3; otherwise the left member of Eq. (1) would be greater than 12.

If $x > 0$ (and $x < 3$), we have from (1),
$-(x - 3) + (2x + 5) + |3x - 1| = 12$ or
$|3x - 1| = 4 - x$; then $x = \frac{5}{4}$ or $x = -\frac{3}{2}$. $\left[\text{Only } \frac{5}{4} \text{ is in the interval } (0, 3).\right]$ The solution set is $\left\{-\frac{3}{2}, \frac{5}{4}\right\}$.

66. $\{x \mid x_0 - \delta < x < x_0 + \delta\}$, or $(x_0 - \delta, x_0 + \delta)$

67. See the answer section in the text.

68. PROVE: $|a + b| \leqslant |a| + |b|$

From Exercise 67 we have $-|a| \leqslant a \leqslant |a|$
and $-|b| \leqslant b \leqslant |b|$.

Adding, we have $-\left[|a| + |b|\right] \leqslant (a + b) \leqslant |a| + |b|$
or $|a + b| \leqslant |a| + |b|$.

69. See the answer section in the text.

70. a) $-\frac{b - a}{2} < x - \frac{a + b}{2} < \frac{b - a}{2}$

$-\frac{b - a}{2} + \frac{a + b}{2} < x < \frac{b - a}{2} + \frac{a + b}{2}$

$\left[\text{Adding } \frac{a + b}{2}\right]$

$\frac{2a}{2} < x < \frac{2b}{2}$

$a < x < b$

b) $|x| < 5$

c) $|x| < 6$

d) $|x - 3| < 4$

e) $|x + 2| < 3$

71. See the answer section in the text.

72. $\{x \mid L - \varepsilon < f(x) < L + \varepsilon\}$, or $(L - \varepsilon, L + \varepsilon)$

73.-82. See Exercises 55-64.

Exercise Set 4.4

1. $(x + 5)(x - 3) > 0$

The solutions of $f(x) = (x + 5)(x - 3) = 0$ are -5 and 3. They divide the real-number line into three intervals as shown:

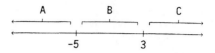

We try a test number in each interval.
A: Test -6, $f(-6) = (-6 + 5)(-6 - 3) = 9$
B: Test 0, $f(0) = (0 + 5)(0 - 3) = -15$
C: Test 4, $f(4) = (4 + 5)(4 - 3) = 9$
Function values are positive in intervals A and C. The solution set is $(-\infty, -5) \cup (3, \infty)$.

2. $(-\infty, -4) \cup (1, \infty)$

3. $(x - 1)(x + 2) \leqslant 0$

The solutions of $f(x) = (x - 1)(x + 2) = 0$ are 1 and -2. They divide the real-number line into three intervals as shown:

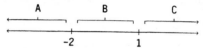

We try a test number in each interval.
A: Test -3, $f(-3) = (-3 - 1)(-3 + 2) = 4$;
B: Test 0, $f(0) = (0 - 1)(0 + 2) = -2$;
C: Test 3, $f(3) = (3 - 1)(3 + 2) = 10$

Function values are negative in interval B. The inequality symbol is $\leqslant$, so we need to include the intercepts. The solution set is $[-2,1]$.

4. $[-5, 3]$

5. $x^2 + x - 2 < 0$
$(x - 1)(x + 2) < 0$ Factoring

See the diagram and test numbers in Exercise 3. The solution set is $\{x \mid -2 < x < 1\}$, or $(-2,1)$.

6. $(-1,2)$

7. $x^2 \geqslant 1$, or $x^2 - 1 \geqslant 0$
$(x + 1)(x - 1) \geqslant 0$ Factoring

The solutions of $f(x) = (x + 1)(x - 1) = 0$ are -1 and 1. They divide the real-number line into three intervals as shown:

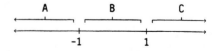

We try a test number in each interval.

A: Test -2, f(-2) = (-2 + 1)(-2 - 1) = 3
B: Test 0, f(0) = (0 + 1)(0 - 1) = -1
C: Test 2, f(2) = (2 + 1)(2 - 1) = 3

Function values are positive in intervals A and C. The inequality symbol is $\geqslant$, so we need to include the intercepts. The solution set is $(-\infty, -1] \cup [1, \infty)$.

8. $(-5, 5)$

9. $9 - x^2 \leqslant 0$

   $(3 + x)(3 - x) \leqslant 0$ Factoring

The solutions of $f(x) = (3 + x)(3 - x) = 0$ are -3 and 3. They divide the real-number line into three intervals as shown:

```
      A         B         C
  <----+----+----+----+----+---->
      -3        3
```

We try a test number in each interval.
A: Test -4, f(-4) = [3 + (-4)][3 - (-4)] = -7
B: Test 0, f(0) = (3 + 0)(3 - 0) = 9
C: Test 4, f(4) = (3 + 4)(3 - 4) = -7

Function values are negative in intervals A and C. The inequality symbol is $\leqslant$, so we need to include the intercepts. The solution set is $(-\infty, -3] \cup [3, \infty)$.

10. $[-2, 2]$

11. $x^2 - 2x + 1 \geqslant 0$

   $(x - 1)^2 \geqslant 0$

The solution of $(x - 1)^2 = 0$ is 1. For all real-number values of x except 1, $(x - 1)^2$ will be positive. Thus the solution set is $\{x | x$ is a real number$\}$, or $(-\infty, \infty)$.

12. $\emptyset$

13. $x^2 + 8 < 6x$
        $x^2 - 6x + 8 < 0$
   $(x - 4)(x - 2) < 0$

The solutions of $f(x) = (x - 4)(x - 2) = 0$ are 4 and 2. They divide the real-number line into three intervals as shown:

```
      A         B         C
  <----+----+----+----+----+---->
      2         4
```

We try a test number in each interval.
A: Test 0, f(0) = (0 - 4)(0 - 2) = 8
B: Test 3, f(3) = (3 - 4)(3 - 2) = -1
C: Test 5, f(5) = (5 - 4)(5 - 2) = 3

Function values are negative in interval B. The solution set is $(2, 4)$.

14. $(-\infty, -2) \cup (6, \infty)$

15. $4x^2 + 7x < 15$
        $4x^2 + 7x - 15 < 0$
   $(4x - 5)(x + 3) < 0$

The solutions of $f(x) = (4x - 5)(x + 3) = 0$ are $\frac{5}{4}$ and -3. They divide the real-number line into three intervals as shown:

```
      A         B         C
  <----+----+----+----+----+---->
      -3        5
                4
```

We try a test number in each interval.
A: Test -4, f(-4) = [4(-4) - 5][-4 + 3] = 21
B: Test 0, f(0) = (4·0 - 5)(0 + 3) = -15
C: Test 2, f(2) = (4·2 - 5)(2 + 3) = 15

Function values are negative in interval B. The solution set is $\left[-3, \frac{5}{4}\right]$.

16. $(-\infty, -3] \cup \left[\frac{5}{4}, \infty\right]$

17. $2x^2 + x > 5$
        $2x^2 + x - 5 > 0$

The solutions of $f(x) = 2x^2 + x - 5 = 0$ are $\frac{-1 \pm \sqrt{41}}{4}$. They divide the real-number line into three intervals as shown:

```
      A         B         C
  <----+----+----+----+----+---->
    -1 - √41    -1 + √41
       4           4
```

We try a test number in each interval. Note that $\frac{-1 - \sqrt{41}}{4} \approx -1.9$ and $\frac{-1 + \sqrt{41}}{4} \approx 1.4$.

A: Test -4, f(-4) = 2(-4)² + (-4) - 5 = 23
B: Test 0, f(0) = 2(0)² + 0 - 5 = -5
C: Test 3, f(3) = 2(3)² + 3 - 5 = 16

Function values are positive in intervals A and C. The solution set is $\left\{x | x < \frac{-1 - \sqrt{41}}{4} \text{ or } x > \frac{-1 + \sqrt{41}}{4}\right\}$, or $\left[-\infty, \frac{-1 - \sqrt{41}}{4}\right] \cup \left[\frac{-1 + \sqrt{41}}{4}, \infty\right]$.

18. $\left[\frac{-1 - \sqrt{17}}{4}, \frac{-1 + \sqrt{17}}{4}\right]$

19. $3x(x + 2)(x - 2) < 0$

The solutions of $f(x) = 3x(x + 2)(x - 2) = 0$ are 0, -2, and 2. They divide the real-number line into four intervals as shown:

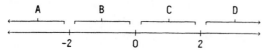

We try a test number in each interval.

A: Test -3, $f(-3) = 3(-3)(-3 + 2)(-3 - 2) = -45$
B: Test -1, $f(-1) = 3(-1)(-1 + 2)(-1 - 2) = 9$
C: Test 1, $f(1) = 3(1)(1 + 2)(1 - 2) = -9$
D: Test 3, $f(3) = 3(3)(3 + 2)(3 - 2) = 45$

Function values are negative in intervals A and C. The solution set is $(-\infty, -2) \cup (0, 2)$.

20. $(-1, 0) \cup (1, \infty)$

21. $(x + 3)(x - 2)(x + 1) > 0$

The solutions of $f(x) = (x + 3)(x - 2)(x + 1) = 0$ are -3, 2, and -1. They divide the real-number line as shown:

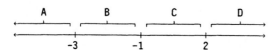

We try a test number in each interval.

A: Test -4, $f(-4) = (-4 + 3)(-4 - 2)(-4 + 1) = -18$
B: Test -2, $f(-2) = (-2 + 3)(-2 - 2)(-2 + 1) = 4$
C: Test 0, $f(0) = (0 + 3)(0 - 2)(0 + 1) = -6$
D: Test 3, $f(3) = (3 + 3)(3 - 2)(3 + 1) = 24$

Function values are positive in intervals B and D. The solution set is $(-3, -1) \cup (2, \infty)$.

22. $(-\infty, -2) \cup (1, 4)$

23. $(x + 3)(x + 2)(x - 1) < 0$

The solutions of $f(x) = (x + 3)(x + 2)(x - 1) = 0$ are -3, -2, and 1. They divide the real-number line into four intervals as shown:

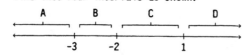

We try a test number in each interval.

A: Test -4, $f(-4) = (-4 + 3)(-4 + 2)(-4 - 1) = -10$

B: Test $-\frac{5}{2}$, $f\left(-\frac{5}{2}\right) =$
$$\left(-\frac{5}{2} + 3\right)\left(-\frac{5}{2} + 2\right)\left(-\frac{5}{2} - 1\right) = \frac{7}{8}$$

C: Test 0, $f(0) = (0 + 3)(0 + 2)(0 - 1) = -6$
D: Test 2, $f(2) = (2 + 3)(2 + 2)(2 - 1) = 20$

Function values are negative in intervals A and C. The solution set is $\{x \mid x < -3 \text{ or } -2 < x < 1\}$,

or $(-\infty, -3) \cup (-2, 1)$.

24. $(-\infty, -1) \cup (2, 3)$

25. $\frac{1}{4 - x} < 0$

We write the related equation by changing the $<$ symbol to $=$:

$$\frac{1}{4 - x} = 0$$

Then we solve the related equation:

$$(4 - x) \cdot \frac{1}{4 - x} = (4 - x) \cdot 0$$
$$1 = 0 \quad \text{False equation}$$

The related equation has no solution. Next we find the replacements that make the denominator 0:

$$4 - x = 0$$
$$4 = x$$

The number 4 divides the number line as shown:

We try test numbers in each interval:

A: Test 0, $\dfrac{1}{4 - x} < 0$

$$\dfrac{1}{4 - 0} \quad\bigg|\quad 0$$
$$\dfrac{1}{4} \quad\bigg|\quad \text{FALSE}$$

The number 0 is not a solution of the inequality, so the interval A is not part of the solution set.

B: Test 5, $\dfrac{1}{4 - x} < 0$

$$\dfrac{1}{4 - 5} \quad\bigg|\quad 0$$
$$-1 \quad\bigg|\quad \text{TRUE}$$

The number 5 is a solution of the inequality, so the interval B is part of the solution set.

The solution set is $(4, \infty)$.

26. $\left(-\infty, -\frac{5}{2}\right]$

27. $3 < \frac{1}{x}$

Solve the related equation.

$$3 = \frac{1}{x}$$
$$x = \frac{1}{3}$$

Find the replacements that are not meaningful.

$$x = 0$$

Use the numbers $\frac{1}{3}$ and 0 to divide the number line into intervals as shown.

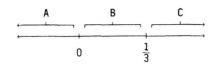

Try test numbers in each interval.

A: Test -1, $3 < \frac{1}{x}$

$$3 \quad \Big| \quad \frac{1}{-1}$$

$$\quad \Big| \quad -1 \quad \text{FALSE}$$

The number -1 is not a solution of the inequality, so the interval A is not in the solution set.

B: Test $\frac{1}{4}$, $3 < \frac{1}{x}$

$$3 \quad \Big| \quad \frac{1}{\frac{1}{4}}$$

$$\quad \Big| \quad 4 \quad \text{TRUE}$$

The number $\frac{1}{4}$ is a solution of the inequality, so the interval B is in the solution set.

C: Test 1, $3 < \frac{1}{x}$

$$3 \quad \Big| \quad \frac{1}{1}$$

$$\quad \Big| \quad 1 \quad \text{FALSE}$$

The number 1 is not a solution of the inequality, so the interval C is not in the solution set.

The solution set is $\left[0, \frac{1}{3}\right]$.

28. $(-\infty, 0) \cup \left[\frac{1}{5}, \infty\right]$

29. $\frac{3x + 2}{x - 3} > 0$

Solve the related equation.

$$\frac{3x + 2}{x - 3} = 0$$
$$3x + 2 = 0$$
$$3x = -2$$
$$x = -\frac{2}{3}$$

Find replacements that are not meaningful.

$$x - 3 = 0$$
$$x = 3$$

Use the numbers $-\frac{2}{3}$ and 3 to divide the number line into intervals as shown.

```
     A        B        C
  ←——————+————————+——————→
        -2/3      3
```

Try test numbers in each interval.

A: Test -1, $\frac{3x + 2}{x - 3} > 0$

$$\frac{3(-1) + 2}{-1 - 3} \quad \Big| \quad 0$$

$$\frac{-1}{-4} \quad \Big|$$

$$\cdot \frac{1}{4} \quad \Big| \quad \text{TRUE}$$

The number -1 is a solution of the inequality, so the interval A is part of the solution set.

B: Test 0, $\frac{3x + 2}{x - 3} > 0$

$$\frac{3 \cdot 0 + 2}{0 - 3} \quad \Big| \quad 0$$

$$\frac{2}{-3} \quad \Big|$$

$$- \frac{2}{3} \quad \Big| \quad \text{FALSE}$$

The number 0 is not a solution of the inequality, so the interval B is not part of the solution set.

C: Test 4, $\frac{3x + 2}{x - 3} > 0$

$$\frac{3 \cdot 4 + 2}{4 - 3} \quad \Big| \quad 0$$

$$14 \quad \Big| \quad \text{TRUE}$$

The number 4 is a solution of the inequality, so the interval C is part of the solution set.

The solution set is $\left(-\infty, -\frac{2}{3}\right] \cup (3, \infty)$.

30. $\left(-\infty, -\frac{3}{4}\right] \cup \left[\frac{5}{2}, \infty\right]$

31. $\frac{x + 2}{x} \leqslant 0$

Solve the related equation.

$$\frac{x + 2}{x} = 0$$
$$x + 2 = 0$$
$$x = -2$$

Find the replacements that are not meaningful.

$$x = 0$$

The numbers -2 and 0 divide the number line into intervals as shown.

```
     A        B        C
  ←——————+————————+——————→
        -2        0
```

Try test numbers in each interval.

A: Test -3, $\frac{x + 2}{x} \leqslant 0$

$$\frac{-3 + 2}{-3} \quad \Big| \quad 0$$

$$\frac{1}{3} \quad \Big| \quad \text{FALSE}$$

The number -3 is not a solution of the inequality, so the interval A is not in the solution set.

B: Test -1, $\dfrac{x + 2}{x} \leqslant 0$

$$\dfrac{\dfrac{-1 + 2}{-1}}{} \Bigg| \begin{array}{l} 0 \\ \\ \end{array}$$

$-1 \Big|$ TRUE

The number -1 is a solution of the inequality, so the interval B is in the solution set.

C: Test 1, $\dfrac{x + 2}{x} \leqslant 0$

$$\dfrac{1 + 2}{1} \Bigg| 0$$

$3 \Big|$ FALSE

The number 1 is not a solution of the inequality, so the interval C is not in the solution set.

The solution set includes the interval B. The number -2 is also included since the inequality symbol is $\leqslant$ and -2 is a solution of the related equation. The number 0 is not included, since it is not a meaningful replacement. The solution set is [-2, 0).

32. $(-\infty, 0] \cup (3, \infty)$

33. $\dfrac{x + 1}{2x - 3} \geqslant 1$

Solve the related equation.

$$\dfrac{x + 1}{2x - 3} = 1$$
$$x + 1 = 2x - 3$$
$$4 = x$$

Find replacements that are not meaningful.

$$2x - 3 = 0$$
$$x = \dfrac{3}{2}$$

Use the numbers 4 and $\dfrac{3}{2}$ to divide the number line as shown.

Try test numbers in each interval.

A: Test 0, $\dfrac{x + 1}{2x - 3} \geqslant 1$

$$\dfrac{0 + 1}{2 \cdot 0 - 3} \Bigg| 1$$

$-\dfrac{1}{3} \Big|$ FALSE

The interval A is not in the solution set.

B: Test 2, $\dfrac{x + 1}{2x - 3} \geqslant 1$

$$\dfrac{2 + 1}{2 \cdot 2 - 3} \Bigg| 1$$

$3 \Big|$ TRUE

The interval B is in the solution set.

C: Test 5, $\dfrac{x + 1}{2x - 3} \geqslant 1$

$$\dfrac{5 + 1}{2 \cdot 5 - 3} \Bigg| 1$$

$\dfrac{6}{7} \Big|$ FALSE

The interval C is not in the solution set. The solution set includes the interval B. The number 4 is also included, since the inequality is $\geqslant$ and 4 is a solution of the related equation. The number $\dfrac{3}{2}$ is not included, since it is not a meaningful replacement. The solution set is $\left[\dfrac{3}{2}, 4\right]$.

34. $\left[2, \dfrac{5}{2}\right]$

35. $\dfrac{x + 1}{x + 2} \leqslant 3$

Solve the related equation.

$$\dfrac{x + 1}{x + 2} = 3$$
$$x + 1 = 3(x + 2)$$
$$x + 1 = 3x + 6$$
$$-5 = 2x$$
$$-\dfrac{5}{2} = x$$

Find the replacements that are not meaningful.

$$x + 2 = 0$$
$$x = -2$$

Use the numbers $-\dfrac{5}{2}$ and -2 to divide the number line as shown.

Try test numbers in each interval.

A: Test -3, $\dfrac{x + 1}{x + 2} \leqslant 3$

$$\dfrac{-3 + 1}{-3 + 2} \Bigg| 3$$

$2 \Big|$ TRUE

The interval A is in the solution set.

B: Test $-\frac{9}{4}$, $\dfrac{x + 1}{x + 2} \leqslant 3$

$$\dfrac{-\frac{9}{4} + 1}{-\frac{9}{4} + 2} \quad \Big| \quad 3$$

$$\dfrac{-\frac{5}{4}}{-\frac{1}{4}}$$

$$5 \quad \Big| \quad \text{FALSE}$$

The interval B is not in the solution set.

C: Test 0, $\dfrac{x + 1}{x + 2} \leqslant 3$

$$\dfrac{0 + 1}{0 + 2} \quad \Big| \quad 3$$

$$\frac{1}{2} \quad \Big| \quad \text{TRUE}$$

The interval C is in the solution set.

The number $-\frac{5}{2}$ is in the solution set since the inequality is $\leqslant$, but -2 is not included since it is not a meaningful replacement. The solution set is $\left[-\infty, -\frac{5}{2}\right] \cup (-2, \infty)$.

<u>36.</u> $\left[-\infty, \frac{3}{2}\right] \cup [4, \infty)$

<u>37.</u> $\dfrac{x - 6}{x} > 1$

Solve the related equation.

$$\dfrac{x - 6}{x} = 1$$

$$x - 6 = x$$

$$-6 = 0$$

The related equation has no solution. We see that 0 is not a meaningful replacement. Use the number 0 to divide the number line as shown.

Try test numbers in each interval.

A: Test -1, $\dfrac{x - 6}{x} > 1$

$$\dfrac{-1 - 6}{-1} \quad \Big| \quad 1$$

$$7 \quad \Big| \quad \text{TRUE}$$

The interval A is in the solution set.

B: Test 1, $\dfrac{x - 6}{x} > 1$

$$\dfrac{1 - 6}{1} \quad \Big| \quad 1$$

$$-5 \quad \Big| \quad \text{FALSE}$$

The interval B is not in the solution set.
The solution set is $(-\infty, 0)$.

<u>38.</u> $(-\infty, -3) \cup \left[-\frac{3}{2}, \infty\right]$

<u>39.</u> $(x + 1)(x - 2) > (x + 3)^2$
Solve the related equation.

$$(x + 1)(x - 2) = (x + 3)^2$$

$$x^2 - x - 2 = x^2 + 6x + 9$$

$$-11 = 7x$$

$$-\frac{11}{7} = x$$

All real numbers are meaningful replacements. Use the number $-\frac{11}{7}$ to divide the number line as shown.

Try test numbers in each interval.

A: Test -2, $(x + 1)(x - 2) > (x + 3)^2$

$$(-2 + 1)(-2 - 2) \quad \Big| \quad (-2 + 3)^2$$

$$4 \quad \Big| \quad 1 \qquad \text{TRUE}$$

The interval A is in the solution set.

B: Test 0, $(x + 1)(x - 2) > (x + 3)^2$

$$(0 + 1)(0 - 2) \quad \Big| \quad (0 + 3)^2$$

$$-2 \quad \Big| \quad 9 \qquad \text{FALSE}$$

The interval B is not in the solution set.
The solution set is $\left[-\infty, -\frac{11}{7}\right]$.

<u>40.</u> $(13, \infty)$

<u>41.</u> $x^3 - x^2 > 0$
$x^2(x - 1) > 0$
The solutions of $f(x) = x^2(x - 1) = 0$ are 0 and 1. They divide the number line into intervals as shown.

Try a test number in each interval.

A: Test -1, $f(-1) = (-1)^3 - (-1)^2 = -2$

B: Test $\frac{1}{2}$, $f\left(\frac{1}{2}\right) = \left(\frac{1}{2}\right)^3 - \left(\frac{1}{2}\right)^2 = -\frac{1}{8}$

C: Test 2, $f(2) = 2^3 - 2^2 = 4$

Function values are positive in interval C. The solution set is $(1, \infty)$.

<u>42.</u> $(-2, 0) \cup (2, \infty)$

43. $x + \frac{4}{x} > 4$

Solve the related equation.

$$x + \frac{4}{x} = 4$$
$$x^2 + 4 = 4x$$
$$x^2 - 4x + 4 = 0$$
$$(x - 2)(x - 2) = 0$$
$$x = 2$$

We see that 0 is not a meaningful replacement. Use the numbers 2 and 0 to divide the number line as shown.

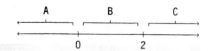

Try test numbers in each interval.

A: Test -1,

$$\begin{array}{c|c} x + \frac{4}{x} > 4 & \\ \hline -1 + \frac{4}{-1} & 4 \\ -5 & \text{FALSE} \end{array}$$

The interval A is not in the solution set.

B: Test 1,

$$\begin{array}{c|c} x + \frac{4}{x} > 4 & \\ \hline 1 + \frac{4}{1} & 4 \\ 5 & \text{TRUE} \end{array}$$

The interval B is in the solution set.

C: Test 4,

$$\begin{array}{c|c} x + \frac{4}{x} > 4 & \\ \hline 4 + \frac{4}{4} & 4 \\ 5 & \text{TRUE} \end{array}$$

The interval C is in the solution set. The solution set is $(0, 2) \cup (2, \infty)$.

44. $(0, 1]$

45. $\frac{1}{x^3} \leqslant \frac{1}{x^2}$

Solve the related equation.

$$\frac{1}{x^3} = \frac{1}{x^2}$$
$$x^3 \cdot \frac{1}{x^3} = x^3 \cdot \frac{1}{x^2}$$
$$1 = x$$

We see that 0 is not a meaningful replacement. Use the numbers 1 and 0 to divide the number line as shown.

Try test numbers in each interval.

A: Test -1,

$$\begin{array}{c|c} \frac{1}{x^3} \leqslant \frac{1}{x^2} & \\ \hline \frac{1}{(-1)^3} & \frac{1}{(-1)^2} \\ -1 & 1 \qquad \text{TRUE} \end{array}$$

The interval A is in the solution set.

B: Test $\frac{1}{2}$,

$$\begin{array}{c|c} \frac{1}{x^3} \leqslant \frac{1}{x^2} & \\ \hline \frac{1}{\left(\frac{1}{2}\right)^3} & \frac{1}{\left(\frac{1}{2}\right)^2} \\ 8 & 4 \qquad \text{FALSE} \end{array}$$

The interval B is not in the solution set.

C: Test 2,

$$\begin{array}{c|c} \frac{1}{x^3} \leqslant \frac{1}{x^2} & \\ \hline \frac{1}{2^3} & \frac{1}{2^2} \\ \frac{1}{8} & \frac{1}{4} \qquad \text{TRUE} \end{array}$$

The interval C is in the solution set. The number 1 is also in the solution set since the inequality is $\leqslant$, but 0 is not included since it is not a meaningful replacement. The solution set is $(-\infty, 0) \cup [1, \infty)$.

46. $(0, 1) \cup (1, \infty)$

47. $\frac{2 + x - x^2}{x^2 + 5x + 6} < 0$

Solve the related equation.

$$\frac{2 + x - x^2}{x^2 + 5x + 6} = 0$$
$$2 + x - x^2 = 0$$
$$(2 - x)(1 + x) = 0$$
$$x = 2 \quad \text{or} \quad x = -1$$

Find the replacements that are not meaningful.

$$x^2 + 5x + 6 = 0$$
$$(x + 3)(x + 2) = 0$$
$$x = -3 \quad \text{or} \quad x = -2$$

Use the numbers -3, -2, -1, and 2 to divide the number line into intervals as shown.

Try test numbers in each interval.

A: Test -4,

$$\begin{array}{c|c} \frac{2 + x - x^2}{x^2 + 5x + 6} < 0 & \\ \hline \frac{2 + (-4) - (-4)^2}{(-4)^2 + 5(-4) + 6} & 0 \\ \frac{-18}{2} & \\ -9 & \text{TRUE} \end{array}$$

The interval A is in the solution set.

B: Test $-\dfrac{5}{2}$,

$$\dfrac{2 + x - x^2}{x^2 + 5x + 6} < 0$$

$$\dfrac{2 + \left(-\dfrac{5}{2}\right) - \left(-\dfrac{5}{2}\right)^2}{\left(-\dfrac{5}{2}\right)^2 + 5\left(-\dfrac{5}{2}\right) + 6} \quad \Bigg| \quad 0$$

$$-\dfrac{27}{4}$$

$$-\dfrac{1}{4}$$

$$27 \quad \Big| \quad \text{FALSE}$$

The interval B is not in the solution set.

C: Test $-\dfrac{3}{2}$,

$$\dfrac{2 + x - x^2}{x^2 + 5x + 6} < 0$$

$$\dfrac{2 + \left(-\dfrac{3}{2}\right) - \left(-\dfrac{3}{2}\right)^2}{\left(-\dfrac{3}{2}\right)^2 + 5\left(-\dfrac{3}{2}\right) + 6} \quad \Bigg| \quad 0$$

$$-\dfrac{7}{4}$$

$$\dfrac{3}{4}$$

$$-\dfrac{7}{3} \quad \Big| \quad \text{TRUE}$$

The interval C is in the solution set.

D: Test 0,

$$\dfrac{2 + x - x^2}{x^2 + 5x + 6} < 0$$

$$\dfrac{2 + 0 - 0^2}{0^2 + 5\cdot 0 + 6} \quad \Bigg| \quad 0$$

$$\dfrac{2}{6}$$

$$\dfrac{1}{3} \quad \Big| \quad \text{FALSE}$$

The interval D is not in the solution set.

E: Test 3,

$$\dfrac{2 + x - x^2}{x^2 + 5x + 6} < 0$$

$$\dfrac{2 + 3 - 3^2}{3^2 + 5\cdot 3 + 6} \quad \Bigg| \quad 0$$

$$-\dfrac{4}{30}$$

$$-\dfrac{2}{15} \quad \Big| \quad \text{TRUE}$$

The interval E is in the solution set.
The solution set is $(-\infty, -3) \cup (-2, -1) \cup (2, \infty)$.

48. $(-2, 0) \cup (0, 2)$

49. $x^4 - 2x^2 \leqslant 0$
$x^2(x^2 - 2) \leqslant 0$

The solutions of $f(x) = x^2(x^2 - 2) = 0$ are 0, $-\sqrt{2}$, and $\sqrt{2}$. They divide the number line as shown.

```
       A         B         C         D
  <----+---------+---------+---------+---->
      -√2        0        √2
```

Try a test number in each interval. Note that $-\sqrt{2} \approx -1.4$ and $\sqrt{2} \approx 1.4$.

A: Test -2, $f(-2) = (-2)^4 - 2(-2)^2 = 8$
B: Test -1, $f(-1) = (-1)^4 - 2(-1)^2 = -1$
C: Test 1, $f(1) = 1^4 - 2(1)^2 = -1$
D: Test 2, $f(2) = 2^4 - 2(2)^2 = 8$

The solution set is $[-\sqrt{2}, \sqrt{2}]$.

50. $(-\infty, -\sqrt{3}) \cup (\sqrt{3}, \infty)$

51. $\left|\dfrac{x + 3}{x - 4}\right| < 2$

$-2 < \dfrac{x + 3}{x - 4} < 2$

$-2 < \dfrac{x + 3}{x - 4}$ and $\dfrac{x + 3}{x - 4} < 2$

First solve $-2 < \dfrac{x + 3}{x - 4}$.

We solve the related equation.

$$-2 = \dfrac{x + 3}{x - 4}$$
$$-2(x - 4) = x + 3$$
$$-2x + 8 = x + 3$$
$$5 = 3x$$
$$\dfrac{5}{3} = x$$

Find the replacements that are not meaningful.
$$x - 4 = 0$$
$$x = 4$$

Use the number $\dfrac{5}{3}$ and 4 to divide the number line as shown.

```
        A         B         C
  <----+---------+---------+---->
       5/3        4
```

Try test numbers in each interval.

A: Test 0,

$$-2 < \dfrac{x + 3}{x - 4}$$

$$-2 \quad \Bigg| \quad \dfrac{0 + 3}{0 - 4}$$

$$\Bigg| \quad -\dfrac{3}{4} \quad \text{TRUE}$$

The interval A is in the solution set.

B: Test 3,

$$-2 < \dfrac{x + 3}{x - 4}$$

$$-2 \quad \Bigg| \quad \dfrac{3 + 3}{3 - 4}$$

$$\Bigg| \quad -6 \quad \text{FALSE}$$

The interval B is not in the solution set.

C: Test 5, $-2 < \dfrac{x + 3}{x - 4}$

$$-2 \ \bigg|\ \dfrac{5 + 3}{5 - 4}$$

$$\bigg|\ 8 \qquad \text{TRUE}$$

The interval C is in the solution set.
The solution set for this portion of the
conjunction is $\left(-\infty, \dfrac{5}{3}\right] \cup (4, \infty)$.

Next solve $\dfrac{x + 3}{x - 4} < 2$.

We solve the related equation.

$$\dfrac{x + 3}{x - 4} = 2$$

$$x + 3 = 2(x - 4)$$

$$x + 3 = 2x - 8$$

$$11 = x$$

As before, 4 is not a meaningful replacement.
Use the numbers 4 and 11 to divide the number
line as shown.

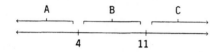

Try test numbers in each interval.

A: Test 0, $\dfrac{x + 3}{x - 4} < 2$

$$\dfrac{0 + 3}{0 - 4} \ \bigg|\ 2$$

$$-\dfrac{3}{4} \ \bigg|\ \text{TRUE}$$

The interval A is in the solution set.

B: Test 5, $\dfrac{x + 3}{x - 4} < 2$

$$\dfrac{5 + 3}{5 - 4} \ \bigg|\ 2$$

$$8 \ \bigg|\ \text{FALSE}$$

The interval B is not in the solution set.

C: Test 12, $\dfrac{x + 3}{x - 4} < 2$

$$\dfrac{12 + 3}{12 - 4} \ \bigg|\ 2$$

$$\dfrac{15}{8} \ \bigg|\ \text{TRUE}$$

The interval C is in the solution set. The
solution set for this portion of the conjunction
is $(-\infty, 4) \cup (11, \infty)$.

The solution set of the original inequality is
the intersection of the solution sets for the
two portions of the conjunction.

$$\left[\left(-\infty, \dfrac{5}{3}\right] \cup (4, \infty)\right] \cap [(-\infty, 4) \cup (11, \infty)] =$$

$$\left(-\infty, \dfrac{5}{3}\right] \cup (11, \infty)$$

<u>52.</u> $[-\sqrt{5}, \sqrt{5}]$

<u>53.</u> $(7 - x)^{-2} < 0$

$$\dfrac{1}{(7 - x)^2} < 0$$

Since $(7 - x)^2 \geqslant 0$ for all values of x,
$\dfrac{1}{(7 - x)^2} > 0$ for all meaningful replacements.
The solution set is $\varnothing$.

<u>54.</u> $(-\infty, 1)$

<u>55.</u> $\left|1 + \dfrac{1}{x}\right| < 3$

$$-3 < 1 + \dfrac{1}{x} < 3$$

$$-3 < 1 + \dfrac{1}{x} \text{ and } 1 + \dfrac{1}{x} < 3$$

First solve $-3 < 1 + \dfrac{1}{x}$

We solve the related equation.

$$-3 = 1 + \dfrac{1}{x}$$

$$-4 = \dfrac{1}{x}$$

$$-4x = 1$$

$$x = -\dfrac{1}{4}$$

We see that 0 is not a meaningful replacement.
Use the numbers $-\dfrac{1}{4}$ and 0 to divide the number
line as shown.

Try test numbers in each interval.

A: Test -1, $-3 < 1 + \dfrac{1}{x}$

$$-3 \ \bigg|\ 1 + \dfrac{1}{-1}$$

$$\bigg|\ 0 \qquad \text{TRUE}$$

The interval A is in the solution set.

B: Test $-\dfrac{1}{8}$, $-3 < 1 + \dfrac{1}{x}$

$$-3 \ \bigg|\ 1 + \dfrac{1}{-\dfrac{1}{8}}$$

$$\bigg|\ 1 - 8$$

$$\bigg|\ -7 \qquad \text{FALSE}$$

The interval B is not in the solution set.

C: Test 1, $-3 < 1 + \frac{1}{x}$

$$\begin{array}{c|c} -3 & 1 + \frac{1}{1} \\ \hline & 2 \qquad \text{TRUE} \end{array}$$

The interval C is in the solution set.

The solution set for this portion of the conjunction is $\left[-\infty, -\frac{1}{4}\right] \cup (0, \infty)$.

Next we solve $1 + \frac{1}{x} < 3$.

We solve the related equation.

$$1 + \frac{1}{x} = 3$$

$$\frac{1}{x} = 2$$

$$1 = 2x$$

$$\frac{1}{2} = x$$

Again, 0 is not a meaningful replacement. We use the numbers $\frac{1}{2}$ and 0 to divide the number line as shown.

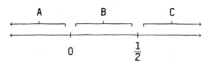

Try test numbers in each interval.

A: Test -1, $1 + \frac{1}{x} < 3$

$$\begin{array}{c|c} 1 + \frac{1}{-1} & 3 \\ \hline 0 & \text{TRUE} \end{array}$$

The interval A is in the solution set.

B: Test $\frac{1}{4}$, $1 + \frac{1}{x} < 3$

$$\begin{array}{c|c} 1 + \frac{1}{\frac{1}{4}} & 3 \\ 1 + 4 & \\ \hline 5 & \text{FALSE} \end{array}$$

The interval B is not in the solution set.

C: Test 1, $1 + \frac{1}{x} < 3$

$$\begin{array}{c|c} 1 + \frac{1}{1} & 3 \\ \hline 2 & \text{TRUE} \end{array}$$

The interval C is in the solution set.

The solution set for this portion of the conjunction is $(-\infty, 0) \cup \left[\frac{1}{2}, \infty\right]$.

The solution set of the original inequality is the intersection of the solution sets for the two portions of the conjunction.

$$\left[\left[-\infty, -\frac{1}{4}\right] \cup (0, \infty)\right] \cap \left[(-\infty, 0) \cup \left[\frac{1}{2}, \infty\right]\right] =$$

$$\left[-\infty, -\frac{1}{4}\right] \cup \left[\frac{1}{2}, \infty\right]$$

56. $(-\infty, -5) \cup (-5, \infty)$

57.

$$|x|^2 - 4|x| + 4 \geqslant 9$$

$$|x|^2 - 4|x| - 5 \geqslant 0$$

$$(|x| - 5)(|x| + 1) \geqslant 0$$

The solutions of $f(x) = (|x| - 5)(|x| + 1) = 0$ are -5 and 5. They divide the number line as shown.

$$\begin{array}{ccc} A & B & C \\ \hline & -5 & 5 \end{array}$$

Try a test number in each interval.

A: Test -6, $f(-6) = |-6|^2 - 4|-6| - 5 = 7$

B: Test 0, $f(0) = |0|^2 - 4|0| - 5 = -5$

C: Test 6, $f(6) = |6|^2 - 4|6| - 5 = 7$

The solution set is $(-\infty, -5] \cup [5, \infty)$.

58. $\left[-\frac{3}{2}, \frac{3}{2}\right]$

59. $\left|2 - \frac{1}{x}\right| \leqslant 2 + \left|\frac{1}{x}\right|$

Note that $\frac{1}{x}$ is not defined when $x = 0$. Thus $x \neq 0$.

Divide the set of real numbers into three intervals:

$x < 0$

$0 < x < \frac{1}{2}$

$x \geqslant \frac{1}{2}$

Find the solution set of $\left|2 - \frac{1}{x}\right| \leqslant 2 + \left|\frac{1}{x}\right|$ for each interval. Then take the union of the three solution sets.

If $x < 0$, then $\left|2 - \frac{1}{x}\right| = 2 - \frac{1}{x}$ and $\left|\frac{1}{x}\right| = -\frac{1}{x}$.

Solve: $x < 0$ and $2 - \frac{1}{x} \leqslant 2 - \frac{1}{x}$

        $x < 0$ and $2 \leqslant 2$

The solution set for this interval is $\{x | x < 0\}$.

If $0 < x < \frac{1}{2}$, then $\left|2 - \frac{1}{x}\right| = -\left(2 - \frac{1}{x}\right)$ and $\left|\frac{1}{x}\right| = \frac{1}{x}$.

Solve: $0 < x < \frac{1}{2}$ and $-\left(2 - \frac{1}{x}\right) \leqslant 2 + \frac{1}{x}$

        $0 < x < \frac{1}{2}$ and $-2 + \frac{1}{x} \leqslant 2 + \frac{1}{x}$

        $0 < x < \frac{1}{2}$ and $-2 \leqslant 2$

The solution set for this interval is $\left\{x \mid 0 < x < \frac{1}{2}\right\}$.

If $x \geqslant \frac{1}{2}$, then $\left|2 - \frac{1}{x}\right| = 2 - \frac{1}{x}$ and $\left|\frac{1}{x}\right| = \frac{1}{x}$.

Solve: $x \geqslant \frac{1}{2}$ and $2 - \frac{1}{x} \leqslant 2 + \frac{1}{x}$

$x \geqslant \frac{1}{2}$ and $-\frac{1}{x} \leqslant \frac{1}{x}$

$x \geqslant \frac{1}{2}$ and $-1 \leqslant 1$

True for all x such that $x \geqslant \frac{1}{2}$. The solution set for this interval is $\left\{x \mid x \geqslant \frac{1}{2}\right\}$.

The <u>union</u> of the above three solution sets is the set of all real numbers except 0, or $(-\infty, 0) \cup (0, \infty)$.

<u>60</u>. $(-2, -1] \cup [3, 4) \cup [2, 2]$

<u>61</u>. $|x^2 + 3x - 1| < 3$

$-3 < x^2 + 3x - 1 < 3$

$-3 < x^2 + 3x - 1$ and $x^2 + 3x - 1 < 3$, or

$0 < x^2 + 3x + 2$ and $x^2 + 3x - 4 < 0$

First we solve $0 < x^2 + 3x + 2$. We solve the related equation.

$0 = x^2 + 3x + 2$

$0 = (x + 2)(x + 1)$

$x = -2$ or $x = -1$

Use the numbers -2 and -1 to divide the number line as shown.

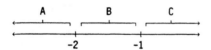

Try test numbers in each interval.

A: Test -3, $f(-3) = (-3)^2 + 3(-3) + 2 = 2$

B: Test $-\frac{3}{2}$, $f\left(-\frac{3}{2}\right) = \left(-\frac{3}{2}\right)^2 + 3\left(-\frac{3}{2}\right) + 2 = -\frac{1}{4}$

C: Test 0, $f(0) = 0^2 + 3\cdot0 + 2 = 2$

The solution set for this portion of the conjunction is $(-\infty, -2) \cup (-1, \infty)$.

Next we solve $x^2 + 3x - 4 < 0$.

Solve the related equation.

$x^2 + 3x - 4 = 0$

$(x + 4)(x - 1) = 0$

$x = -4$ or $x = 1$

Use the numbers -4 and 1 to divide the number line as shown.

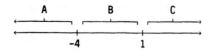

Try test numbers in each interval.

A: Test -5, $f(-5) = (-5)^2 + 3(-5) - 4 = 6$

B: Test 0, $f(0) = 0^2 + 3\cdot0 - 4 = -4$

C: Test 2, $f(2) = 2^2 + 3\cdot2 - 4 = 6$

The solution set for this portion of the conjunction is $(-4, 1)$.

The solution set for the original inequality is the intersection of the solution sets for the two portions of the disjunction.

$[(-\infty, -2) \cup (-1, \infty)] \cap (-4, 1) = (-4, -2) \cup (-1, 1)$

<u>62</u>. $(-\infty, -1] \cup [1, 4] \cup [6, \infty)$

<u>63</u>. Familiarize: We first make a drawing.

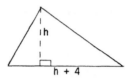

We let h represent the height and $h + 4$ the base.

Translate: The area is $\frac{1}{2}(h + 4)h$. We now have an inequality:

$\frac{1}{2} h(h + 4) > 10$

$h(h + 4) > 20$

$h^2 + 4h > 20$

$h^2 + 4h - 20 > 0$

Carry out: Using the quadratic formula, we find that the solutions of $f(x) = h^2 + 4h - 20$ are $-2 \pm 2\sqrt{6}$. Since h must be positive we use $-2 + 2\sqrt{6}$ to divide the interval $(0, \infty)$ as shown.

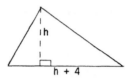

Try a test number in each interval. Note that $-2 + 2\sqrt{6} \approx 2.9$.

A: Test 1, $f(1) = 1^2 + 4\cdot1 - 20 = -15$

B: Test 3, $f(3) = 3^2 + 4\cdot3 - 20 = 1$

Function values are positive in interval B.

Check: We go over our work.

State: The area of the triangle will be greater than 10 cm² for $\{h \mid h > -2 + 2\sqrt{6}$ cm$\}$.

<u>64</u>. $\left\{w \mid w > \frac{-3 + \sqrt{69}}{2} \text{ m}\right\}$

65. a) $-3x^2 + 630x - 6000 > 0$

    $x^2 - 210x + 2000 < 0$ $\left[\text{Multiplying by } -\tfrac{1}{3}\right]$

    $(x - 200)(x - 10) < 0$

The solutions of $f(x) = (x - 200)(x - 10) = 0$ are 200 and 10. Since x must be positive, we use 10 and 200 to divide the interval $(0, \infty)$ as shown.

```
        A         B           C
    |-------|----------|------------->
    0      10         200
```

A: Test 1, $f(1) = (1 - 200)(1 - 10) = 1791$

B: Test 100, $f(100) = (100 - 200)(100 - 10) = -9000$

C: Test 201, $f(201) = (201 - 200)(201 - 10) = 191$

Note that $P(x) > 0$ is equivalent to $f(x) < 0$. Then the solution set is $\{x \mid 10 < x < 200\}$.

b) Note that $P(x) < 0$ for $f(x) > 0$. Then the solution set is $\{x \mid 0 < x < 10 \text{ or } x > 200\}$.

66. a) $\{t \mid 0 \text{ sec} < t < 2 \text{ sec}\}$

b) $\{t \mid t > 10 \text{ sec}\}$

67. a) Solve $R(x) = C(x)$.

    $50x - x^2 = 5x + 350$

    $0 = x^2 - 45x + 350$

    $0 = (x - 35)(x - 10)$

    $x = 35$ or $x = 10$

The break even values are 10 units and 35 units.

b) $R(x) > C(x)$

    $50x - x^2 > 5x + 350$

    $0 > x^2 - 45x + 350$

    $0 > (x - 35)(x - 10)$

The solutions of $f(x) = (x - 35)(x - 10) = 0$ are 35 and 10. We will restrict x to positive values.

```
        A         B         C
    |-------|----------|------------->
    0      10         35
```

A: Test 1, $f(1) = 1^2 - 45 \cdot 1 + 350 = 306$

B: Test 11, $f(11) = 11^2 - 45 \cdot 11 + 350 = -24$

C: Test 36, $f(36) = 36^2 - 45 \cdot 36 + 350 = 26$

Note that $R(x) > C(x)$ corresponds to $f(x) < 0$. There is a profit for $\{x \mid 10 < x < 35\}$.

c) $R(x) < C(x)$

    $50x - x^2 < 5x + 350$

    $0 < x^2 - 45x + 350$

    $0 < (x - 35)(x - 10)$

From part (b) we see that the function values of $f(x) = (x - 35)(x - 10)$ are greater than zero for $\{x \mid 0 < x < 10 \text{ or } x > 35\}$.

68. a) 10, 60

b) $\{x \mid 10 < x < 60\}$

c) $\{x \mid x < 10 \text{ or } x > 60\}$ (Of course, we would expect x to be positive as well.)

69. The discriminant of $x^2 + kx + 1 = 0$ is $k^2 - 4$.

a) There are two real-number solutions when the discriminant is positive, so we solve $k^2 - 4 > 0$, or $(k + 2)(k - 2) > 0$.

The solutions of $f(k) = (k + 2)(k - 2) = 0$ are -2 and 2. They divide the real-number line as shown.

```
        A         B           C
    |-------|----------|------------->
          -2          2
```

A: Test -3, $f(-3) = (-3 + 2)(-3 - 2) = 5$

B: Test 0, $f(0) = (0 + 2)(0 - 2) = -4$

C: Test 3, $f(3) = (3 + 2)(3 - 2) = 5$

The solution set is $\{k \mid k < -2 \text{ or } k > 2\}$, or $(-\infty, -2) \cup (2, \infty)$.

b) There is no real-number solution when the discriminant is negative, so we solve $k^2 - 4 < 0$. Using the test points in part (a), we see that the solution set is $\{k \mid -2 < k < 2\}$, or $(-2, 2)$.

70. a) $\left\{k \mid k < -2\sqrt{2} \text{ or } k > 2\sqrt{2}\right\}$

b) $\left\{k \mid -2\sqrt{2} < k < 2\sqrt{2}\right\}$

71. $f(x) = \sqrt{1 - x^2}$

The radicand must be nonnegative, so we solve $1 - x^2 \geq 0$, or $x^2 - 1 \leq 0$ (multiplying by -1). Using the test points in Exercise 7, we see that the domain is $\{x \mid -1 \leq x \leq 1\}$, or $[-1, 1]$.

72. $\{x \mid -1 < x < 1\}$, or $(-1, 1)$

73. $g(x) = \sqrt{x^2 + 2x - 3}$

The radicand must be nonnegative, so we solve $x^2 + 2x - 3 \geq 0$, or $(x + 3)(x - 1) \geq 0$. The solutions of $f(x) = (x + 3)(x - 1) = 0$ are -3 and 1. They divide the real-number line as shown.

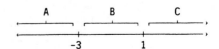

A: Test -4, $f(-4) = (-4)^2 + 2(-4) - 3 = 5$

B: Test 0, $f(0) = 0^2 + 2 \cdot 0 - 3 = -3$

C: Test 2, $f(2) = 2^2 + 2 \cdot 2 - 3 = 5$

The domain is $\{x \mid x \leq -3 \text{ or } x \geq 1\}$, or $(-\infty, -3] \cup [1, \infty)$.

<u>74</u>. {x I -2 ⩽ x ⩽ 2}, or [-2, 2]

75. From the graph we determine the following:

The solutions of f(x) = 0 are -2, 1, and 3.

The solution of f(x) < 0 is
{x I x < -2 or 1 < x < 3}, or (-∞,-2) ∪ (1,3).

The solution of f(x) > 0 is
{x I -2 < x < 1 or x > 3}, or (-2,1) ∪ (3,∞).

<u>76</u>. Roots: -2, 1; f(x) < 0: {x I x < -2}, or (-∞,-2);

f(x) > 0: {x I -2 < x < 1 or x > 1}, or
(-2,1) ∪ (1,∞)

<u>77</u>. From the graph we determine the following:

f(x) has no zeros.

The solutions f(x) < 0 are {x I x < 0}, or
(-∞,0).

The solutions of f(x) > 0 are {x I x > 0}, or
(0,∞).

<u>78</u>. Roots: 0, 1; f(x) < 0: {x I 0 < x < 1}, or (0,1);

f(x) > 0: {x I x > 1}, or (1,∞)

<u>79</u>. From the graph we determine the following:

The solutions of f(x) = 0 are -2, 1, 2, and 3.

The solutions of f(x) < 0 are
{x I -2 < x < 1 or 2 < x < 3}, or (-2,1) ∪ (2,3).

The solutions of f(x) > 0 are
{x I x < -2 or 1 < x < 2 or x > 3}, or
(-∞,-2) ∪ (1,2) ∪ (3,∞).

<u>80</u>. Roots: -2, 0, 1; f(x) < 0:
{x I x < -3 or -2 < x < 0 or 1 < x < 2}, or
(-∞,-3) ∪ (-2,0) ∪ (1,2);

f(x) > 0:
{x I -3 < x < -2 or 0 < x < 1 or x > 2}, or
(-3,-2) ∪ (0,1) ∪ (2,∞)

Exercise Set 4.5

<u>1</u>. $f(x) = \frac{1}{3}x^6$

The graph of $f(x) = \frac{1}{3}x^6$ has the same general shape as $f(x) = x^2$. This is an even function, so it is symmetric with respect to the y-axis. We can compute some function values and use symmetry to find others.

x	0	1	2	3	-1	-2	-3
f(x)	0	$\frac{1}{3}$	$\frac{64}{3}$	243	$\frac{1}{3}$	$\frac{64}{3}$	243

Found using symmetry

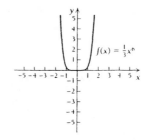
$f(x) = \frac{1}{3}x^6$

2.

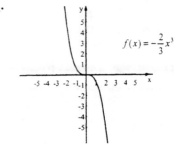
$f(x) = -\frac{2}{3}x^3$

<u>3</u>. $f(x) = -0.6x^5$

The graph of $f(x) = -0.6x^5$ has the same general shape as $f(x) = x^3$ reflected across the x-axis (since -0.6 < 0). It is an odd function, so it is symmetric with respect to the origin. We can compute some function values and use symmetry to find others.

x	0	1	2	-1	-2
f(x)	0	-0.6	-19.2	0.6	19.2

Found using symmetry

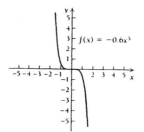
$f(x) = -0.6x^5$

<u>4</u>.

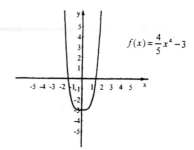
$f(x) = \frac{4}{5}x^4 - 3$

5. f(x) = (x + 1)⁵ - 4

The graph of f(x) = (x + 1)⁵ - 4 is a
transformation of f(x) = x⁵, so its graph will
have the same general shape as f(x) = x³. We
compute some function values.

x	-3	-2	-1	0	1	2
f(x)	-36	-5	-4	-3	28	239

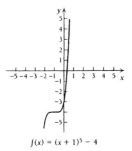

$f(x) = (x + 1)^5 - 4$

6.

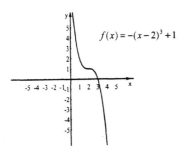

$f(x) = -(x-2)^3 + 1$

7. f(x) = $\frac{1}{4}$ (x + 1)⁴

This function is a transformation of f(x) = x⁴,
so its graph will have the same general shape as
f(x) = x². We compute some function values.

x	-4	-3	-2	-1	0	1	2
f(x)	$\frac{81}{4}$	4	$\frac{1}{4}$	0	$\frac{1}{4}$	4	$\frac{81}{4}$

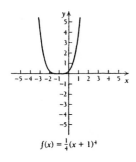

$f(x) = \frac{1}{4}(x + 1)^4$

8.

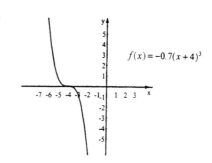

$f(x) = -0.7(x + 4)^3$

9. f(x) = (x + 3)(x - 2)(x + 1)

The zeros of this function are -3, 2, and -1.
They divide the real-number line into 4 open
intervals as indicated in the table below. We
try a test value in each interval and determine
the sign of the function for values of x in the
interval.

Interval	(-∞, -3)	(-3, -1)	(-1, 2)	(2, ∞)
Test value	f(-4) = -18	f(-2) = 4	f(0) = -6	f(3) = 24
Sign of f(x)	-	+	-	+
Location of points on graph	Below x-axis	Above x-axis	Below x-axis	Above x-axis

We use the information in the table, calculate
some additional function values if needed, plot
points, and sketch the graph.

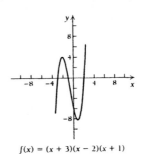

$f(x) = (x + 3)(x - 2)(x + 1)$

10.

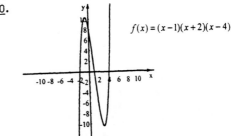

$f(x) = (x - 1)(x + 2)(x - 4)$

11. $f(x) = 9x^2 - x^4$

     $\quad\quad = x^2(9 - x^2)$

     $\quad\quad = x^2(3 + x)(3 - x)$

The zeros of the function are 0, -3, and 3.

Interval	$(-\infty, -3)$	$(-3, 0)$	$(0, 3)$	$(3, \infty)$
Test value	$f(-4) = -112$	$f(-1) = 8$	$f(1) = 8$	$f(4) = -112$
Sign of $f(x)$	-	+	+	-
Location of points on graph	Below x-axis	Above x-axis	Above x-axis	Below x-axis

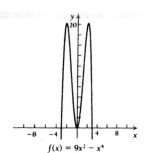

$f(x) = 9x^2 - x^4$

12.

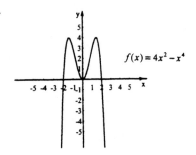

$f(x) = 4x^2 - x^4$

13. $f(x) = x^4 - x^3$

     $\quad\quad = x^3(x - 1)$

The zeros of the function are 0 and 1.

Interval	$(-\infty, 0)$	$(0, 1)$	$(1, \infty)$
Test value	$f(-1) = 2$	$f\left(\frac{1}{2}\right) = -\frac{1}{16}$	$f(2) = 8$
Sign of $f(x)$	+	-	+
Location of points on graph	Above x-axis	Below x-axis	Above x-axis

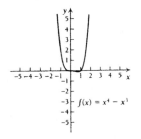

$f(x) = x^4 - x^3$

14.

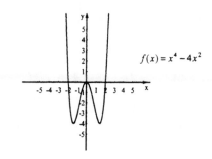

$f(x) = x^4 - 4x^2$

15. $f(x) = x^3 - 4x$

     $\quad\quad = x(x^2 - 4)$

     $\quad\quad = x(x + 2)(x - 2)$

The zeros of the function are 0, -2, and 2.

Interval	$(-\infty, -2)$	$(-2, 0)$	$(0, 2)$	$(2, \infty)$
Test value	$f(-3) = -15$	$f(-1) = 3$	$f(1) = -3$	$f(3) = 15$
Sign of $f(x)$	-	+	-	+
Location of points on graph	Below x-axis	Above x-axis	Below x-axis	Above x-axis

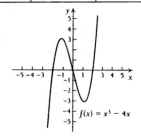

$f(x) = x^3 - 4x$

16.

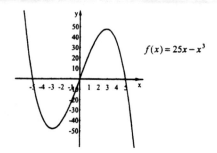

$f(x) = 25x - x^3$

17. $f(x) = x^3 + x^2 - 2x$

$\qquad = x(x^2 + x - 2)$

$\qquad = x(x + 2)(x - 1)$

The zeros of the function are 0, -2, and 1.

Interval	$(-\infty, -2)$	$(-2, 0)$	$(0, 1)$	$(1, \infty)$
Test value	$f(-3) = -12$	$f(-1) = 2$	$f\left(\frac{1}{2}\right) = -\frac{5}{8}$	$f(2) = 8$
Sign of $f(x)$	−	+	−	+
Location of points on graph	Below x-axis	Above x-axis	Below x-axis	Above x-axis

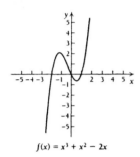

$f(x) = x^3 + x^2 - 2x$

18.

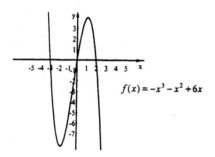

$f(x) = -x^3 - x^2 + 6x$

19. $f(x) = x^4 - 9x^2 + 20$

$\qquad = (x^2 - 4)(x^2 - 5)$

$\qquad = (x + 2)(x - 2)(x + \sqrt{5})(x - \sqrt{5})$

The zeros of the function are -2, 2, $-\sqrt{5}$, and $\sqrt{5}$.

Interval	$(-\infty, -\sqrt{5})$	$(-\sqrt{5}, -2)$	$(-2, 2)$	$(2, \sqrt{5})$	$(\sqrt{5}, \infty)$
Test value	$f(-3) =$ 20	$f(-2.1) =$ -0.2419	$f(0) =$ 20	$f(2.1) =$ -0.2419	$f(3) =$ 20
Sign of $f(x)$	+	−	+	−	+
Location of points on graph	Above x-axis	Below x-axis	Above x-axis	Below x-axis	Above x-axis

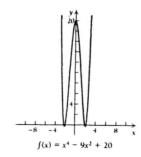

$f(x) = x^4 - 9x^2 + 20$

20.

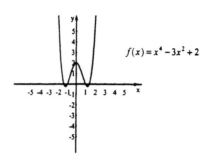

$f(x) = x^4 - 3x^2 + 2$

21. $f(x) = x^3 - 3x^2 - 4x + 12$

$\qquad = x^2(x - 3) - 4(x - 3)$

$\qquad = (x^2 - 4)(x - 3)$

$\qquad = (x + 2)(x - 2)(x - 3)$

The zeros of the function are -2, 2, and 3.

Interval	$(-\infty, -2)$	$(-2, 2)$	$(2, 3)$	$(3, \infty)$
Test value	$f(-3) = -30$	$f(0) = 12$	$f\left(\frac{5}{2}\right) = -\frac{9}{8}$	$f(4) = 12$
Sign of $f(x)$	−	+	−	+
Location of points on graph	Below x-axis	Above x-axis	Below x-axis	Above x-axis

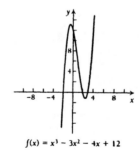

$f(x) = x^3 - 3x^2 - 4x + 12$

22.

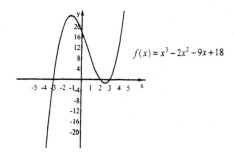

$f(x) = x^3 - 2x^2 - 9x + 18$

23. $f(x) = -x^4 - 3x^3 - 3x^2$

   $= -x^2(x^2 + 3x + 3)$

The discriminant of $x^2 + 3x + 3 = 0$ is negative, so the only zero of the function is 0.

Interval	$(-\infty, 0)$	$(0, \infty)$
Test value	$f(-1) = -1$	$f(1) = -7$
Sign of $f(x)$	–	–
Location of points on graph	Below x-axis	Below x-axis

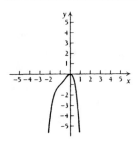

$f(x) = -x^4 - 3x^3 - 3x^2$

24.

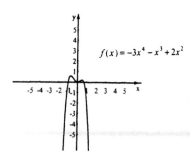

$f(x) = -3x^4 - x^3 + 2x^2$

25. $f(x) = x(x - 2)(x + 1)(x + 3)$

   The zeros of the function are 0, 2, -1, and -3.

Interval	$(-\infty,-3)$	$(-3,-1)$	$(-1,0)$	$(0,2)$	$(2,\infty)$
Test value	$f(-4) =$ 72	$f(-2) =$ -8	$f\left(-\frac{1}{2}\right) =$ $\frac{25}{16}$	$f(1) =$ -8	$f(3) =$ 72
Sign of f(x)	+	–	+	–	+
Location of points on graph	Above x-axis	Below x-axis	Above x-axis	Below x-axis	Above x-axis

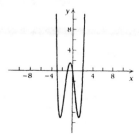

$f(x) = x(x - 2)(x + 1)(x + 3)$

26.

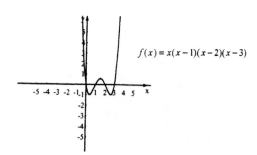

$f(x) = x(x-1)(x-2)(x-3)$

27. a)   1 unit: $y = \frac{1}{13}(1)^3 - \frac{1}{14}(1) \approx 0.0055$

   2 units: $y = \frac{1}{13}(2)^3 - \frac{1}{14}(2) \approx 0.4725$

   3 units: $y = \frac{1}{13}(3)^3 - \frac{1}{14}(3) \approx 1.8626$

   6 units: $y = \frac{1}{13}(6)^3 - \frac{1}{14}(6) \approx 16.1868$

   8 units: $y = \frac{1}{13}(8)^3 - \frac{1}{14}(8) \approx 38.8132$

   9 units: $y = \frac{1}{13}(9)^3 - \frac{1}{14}(9) \approx 55.4341$

   10 units: $y = \frac{1}{13}(10)^3 - \frac{1}{14}(10) \approx 76.2088$

   b)

$y = \frac{1}{13}x^3 - \frac{1}{14}x$

28. a) 4.03, 5.79, 7.00, 7.38, 6.65, 4.54, 0.78

    b) $N(t) = -0.046t^3 + 2.08t + 2$

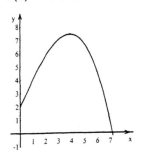

29. a) 5 ft 7 in. = 67 in.

    $W(67) = \left(\dfrac{67}{12.3}\right)^3 \approx 161.6$ lb

    5 ft 10 in. = 70 in.

    $W(70) = \left(\dfrac{70}{12.3}\right)^3 \approx 184.3$ lb

    b) 6 ft 1 in. = 73 in.

    $W(73) = \left(\dfrac{73}{12.3}\right)^3 \approx 209.1$ lb

    He should watch his weight.

30. a) 7.68 watts, 15 watts, 50.625 watts

    b) 20 mph

31. a) We make a drawing.

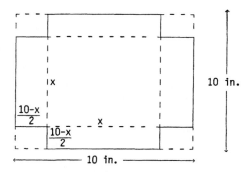

The length of the box is x, the width is x, the height is $\dfrac{10 - x}{2}$. Thus, the volume is given by

$V(x) = x \cdot x \cdot \left(\dfrac{10 - x}{2}\right) = x^2 \left(\dfrac{10 - x}{2}\right)$, or

$V(x) = 5x^2 - \dfrac{x^3}{2}$.

b) The zeros of V(x) are 0 and 10.

Interval	$(-\infty, 0)$	$(0, 10)$	$(10, \infty)$
Test value	$V(-1) = \dfrac{11}{2}$	$V(1) = \dfrac{9}{2}$	$V(11) = -\dfrac{121}{2}$
Sign of f(x)	+	+	−
Location of points on graph	Above x-axis	Above x-axis	Below x-axis

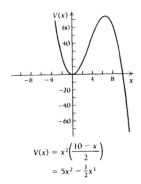

$V(x) = x^2\left(\dfrac{10 - x}{2}\right)$
$= 5x^2 - \dfrac{1}{2}x^3$

c) Using the table in part b), we see that the function is positive over $(-\infty, 0)$, and $(0, 10)$.

32. a) $V(x) = x^2\left(\dfrac{8 - x}{2}\right)$, or $V(x) = 4x^2 - \dfrac{x^3}{2}$

    b)

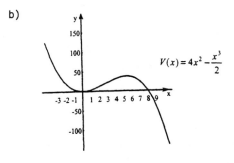

$V(x) = 4x^2 - \dfrac{x^3}{2}$

    c) $(-\infty, 0) \cup (0, 8)$

33. The functions for which f(-x) = f(x) are those in Exercises 1, 4, 11, 12, 14, 19, and 20.

34. Exercise 2, 3, 15 and 16

35. Every exponent of the polynomial function must be even or f(x) = c.

36. Every exponent of the polynomial function must be odd and the constant term must be 0.

37. The zeros of the function are -3, -2, and 1.

38. -3.337

39. The function has zeros at -1 and 2 and at about
     -1.41421 and 1.41421.

40. -2, 0, 3, -1.73215, 1.73215

41. a) $f(x) = -0.046x^3 + 2.08x + 2$

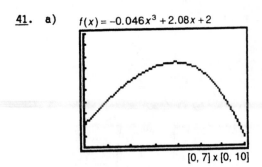

[0, 7] x [0, 10]

   b) The zeros of the function are approximately
       -6.17908, -0.98251, and 7.16159.

   c) The concentration will be 0 after about
       7.2 hr.

Exercise Set 4.6

1. $P(x) = x^3 - 9x^2 + 14x + 24$

    $P(4) = 4^3 - 9 \cdot 4^2 + 14 \cdot 4 + 24 = 0$
    Since $P(4) = 0$, 4 is a root of $P(x)$.

    $P(5) = 5^3 - 9 \cdot 5^2 + 14 \cdot 5 + 24 = -6$
    Since $P(5) \neq 0$, 5 is not a root of $P(x)$.

    $P(-2) = (-2)^3 - 9(-2)^2 + 14(-2) + 24 = -48$
    Since $P(-2) \neq 0$, -2 is not a root of $P(x)$.

2. All no

3. Using the techniques in subsection ⓑ, we divide
    to determine whether each binomial is a factor of
    $P(x)$.

    a)
$$
\begin{array}{r}
x^2 - 5x - 6 \\
x - 4 \overline{\smash{)}\ x^3 - 9x^2 + 14x + 24} \\
\underline{x^3 - 4x^2} \\
-5x^2 + 14x \\
\underline{-5x^2 + 20x} \\
-6x + 24 \\
\underline{-6x + 24} \\
0
\end{array}
$$

    Since the remainder is 0, x - 4 is a factor
    of $P(x)$.

b)
$$
\begin{array}{r}
x^2 - 4x - 6 \\
x - 5 \overline{\smash{)}\ x^3 - 9x^2 + 14x + 24} \\
\underline{x^3 - 5x} \\
-4x^2 + 14x \\
\underline{-4x^2 + 20x} \\
-6x + 24 \\
\underline{-6x + 30} \\
-6
\end{array}
$$

Since the remainder is not 0, x - 5 is not a
factor of $P(x)$.

c)
$$
\begin{array}{r}
x^2 - 11x + 36 \\
x + 2 \overline{\smash{)}\ x^3 - 9x^2 + 14x + 24} \\
\underline{x^3 + 2x^2} \\
-11x^2 + 14x \\
\underline{-11x^2 - 22x} \\
36x + 24 \\
\underline{36x + 72} \\
-48
\end{array}
$$

Since the remainder is not 0, x + 2 is not a
factor of $P(x)$.

4. No; b) No; c) No

5.
$$
\begin{array}{r}
x^2 + 8x + 15 \\
x - 2 \overline{\smash{)}\ x^3 + 6x^2 - x - 30} \\
\underline{x^3 - 2x^2} \\
8x^2 - x \\
\underline{8x^2 - 16x} \\
15x - 30 \\
\underline{15x - 30} \\
0
\end{array}
$$

    $x^3 + 6x^2 - x - 30 = (x - 2)(x^2 + 8x + 15) + 0$

6. $2x^3 - 3x^2 + x - 1 = (x - 2)(2x^2 + x + 3) + 5$

7.
$$
\begin{array}{r}
x^2 + 9x + 26 \\
x - 3 \overline{\smash{)}\ x^3 + 6x^2 - x - 30} \\
\underline{x^3 - 3x^2} \\
9x^2 - x \\
\underline{9x^2 - 27x} \\
26x - 30 \\
\underline{26x - 78} \\
48
\end{array}
$$

    $x^3 + 6x^2 - x - 30 = (x - 3)(x^2 + 9x + 26) + 48$

8. $2x^3 - 3x^2 + x - 1 = (x - 3)(2x^2 + 3x + 10) + 29$

9.
$$
\begin{array}{r}
x^2 - 2x + 4 \\
x + 2 \overline{)\ x^3 + 0x^2 + 0x - 8} \\
\underline{x^3 + 2x^2} \\
-2x^2 + 0x \\
\underline{-2x^2 - 4x} \\
4x - 8 \\
\underline{4x + 8} \\
-16
\end{array}
$$

$x^3 - 8 = (x + 2)(x^2 - 2x + 4) - 16$

10. $x^3 + 27 = (x + 1)(x^2 - x + 1) + 26$

11.
$$
\begin{array}{r}
x^2 + 5 \\
x^2 + 4 \overline{)\ x^4 + 9x^2 + 20} \\
\underline{x^4 + 4x^2} \\
5x^2 + 20 \\
\underline{5x^2 + 20} \\
0
\end{array}
$$

$x^4 + 9x^2 + 20 = (x^2 + 4)(x^2 + 5) + 0$

12. $x^4 + x^2 + 2 = (x^2 + x + 1)(x^2 - x + 1) + 1$

13.
$$
\frac{5}{2}x^5 + \frac{5}{4}x^4 - \frac{5}{8}x^3 - \frac{39}{16}x^2 - \frac{29}{32}x + \frac{113}{64}
$$
$$2x^2-x+1\overline{)5x^7 + 0x^6 + 0x^5 - 3x^4 + 0x^3 + 2x^2 + 0x - 3}$$

$$
\begin{array}{l}
\underline{5x^7 - \frac{5}{2}x^6 + \frac{5}{2}x^5} \\
\frac{5}{2}x^6 - \frac{5}{2}x^5 - 3x^4 \\
\underline{\frac{5}{2}x^6 - \frac{5}{4}x^5 + \frac{5}{4}x^4} \\
-\frac{5}{4}x^5 - \frac{17}{4}x^4 + 0x^3 \\
\underline{-\frac{5}{4}x^5 + \frac{5}{8}x^4 - \frac{5}{8}x^3} \\
-\frac{39}{8}x^4 + \frac{5}{8}x^3 + 2x^2 \\
\underline{-\frac{39}{8}x^4 + \frac{39}{16}x^3 - \frac{39}{16}x^2} \\
-\frac{29}{16}x^3 + \frac{71}{16}x^2 + 0x \\
\underline{-\frac{29}{16}x^3 + \frac{29}{32}x^2 - \frac{29}{32}x} \\
\frac{113}{32}x^2 + \frac{29}{32}x - 3 \\
\underline{\frac{113}{32}x^2 - \frac{113}{64}x + \frac{113}{64}} \\
\frac{171}{64}x - \frac{305}{64}
\end{array}
$$

$5x^7 - 3x^4 + 2x^2 - 3$

$= (2x^2 - x + 1)\left(\frac{5}{2}x^5 + \frac{5}{4}x^4 - \frac{5}{8}x^3 - \frac{39}{16}x^2 - \frac{29}{32}x + \frac{113}{64}\right) + \left(\frac{171}{64}x - \frac{305}{64}\right)$

14. $6x^5 + 4x^4 - 3x^2 + x - 2 =$
$(3x^2 + 2x - 1)\left(2x^3 + \frac{2}{3}x - \frac{13}{9}\right) + \left(\frac{41}{9}x - \frac{31}{9}\right)$

15. $(2x^4 + 7x^3 + x - 12) \div (x + 3)$
$= (2x^4 + 7x^3 + 0x^2 + x - 12) \div [x - (-3)]$

$$
\begin{array}{r|rrrrr}
-3 & 2 & 7 & 0 & 1 & -12 \\
   &   & -6 & -3 & 9 & -30 \\
\hline
   & 2 & 1 & -3 & 10 & -42
\end{array}
$$

The quotient is $2x^3 + x^2 - 3x + 10$.
The remainder is -42.

16. $Q(x) = x^2 - 5x + 3, R(x) = 9$

17. $(x^3 - 2x^2 - 8) \div (x + 2)$
$= (x^3 - 2x^2 + 0x - 8) \div [x - (-2)]$

$$
\begin{array}{r|rrrr}
-2 & 1 & -2 & 0 & -8 \\
   &   & -2 & 8 & -16 \\
\hline
   & 1 & -4 & 8 & -24
\end{array}
$$

The quotient is $x^2 - 4x + 8$.
The remainder is -24.

18. $Q(x) = x^2 + 2x + 1, R(x) = 12$

19. $(x^4 - 1) \div (x - 1)$
$= (x^4 + 0x^3 + 0x^2 + 0x - 1) \div (x - 1)$

$$
\begin{array}{r|rrrrr}
1 & 1 & 0 & 0 & 0 & -1 \\
  &   & 1 & 1 & 1 & 1 \\
\hline
  & 1 & 1 & 1 & 1 & 0
\end{array}
$$

The quotient is $x^3 + x^2 + x + 1$.
The remainder is 0.

20. $Q(x) = x^4 - 2x^3 + 4x^2 - 8x + 16, R(x) = 0$

21. $(2x^4 + 3x^2 - 1) \div \left(x - \frac{1}{2}\right)$
$= (2x^4 + 0x^3 + 3x^2 + 0x - 1) \div \left(x - \frac{1}{2}\right)$

$$
\begin{array}{r|rrrrr}
\frac{1}{2} & 2 & 0 & 3 & 0 & -1 \\
   &   & 1 & \frac{1}{2} & \frac{7}{4} & \frac{7}{8} \\
\hline
   & 2 & 1 & \frac{7}{2} & \frac{7}{4} & -\frac{1}{8}
\end{array}
$$

The quotient is $2x^3 + x^2 + \frac{7}{2}x + \frac{7}{4}$.
The remainder is $-\frac{1}{8}$.

22. $Q(x) = 3x^3 + \frac{3}{4}x^2 - \frac{29}{16}x - \frac{29}{64}, R(x) = \frac{483}{256}$

23. $(x^4 - y^4) \div (x - y)$

$= (x^4 + 0x^3 + 0x^2 + 0x - y^4) \div (x - y)$

$$
\begin{array}{r|rrrr}
y & 1 & 0 & 0 & 0 & -y^4 \\
  &   & y & y^2 & y^3 & y^4 \\
\hline
  & 1 & y & y^2 & y^3 & 0 \\
\end{array}
$$

The quotient is $x^3 + x^2y + xy^2 + y^3$.
The remainder is 0.

24. $Q(x) = x^2 + 2ix + (2 - 4i)$, $R(x) = -6 - 2i$

25. $P(x) = x^3 - 6x^2 + 11x - 6$

Find $P(1)$.

$$
\begin{array}{r|rrrr}
1 & 1 & -6 & 11 & -6 \\
  &   & 1 & -5 & 6 \\
\hline
  & 1 & -5 & 6 & 0 \\
\end{array}
$$

$P(1) = 0$

Find $P(-2)$.

$$
\begin{array}{r|rrrr}
-2 & 1 & -6 & 11 & -6 \\
   &   & -2 & 16 & -54 \\
\hline
   & 1 & -8 & 27 & -60 \\
\end{array}
$$

$P(-2) = -60$

Find $P(3)$.

$$
\begin{array}{r|rrrr}
3 & 1 & -6 & 11 & -6 \\
  &   & 3 & -9 & 6 \\
\hline
  & 1 & -3 & 2 & 0 \\
\end{array}
$$

$P(3) = 0$

26. $P(-3) = 69$, $P(-2) = 41$, $P(1) = -7$

27. $P(x) = 2x^5 - 3x^4 + 2x^3 - x + 8$

Find $P(20)$.

$$
\begin{array}{r|rrrrrr}
20 & 2 & -3 & 2 & 0 & -1 & 8 \\
   &   & 40 & 740 & 14{,}840 & 296{,}800 & 5{,}935{,}980 \\
\hline
   & 2 & 37 & 742 & 14{,}840 & 296{,}799 & 5{,}935{,}988 \\
\end{array}
$$

$P(20) = 5{,}935{,}988$

Find $P(-3)$.

$$
\begin{array}{r|rrrrrr}
-3 & 2 & -3 & 2 & 0 & -1 & 8 \\
   &   & -6 & 27 & -87 & 261 & -780 \\
\hline
   & 2 & -9 & 29 & -87 & 260 & -772 \\
\end{array}
$$

$P(-3) = -772$

28. $P(-10) = -220{,}050$, $P(5) = -750$

29. $P(x) = x^4 - 16$

Find $P(2)$.

$$
\begin{array}{r|rrrrr}
2 & 1 & 0 & 0 & 0 & -16 \\
  &   & 2 & 4 & 8 & 16 \\
\hline
  & 1 & 2 & 4 & 8 & 0 \\
\end{array}
$$

$P(2) = 0$

Find $P(-2)$.

$$
\begin{array}{r|rrrrr}
-2 & 1 & 0 & 0 & 0 & -16 \\
   &   & -2 & 4 & -8 & 16 \\
\hline
   & 1 & -2 & 4 & -8 & 0 \\
\end{array}
$$

$P(-2) = 0$.

Find $P(3)$.

$$
\begin{array}{r|rrrrr}
3 & 1 & 0 & 0 & 0 & -16 \\
  &   & 3 & 9 & 27 & 81 \\
\hline
  & 1 & 3 & 9 & 27 & 65 \\
\end{array}
$$

$P(3) = 65$.

30. $P(2) = 64$, $P(-2) = 0$, $P(3) = 275$

31. $P(x) = 3x^3 + 5x^2 - 6x + 18$

If -3 is a root of $P(x)$, then $P(-3) = 0$.
Find $P(-3)$ using synthetic division.

$$
\begin{array}{r|rrrr}
-3 & 3 & 5 & -6 & 18 \\
   &   & -9 & 12 & -18 \\
\hline
   & 3 & -4 & 6 & 0 \\
\end{array}
$$

Since $P(-3) = 0$, -3 is a root of $P(x)$.

If 2 is a root of $P(x)$, then $P(2) = 0$.
Find $P(2)$ using synthetic division.

$$
\begin{array}{r|rrrr}
2 & 3 & 5 & -6 & 18 \\
  &   & 6 & 22 & 32 \\
\hline
  & 3 & 11 & 16 & 50 \\
\end{array}
$$

Since $P(2) \neq 0$, 2 is not a root of $P(x)$.

32. -4 yes, 2 no

33. $P(x) = x^3 - \frac{7}{2}x^2 + x - \frac{3}{2}$

If -3 is a root of $P(x)$, then $P(-3) = 0$.
Find $P(-3)$ using synthetic division.

$$
\begin{array}{r|rrrr}
-3 & 1 & -\frac{7}{2} & 1 & -\frac{3}{2} \\
   &   & -3 & \frac{39}{2} & -\frac{123}{2} \\
\hline
   & 1 & -\frac{13}{2} & \frac{41}{2} & -63 \\
\end{array}
$$

Since $P(-3) \neq 0$, -3 is not a root of $P(x)$.

If $\frac{1}{2}$ is a root of $P(x)$, then $P\left(\frac{1}{2}\right) = 0$.
Find $P\left(\frac{1}{2}\right)$ using synthetic division.

$$\frac{1}{2} \bigg\rfloor \quad 1 \quad -\frac{7}{2} \quad 1 \quad -\frac{3}{2}$$

$$\frac{1}{2} \quad -\frac{3}{2} \quad -\frac{1}{4}$$

$$1 \quad -3 \quad -\frac{1}{2} \bigg| \quad -\frac{7}{4}$$

Since $P\left(\frac{1}{2}\right) \neq 0$, $\frac{1}{2}$ is not a root of $P(x)$.

34. All yes

35. $P(x) = x^3 + 4x^2 + x - 6$

Try $x - 1$. Use synthetic division to see whether $P(1) = 0$.

$$1 \,\rfloor \quad 1 \quad 4 \quad 1 \quad -6$$
$$1 \quad 5 \quad 6$$
$$1 \quad 5 \quad 6 \,\big| \quad 0$$

Since $P(1) = 0$, $x - 1$ is a factor of $P(x)$.
Thus $P(x) = (x - 1)(x^2 + 5x + 6)$.

Factoring the trinomial we get
$P(x) = (x - 1)(x + 2)(x + 3)$.

To solve the equation $P(x) = 0$, use the principle of zero products.

$(x - 1)(x + 2)(x + 3) = 0$

$x - 1 = 0$ or $x + 2 = 0$ or $x + 3 = 0$
$\quad$ $x = 1$ or $\quad$ $x = -2$ or $\quad$ $x = -3$

The solutions are 1, -2, and -3.

36. $P(x) = (x - 2)(x + 3)(x + 4)$; 2, -3, and -4

37. $P(x) = x^3 - 6x^2 + 3x + 10$

Try $x - 1$. Use synthetic division to see whether $P(1) = 0$.

$$1 \,\rfloor \quad 1 \quad -6 \quad 3 \quad 10$$
$$1 \quad -5 \quad -2$$
$$1 \quad -5 \quad -2 \,\big| \quad 8$$

Since $P(1) \neq 0$, $x - 1$ is not a factor of $P(x)$.

Try $x + 1$. Use synthetic division to see whether $P(-1) = 0$.

$$-1 \,\rfloor \quad 1 \quad -6 \quad 3 \quad 10$$
$$-1 \quad 7 \quad -10$$
$$1 \quad -7 \quad 10 \,\big| \quad 0$$

Since $P(-1) = 0$, $x + 1$ is a factor of $P(x)$.
Thus $P(x) = (x + 1)(x^2 - 7x + 10)$.

Factoring the trinomial we get
$P(x) = (x + 1)(x - 2)(x - 5)$.

To solve the equation $P(x) = 0$, use the principle of zero products.

$(x + 1)(x - 2)(x - 5) = 0$

$x + 1 = 0$ or $x - 2 = 0$ or $x - 5 = 0$
$\quad$ $x = -1$ or $\quad$ $x = 2$ or $\quad$ $x = 5$

The solutions are -1, 2, and 5.

38. $P(x) = (x - 1)(x - 2)(x + 5)$; 1, 2, and -5

39. $P(x) = x^3 - x^2 - 14x + 24$

Try $x + 1$, $x - 1$, and $x + 2$. Using synthetic division we find that $P(-1) \neq 0$, $P(1) \neq 0$, and $P(-2) \neq 0$. Thus $x + 1$, $x - 2$, and $x + 2$ are not factors of $P(x)$.

Try $x - 2$. Use synthetic division to see whether $P(2) = 0$.

$$2 \,\rfloor \quad 1 \quad -1 \quad -14 \quad 24$$
$$2 \quad 2 \quad -24$$
$$1 \quad 1 \quad -12 \,\big| \quad 0$$

Since $P(2) = 0$, $x - 2$ is a factor of $P(x)$.
Thus $P(x) = (x - 2)(x^2 + x - 12)$.

Factoring the trinomial we get
$P(x) = (x - 2)(x + 4)(x - 3)$.

To solve the equation $P(x) = 0$, use the principle of zero products.

$(x - 2)(x + 4)(x - 3) = 0$

$x - 2 = 0$ or $x + 4 = 0$ or $x - 3 = 0$
$\quad$ $x = 2$ or $\quad$ $x = -4$ or $\quad$ $x = 3$

The solutions are 2, -4, and 3.

40. $P(x) = (x - 2)(x - 4)(x + 3)$; 2, 4, and -3

41. $P(x) = x^4 - x^3 - 19x^2 + 49x - 30$

Try $x - 1$. Use synthetic division to see whether $P(1) = 0$.

$$1 \,\rfloor \quad 1 \quad -1 \quad -19 \quad 49 \quad -30$$
$$1 \quad 0 \quad -19 \quad 30$$
$$1 \quad 0 \quad -19 \quad 30 \,\big| \quad 0$$

Since $P(1) = 0$, $x - 1$ is a factor of $P(x)$.
Thus $P(x) = (x - 1)(x^3 - 19x + 30)$.

We continue to use synthetic division to factor $Q(x) = x^3 - 19x + 30$. Trying $x - 1$, $x + 1$, and $x + 2$ we find that $Q(1) \neq 0$, $Q(-1) \neq 0$, and $Q(-2) \neq 0$. Thus $x - 1$, $x + 1$, and $x + 2$ are not factors of $x^3 - 19x + 30$. Try $x - 2$.

$$2 \,\rfloor \quad 1 \quad 0 \quad -19 \quad 30$$
$$2 \quad 4 \quad -30$$
$$1 \quad 2 \quad -15 \,\big| \quad 0$$

Since $Q(2) = 0$, $x - 2$ is a factor of $x^3 - 19x + 30$.

Thus $P(x) = (x - 1)(x - 2)(x^2 + 2x - 15)$.

Factoring the trinomial we get
$P(x) = (x - 1)(x - 2)(x - 3)(x + 5)$.

To solve the equation $P(x) = 0$, use the principle of zero products.

$$(x - 1)(x - 2)(x - 3)(x + 5) = 0$$

$x - 1 = 0$ or $x - 2 = 0$ or $x - 3 = 0$ or $x + 5 = 0$

$\quad x = 1$ or $\quad x = 2$ or $\quad x = 3$ or $\quad x = -5$

The solutions are 1, 2, 3, and -5.

42. $P(x) = (x + 1)(x + 2)(x + 3)(x + 5)$; -1, -2, -3, and -5

43. We solve $f(n) = 0$.

$$\frac{1}{2}(n^2 - n) = 0$$

$\quad n^2 - n = 0$ (Multiplying by 2)

$\quad n(n - 1) = 0$

$n = 0$ or $n = 1$

The zeros of the function are 0 and 1.

44. 7.16

45. We solve $W(h) = 0$, $0 \leqslant h < \infty$.

$$\left(\frac{h}{12.3}\right)^3 = 0$$

$$\frac{h^3}{(12.3)^3} = 0$$

$\quad h^3 = 0$ [Multiplying by $(12.3)^3$]

$\quad h = 0$

The root of the polynomial is 0.

46. 0, 0.9636

47. $\dfrac{6x^2}{x^2 + 11} + \dfrac{60}{x^3 - 7x^2 + 11x - 77} = \dfrac{1}{x - 7}$

LCM is $(x^2 + 11)(x - 7)$.

$$(x^2+11)(x-7)\left[\frac{6x^2}{x^2+11} + \frac{60}{(x^2+11)(x-7)}\right] = (x^2+11)(x-7)\cdot\frac{1}{x-7}$$

$$6x^2(x - 7) + 60 = x^2 + 11$$

$$6x^3 - 42x^2 + 60 = x^2 + 11$$

$$6x^3 - 43x^2 + 49 = 0$$

Use synthetic division to find factors of $P(x) = 6x^3 - 43x^2 + 49$. Try $x + 1$. Use synthetic division to see whether $P(-1) = 0$.

```
-1 | 6   -43    0    49
   |      -6   49   -49
   ------------------------
     6   -49   49  |  0
```

Since $P(-1) = 0$, $x + 1$ is a factor of $P(x)$. Thus $P(x) = (x + 1)(6x^2 - 49x + 49)$. Factoring the trinomial we get $P(x) = (x + 1)(6x - 7)(x - 7)$.

To solve $P(x) = 0$, use the principle of zero products.

$$(x + 1)(6x - 7)(x - 7) = 0$$

$x + 1 = 0$ or $6x - 7 = 0$ or $x - 7 = 0$

$\quad x = -1$ or $\quad 6x = 7$ or $\quad x = 7$

$\quad x = -1$ or $\quad x = \frac{7}{6}$ or $\quad x = 7$

The value $x = 7$ does not check, but $x = -1$ and $x = \frac{7}{6}$ do. The solutions are -1 and $\frac{7}{6}$.

48. $-1 \pm \sqrt{7}$

49. To solve $x^3 + 2x^2 - 13x + 10 > 0$ we first find the solutions of $f(x) = x^3 + 2x^2 - 13x + 10 = 0$. Using synthetic division we find the solutions to be -5, 1, and 2. They divide the real-number line as shown.

Try a test number in each interval.

A: $f(-6) = (-6)^3 + 2(-6)^2 - 13(-6) + 10 = -56$

B: $f(0) = 0^3 + 2\cdot0^2 - 13\cdot0 + 10 = 10$

C: $f\left(\frac{3}{2}\right) = \left(\frac{3}{2}\right)^3 + 2\left(\frac{3}{2}\right)^2 - 13\left(\frac{3}{2}\right) + 10 = -\frac{13}{8}$

D: $f(3) = 3^3 + 2\cdot3^2 - 13\cdot3 + 10 = 16$

The solution set is $\{x \mid -5 < x < 1$ or $x > 2\}$.

50. $\{x \mid -5 < x < 1$ or $2 < x < 3\}$

51. $P(x) = x^3 - kx^2 + 3x + 7k$

Think of $x + 2$ as $x - (-2)$.
Find $P(-2)$.

```
-2 | 1    -k       3        7k
   |      -2    2k + 4   -4k - 14
   -----------------------------------
     1  -k - 2  2k + 7 |  3k - 14
```

Thus $P(-2) = 3k - 14$.
We know that if $x + 2$ is a factor of $P(x)$, then $P(-2) = 0$.

We solve $0 = 3k - 14$ for k.

$0 = 3k - 14$

$14 = 3k$

$\dfrac{14}{3} = k$

52. a) -485.1587,   b) -485.1587

53. Divide $x^2 + kx + 4$ by $x - 1$.

```
1 | 1    k       4
  |      1     k + 1
  ---------------------
    1  k + 1 | k + 5
```

The remainder is $k + 5$.

Divide $x^2 + kx + 4$ by $x + 1$.

```
-1 | 1    k       4
   |     -1    -k + 1
   ─────────────────────
     1   k - 1 | -k + 5
```

The remainder is $-k + 5$.

Set $k + 5 = -k + 5$ and solve for k.

$k + 5 = -k + 5$

$2k = 0$

$k = 0$

54. $-\dfrac{3}{2}$

55. See the answer section in the text.

56. Let $f(x) = x^n - a^n$. Then $f(a) = 0$ so that $x - a$ is a factor.

Exercise Set 4.7

1. $P(x) = (x + 3)^2(x - 1) = (x + 3)(x + 3)(x - 1)$

The factor $x + 3$ occurs twice. Thus the root -3 has a multiplicity of two.

The factor $x - 1$ occurs only one time. Thus the root 1 has a mulitplicity of one.

2. 3, Multiplicity 2; -4, Multiplicity 3; 0, Multiplicity 4

3. $x^3(x - 1)^2(x + 4) = 0$

$x \cdot x \cdot x(x - 1)(x - 1)(x + 4) = 0$

The factor x occurs three times. Thus the root 0 has a multiplicity of three.

The factor $x - 1$ occurs twice. Thus the root 1 has a multiplicity of two.

The factor $x + 4$ occurs only one time. Thus the root -4 has a multiplicity of one.

4. 3, Multiplicity 2; 2, Multiplicity 2

5. $P(x) = x^4 - 4x^2 + 3$

We factor as follows:

$P(x) = (x^2 - 3)(x^2 - 1)$

$= (x - \sqrt{3})(x + \sqrt{3})(x - 1)(x + 1)$

The roots of the polynomial are $\sqrt{3}$, $-\sqrt{3}$, 1, and -1. Each has multiplicity of one.

6. ± 3, ± 1; each has multiplicity of 1

7. $P(x) = x^3 + 3x^2 - x - 3$

We factor by grouping:

$P(x) = x^2(x + 3) - (x + 3)$

$= (x^2 - 1)(x + 3)$

$= (x - 1)(x + 1)(x + 3)$

The roots of the polynomial are 1, -1, and -3. Each has multiplicity of one.

8. $\sqrt{2}$, $-\sqrt{2}$, 1; each has multiplicity of 1

9. Find a polynomial of degree 3 with -2, 3, and 5 as roots.

By Theorem 6 such a polynomial has factors $x + 2$, $x - 3$, and $x - 5$, so we have

$P(x) = a_n(x + 2)(x - 3)(x - 5)$.

The number a_n can be any nonzero number. The simplest polynomial will be obtained if we let it be 1. Multiplying the factors, we obtain

$P(x) = (x + 2)(x - 3)(x - 5)$

$= (x^2 - x - 6)(x - 5)$

$= x^3 - 6x^2 - x + 30$

10. $x^3 - 2x^2 + x - 2$

11. Find a polynomial of degree 3 with -3, $2i$, and $-2i$ as roots.

By Theorem 6 such a polynomial has factors $x + 3$, $x - 2i$, and $x + 2i$, so we have

$P(x) = a_n(x + 3)(x - 2i)(x + 2i)$.

The number a_n can be any nonzero number. The simplest polynomial will be obtained if we let it be 1. Multiplying the factors, we obtain

$P(x) = (x + 3)(x - 2i)(x + 2i)$

$= (x + 3)(x^2 + 4)$

$= x^3 + 3x^2 + 4x + 12$

12. $x^3 - x^2 + 15x + 17$

13. Find a polynomial of degree 3 with $\sqrt{2}$, $-\sqrt{2}$, and $\sqrt{3}$ as roots.

By Theorem 6 such a polynomial has factors $x - \sqrt{2}$, $x + \sqrt{2}$, and $x - \sqrt{3}$, so we have

$P(x) = a_n(x - \sqrt{2})(x + \sqrt{2})(x - \sqrt{3})$.

The number a_n can be any nonzero number. The simplest polynomial will be obtained if we let it be 1. Multiplying the factors, we obtain

$P(x) = (x - \sqrt{2})(x + \sqrt{2})(x - \sqrt{3})$

$= (x^2 - 2)(x - \sqrt{3})$

$= x^3 - \sqrt{3}\, x^2 - 2x + 2\sqrt{3}$

14. $x^4 - 3x^3 - 7x^2 + 15x + 18$

15. A polynomial or polynomial equation of degree 5 has at most 5 roots. Three of the roots are 6, $-3 + 4i$, and $4 - \sqrt{5}$. Using Theorems 10 and 11 we know that $-3 - 4i$ and $4 + \sqrt{5}$ are also roots.

16. $1 + i$

17. Find a polynomial of lowest degree with rational coefficients that has $1 + i$ and 2 as some of its roots.

    $1 - i$ is also a root.   (Theorem 10)

    Thus the polynomial is

    $a_n(x - 2)[x - (1 + i)][x - (1 - i)]$.

    If we let $a_n = 1$, we obtain

    $(x - 2)[(x - 1) - i][(x - 1) + i]$

    $= (x - 2)[(x - 1)^2 - i^2]$

    $= (x - 2)(x^2 - 2x + 1 + 1)$

    $= (x - 2)(x^2 - 2x + 2)$

    $= x^3 - 4x^2 + 6x - 4$

18. $x^3 - 3x^2 + x + 5$

19. Find a polynomial of lowest degree with rational coefficients that has $-4i$ and 5 as some of its roots.

    $4i$ is also a root.   (Theorem 10)

    Thus the polynomial is

    $a_n(x - 5)(x + 4i)(x - 4i)$.

    If we let $a_n = 1$, we obtain

    $(x - 5)[x^2 - (4i)^2]$

    $= (x - 5)(x^2 + 16)$

    $= x^3 - 5x^2 + 16x - 80$

20. $x^4 - 6x^3 + 11x^2 - 10x + 2$

21. Find a polynomial of lowest degree with rational coefficients that has $\sqrt{5}$ and $-3i$ as some of its roots.

    $-\sqrt{5}$ is also a root.   (Theorem 11)

    $3i$ is also a root.   (Theorem 10)

    Thus the polynomial is

    $a_n(x - \sqrt{5})(x + \sqrt{5})(x + 3i)(x - 3i)$.

    If we let $a_n = 1$, we obtain

    $(x^2 - 5)(x^2 + 9)$

    $= x^4 + 4x^2 - 45$

22. $x^4 + 14x^2 - 32$

23. If $-i$ is a root of $x^4 - 5x^3 + 7x^2 - 5x + 6$, i is also a root (Theorem 10). Thus $x + i$ and $x - i$ or $(x + i)(x - i)$ which is $x^2 + 1$ are factors of the polynomial. Divide $x^4 - 5x^3 + 7x^2 - 5x + 6$ by $x^2 + 1$.

$$
\begin{array}{r}
x^2 - 5x + 6 \\
x^2 + 1 \overline{\smash{\big)}\ x^4 - 5x^3 + 7x^2 - 5x + 6} \\
\underline{x^4 + x^2} \\
-5x^3 + 6x^2 - 5x \\
\underline{-5x^3 - 5x} \\
6x^2 + 6 \\
\underline{6x^2 + 6} \\
0
\end{array}
$$

Thus

$x^4 - 5x^3 + 7x^2 - 5x + 6 = (x + i)(x - i)(x^2 - 5x + 6)$

$= (x + i)(x - i)(x - 2)(x - 3)$

Using the principle of zero products we find the other roots to be i, 2, and 3.

24. $-2i$, -2, 2

25. $x^3 - 6x^2 + 13x - 20 = 0$

    If 4 is a root, then $x - 4$ is a factor. Use synthetic division to find another factor.

$$
\begin{array}{r}
4 \ \big|\ \begin{array}{rrrr} 1 & -6 & 13 & -20 \end{array} \\
\begin{array}{rrrr} & 4 & -8 & 20 \end{array} \\
\hline
\begin{array}{rrr|r} 1 & -2 & 5 & 0 \end{array}
\end{array}
$$

    $(x - 4)(x^2 - 2x + 5) = 0$

    $x - 4 = 0$ or $x^2 - 2x + 5 = 0$ (Principle of zero products)

    $x = 4$ or $x = \dfrac{2 \pm \sqrt{4 - 20}}{2}$ (Quadratic formula)

    $x = 4$ or $x = \dfrac{2 \pm 4i}{2} = 1 \pm 2i$

    The other roots are $1 + 2i$ and $1 - 2i$.

26. $-1 + i\sqrt{3}$, $-1 - i\sqrt{3}$

27. $P(x) = x^5 - 3x^2 + 1$

    Since the leading coefficient is 1, the only possibilities for rational roots (Theorem 12) are the factors of the last coefficient 1: 1 and -1.

28. $\pm (1, 2, 3, 4, 6, 12)$

29. $P(x) = 15x^6 + 47x^2 + 2$

    By Theorem 12, if c/d, a rational number in lowest terms, is a root of $P(x)$, then c must be a factor of 2 and d must be a factor of 15.

    The possibilities for c and d are

    c: 1, -1, 2, -2   d: 1, -1, 3, -3, 5, -5, 15, -15

    The resulting possibilities for c/d are

$\frac{c}{d}$: 1, -1, $\frac{1}{3}$, -$\frac{1}{3}$, $\frac{1}{5}$, -$\frac{1}{5}$, $\frac{1}{15}$, -$\frac{1}{15}$

2, -2, $\frac{2}{3}$, -$\frac{2}{3}$, $\frac{2}{5}$, -$\frac{2}{5}$, $\frac{2}{15}$, -$\frac{2}{15}$

30. $\pm \left[1, 2, 3, 6, \frac{1}{2}, \frac{3}{2}, \frac{1}{5}, \frac{2}{5}, \frac{3}{5}, \frac{6}{5}, \frac{1}{10}, \frac{3}{10} \right]$

31. $P(x) = x^3 + 3x^2 - 2x - 6$

Using Theorem 12, the only possible rational roots are 1, -1, 2, -2, 3, -3, 6 and -6. Of these eight possibilities we know that at most three of them can be roots because $P(x)$ is of degree 3. Use synthetic division to determine which are roots.

```
1 | 1   3  -2  -6
  |     1   4   2
    1   4   2 | -4
```
$P(1) \neq 0$, so 1 is not a root.

```
-1 | 1   3  -2  -6
   |    -1  -2   4
     1   2  -4 | -2
```
$P(-1) \neq 0$, so -1 is not a root.

```
2 | 1   3  -2  -6
  |     2  10  16
    1   5   8 | 10
```
$P(2) \neq 0$, so 2 is not a root.

```
-2 | 1   3  -2  -6
   |    -2  -2   8
     1   1  -4 | 2
```
$P(-2) \neq 0$, so -2 is not a root.

```
3 | 1   3  -2  -6
  |     3  18  48
    1   6  16 | 42
```
$P(3) \neq 0$, so 3 is not a root.

```
-3 | 1   3  -2  -6
   |    -3   0   6
     1   0  -2 | 0
```
$P(-3) = 0$, so -3 is a root.

We can now express $P(x)$ as follows:

$P(x) = (x + 3)(x^2 - 2)$.

To find the other roots we solve the quadratic equation $x^2 - 2 = 0$. The other roots are $\sqrt{2}$ and $-\sqrt{2}$. These are irrational numbers. Thus the only rational root is -3.

We write the polynomial in factored form:

$P(x) = (x + 3)(x - \sqrt{2})(x + \sqrt{2})$

(Note that we could have also used factoring by grouping in this exercise.)

32. 1, $\sqrt{3}$, and $-\sqrt{3}$;
$(x - 1)(x - \sqrt{3})(x + \sqrt{3}) = 0$

33. $x^3 - 3x + 2 = 0$

Using Theorem 12, the only possible rational roots are 1, -1, 2, and -2. At most three of these can be roots because the polynomial is of degree 3. Use synthetic division to determine which are roots.

```
1 | 1   0  -3   2
  |     1   1  -2
    1   1  -2 | 0
```
1 is a root

We can now express the equation as follows:

$(x - 1)(x^2 + x - 2) = 0$

To find the other roots we solve the following quadratic equation:

$x^2 + x - 2 = 0$

$(x + 2)(x - 1) = 0$

$x = -2$ or $x = 1$

The rational roots are 1 and -2. (One is a root with multiplicity of 2.)

We write the polynomial in factored form:

$(x - 1)^2(x + 2) = 0$

34. No rational roots

35. $P(x) = x^3 - 5x^2 + 11x - 19$

Using Theorem 12, the only possible rational roots are 1, -1, 19, and -19. At most three of these can be roots because $P(x)$ is of degree 3. Use synthetic division to determine which are roots.

```
1 | 1  -5   11  -19
  |     1  -4    7
    1  -4   7 | -12
```
$P(1) \neq 0$, so 1 is not a root.

```
-1 | 1  -5   11  -19
   |    -1    6  -17
     1  -6   17 | -36
```
$P(-1) \neq 0$, so -1 is not a root.

```
19 | 1   -5    11    -19
   |      19   266   5263
     1    14   277 | 5244
```
$P(19) \neq 0$, so 19 is not a root.

```
-19 | 1   -5    11    -19
    |     -19   456  -8873
      1  -24   467 | -8892
```
$P(19) \neq 0$, so -19 is not a root.

There are no rational roots.

36. No rational roots

37. $P(x) = 5x^4 - 4x^3 + 19x^2 - 16x - 4$

Using Theorem 12, the only possible rational roots are $\pm \left[1, 2, 4, \frac{1}{5}, \frac{2}{5}, \frac{4}{5} \right]$. Of these twelve possibilities we know that at most four of them can be roots because $P(x)$ is of degree 4. Using synthetic division we determine that the only rational roots are 1 and $-\frac{1}{5}$.

```
1 | 5  -4   19  -16  -4
  |    5    1   20   4
    5   1   20    4 | 0
```

$P(1) = 0$, so 1 is a root.

We can now express P(x) as follows:

P(x) = (x - 1)(5x³ + x² + 20x + 4).

We now use Q(x) = 5x³ + x² + 20x + 4 and check to see if 1 is a double root and check for other possible rational roots. The only other rational root is $-\frac{1}{5}$.

$$-\frac{1}{5} \begin{array}{|rrrr} 5 & 1 & 20 & 4 \\ & -1 & 0 & -4 \\ \hline 5 & 0 & 20 & \big| \ 0 \end{array}$$

$Q\left(-\frac{1}{5}\right) = 0$, so $-\frac{1}{5}$ is a root.

We can now express P(x) as follows:

$P(x) = (x - 1)\left[x + \frac{1}{5}\right](5x^2 + 20)$

To find the other roots we solve the quadratic equation 5x² + 20 = 0.

5x² + 20 = 0

    5x² = -20

     x² = -4

      x = ± 2i

The other roots are 2i and -2i, neither of which is rational. The only rational roots are 1 and $-\frac{1}{5}$.

We write the polynomial in factored form:

$P(x) = (x - 1)\left[x + \frac{1}{5}\right](x - 2i)(x + 2i)$

__38.__ $-1, \frac{1}{3}, 1 \pm i$;

$3(x + 1)\left[x - \frac{1}{3}\right](x - 1 - i)(x - 1 + i)$

__39.__ P(x) = x⁴ - 3x³ - 20x² - 24x - 8

Using Theorem 12, the only possible rational roots are 1, -1, 2, -2, 4, -4, 8, and -8. Of these eight possibilities, we know that at most four of them can be roots because P(x) is of degree 4. Using synthetic division we determine that the only rational roots are -1 and -2.

$$-1 \begin{array}{|rrrrr} 1 & -3 & -20 & -24 & -8 \\ & -1 & 4 & 16 & 8 \\ \hline 1 & -4 & -16 & -8 & \big| \ 0 \end{array}$$

P(-1) = 0, so -1 is a root.

We can now express P(x) as follows:

P(x) = (x + 1)(x³ - 4x² - 16x - 8).

We now use Q(x) = x³ - 4x² - 16x - 8 and check to see if -1 is a double root and check for other possible rational roots. The only other rational root is -2.

$$-2 \begin{array}{|rrrr} 1 & -4 & -16 & -8 \\ & -2 & 12 & 8 \\ \hline 1 & -6 & -4 & \big| \ 0 \end{array}$$

Q(-2) = 0, so -2 is a root.

We can now express P(x) as follows:

P(x) = (x + 1)(x + 2)(x² - 6x - 4).

Since the factor x² - 6x - 4 is quadratic, use the quadratic formula to find the other roots, $3 + \sqrt{13}$ and $3 - \sqrt{13}$. These are irrational numbers. The only rational roots are -1 and -2.

We write the polynomial in factored form:

$P(x) = (x + 1)(x + 2)(x - 3 - \sqrt{13})(x - 3 + \sqrt{13})$

__40.__ $1, 2, -4 \pm \sqrt{21}$;

$(x - 1)(x - 2)(x + 4 - \sqrt{21})(x + 4 + \sqrt{21})$

__41.__ P(x) = x³ - 4x² + 2x + 4

Using Theorem 12, the only possible rational roots are 1, -1, 2, -2, 4, and -4. Of these six possibilities, we know that at most three of them can be roots, because P(x) is of degree 3. Use synthetic division to determine which are roots.

$$1 \begin{array}{|rrrr} 1 & -4 & 2 & 4 \\ & 1 & -3 & -1 \\ \hline 1 & -3 & -1 & \big| \ 3 \end{array}$$ P(1) ≠ 0, so 1 is not a root.

$$-1 \begin{array}{|rrrr} 1 & -4 & 2 & 4 \\ & -1 & 5 & -7 \\ \hline 1 & -5 & 7 & \big| \ -3 \end{array}$$ P(-1) ≠ 0, so -1 is not a root.

$$2 \begin{array}{|rrrr} 1 & -4 & 2 & 4 \\ & 2 & -4 & -4 \\ \hline 1 & -2 & -2 & \big| \ 0 \end{array}$$ P(2) = 0, so 2 __is__ a root.

We can now express P(x) as

P(x) = (x - 2)(x² - 2x - 2).

To find the other roots we use the quadratic formula to solve the equation x² - 2x - 2 = 0. The other roots are $1 + \sqrt{3}$ and $1 - \sqrt{3}$. These are irrational roots.

The only rational root is 2.

We write the polynomial in factored form:

$P(x) = (x - 2)(x - 1 - \sqrt{3})(x - 1 + \sqrt{3})$

__42.__ 4 and $2 \pm \sqrt{3}$;

$(x - 4)(x - 2 - \sqrt{3})(x - 2 + \sqrt{3})$

__43.__ P(x) = x³ + 8

Using Theorem 12, the only possible rational roots are 1, -1, 2, -2, 4, -4, 8, and -8. Of these eight possibilities we know that at most three of them can be roots because P(x) is of degree 3. Using synthetic division we determine that -2 is the only rational root.

```
-2 | 1    0    0    8
   |     -2    4   -8
   --------------------
     1   -2    4 |  0
```

P(-2) = 0, so -2 is a root.

We can now express P(x) as follows:

P(x) = (x + 2)(x² - 2x + 4)

Since the factor x² - 2x + 4 is quadratic, use the quadratic formula to find the other roots, 1 + i√3 and 1 - i√3. These are irrational numbers. Thus the only rational root is -2.

We write the polynomial in factored form:

P(x) = (x + 2)(x - 1 - i√3)(x - 1 + i√3)

44. 2 and -1 ± i√3;
(x - 2)(x + 1 - i√3)(x + 1 + i√3) = 0

45. P(x) = $\frac{1}{3}$ x³ - $\frac{1}{2}$ x² - $\frac{1}{6}$ x + $\frac{1}{6}$

6P(x) = 2x³ - 3x² - x + 1

We multiplied by 6, the LCM of the denominators. This equation is equivalent to the first, and all coefficients on the right are integers. Thus any root of 6P(x) is a root of P(x).

The possible rational roots of 2x³ - 3x² - x + 1 are 1, -1, $\frac{1}{2}$, and - $\frac{1}{2}$. Of these four possibilities we know that at most three of them can be roots because the degree of 6P(x) is 3. Using synthetic division we determine that $\frac{1}{2}$ is a root.

```
1/2 | 2   -3   -1    1
    |      1   -1   -1
    -------------------
      2   -2   -2 |  0
```

We can now express 6P(x) as follows:

6P(x) = $\left(x - \frac{1}{2}\right)$(2x² - 2x - 2)

= 2$\left(x - \frac{1}{2}\right)$(x² - x - 1)

Use the quadratic formula to find the other roots, $\frac{1 + \sqrt{5}}{2}$ and $\frac{1 - \sqrt{5}}{2}$. These are irrational numbers. Thus the only rational root is $\frac{1}{2}$.

We write the polynomial in factored form:

6P(x) = 2$\left(x - \frac{1}{2}\right)\left(x - \frac{1 + \sqrt{5}}{2}\right)\left(x - \frac{1 - \sqrt{5}}{2}\right)$, so

P(x) = $\frac{1}{3}\left(x - \frac{1}{2}\right)\left(x - \frac{1 + \sqrt{5}}{2}\right)\left(x - \frac{1 - \sqrt{5}}{2}\right)$

46. $\frac{3}{4}$, ± i; $\frac{2}{3}\left(x - \frac{3}{4}\right)$(x - i)(x + i)

47. P(x) = x⁴ + 32

Using Theorem 12, the only possible rational roots are ± (1, 2, 4, 8, 16, 32). Of these twelve possibilities we know that at most four of them can be roots because P(x) is of degree 4. Use synthetic division to check each possibility.

```
1 | 1   0   0   0   32      -1 | 1   0   0   0   32
  |     1   1   1    1         |    -1   1  -1    1
  ----------------------       ----------------------
    1   1   1   1 | 33           1  -1   1  -1 | 33

2 | 1   0   0   0   32      -2 | 1   0   0   0   32
  |     2   4   8   16         |    -2   4  -8   16
  ----------------------       ----------------------
    1   2   4   8 | 48           1  -2   4  -8 | 48
```

P(1) ≠ 0, P(-1) ≠ 0, P(2) ≠ 0, P(-2) ≠ 0; therefore 1, -1, 2, and -2 are not roots. Similarly we can show that ± (4, 8, 16, 32) are not roots. Thus there are no rational roots.

48. No rational roots

49. x³ - x² - 4x + 3 = 0

The possible rational roots are 1, -1, 3, and -3. Of these four possibilities, at most three of them can be roots because the polynomial equation is of degree 3. Use synthetic division to check each possibility.

```
1 | 1  -1  -4   3      -1 | 1  -1  -4    3
  |     1   0  -4         |    -1   2    2
  ------------------       -------------------
    1   0  -4 | -1          1  -2  -2 |  5

3 | 1  -1  -4   3      -3 | 1  -1   -4    3
  |     3   6   6         |    -3   12  -24
  ------------------       -------------------
    1   2   2 |  9          1  -4    8 | -21
```

P(1) ≠ 0, P(-1) ≠ 0, P(3) ≠ 0, P(-3) ≠ 0; therefore 1, -1, 3, and -3 are not roots. Thus there are no rational roots.

50. - $\frac{3}{2}$

51. x⁴ + 2x³ + 2x² - 4x - 8 = 0

The possible rational roots are ± (1, 2, 4, 8). Of these eight possibilities, at most four of them can be roots because the polynomial equation is of degree 4. Use synthetic division to check each possibility.

```
 1 | 1   2   2   -4   -8
   |     1   3    5    1
   -----------------------
     1   3   5    1 | -7

-1 | 1   2   2   -4   -8
   |    -1  -1   -1    5
   -----------------------
     1   1   1   -5 | -3

 2 | 1   2   2   -4   -8
   |     2   8   20   32
   -----------------------
     1   4  10   16 | 24
```

```
-2 | 1    2    2   -4   -8
   |     -2    0   -4   16
     1    0    2   -8 | 8
```

$P(1) \neq 0$, $P(-1) \neq 0$, $P(2) \neq 0$, $P(-2) \neq 0$; therefore 1, -1, 2, and -2 are not roots. Similarly we can show that 4, -4, 8, and -8 are not roots. Thus there are <u>no</u> rational roots.

<u>52.</u> No rational roots

<u>53.</u> $P(x) = x^5 - 5x^4 + 5x^3 + 15x^2 - 36x + 20$

The possible rational roots are $\pm (1, 2, 4, 5, 10, 20)$. Of these twelve possibilities at most five of them can be roots because $P(x)$ is of degree 5. Using synthetic division we determine the only rational roots are -2, 1, and 2.

```
-2 | 1   -5    5   15  -36   20
   |     -2   14  -38   46  -20
     1   -7   19  -23   10 | 0
```

$P(-2) = 0$, so -2 is a root.

We now use $x^4 - 7x^3 + 19x^2 - 23x + 10$ and check other possible roots. Synthetic division shows that -2 is not a double root. We try 1.

```
1 | 1   -7   19  -23   10
  |      1   -6   13  -10        1 is a root
    1   -6   13  -10 | 0
```

We now use $x^3 - 6x^2 + 13x - 10$ and check other possible roots. Synthetic division shows that 1 is not a double root. We try 2.

```
2 | 1   -6   13  -10
  |      2   -8   10            2 is a root
    1   -4    5 | 0
```

$P(x) = (x + 2)(x - 1)(x - 2)(x^2 - 4x + 5)$

Using the discriminant, we determine that $x^2 - 4x + 5$ has two nonreal roots. Thus the rational roots are -2, 1, and 2.

<u>54.</u> 2, 3, -2

<u>55.</u> $x^3 - 4x^2 + x - 4 = 0$

$x^2(x - 4) + (x - 4) = 0$ Factoring by grouping

$(x^2 + 1)(x - 4) = 0$

$x - 4 = 0$ or $x^2 + 1 = 0$

$x = 4$ or $\quad x^2 = -1$

$x = 4$ or $\quad x = \pm i$

The solutions are 4, i, and -i.

<u>56.</u> -3, 2 $\pm$ i

<u>57.</u> $x^4 - 2x^3 - 2x - 1 = 0$

The possible rational roots are ± 1. Synthetic division shows that neither of these is a root.

Using synthetic division we find that i is a root.

```
i | 1    -2      0      -2     -1
  |       i   -2i - 1   2 - i    1
    1  -2 + i  -2i - 1   -i  | 0
```

Using Theorem 10 we know that if i is a root then -i is also a root. Thus $(x - i)(x + i)$, or $x^2 + 1$, is a factor of $x^4 - 2x^3 - 2x - 1$. Using division we find that $x^2 - 2x - 1$ is also a factor.

$$
\begin{array}{r}
x^2 - 2x - 1 \\
x^2 + 1 \overline{\smash{\big)}\ x^4 - 2x^3 + 0x^2 - 2x - 1} \\
\underline{x^4 \qquad + x^2} \\
-2x^3 - x^2 - 2x \\
\underline{-2x^3 \qquad - 2x} \\
-x^2 \qquad - 1 \\
\underline{-x^2 \qquad - 1} \\
0
\end{array}
$$

Therefore, $(x - i)(x + i)(x^2 - 2x - 1) = 0$. Using the principle of zero products and the quadratic formula, we find that the roots are i, -i, $1 + \sqrt{2}$, and $1 - \sqrt{2}$.

<u>58.</u> $\pm 4i$, $\pm \sqrt{7}$

<u>59.</u> If the equation has a double root, the discriminant must be 0.

$(2a)^2 - 4 \cdot 1 \cdot b = 0$ gives $b = a^2$.

Then $x^2 + 2ax + a^2 = 0$ has the double root -a.

<u>60.</u> For $P(x) = a_n x^n + a_{n-1} x^{n-1} + \ldots + a_0$ with a_i positive, consider any positive x. Every term will be positive, hence $P(x) > 0$.

<u>61.</u> Solve $x^3 - 64 = 0$

One root is 4. We use synthetic division and express the polynomial in factored form.

```
4 | 1    0    0   -64
  |      4   16    64
    1    4   16 | 0
```

$x^3 - 64 = (x - 4)(x^2 + 4x + 16) = 0$

The discriminant of $x^2 + 4x + 16$ is -48, so the other roots of the polynomial are not real numbers. Thus, the length of a side is 4 cm.

<u>62.</u> 5 cm

63.

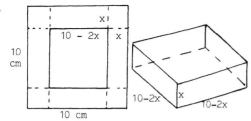

$$V = \ell wh \qquad \text{(Volume formula for rectangular solid)}$$

$$48 = (10 - 2x)(10 - 2x)(x)$$

(Substituting $10 - 2x$ for ℓ and w, x for h, and 48 for V)

$$48 = 100x - 40x^2 + 4x^3$$

$$12 = 25x - 10x^2 + x^3$$

$$0 = x^3 - 10x^2 + 25x - 12$$

The possible rational roots of this equation are $\pm (1, 2, 3, 4, 6, 12)$. Using synthetic division we find that 3 is the only rational root.

```
3 | 1  -10   25  -12
  |       3  -21   12
  -----------------------
    1   -7    4 |  0
```

We can now express the polynomial equation as follows:

$$0 = (x - 3)(x^2 - 7x + 4)$$

Since the factor $x^2 - 7x + 4$ is quadratic, we use the quadratic formula to find the other roots, $\frac{7 + \sqrt{33}}{2}$ and $\frac{7 - \sqrt{33}}{2}$. Since the length, x, must be positive and less then 5, $\frac{7 + \sqrt{33}}{2}$ is not a possible solution. The length of a side of the square can be 3 cm or $\frac{7 - \sqrt{33}}{2}$ cm.

64. 5 cm or $\frac{15 - 5\sqrt{5}}{2}$ cm

65. See the answer section in the text.

66. Each pair of nonreal roots, $a \pm bi$, gives a quadratic factor $x^2 - 2ax + a^2 + b^2$ with real coefficients. Each real root c gives a linear factor $(x - c)$. Hence P(x) is composed of linear and quadratic factors with real coefficients.

67. See the answer section in the text.

68. $\sqrt{N}$ is rational if and only if $x^2 - N = 0$ has rational roots, where N is a positive integer.

69. $x^3 + 7 = 0$

The possible rational roots are 1, -1, 7, and -7. Synthetic division shows that none of these is a root. Since $x^3 + 7 = 0$, we know $x^3 = -7$ and one root is $x = -\sqrt[3]{7}$. We use synthetic division.

```
-∛7 | 1    0      0      7
    |    -∛7    ∛7²     -7
    ------------------------
      1  -∛7    ∛7²  |   0
```

We can write the equation in factored form.

$$(x + \sqrt[3]{7})(x^2 - \sqrt[3]{7}x + \sqrt[3]{7^2}) = 0$$

Solve $x^2 - \sqrt[3]{7}x + \sqrt[3]{7^2}$ using the quadratic formula.

$$x = \frac{\sqrt[3]{7} \pm \sqrt{\sqrt[3]{7^2} - 4\sqrt[3]{7^2}}}{2}$$

$$x = \frac{\sqrt[3]{7} \pm \sqrt{-3\sqrt[3]{7^2}}}{2} = \frac{\sqrt[3]{7} \pm \sqrt{-3 \cdot 7^{2/3}}}{2}$$

$$x = \frac{\sqrt[3]{7} \pm \sqrt[3]{7}\sqrt{-3}}{2} \qquad (\sqrt{7^{2/3}} = (7^{2/3})^{1/2} = 7^{1/3})$$

$$x = \sqrt[3]{7}\left[\frac{1 \pm i\sqrt{3}}{2}\right]$$

The roots are $-\sqrt[3]{7}$, $\sqrt[3]{7}\left[\frac{1}{2} + \frac{\sqrt{3}}{2}i\right]$, and $\sqrt[3]{7}\left[\frac{1}{2} - \frac{\sqrt{3}}{2}i\right]$.

70. $\sqrt[3]{5}$, $\sqrt[3]{5}\left[-\frac{1}{2} \pm \frac{\sqrt{3}}{2}i\right]$

71. $P(x) = x^3 - 31x^2 + 230x - 200$

a) Solve $x^3 - 31x^2 + 230x - 200 = 0$

The possible rational roots are ± 1, ± 2, ± 4, ± 5, ± 8, ± 10, ± 20, ± 25, ± 40, ± 50, ± 100, ± 200.

```
1 | 1  -31   230  -200
  |       1   -30   200
  -----------------------
    1  -30   200 |   0
```

We see that 1 is a root.

Then $(x - 1)(x^2 - 30x + 200) = 0$.

Solve $x^2 - 30x + 200 = 0$.

$$(x - 10)(x - 20) = 0$$

$$x = 10 \quad \text{or} \quad x = 20$$

The break-even values are 1, 10, and 20.

b) Solve $x^3 - 31x^2 + 230x - 200 > 0$, $x \geqslant 0$.

The numbers 1, 10, and 20 divide the nonnegative portion of the real number line as shown.

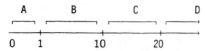

```
     A        B         C          D
   ⌐‾‾⌐  ⌐‾‾‾‾‾‾⌐  ⌐‾‾‾‾‾‾‾⌐  ⌐‾‾‾‾‾‾
 ──┴──┴──────┴────────┴────────
   0   1        10        20
```

Try a test number in each interval.

A: $P\left(\frac{1}{2}\right) = \left(\frac{1}{2}\right)^3 - 31\left(\frac{1}{2}\right)^2 + 230\left(\frac{1}{2}\right) - 200 =$

$-\frac{741}{8}$

B: $P(2) = 2^3 - 31(2)^2 + 230 \cdot 2 - 200 = 144$

C: $P(11) = 11^3 - 31(11)^2 + 230 \cdot 11 - 200 = -90$

D: $P(21) = 21^3 - 31(21)^2 + 230 \cdot 21 - 200 = 220$

The nonnegative values of x for which the company makes a profit are $\{x | 1 < x < 10$ or $x > 20\}$.

c) Using the results of part (b) we see that the nonnegative values of x for which the company has a loss are $\{x | 0 \leqslant x < 1$ or $10 < x < 20\}$.

72. $\frac{9}{2}$ m $\times \frac{3}{2}$ m $\times \frac{1}{2}$ m

---

Exercise Set 4.8

1. $P(x) = 3x^5 - 2x^2 + x - 1$

```
           ⌐‾⌐ ⌐‾⌐ ⌐‾⌐
           a)  b)  c)
```

a) From positive to negative: a variation
b) From negative to positive: a variation
c) From positive to negative: a variation

The number of variations of sign is three. Therefore, the number of positive real roots is either 3 or 1.

$P(-x) = 3(-x)^5 - 2(-x)^2 + (-x) - 1$

$= -3x^5 - 2x^2 - x - 1$

```
           ⌐‾⌐ ⌐‾⌐ ⌐‾⌐
           a)  b)  c)
```

a) From negative to negative: no variation
b) From negative to negative: no variation
c) From negative to negative: no variation

There are no variations of sign. Thus, there are no negative real roots of P(x).

2. 3 or 1; 0

---

3. $P(x) = 6x^7 + 2x^2 + 5x + 4$

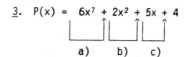

```
       ⌐‾⌐  ⌐‾⌐  ⌐‾⌐
       a)   b)   c)
```

a) From positive to positive: no variation
b) From positive to positive: no variation
c) From positive to positive: no variation

There are no variations of sign. Thus, there are no positive real roots.

$P(-x) = 6(-x)^7 + 2(-x)^2 + 5(-x) + 4$

$= -6x^7 + 2x^2 - 5x + 4$

```
       ⌐‾⌐  ⌐‾⌐  ⌐‾⌐
       a)   b)   c)
```

a) From negative to positive: a variation
b) From positive to negative: a variation
c) From negative to positive: a variation

The number of variations of sign is three. Therefore, the number of negative real roots is 3 or 1.

4. 0; 1

5. $P(p) = 3p^{18} + 2p^4 - 5p^2 + p + 3$

```
       ⌐‾⌐  ⌐‾⌐  ⌐‾⌐  ⌐‾⌐
       a)   b)   c)   d)
```

a) From positive to positive: no variation
b) From positive to negative: a variation
c) From negative to positive: a variation
d) From positive to positive: no variation

The number of variations of sign is two. Therefore, the number of positive real roots is either 2 or 0.

$P(-p) = 3(-p)^{18} + 2(-p)^4 - 5(-p)^2 + (-p) + 3$

$= 3p^{18} + 2p^4 - 5p^2 - p + 3$

```
       ⌐‾⌐  ⌐‾⌐  ⌐‾⌐  ⌐‾⌐
       a)   b)   c)   d)
```

a) From positive to positive: no variation
b) From positive to negative: a variation
c) From negative to negative: no variation
d) From negative to positive: a variation

The number of variations of sign is two. Therefore, the number of negative real roots of P(p) is 2 or 0.

6. 2 or 0; 4, 2, or 0

7. $P(x) = 7x^6 + 3x^4 - x - 10$

The number of variations of sign is 1. Therefore, the number of positive real roots is 1.

$P(-x) = 7(-x)^6 + 3(-x)^4 - (-x) - 10$

$= 7x^6 + 3x^4 + x - 10$

The number of variations of sign is 1. Therefore, the number of negative real roots is 1.

8. 2 or 0; 0

9. $P(t) = -4t^5 - t^3 + 2t^2 + 1$

The number of variations of sign is 1.
Therefore, the number of positive real roots is 1.

$P(-t) = -4(-t)^5 - (-t)^3 + 2(-t)^2 + 1$

$\qquad = 4t^5 + t^3 + 2t^2 + 1$

There are no variations of sign. Therefore, there are no negative real roots.

10.   1; 1

11. $P(y) = y^4 + 13y^3 - y + 5$

The number of variations of sign is two.
Therefore, the number of positive real roots is 2 or 0.

$P(-y) = (-y)^4 + 13(-y)^3 - (-y) + 5$

$\qquad = y^4 - 13y^3 + y + 5$

The number of variations of sign is two.
Therefore, the number of negative real roots is 2 or 0.

12.   3 or 1; 2 or 0

13. $P(x) = 3x^4 - 15x^2 + 2x - 3$

Try 1:

```
1 | 3    0   -15    2    -3
  |      3     3  -12   -10
  ----------------------------
    3    3   -12  -10  | -13
```

Since some of the coefficients of the quotient and the remainder are negative, there is no guarantee that 1 is an upper bound.

Try 2:

```
2 | 3    0   -15    2    -3
  |      6    12   -6    -8
  ----------------------------
    3    6    -3   -4  | -11
```

Again there is no guarantee that 2 is an upper bound.

Try 3:

```
3 | 3    0   -15    2    -3
  |      9    27   36   114
  ----------------------------
    3    9    12   38  | 111
```

Since all of the numbers in the bottom row are nonnegative, 3 is an upper bound. Thus, 3 is the smallest positive integer that is guaranteed by Theorem 15 to be an upper bound to the roots of $P(x)$.

Next we choose a negative number. Try -1:

```
-1 | 3    0   -15    2    -3
   |      -3     3   12  -14
   ----------------------------
     3   -3   -12   14  | -17
```

Since the numbers in the bottom row are not alternately positive and negative, -1 is not guaranteed to be a lower bound.

Try -2:

```
-2 | 3    0   -15    2    -3
   |      -6    12    6  -16
   ----------------------------
     3   -6    -3    8  | -19
```

By Theorem 15, -2 is not guaranteed to be a lower bound.

Try -3:

```
-3 | 3    0   -15    2    -3
   |      -9    27  -36   102
   ----------------------------
     3   -9    12  -34  | 99
```

Since the numbers in the bottom row are alternately positive and negative, -3 is a lower bound. Thus, -3 is the largest negative integer that is guaranteed by Theorem 15 to be a lower bound to the roots of $P(x)$.

14.   2, -3

15. $P(x) = 6x^3 - 17x^2 - 3x - 1$

Try 2:

```
2 | 6   -17    -3    -1
  |      12   -10   -26
  -------------------------
    6    -5   -13  | -27
```

Since some of the coefficients of the quotient and the remainder are negative, there is no guarantee that 2 is an upper bound.

Try 3:

```
3 | 6   -17    -3    -1
  |      18     3     0
  -------------------------
    6     1     0  | -1
```

Since the remainder is negative, there is no guarantee that 3 is an upper bound.

Try 4:

```
4 | 6   -17    -3    -1
  |      24    28   100
  -------------------------
    6     7    25  | 99
```

Since all of the numbers in the bottom row are nonnegative, 4 is an upper bound. Thus, 4 is the smallest positive integer that is guaranteed by Theorem 15 to be an upper bound to the roots of $P(x)$.

Next we choose a negative number. Try -1:

```
-1 | 6   -17    -3    -1
   |      -6    23   -20
   -------------------------
     6   -23    20  | -21
```

Since the numbers in the bottom row are alternately positive and negative, -1 is a lower bound. Since -1 is the largest negative integer, then -1 is the largest negative integer that is guaranteed by Theorem 15 to be a lower bound to the roots of $P(x)$.

16.   3, -1

17. $P(x) = -3x^4 + 15x^3 - 2x + 3$

$-P(x) = -(-3x^4 + 15x^3 - 2x + 3)$

$= 3x^4 - 15x^3 + 2x - 3$

Since $P(x) = 0$ and $-P(x) = 0$ are equivalent equations, we will consider $-P(x)$.

Try 4:

```
4 | 3   -15    0     2     -3
  |      12   -12   -48   -184
    3    -3   -12   -46 | -187
```

Since some of the numbers in the bottom row are negative, 4 is not guaranteed to be an upper bound.

Try 5:

```
5 | 3   -15    0     2    -3
  |      15    0     0    10
    3     0    0     2 |   7
```

Since all of the numbers in the bottom row are nonnegative, 5 is an upper bound. Thus, 5 is the smallest positive integer that is guaranteed by Theorem 15 to be an upper bound to the roots of $P(x)$.

Next we choose a negative number. Try -1:

```
-1 | 3   -15    0     2    -3
   |      -3   18   -18   16
     3   -18   18   -16 | 13
```

Since the numbers in the bottom row are alternately positive and negative, -1 is a lower bound. Since -1 is the largest negative integer, -1 is the largest negative integer guaranteed by Theorem 15 to be a lower bound to the roots of $P(x)$.

18.   5, -1

19. $P(x) = 6x^3 + 15x^2 + 3x - 1$

Try 1:

```
1 | 6   15    3    -1
  |       6   21    24
    6   21   24 |   23
```

Since all of the numbers in the bottom row are nonnegative, 1 is an upper bound. Since 1 is the smallest positive integer, then 1 is the smallest positive integer guaranteed by Theorem 15 to be an upper bound to the roots of $P(x)$.

Next we choose a negative number. Try -2:

```
-2 | 6   15    3    -1
   |     -12   -6    6
     6    3   -3 |   5
```

-2 is not guaranteed by Theorem 15 to be a lower bound.

Try -3:

```
-3 | 6   15    3    -1
   |     -18    9   -36
     6   -3   12 | -37
```

Since the numbers in the bottom row are alternately positive and negative, -3 is a lower bound. Thus, -3 is the largest negative integer guaranteed by Theorem 15 to be a lower bound to the roots of $P(x)$.

20.   1, -2

21. $P(x) = x^4 - 2x^2 + 12x - 8$

There are three variations of sign, so the number of positive real roots is 3 or 1.

$P(-x) = (-x)^4 - 2(-x)^2 + 12(-x) - 8$

$= x^4 - 2x^2 - 12x - 8$

There is one variation of sign, so the number of negative real roots of $P(x)$ is 1.

Look for an upper bound. Use Theorem 15.

Try 1:

```
1 | 1    0   -2   12    -8
  |       1    1   -1    11
    1    1   -1   11 |    3
```

Since one of the coefficients of the quotient is negative, 1 is not guaranteed to be a lower bound.

Try 2:

```
2 | 1    0   -2   12    -8
  |       2    4    4    32
    1    2    2   16 |   24
```

Since all of the numbers in the bottom row are nonnegative, 2 is an upper bound.

Look for a lower bound. Use Theorem 15.

Try -2:

```
-2 | 1    0   -2   12    -8
   |      -2    4   -4   -16
     1   -2    2    8 |  -24
```

-2 is not guaranteed to be a lower bound.

Try -3:

```
-3 | 1    0   -2   12    -8
   |      -3    9  -21    27
     1   -3    7   -9 |   19
```

Since the numbers in the bottom row are alternately positive and negative, -3 is a lower bound.

22.   3 or 1 positive; 1 negative; upper bound: 3; lower bound: -4

23. $P(x) = x^4 - 2x^2 - 8$

There is one variation of sign, so the number of positive real roots is 1.

$P(-x) = (-x)^4 - 2(-x)^2 - 8$

$\quad\quad = x^4 - 2x^2 - 8$

There is one variation of sign, so the number of negative real roots is 1. Since the degree of the equation is 4 and there is only one positive real root and one negative real root, there must be two nonreal roots.

Look for an upper bound. Use Theorem 15.

Try 1:

```
1 | 1   0   -2    0   -8
  |     1    1   -1   -1
  ----------------------
    1   1   -1   -1 | -9
```

Since some of the coefficients of the quotient and the remainder are negative, 1 is not guaranteed to be an upper bound.

Try 2:

```
2 | 1   0   -2    0   -8
  |     2    4    4    8
  ----------------------
    1   2    2    4 |  0
```

Since all of the numbers in the bottom row are nonnegative, 2 is an upper bound. Also note that 2 is a root, $P(2) = 0$.

Look for a lower bound. Use Theorem 15.

Try -1:

```
-1 | 1   0   -2    0   -8
   |    -1    1    1   -1
   ----------------------
     1  -1   -1    1 | -9
```

-1 is not guaranteed to be a lower bound.

Try -2:

```
-2 | 1   0   -2    0   -8
   |    -2    4   -4    8
   ----------------------
     1  -2    2   -4 |  0
```

Since the numbers in the bottom row are alternately positive and negative, -2 is a lower bound.

24. 1 positive; 1 negative; 2 nonreal; upper bound: 2; lower bound: -2

25. $P(x) = x^4 - 9x^2 - 6x + 4$

There are two variations of sign, so the number of positive real roots is 2 or 0.

$P(-x) = (-x)^4 - 9(-x)^2 - 6(-x) + 4$

$\quad\quad = x^4 - 9x^2 + 6x + 4$

There are two variations of sign, so the number of negative real roots of $P(x)$ is 2 or 0.

Look for an upper bound. Use Theorem 15.

Try 3:

```
3 | 1   0   -9   -6    4
  |     3    9    0  -18
  -----------------------
    1   3    0   -6 | -14
```

Since one of the coefficients of the quotient and the remainder are negative, 3 is not guaranteed to be an upper bound.

Try 4:

```
4 | 1   0   -9   -6    4
  |     4   16   28   88
  -----------------------
    1   4    7   22 | 92
```

Since all of the numbers in the bottom row are nonnegative, 4 is an upper bound.

Look for a lower bound. Use Theorem 15.

Try -2:

```
-2 | 1   0   -9   -6    4
   |    -2    4   10   -8
   -----------------------
     1  -2   -5    4 | -4
```

-2 is not guaranteed to be a lower bound.

Try -3:

```
-3 | 1   0   -9   -6    4
   |    -3    9    0   18
   -----------------------
     1  -3    0   -6 | 22
```

Since the numbers in the bottom row are alternately positive and negative, -3 is a lower bound.

26. 2 or 0 positive; 2 or 0 negative; upper bound: 5; lower bound: -5

27. $P(x) = x^4 + 3x^2 + 2$

There are no variations of sign. Thus there are no positive real roots.

$P(-x) = (-x)^4 + 3(-x)^2 + 2$

$\quad\quad = x^4 + 3x^2 + 2$

There are no variations of sign. Thus there are no negative real roots. Since the degree of the equation is 4, there are 4 nonreal roots.

Since there are no real roots, there is no upper bound and no lower bound.

28. 0 positive; 0 negative; 4 nonreal

29. $P(x) = x^3 - 3x - 2$

The possible rational roots are 1, -1, 2, and -2. Synthetic division shows that -1 is a root with multiplicity of 2 and 2 is also a root. Thus all the roots are rational.

30. -0.9, 1.3, 2.5

31. $x^3 - 3x - 4 = 0$

    The possible rational roots are 1, -1, 2, -2, 4, and -4. Synthetic division shows that none of these is a root. Sketch a graph of the polynomial $P(x) = x^3 - 3x - 4$.

    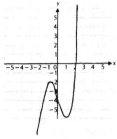

    We see from the graph that there seems to be a root between 2 and 3. Checking function values confirms this: $P(2) = -2$ and $P(3) = 14$. To find a better approximation we compute additional function values.

x	P(x)
2	-2
2.1	-1.039
2.2	0.048

    Since $P(2.1) < 0$, there is no root between 2 and 2.1.

    Since $P(2.1) < 0$ and $P(2.2) > 0$, there is a root between 2.1 and 2.2.

    We have the accuracy desired, so we stop. Since 0.048 is closer to 0 than -1.039, we accept 2.2 as a better approximation to the nearest tenth.

32. -1.1

33. $x^4 + x^2 + 1 = 0$

    $P(x) = x^4 + x^2 + 1$ has no variations in sign and hence no positive real roots. $P(-x) = (-x)^4 + (-x)^2 + 1 = x^4 + x^2 + 1$ also has no variations in sign and hence $P(x)$ has no negative real roots. Thus the equation has no real roots.

34. No real roots

35. $x^4 - 6x^2 + 8 = 0$

    The possible rational roots are ± 1, ± 2, ± 4, and ± 8. Using synthetic division we find that 2 and -2 are roots and we can factor the polynomial as follows: $(x - 2)(x + 2)(x^2 - 2)$. Solving $x^2 - 2 = 0$, we find that the irrational roots are $\sqrt{2}$ and $-\sqrt{2}$, or 1.4 and -1.4 to the nearest tenth.

36. -1.8, -0.8, 0.8, 1.8

37. $x^5 + x^4 - x^3 - x^2 - 2x - 2 = 0$

    The possible rational roots are ± 1 and ± 2. Using synthetic division we find that -1 is a root and we can factor the polynomial as follows: $(x + 1)(x^4 - x^2 - 2)$. We can solve $x^4 - x^2 - 2 = 0$ by substituting u for x^2, factoring, and solving for u and then for x. We get $x = \pm \sqrt{2}$ or $x = \pm i$. The irrational roots are $\sqrt{2}$ and $-\sqrt{2}$, or 1.4 and -1.4 to the nearest tenth.

38. ± 1.7

39. $P(x) = 2x^5 + 2x^3 - x^2 - 1$

x	P(x)
0	-1
1	2
0.5	-0.9375
0.7	-0.46786
0.8	0.03936
0.75	-0.24414
0.79	-0.02261

    Since $P(0.5) < 0$, there is a root between 0.5 and 1.

    Since $P(0.7) < 0$, there is a root between 0.7 and 1.

    Since $P(0.8) > 0$, there is a root between 0.7 and 0.8.

    Since $P(0.75) < 0$, there is a root between 0.75 and 0.8.

    Since $P(0.79) < 0$, there is a root between 0.79 and 0.8.

    We have the accuracy we desire, so we stop. Since -0.02261 is closer to 0 than 0.03936, we accept 0.79 as a better approximation to the nearest hundredth.

40. 1.41

41. $P(x) = x^3 - 2x^2 - x + 4$

    The possible rational roots are ± 1, ± 2, and ± 4. Synthetic division shows that none of these is a root. Sketch a graph of the polynomial.

    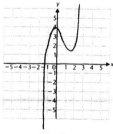

    The graph shows that there seems to be a root between -2 and -1. We check function values to confirm this: $P(-2) = -10$, $P(-1) = 2$. We do further computations to find a better approximation.

x	P(x)
-1	2
-1.1	1.349
-1.2	0.592
-1.3	-0.277
-1.25	0.1719
-1.27	-0.0042
-1.26	0.0844

Since P(-1.1) > 0, there is no root between -1.1 and -1.

Since P(-1.2) > 0, there is no root between -1.2 and -1.1.

Since P(-1.3) < 0, there is a root between -1.3 and -1.2.

Since P(-1.25) > 0, there is a root between -1.3 and -1.25.

Since P(-1.27) < 0, there is a root between -1.27 and -1.25.

Since P(-1.26) > 0, there is a root between -1.27 and -1.26.

We have the desired accuracy, so we stop. Since -0.0042 is closer to 0 than 0.0844, we accept -1.27 as the better approximation to the nearest hundredth.

42. -0.70, 1.24, and 3.46

43. See the answer section in the text.

44. f(x) has one variation in sign; hence one positive root. f(-x) has no variation in sign; hence no negative roots. Zero is not a root. Hence there is exactly one real root.

45. -1.3

46. -1.88, 0.35, 1.53

47. 1.5, 5.7

48. -7, -5, -2, 1, 3, 6

49. 7.16 hr

50. 0.08, or 8%

Exercise Set 4.9

1. $f(x) = \dfrac{1}{x - 3}$

   a) The function is neither even nor odd and no symmetries are apparent.

   b) The degree of the numerator is less than the degree of the denominator, so the x-axis is an asymptote.

   c) The zero of the denominator is 3, so x = 3 is a vertical asymptote. The function has no zeros.

d) The vertical asymptote divides the x-axis into intervals. To find where the function is positive or negative, we try a test point in each interval.

Interval	$(-\infty, 3)$	$(3, \infty)$
Test value	$f(0) = -\dfrac{1}{3}$	$f(4) = 1$
Sign of f(x)	-	+
Location of points on graph	Below x-axis	Above x-axis

e) We make a table of function values.

x	$3\frac{1}{8}$	$3\frac{1}{2}$	4	5	8	$2\frac{3}{4}$	2	1	0	-2
f(x)	8	2	1	$\frac{1}{2}$	$\frac{1}{5}$	-4	-1	$-\frac{1}{2}$	$-\frac{1}{3}$	$-\frac{1}{5}$

f) Using all available information, we draw the graph.

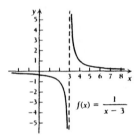

$$f(x) = \frac{1}{x - 3}$$

2.

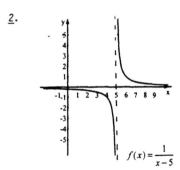

$$f(x) = \frac{1}{x - 5}$$

3. $f(x) = \dfrac{-2}{x - 5}$

   a) The function is neither even nor odd and no symmetries are apparent.

   b) The degree of the numerator is less than the degree of the denominator, so the x-axis is an asymptote.

   c) The zero of the denominator is 5, so x = 5 is a vertical asymptote. The function has no zeros.

d) The vertical asymptote divides the x-axis into intervals. To find where the function is positive or negative, we try a test point in each interval.

Interval	$(-\infty, 5)$	$(5, \infty)$
Test value	$f(0) = \frac{2}{5}$	$f(6) = -2$
Sign of f(x)	+	-
Location of points on graph	Above x-axis	Below x-axis

e) We make a table of function values.

x	-3	0	2	4	$4\frac{2}{3}$	$5\frac{1}{3}$	6	8	11
f(x)	$\frac{1}{4}$	$\frac{2}{5}$	$\frac{2}{3}$	2	6	-6	-2	$-\frac{2}{3}$	$-\frac{1}{3}$

f) Using all available information, we draw the graph.

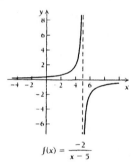

$$f(x) = \frac{-2}{x - 5}$$

4.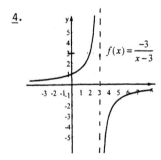

$$f(x) = \frac{-3}{x-3}$$

5. $f(x) = \dfrac{2x + 1}{x}$

a) The function is neither even nor odd and no symmetries are apparent.

b) Since the numerator and denominator have the same degree, the horizontal asymptote can be determined by dividing the leading coefficients of the two polynomials. Thus $2 \div 1 = 2$, and we know that $y = 2$ is a horizontal asymptote.

c) The zero of the denominator is 0, so x = 0 (the y-axis) is a vertical asymptote. The zero of the function is $-\frac{1}{2}$.

d) The vertical asymptote and the zero of the function divide the x-axis into intervals. To find where the function is positive or negative, we try a test number in each interval.

Interval	$\left(-\infty, -\frac{1}{2}\right]$	$\left[-\frac{1}{2}, 0\right)$	$(0, \infty)$
Test value	$f(-1) = 1$	$f\left(-\frac{1}{4}\right) = -8$	$f(1) = 3$
Sign of f(x)	+	-	+
Location of points on graph	Above x-axis	Below x-axis	Above x-axis

e) We make a table of function values.

x	-100	-10	-3	$-\frac{1}{2}$	$-\frac{1}{10}$	1	4	10	100
f(x)	$1\frac{99}{100}$	$1\frac{9}{10}$	$1\frac{2}{3}$	0	-8	3	$2\frac{1}{4}$	$2\frac{1}{10}$	$2\frac{1}{100}$

f) Using all available information, we draw the graph.

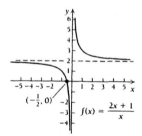

$$f(x) = \frac{2x + 1}{x}$$

6.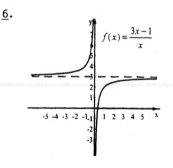

$$f(x) = \frac{3x - 1}{x}$$

7. $f(x) = \dfrac{1}{(x-2)^2}$

a) The function is neither even nor odd and no symmetries are apparent.

b) The degree of the numerator is less than the degree of the denominator, so the x-axis is an asymptote.

c) The zero of the denominator is 2, so x = 2 is a vertical asymptote. The function has no zeros.

d) $(x-2)^2 > 0$ for all x ≠ 2, so f(x) is positive at each point in the domain.

e) We make a table of function values.

x	$2\frac{1}{2}$	3	4	5	$1\frac{1}{2}$	1	0	-1
f(x)	4	1	$\frac{1}{4}$	$\frac{1}{9}$	4	1	$\frac{1}{4}$	$\frac{1}{9}$

f) Using all available information, we draw the graph.

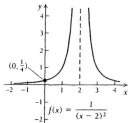

8.

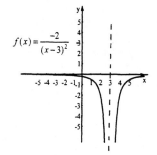

9. $f(x) = \dfrac{2}{x^2}$

a) The function is even, so it is symmetric with respect to the y-axis.

b) The degree of the numerator is less than the degree of the denominator, so the x-axis is an asymptote.

c) The zero of the denominator is 0, so x = 0 (the y-axis) is a vertical asymptote. The function has no zeros.

d) $x^2 > 0$ for all x ≠ 0, so f(x) is positive at each point in the domain.

e) We make a table of function values.

x	$\pm\frac{1}{2}$	± 1	± 2	± 3	± 4
f(x)	8	2	$\frac{1}{2}$	$\frac{2}{9}$	$\frac{1}{8}$

f) Using all available information, we draw the graph.

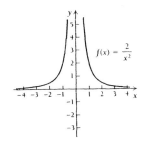

10.

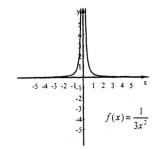

11. $f(x) = \dfrac{1}{x^2 + 3}$

a) The function is even, so the graph is symmetric with respect to the y-axis.

b) The degree of the numerator is less than the degree of the denominator, so the x-axis is an asymptote.

c) The denominator has no zeros, so there are no vertical asymptotes. The function has no zeros.

d) $x^2 + 3 > 0$ for all real numbers x, so f(x) is positive for all real numbers x.

e) We make a table of function values.

x	0	± 1	± 2	± 3
f(x)	$\frac{1}{3}$	$\frac{1}{4}$	$\frac{1}{7}$	$\frac{1}{12}$

f) Using all available information, we draw the graph.

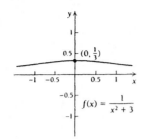

$$f(x) = \frac{1}{x^2 + 3}$$

f) Using all available information, we draw the graph.

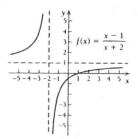

$$f(x) = \frac{x-1}{x+2}$$

12.

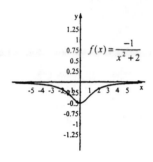

$$f(x) = \frac{-1}{x^2 + 2}$$

14.

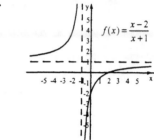

$$f(x) = \frac{x-2}{x+1}$$

13. $f(x) = \frac{x-1}{x+2}$

a) The function is neither even nor odd and no symmetries are apparent.

b) Since the numerator and denominator have the same degree, the horizontal asymptote can be determined by dividing the leading coefficients of the two polynomials. Thus $1 \div 1 = 1$, and we know that $y = 1$ is a horizontal asymptote.

c) The zero of the denominator is -2, so $x = -2$ is a vertical asymptote. The zero of the function is 1.

d) The vertical asymptote and the zero of the function divide the x-axis into intervals. To find where the function is positive or negative, we try a test point in each interval.

Interval	$(-\infty, -2)$	$(-2, 1)$	$(1, \infty)$
Test value	$f(-3) = 4$	$f(0) = -\frac{1}{2}$	$f(2) = \frac{1}{4}$
Sign of $f(x)$	+	-	+
Location of points on graph	Above x-axis	Below x-axis	Above x-axis

e) We make a table of function values.

x	-6	-4	-3	$-2\frac{1}{2}$	$-1\frac{1}{2}$	-1	0	1	2	4	6	20
$f(x)$	$\frac{7}{4}$	$\frac{5}{2}$	4	7	-5	-2	$-\frac{1}{2}$	0	$\frac{1}{4}$	$\frac{1}{2}$	$\frac{5}{8}$	$\frac{19}{22}$

15. $f(x) = \frac{3x}{x^2 + 5x + 4} = \frac{3x}{(x+4)(x+1)}$

a) The function is neither even nor odd and no symmetries are apparent.

b) The degree of the denominator is greater than that of the numerator. Thus the x-axis is an asymptote.

c) The zeros of the denominator are -4 and -1, so $x = -4$ and $x = -1$ are vertical asymptotes. The zero of the function is 0.

d) The vertical asymptotes and the zero of the function divide the x-axis into intervals. To find where the function is positive or negative, we try a test point in each interval.

Interval	Test value	Sign of $f(x)$	Location of points on graph
$(-\infty, -4)$	$f(-5) = -\frac{15}{4}$	-	Below x-axis
$(-4, -1)$	$f(-2) = 3$	+	Above x-axis
$(-1, 0)$	$f\left(-\frac{1}{2}\right) = -\frac{6}{7}$	-	Below x-axis
$(0, \infty)$	$f(1) = \frac{3}{10}$	+	Above x-axis

e) We make a table of function values.

x	-6	-5	$-\frac{9}{2}$	$-\frac{7}{2}$	-3	-2	$-\frac{3}{2}$
f(x)	$-\frac{9}{5}$	$-\frac{15}{4}$	$-\frac{54}{7}$	$\frac{42}{5}$	$\frac{9}{2}$	3	$\frac{18}{5}$

x	$-\frac{3}{4}$	$-\frac{1}{2}$	0	1	2	3
f(x)	$-\frac{36}{13}$	$-\frac{6}{7}$	0	$\frac{3}{10}$	$\frac{1}{3}$	$\frac{9}{28}$

f) Using all the information, we draw the graph.

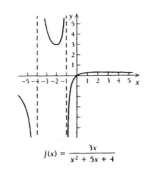

$$f(x) = \frac{3x}{x^2 + 5x + 4}$$

16.

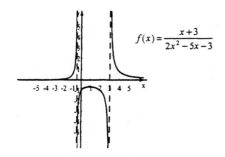

$$f(x) = \frac{x + 3}{2x^2 - 5x - 3}$$

17. $f(x) = \frac{x^2 - 4}{x - 1} = \frac{(x + 2)(x - 2)}{x - 1}$

a) The function is neither even nor odd and no symmetries are apparent.

b) The degree of the numerator is one greater than that of the denominator. Dividing numerator by denominator will show that y = x + 1 is an oblique asymptote.

c) The zero of the denominator is 1, so x = 1 is a vertical asymptote. The zeros of the function are -2 and 2.

d) The vertical asymptote and the zeros of the function divide the x-axis into intervals. To find where the function is positive or negative, we try a test point in each interval.

Interval	Test value	Sign of f(x)	Location of points on graph
$(-\infty,\ -2)$	$f(-3) = -\frac{5}{4}$	-	Below x-axis
$(-2,\ 1)$	$f(0) = 4$	+	Above x-axis
$(1,\ 2)$	$f\left[\frac{3}{2}\right] = -\frac{7}{2}$	-	Below x-axis
$(2,\ \infty)$	$f(3) = \frac{5}{2}$	+	Above x-axis

e) We make a table of function values.

x	-3	-2	-1	0	$\frac{1}{2}$	$\frac{3}{2}$	2	3
f(x)	$-1\frac{1}{4}$	0	$1\frac{1}{2}$	4	$7\frac{1}{2}$	$-3\frac{1}{2}$	0	$2\frac{1}{2}$

f) Using all available information, we draw the graph.

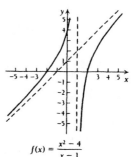

$$f(x) = \frac{x^2 - 4}{x - 1}$$

18.

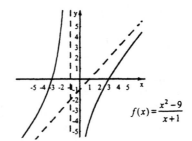

$$f(x) = \frac{x^2 - 9}{x + 1}$$

19. $f(x) = \frac{x^2 + x - 2}{2x^2 + 1} = \frac{(x + 2)(x - 1)}{2x^2 + 1}$

a) The function is neither even nor odd and no symmetries are apparent.

b) Since the numerator and denominator have the same degree, the horizontal asymptote can be determined by dividing the leading coefficients of the two polynomials. Thus 1 ÷ 2 = 1/2, and we know that y = 1/2 is a horizontal asymptote.

c) The denominator has no zeros, so there are no vertical asymptotes. The zeros of the function are -2 and 1.

d) The zeros of the function divide the x-axis into intervals. To find where the function is positive or negative, we try a test point in each interval.

Interval	$(-\infty, -2)$	$(-2, 1)$	$(1, \infty)$
Test value	$f(-3) = \frac{4}{19}$	$f(0) = -2$	$f(2) = \frac{4}{9}$
Sign of f(x)	+	−	+
Location of points on graph	Above x-axis	Below x-axis	Above x-axis

e) We make a table of function values.

x	−5	−3	−2	−1	0	1	2	3
f(x)	$\frac{6}{17}$	$\frac{4}{19}$	0	$-\frac{2}{3}$	−2	0	$\frac{4}{9}$	$\frac{10}{19}$

f) Using all the information, we draw the graph.

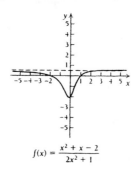

$$f(x) = \frac{x^2 + x - 2}{2x^2 + 1}$$

20.

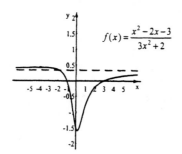

$$f(x) = \frac{x^2 - 2x - 3}{3x^2 + 2}$$

21. $f(x) = \dfrac{x - 1}{x^2 - 2x - 3} = \dfrac{x - 1}{(x - 3)(x + 1)}$

a) The function is neither even nor odd and no symmetries are apparent.

b) The degree of the denominator is greater than the degree of the numerator. Thus the x-axis is an asymptote.

c) The zeros of the denominator are 3 and −1, so x = 3 and x = −1 are vertical asymptotes. The zero of the function is 1.

d) The vertical asymptotes and the zeros of the function divide the x-axis into intervals. To find where the function is positive or negative, we try a test point in each interval.

Interval	Test value	Sign of f(x)	Location of points on graph
$(-\infty, -1)$	$f(-2) = -\frac{3}{5}$	−	Below x-axis
$(-1, 1)$	$f(0) = \frac{1}{3}$	+	Above x-axis
$(1, 3)$	$f(2) = -\frac{1}{3}$	−	Below x-axis
$(3, \infty)$	$f(4) = \frac{3}{5}$	+	Above x-axis

e) We make a table of function values.

x	−5	−3	−2	$-\frac{3}{2}$	$-\frac{1}{2}$	0	1
f(x)	$-\frac{3}{16}$	$-\frac{1}{3}$	$-\frac{3}{5}$	$-\frac{10}{9}$	$\frac{6}{7}$	$\frac{1}{3}$	0

x	2	$\frac{5}{2}$	4	5
f(x)	$-\frac{1}{3}$	$-\frac{6}{7}$	$\frac{3}{5}$	$\frac{1}{3}$

f) Using all available information, we draw the graph.

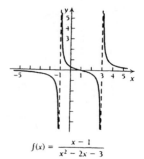

$$f(x) = \frac{x - 1}{x^2 - 2x - 3}$$

22.

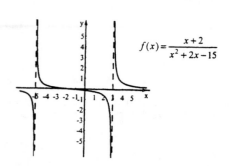

$$f(x) = \frac{x + 2}{x^2 + 2x - 15}$$

23. $f(x) = \dfrac{x + 2}{(x - 1)^3}$

   a) The function is neither even nor odd and no symmetries are apparent.

   b) The degree of the denominator is greater than the degree of the numerator. Thus the x-axis is an asymptote.

   c) The zero of the denominator is 1, so x = 1 is a vertical asymptote. The zero of the function is -2.

   d) The vertical asymptote and the zero of the function divide the x-axis into intervals. To find where the function is positive or negative, we try a test point in each interval.

Interval	$(-\infty, -2)$	$(-2, 1)$	$(1, \infty)$
Test value	$f(-3) = \frac{1}{64}$	$f(0) = -2$	$f(2) = 4$
Sign of f(x)	+	−	+
Location of points on graph	Above x-axis	Below x-axis	Above x-axis

   e) Make a table of function values.

x	-5	-4	-3	-2	-1	0	$\frac{1}{2}$
f(x)	$\frac{1}{72}$	$\frac{2}{125}$	$\frac{1}{64}$	0	$-\frac{1}{8}$	-2	-20

x	$\frac{3}{2}$	2	3
f(x)	28	4	$\frac{5}{8}$

   f) Using all available information, we draw the graph.

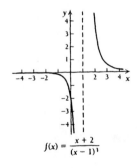

$f(x) = \dfrac{x + 2}{(x - 1)^3}$

24.

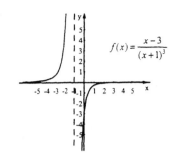

$f(x) = \dfrac{x - 3}{(x + 1)^3}$

25. $f(x) = \dfrac{x^3 + 1}{x} = \dfrac{(x + 1)(x^2 - x + 1)}{x}$

   a) The function is neither even nor odd and no symmetries are apparent.

   b) The degree of the numerator is two greater than the denominator. There are no horizontal or oblique asymptotes.

   c) The zero of the denominator is 0, so x = 0 (the y-axis) is a vertical asymptote. The only real-number zero of the function is -1.

   d) The vertical asymptote and the real zero of the function divide the x-axis into intervals. To find where the function is positive or negative, we try a test point in each interval.

Interval	$(-\infty, -1)$	$(-1, 0)$	$(0, \infty)$
Test value	$f(-2) = \frac{7}{2}$	$f\left(-\frac{1}{2}\right) = -\frac{7}{4}$	$f(1) = 2$
Sign of f(x)	+	−	+
Location of points on graph	Above x-axis	Below x-axis	Above x-axis

   e) Make a table of function values.

x	-3	-2	-1	$-\frac{1}{2}$	$\frac{1}{2}$	1	2	3
f(x)	$\frac{26}{3}$	$\frac{7}{2}$	0	$-\frac{7}{4}$	$\frac{9}{4}$	2	$\frac{9}{2}$	$\frac{28}{3}$

   f) Draw the graph using all the known information.

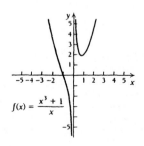

$f(x) = \dfrac{x^3 + 1}{x}$

26.

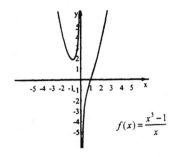

$$f(x) = \frac{x^3 - 1}{x}$$

27. $f(x) = \dfrac{x^3 + 2x^2 - 15x}{x^2 - 5x - 14} = \dfrac{x(x + 5)(x - 3)}{(x + 2)(x - 7)}$

a) The function is neither even nor odd and no symmetries are apparent.

b) The degree of the numerator is one greater than that of the denominator. Dividing numerator by denominator will show that y = x + 7 is an oblique asymptote.

c) The zeros of the denominator are -2 and 7, so x = -2 and x = 7 are vertical asymptotes. The zeros of the function are 0, -5, and 3.

d) The vertical asymptotes and the zeros of the function divide the x-axis into intervals. To find where the function is positive or negative, we try a test value in each interval.

Interval	Test value	Sign of f(x)	Location of points on graph
$(-\infty, -5)$	$f(-6) = -\dfrac{27}{26}$	$-$	Below x-axis
$(-5, -2)$	$f(-3) = \dfrac{18}{5}$	$+$	Above x-axis
$(-2, 0)$	$f(-1) = -2$	$-$	Below x-axis
$(0, 3)$	$f(1) = \dfrac{2}{3}$	$+$	Above x-axis
$(3, 7)$	$f(4) = -2$	$-$	Below x-axis
$(7, \infty)$	$f(8) = 52$	$+$	Above x-axis

e) Make a table of function values.

x	-100	-10	-8	-7	-6	-5	-4	-3	-2.5
f(x)	-93.3	-4.8	-2.9	-2	-1.0	0	1.3	3.6	7.2

x	-2.1	-1.9	-1.5	-1	0	1	2	3	4	5
f(x)	34.1	-32.4	-5.6	-2	0	0.7	0.7	0	-2	-7.1

x	6	7.1	7.5	8	10	20	100
f(x)	-24.8	387	88.8	52	29.2	29.7	107

f) Draw the graph using all available information.

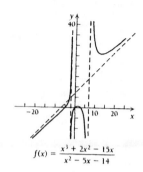

$$f(x) = \frac{x^3 + 2x^2 - 15x}{x^2 - 5x - 14}$$

28.

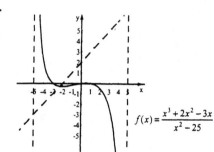

$$f(x) = \frac{x^3 + 2x^2 - 3x}{x^2 - 25}$$

29. $f(x) = \dfrac{5x^4}{x^4 + 1}$

a) The function is even, so the graph is symmetric about the y-axis.

b) The degree of the numerator equals the degree of the denominator, so $y = \dfrac{5}{1}$, or y = 5, is a horizontal asymptote.

c) The denominator has no zeros, so there are no vertical asymptotes. The zero of the function is 0.

d) $5x^4 \geq 0$ for all real x and $x^4 + 1 > 0$ for all real x, so $f(x) \geq 0$ for all real values of x.

e) We make a table of function values.

x	± 5	± 3	± 1	0
f(x)	4.99	4.94	2.5	0

f) Using all available information, we draw the graph.

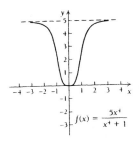

$$f(x) = \frac{5x^4}{x^4 + 1}$$

30.

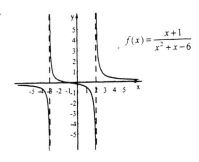

$$f(x) = \frac{x+1}{x^2 + x - 6}$$

31. $f(x) = \dfrac{x^2 - x - 2}{x + 2} = \dfrac{(x - 2)(x + 1)}{x + 2}$

a) The function is neither even nor odd, and no symmetries are apparent.

b) The degree of the numerator is one more than the degree of the denominator. Dividing the numerator by the denominator will show that the line y = x - 3 is an oblique asymptote.

c) The zero of the denominator is -2, so x = -2 is a vertical asymptote. The zeros of the function are 2 and -1.

d) The vertical asymptote and the zeros of the function divide the x-axis into intervals. To find where the function is positive or negative, we try a test point in each interval.

Interval	Test value	Sign of $f(x)$	Location of points on graph
$(-\infty, -2)$	$f(-3) = -10$	−	Below x-axis
$(-2, -1)$	$f\left(-\frac{3}{2}\right) = \frac{7}{2}$	+	Above x-axis
$(-1, 2)$	$f(0) = -1$	−	Below x-axis
$(2, \infty)$	$f(3) = \frac{4}{5}$	+	Above x-axis

e) Make a table of function values.

x	-5	-3	$-\frac{3}{2}$	-1	0	1	2
f(x)	$-\frac{28}{3}$	-10	$\frac{7}{2}$	0	-1	$-\frac{2}{3}$	0

x	3	5
f(x)	$\frac{4}{5}$	$\frac{18}{7}$

f) Using all available information, we draw the graph.

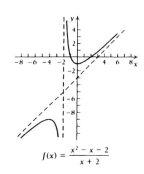

$$f(x) = \frac{x^2 - x - 2}{x + 2}$$

32.

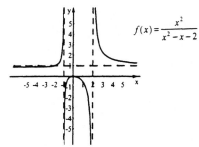

$$f(x) = \frac{x^2}{x^2 - x - 2}$$

33. a) We use the formula d = rt:

   d = rt

   $t = \dfrac{d}{r}$ (Solving for t)

   $t = \dfrac{500}{r}$ (Substituting 500 for d)

b) The graph over (0, ∞) is one branch of a hyperbola.

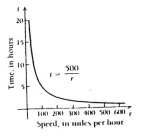

173

<u>34.</u> a) $R = \dfrac{10r}{10 + r}$

b)

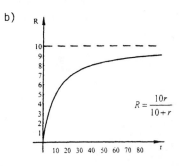

$R = \dfrac{10r}{10+r}$

c) Domain: $(0, \infty)$, range: $(0, 10)$

d) 10

<u>35.</u> $f(x) = \dfrac{x^3 + 4x^2 + x - 6}{x^2 - x - 2} = \dfrac{(x + 3)(x + 2)(x - 1)}{(x - 2)(x + 1)}$

(We can use the rational roots theorem and synthetic division to factor the numerator.)

a) The function is neither even nor odd, and no symmetries are apparent.

b) The degree of the numerator is one more than the degree of the denominator. Dividing the numerator by the denominator will show that the line $y = x + 5$ is an oblique asymptote.

c) The zeros of the denominator are 2 and -1, so $x = 2$ and $x = -1$ are vertical asymptotes. The zeros of the function are -3, -2, and 1.

d) We find where the function is positive or negative.

Interval	Test value	Sign of $f(x)$	Location of points on graph
$(-\infty, -3)$	$f(-4) = -\dfrac{5}{9}$	$-$	Below x-axis
$(-3, -2)$	$f\left(-\dfrac{5}{2}\right) = \dfrac{7}{54}$	$+$	Above x-axis
$(-2, -1)$	$f\left(-\dfrac{3}{2}\right) = -\dfrac{15}{14}$	$-$	Below x-axis
$(-1, 1)$	$f(0) = 3$	$+$	Above x-axis
$(1, 2)$	$f\left(\dfrac{3}{2}\right) = -\dfrac{63}{10}$	$-$	Below x-axis
$(2, \infty)$	$f(3) = 15$	$+$	Above x-axis

e) We make a table of function values.

x	-5	-4	-3	$-\dfrac{5}{2}$	-2	$-\dfrac{3}{2}$	0
$f(x)$	$-\dfrac{9}{7}$	$-\dfrac{5}{9}$	0	$\dfrac{7}{54}$	0	$-\dfrac{15}{14}$	3

x	1	$\dfrac{3}{2}$	3	5	7	9
$f(x)$	0	$-\dfrac{63}{10}$	15	$\dfrac{112}{9}$	$\dfrac{27}{2}$	$\dfrac{528}{35}$

f) Using all available information, we draw the graph.

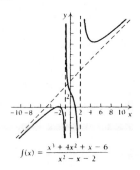

$f(x) = \dfrac{x^3 + 4x^2 + x - 6}{x^2 - x - 2}$

<u>36.</u>

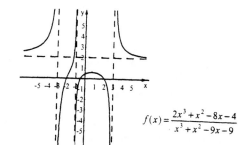

$f(x) = \dfrac{2x^3 + x^2 - 8x - 4}{x^3 + x^2 - 9x - 9}$

<u>37.</u> a) The distance of the round trip is $2 \cdot 300$, or 600 km, so $t_0 = \dfrac{600 \text{ km}}{200 \text{ km/h}} = 3$ hr.

We substitute:

$$T = \dfrac{(200)^2(3)}{(200)^2 - w^2} = \dfrac{120{,}000}{40{,}000 - w^2}$$

b) For $w = 5$ km/h:

$$T = \dfrac{120{,}000}{40{,}000 - 5^2} \approx 3.001876 \text{ hr}$$

For $w = 10$ km/h:

$$T = \dfrac{120{,}000}{40{,}000 - 10^2} \approx 3.007519 \text{ hr}$$

For $w = 20$ km/h:

$$T = \dfrac{120{,}000}{40{,}000 - 20^2} \approx 3.030303 \text{ hr}$$

c) The wind speed must be nonnegative, and it must also be less than the speed of the plane in still air. Thus the domain is $\{w \mid 0 \leqslant w < 200\}$.

For $w = 0$, $T = 3$; for other values of w in the domain, $T > 3$. Thus the range is $\{T \mid T \geqslant 3\}$.

d)

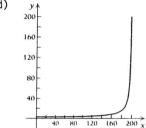

38. a) $S = x^2 + \dfrac{432}{x}$

    b)

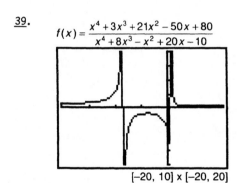

$S = x^2 + \dfrac{432}{x}$

    c) 108 cm²; 6 cm

39.

$f(x) = \dfrac{x^4 + 3x^3 + 21x^2 - 50x + 80}{x^4 + 8x^3 - x^2 + 20x - 10}$

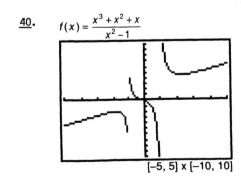

[−20, 10] x [−20, 20]

40. $f(x) = \dfrac{x^3 + x^2 + x}{x^2 - 1}$

[−5, 5] x [−10, 10]

41. $f(x) = \dfrac{x^3}{x^2 - 1}$

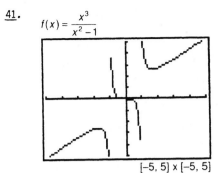

[−5, 5] x [−5, 5]

42. $f(x) = \dfrac{x^3 + 4}{x}$

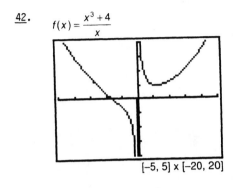

[−5, 5] x [−20, 20]

Exercise Set 5.1

1. We interchange the first and second coordinates of each ordered pair to find the inverse of the relation. It is

   {(1, 0), (6, 5), (-4, -2)}.

2. {(3, -1), (5, 2), (5, -3), (0, 2)}

3. We interchange the first and second coordinates of each ordered pair to find the inverse of the relation. It is

   {(8, 7), (8, -2), (-4, 3), (-8, 8)}.

4. {(-1, -1), (4, -3)}

5. Interchange x and y.

   $y = 4x - 5$
   $\downarrow \quad \downarrow$
   $x = 4y - 5$ (Equation of the inverse relation)

6. $x = 3y + 5$

7. Interchange x and y.

   $x^2 - 3y^2 = 3$
   $\downarrow \quad \downarrow$
   $y^2 - 3x^2 = 3$ (Equation of the inverse relation)

8. $2y^2 + 5x^2 = 4$

9. Interchange x and y.

   $y = 3x^2 + 2$
   $\downarrow \quad \downarrow$
   $x = 3y^2 + 2$ (Equation of the inverse relation)

10. $x = 5y^2 - 4$

11. Interchange x and y.

    $xy = 7$
    $\downarrow\downarrow$
    $yx = 7$ (Equation of the inverse relation)

12. $yx = -5$

13. Graph $y = x^2 + 1$. A few solutions of the equation are (0, 1), (1, 2), (-1, 2), (2, 5), and (-2, 5). Plot these points and draw the curve. Then reflect the graph of $y = x^2 + 1$ across the line y = x (----). A few solutions of the inverse relation, $x = y^2 + 1$, are (1, 0), (2, 1), (2, -1), (5, 2), and (5, -2). Both graphs are shown below.

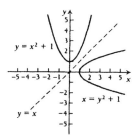

14.

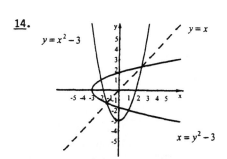

15. Graph x = |y|. A few solutions of the equation are (0, 0), (2, 2), (2, -2), (5, 5), and (5, -5). Plot these points and draw the graph. Then reflect the graph across the line y = x (----). A few solutions of the inverse relation, y = |x|, are (0, 0), (2, 2), (-2, 2), (5, 5), and (-5, 5). Both graphs are shown below.

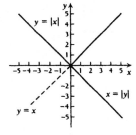

16.

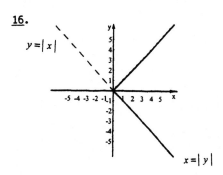

17. The graph of f(x) = 5x - 8 is shown below.

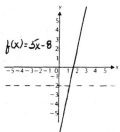

Since there is no horizontal line that crosses the graph more than once, the function is one-to-one.

18. Yes

19. The graph of f(x) = x² - 7 is shown below.

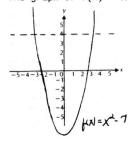

There are many horizontal lines that cross the graph more than once. In particular, the line y = 4 crosses the graph more than once. The function is not one-to-one.

20. No

21. The graph of g(x) = 3 is shown below.

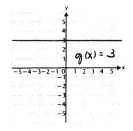

Since the horizontal line y = 3 crosses the graph more than once, the function is not one-to-one.

22. No

23. The graph of g(x) = ᴵxᴵ is shown below.

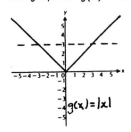

There are many horizontal lines that cross the graph more than once. In particular, the line y = 3 crosses the graph more than once. The function is not one-to-one.

24. No

25. The graph of f(x) = ᴵx + 1ᴵ is shown below.

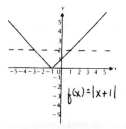

The line y = 2 is one of many lines that cross the graph more than once. The function is not one-to-one.

26. No

27. The graph of g(x) = $\frac{-4}{x}$ is shown below.

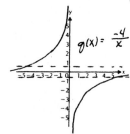

Since there is no horizontal line that crosses the graph more than once, the function is one-to-one.

28. Yes

29. a) The graph of f(x) = x + 4 is shown below. It passes the horizontal line test, so it is one-to-one.

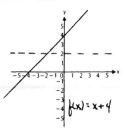

b) Replace f(x) by y: y = x + 4
    Interchange x and y: x = y + 4
    Solve for y: x - 4 = y
    Replace y by f⁻¹(x): f⁻¹(x) = x - 4

30. a) Yes; b) f⁻¹(x) = x - 5

31. a) The graph of f(x) = 5 - x is shown below. It passes the horizontal line test, so the function is one-to-one.

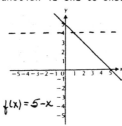

b) Replace f(x) by y: y = 5 - x

Interchange x and y: x = 5 - y

Solve for y: y = 5 - x

Replace y by $f^{-1}(x)$: $f^{-1}(x)$ = 5 - x

32. a) Yes; b) $f^{-1}(x)$ = 7 - x

33. a) The graph of g(x) = x - 3 is shown below. It passes the horizontal line test, so the function is one-to-one.

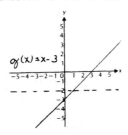

b) Replace g(x) by y: y = x - 3

Interchange x and y: x = y - 3

Solve for y: x + 3 = y

Replace y by $g^{-1}(x)$: $g^{-1}(x)$ = x + 3

34. a) Yes; b) $g^{-1}(x)$ = x + 10

35. a) The graph of f(x) = 2x is shown below. It passes the horizontal line test, so the function is one-to-one.

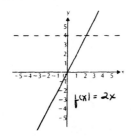

b) Replace f(x) by y: y = 2x

Interchange x and y: x = 2y

Solve for y: $\frac{x}{2}$ = y

Replace y by $f^{-1}(x)$: $f^{-1}(x)$ = $\frac{x}{2}$

36. a) Yes; b) $f^{-1}(x)$ = $\frac{x}{5}$

37. a) The graph of g(x) = 2x + 5 is shown below. It passes the horizontal line test, so the function is one-to-one.

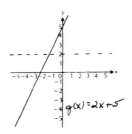

b) Replace g(x) by y: y = 2x + 5

Interchange variables: x = 2y + 5

Solve for y: x - 5 = 2y

$\frac{x - 5}{2}$ = y

Replace y by $g^{-1}(x)$: $g^{-1}(x)$ = $\frac{x - 5}{2}$

38. a) Yes; b) $g^{-1}(x)$ = $\frac{x - 8}{5}$

39. a) The graph of h(x) = $\frac{4}{x + 7}$ is shown below. It passes the horizontal line test, so the function is one-to-one.

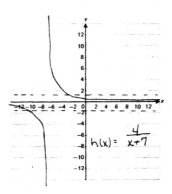

b) Replace h(x) by y: y = $\frac{4}{x + 7}$

Interchange x and y: x = $\frac{4}{y + 7}$

Solve for y: x(y + 7) = 4

y + 7 = $\frac{4}{x}$

y = $\frac{4}{x}$ - 7

Replace y by $h^{-1}(x)$: $h^{-1}(x)$ = $\frac{4}{x}$ - 7

40. a) Yes; b) $h^{-1}(x)$ = $\frac{1}{x}$ + 6

41. a) The graph of $f(x) = \frac{1}{x}$ is shown below. It
         passes the horizontal line test, so the
         function is one-to-one.

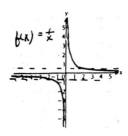

     b) Replace f(x) by y: $y = \frac{1}{x}$

         Interchange x and y: $x = \frac{1}{y}$

         Solve for y: $xy = 1$

                       $y = \frac{1}{x}$

         Replace y by $f^{-1}(x)$: $f^{-1}(x) = \frac{1}{x}$

42. a) Yes; b) $f^{-1}(x) = -\frac{4}{x}$

43. a) The graph of $f(x) = \frac{2x + 3}{4}$ is shown below.
         It passes the horizontal line test, so the
         function is one-to-one.

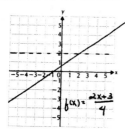

     b) Replace f(x) by y: $y = \frac{2x + 3}{4}$

         Interchange x and y: $x = \frac{2y + 3}{4}$

         Solve for y: $4x = 2y + 3$

                       $4x - 3 = 2y$

                       $\frac{4x - 3}{2} = y$

         Replace y by $f^{-1}(x)$: $f^{-1}(x) = \frac{4x - 3}{2}$

44. a) Yes; b) $f^{-1}(x) = \frac{4x + 5}{3}$

45. a) The graph of $g(x) = \frac{x + 4}{x - 3}$ is shown below. It
         passes the horizontal line test, so the
         function is one-to-one.

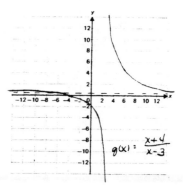

     b) Replace g(x) by y: $y = \frac{x + 4}{x - 3}$

         Interchange x and y: $x = \frac{y + 4}{y - 3}$

         Solve for y: $(y - 3)x = y + 4$

                       $xy - 3x = y + 4$

                       $xy - y = 3x + 4$

                       $y(x - 1) = 3x + 4$

                       $y = \frac{3x + 4}{x - 1}$

         Replace y by $g^{-1}(x)$: $g^{-1}(x) = \frac{3x + 4}{x - 1}$

46. a) Yes; b) $g^{-1}(x) = \frac{x + 3}{5 - 2x}$

47. a) The graph of $f(x) = x^3 - 1$ is shown below.
         It passes the horizontal line test, so the
         function is one-to-one.

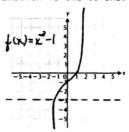

     b) Replace f(x) by y: $y = x^3 - 1$
         Interchange x and y: $x = y^3 - 1$
         Solve for y: $x + 1 = y^3$

                       $\sqrt[3]{x + 1} = y$

         Replace y by $f^{-1}(x)$: $f^{-1}(x) = \sqrt[3]{x + 1}$

48. a) Yes; b) $f^{-1}(x) = \sqrt[3]{x - 7}$

49. a) The graph of G(x) = (x - 4)³ is shown below. It passes the horizontal line test, so the function is one-to-one.

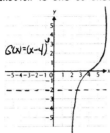

b) Replace G(x) by y: y = (x - 4)³

Interchange x and y: x = (y - 4)³

Solve for y: $\sqrt[3]{x}$ = y - 4

$\sqrt[3]{x}$ + 4 = y

Replace y by G⁻¹(x): G⁻¹(x) = $\sqrt[3]{x}$ + 4

50. a) Yes; b) G⁻¹(x) = $\sqrt[3]{x}$ - 5

51. a) The graph of f(x) = $\sqrt[3]{x}$ is shown below. It passes the horizontal line test, so the function is one-to-one.

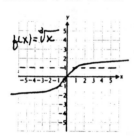

b) Replace f(x) by y: y = $\sqrt[3]{x}$

Interchange x and y: x = $\sqrt[3]{y}$

Solve for y: x³ = y

Replace y by f⁻¹(x): f⁻¹(x) = x³

52. a) Yes; b) f⁻¹(x) = x³ + 8

53. a) The graph of f(x) = 4x² + 3, x > 3, is shown below. It passes the horizontal line test, so the function is one-to-one.

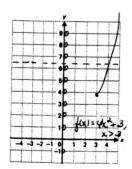

b) Replace f(x) by y: y = 4x² + 3

Interchange x and y: x = 4y² + 3

Solve for y: x - 3 = 4y²

$\dfrac{x - 3}{4}$ = y²

$\dfrac{\sqrt{x - 3}}{2}$ = y

(We take the principal square root since x > 3 in the original function.)

Replace y by f⁻¹(x): f⁻¹(x) = $\dfrac{\sqrt{x - 3}}{2}$ for all x in the range of f(x), or

f⁻¹(x) = $\dfrac{\sqrt{x - 3}}{2}$, x > 39

54. a) Yes; b) f⁻¹(x) = $\sqrt{\dfrac{x + 2}{5}}$, x > 18

55. a) The graph of f(x) = $\sqrt{x + 1}$ is shown below. It passes the horizontal line test, so the function is one-to-one.

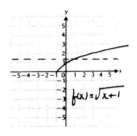

b) Replace f(x) by y: y = $\sqrt{x + 1}$

Interchange x and y: x = $\sqrt{y + 1}$

Solve for y: x² = y + 1

x² - 1 = y

Replace y by f⁻¹(x): f⁻¹(x) = x² - 1 for all x in the range of f(x), or f⁻¹(x) = x² - 1, x ⩾ 0.

56. a) Yes; b) g⁻¹(x) = $\dfrac{x² + 3}{2}$, x ⩾ 0

57. Only c) passes the horizontal line test, and thus has an inverse that is a function.

58. (d)

59. First graph f(x) = $\frac{1}{2}$x - 4. Then graph the inverse by flipping the graph of f(x) = $\frac{1}{2}$x - 4 over the line y = x. The graph of the inverse function can also be found by first finding a formula for the inverse and then substituting to find function values.

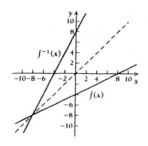

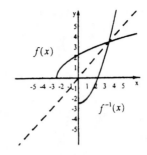

64.

60.

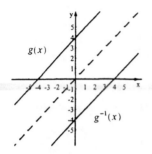

65. Use the procedure described in Exercise 59 to graph the function and its inverse.

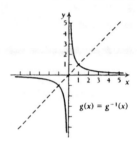

61. Use the procedure described in Exercise 59 to graph the function and its inverse.

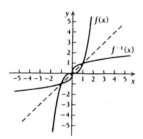

66.

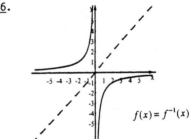

62.

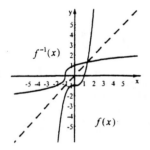

67. Use the procedure described in Exercise 59 to graph the function and its inverse.

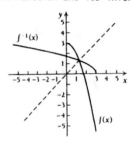

63. Use the procedure described in Exercise 59 to graph the function and its inverse.

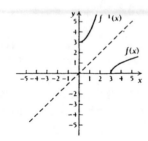

68.

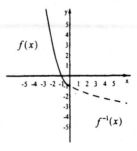

69. We find $f^{-1} \circ f(x)$ and $f \circ f^{-1}(x)$ and check to see that each is x.

   a) $f^{-1} \circ f(x) = f^{-1}(f(x)) = f^{-1}\left[\frac{7}{8}x\right] = \frac{8}{7}\left[\frac{7}{8}x\right] = x$

   b) $f \circ f^{-1}(x) = f(f^{-1}(x)) = f\left[\frac{8}{7}x\right] = \frac{7}{8}\left[\frac{8}{7}x\right] = x$

70. a) $f^{-1} \circ f(x) = 4\left[\frac{x+5}{4}\right] - 5 = x + 5 - 5 = x$

   b) $f \circ f^{-1}(x) = \frac{4x - 5 + 5}{4} = \frac{4x}{4} = x$

71. We find $f^{-1} \circ f(x)$ and $f \circ f^{-1}(x)$ and check to see that each is x.

   a) $f^{-1} \circ f(x) = f^{-1}(f(x)) = f^{-1}\left[\frac{1-x}{x}\right] =$

   $\dfrac{1}{\frac{1-x}{x} + 1} = \dfrac{1}{\frac{1-x+x}{x}} = \dfrac{1}{\frac{1}{x}} = x$

   b) $f \circ f^{-1}(x) = f(f^{-1}(x)) = f\left[\frac{1}{x+1}\right] =$

   $\dfrac{1 - \frac{1}{x+1}}{\frac{1}{x+1}} = \dfrac{\frac{x+1-1}{x+1}}{\frac{1}{x+1}} = \dfrac{\frac{x}{x+1}}{\frac{1}{x+1}} = x$

72. a) $f^{-1} \circ f(x) = \sqrt[3]{x^3 - 4 + 4} = \sqrt[3]{x^3} = x$

   b) $f \circ f^{-1}(x) = (\sqrt[3]{x+4})^3 - 4 = x + 4 - 4 = x$

73. Using Theorem 1 we have
   $f^{-1}(f(3)) = 3$ and $f(f^{-1}(-125)) = -125$.

74. 5; -12

75. Using Theorem 1, we have
   $f^{-1}(f(12,053)) = 12,053$ and $f(f^{-1}(-17,243)) = -17,243$.

76. 489; -17,422

77. a) $f(8) = 8 + 32 = 40$
   Size 40 in France corresponds to size 8 in the U.S.

   $f(10) = 10 + 32 = 42$
   Size 42 in France corresponds to size 10 in the U.S.

   $f(14) = 14 + 32 = 46$
   Size 46 in France corresponds to size 14 in the U.S.

   $f(18) = 18 + 32 = 50$
   Size 50 in France corresponds to size 18 in the U.S.

   b) The graph of $f(x) = x + 32$ is shown below.

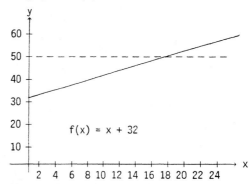

   It passes the horizontal line test, so the function is one-to-one and, hence, has an inverse that is a function. We now find a formula for the inverse.

   Replace f(x) by y: $y = x + 32$

   Interchange x and y: $x = y + 32$

   Solve for y: $x - 32 = y$

   Replace y by $f^{-1}(x)$: $f^{-1}(x) = x - 32$

   c) $f^{-1}(40) = 40 - 32 = 8$
   Size 8 in the U.S. corresponds to size 40 in France.

   $f^{-1}(42) = 42 - 32 = 10$
   Size 10 in the U.S. corresponds to size 42 in France.

   $f^{-1}(46) = 46 - 32 = 14$
   Size 14 in the U.S. corresponds to size 46 in France.

   $f^{-1}(50) = 50 - 32 = 18$
   Size 18 in the U.S. corresponds to size 50 in France.

78. a) 40, 44, 52, 60; b) yes, $g^{-1}(x) = \frac{x-24}{2}$, or $\frac{x}{2} - 12$;

   c) 8, 10, 14, 18

79. The graph of $f(x) = 5$ is shown below. Since the horizontal line $y = 5$ crosses the graph in more than one place, the function does not have an inverse that is a function.

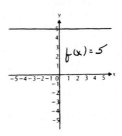

80. $C^{-1}(x) = \frac{100}{x-5}$

   $C^{-1}(x)$ gives the number of people in the group, where x is the cost per person, in dollars.

81. Check to see if g ∘ f(x) = x and f ∘ g(x) = x.

   a) g ∘ f(x) = g(f(x)) = g$\left(\frac{2}{3}\right)$ = $\frac{3}{2}$

      Since g ∘ f(x) ≠ x, the functions are not inverses of each other.

82. Yes

83. Check to see if g ∘ f(x) = x and f ∘ g(x) = x.

   a) g ∘ f(x) = g(f(x)) = g($\sqrt[4]{x}$) = ($\sqrt[4]{x}$)^4 = x

   b) f ∘ g(x) = f(g(x)) = f(x^4) = $\sqrt[4]{x^4}$ = |x|

      For x < 0, |x| = -x and f ∘ g(x) ≠ x.

   The functions are not inverses of each other.

84. No

85. Answers may vary. f(x) = $\frac{3}{x}$, f(x) = 1 - x,
   f(x) = x

86. Consider any two inputs a and b, a < b, of an increasing function f(x). Then f(a) < f(b). Thus, if a ≠ b, f(a) ≠ f(b) and the function is one-to-one.

87. Graph y = $\frac{1}{x^2}$.

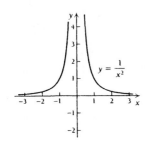

   Test y = $\frac{1}{x^2}$ for symmetry with respect to the x-axis.

   Replace y by -y: -y = $\frac{1}{x^2}$

   Since y = $\frac{1}{x^2}$ is not equivalent to -y = $\frac{1}{x^2}$, y = $\frac{1}{x^2}$ is not symmetric with respect to the x-axis.

   Test y = $\frac{1}{x^2}$ for symmetry with respect to the y-axis.

   Replace x by -x: y = $\frac{1}{(-x)^2}$ = $\frac{1}{x^2}$

   Since the equations are equivalent, y = $\frac{1}{x^2}$ is symmetric with respect to the y-axis.

   Test y = $\frac{1}{x^2}$ for symmetry with respect to the origin.

   Replace x by -x and y by -y: -y = $\frac{1}{(-x)^2}$ = $\frac{1}{x^2}$

   Since y = $\frac{1}{x^2}$ is not equivalent to -y = $\frac{1}{x^2}$, y = $\frac{1}{x^2}$ is not symmetric with respect to the origin.

88. C and F are inverses.

89.

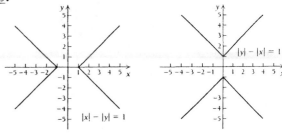

   Test |x| - |y| = 1 for symmetry with respect to the x-axis.

   Replace y by -y: |x| - |-y| = 1
                     |x| - |y| = 1

   Since the equations are equivalent, |x| - |y| = 1 is symmetric with respect to the x-axis.

   Test |x| - |y| = 1 for symmetry with respect to the y-axis.

   Replace x by -x: |-x| - |y| = 1
                     |x| - |y| = 1

   Since the equations are equivalent, |x| - |y| = 1 is symmetric with respect to the y-axis.

   Test |x| - |y| = 1 for symmetry with respect to the origin:

   Replace x by -x and y by -y: |-x| - |-y| = 1
                                 |x| - |y| = 1

   Since the equations are equivalent, |x| - |y| = 1 is symmetric with respect to the origin.

90. a) The inverse of the given function y = $\frac{|x|}{x}$ is
      x = $\frac{|y|}{y}$.

   The graphs are as shown:

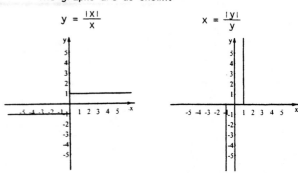

b) Each of the two graphs is symmetric with respect to the origin. Neither is symmetric with respect to the x-axis or the y-axis.

91. The function fails the horizontal line test. Thus it is not one-to-one and does not have an inverse that is a function.

92. $f(x) = (3x - 9)^3$

$f^{-1}(x) = \dfrac{\sqrt[3]{x} + 9}{3}$

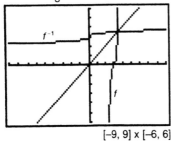

[-9, 9] x [-6, 6]

Inverse function: $y = \dfrac{\sqrt[3]{x} + 9}{3}$

93. $f(x) = \sqrt{2x - 1}$

$f^{-1}(x) = \dfrac{x^2 + 1}{2}$

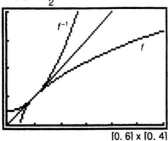

[0, 6] x [0, 4]

The function passes the horizontal line test, so it is one-to-one. We find the inverse function.

$$y = \sqrt{2x - 1}$$
$$x = \sqrt{2y - 1} \qquad \text{Interchange } x \text{ and } y.$$
$$x^2 = 2y - 1 \qquad \text{Solve for } y.$$
$$x^2 + 1 = 2y$$
$$\frac{x^2 + 1}{2} = y, \; x \geqslant 0 \qquad \text{Restrict the domain of the inverse function to the range of the original function.}$$

94. Inverse function: $y = \dfrac{x^4}{90}, \; x \geqslant 0$

Exercise Set 5.2

1. Graph: $y = 2^x$

We compute some function values, thinking of y as f(x), and keep the results in a table.

$f(0) = 2^0 = 1$

$f(1) = 2^1 = 2$

$f(2) = 2^2 = 4$

$f(-1) = 2^{-1} = \dfrac{1}{2^1} = \dfrac{1}{2}$

$f(-2) = 2^{-2} = \dfrac{1}{2^2} = \dfrac{1}{4}$

x	y, or f(x)
0	1
1	2
2	4
-1	$\frac{1}{2}$
-2	$\frac{1}{4}$

Next we plot these points and connect them with a smooth curve.

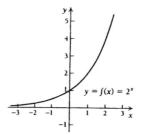

2.

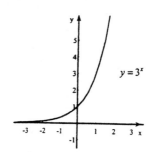

3. Graph: $y = 5^x$

We compute some function values, thinking of y as f(x), and keep the results in a table.

$f(0) = 5^0 = 1$

$f(1) = 5^1 = 5$

$f(2) = 5^2 = 25$

$f(-1) = 5^{-1} = \dfrac{1}{5^1} = \dfrac{1}{5}$

$f(-2) = 5^{-2} = \dfrac{1}{5^2} = \dfrac{1}{25}$

x	y, or f(x)
0	1
1	5
2	25
-1	$\frac{1}{5}$
-2	$\frac{1}{25}$

Next we plot these points and connect them with a smooth curve.

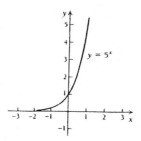

$y = 5^x$

4.

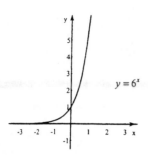

$y = 6^x$

5.   Graph: $y = 2^{x+1}$

We compute some function values, thinking of y as f(x), and keep the results in a table.

$f(0) = 2^{0+1} = 2^1 = 2$

$f(-1) = 2^{-1+1} = 2^0 = 1$

$f(-2) = 2^{-2+1} = 2^{-1} = \frac{1}{2^1} = \frac{1}{2}$

$f(-3) = 2^{-3+1} = 2^{-2} = \frac{1}{2^2} = \frac{1}{4}$

$f(1) = 2^{1+1} = 2^2 = 4$

$f(2) = 2^{2+1} = 2^3 = 8$

x	y, or f(x)
0	2
-1	1
-2	$\frac{1}{2}$
-3	$\frac{1}{4}$
1	4
2	8

Next we plot these points and connect them with a smooth curve.

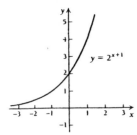

$y = 2^{x+1}$

6.

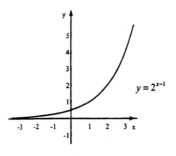

$y = 2^{x-1}$

7.   Graph: $y = 3^{x-2}$

We compute some function values, thinking of y as f(x), and keep the results in a table.

$f(0) = 3^{0-2} = 3^{-2} = \frac{1}{3^2} = \frac{1}{9}$

$f(1) = 3^{1-2} = 3^{-1} = \frac{1}{3^1} = \frac{1}{3}$

$f(2) = 3^{2-2} = 3^0 = 1$

$f(3) = 3^{3-2} = 3^1 = 3$

$f(4) = 3^{4-2} = 3^2 = 9$

$f(-1) = 3^{-1-2} = 3^{-3} = \frac{1}{3^3} = \frac{1}{27}$

$f(-2) = 3^{-2-2} = 3^{-4} = \frac{1}{3^4} = \frac{1}{81}$

x	y, or f(x)
0	$\frac{1}{9}$
1	$\frac{1}{3}$
2	1
3	3
4	9
-1	$\frac{1}{27}$
-2	$\frac{1}{81}$

Next we plot these points and connect them with a smooth curve.

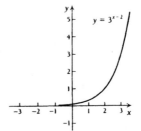

$y = 3^{x-2}$

8.

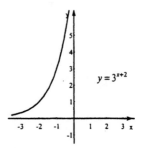

$y = 3^{x+2}$

9.   Graph: $y = 2^x - 3$

We construct a table of values, thinking of y as f(x). Then we plot the points and connect them with a smooth curve.

$f(0) = 2^0 - 3 = 1 - 3 = -2$

$f(1) = 2^1 - 3 = 2 - 3 = -1$

$f(2) = 2^2 - 3 = 4 - 3 = 1$

$f(3) = 2^3 - 3 = 8 - 3 = 5$

$f(-1) = 2^{-1} - 3 = \frac{1}{2} - 3 = -\frac{5}{2}$

$f(-2) = 2^{-2} - 3 = \frac{1}{4} - 3 = -\frac{11}{4}$

x	y, or f(x)
0	-2
1	-1
2	1
3	5
-1	$-\frac{5}{2}$
-2	$-\frac{11}{4}$

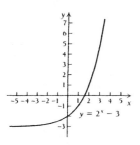

$y = 2^x - 3$

10.

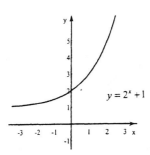

$y = 2^x + 1$

11. Graph: $y = 5^{x+3}$

We construct a table of values, thinking of y as f(x). Then we plot the points and connect them with a smooth curve.

$f(0) = 5^{0+3} = 5^3 = 125$

$f(-1) = 5^{-1+3} = 5^2 = 25$

$f(-2) = 5^{-2+3} = 5^1 = 5$

$f(-3) = 5^{-3+3} = 5^0 = 1$

$f(-4) = 5^{-4+3} = 5^{-1} = \frac{1}{5}$

$f(-5) = 5^{-5+3} = 5^{-2} = \frac{1}{25}$

x	y, or f(x)
0	125
-1	25
-2	5
-3	1
-4	$\frac{1}{5}$
-5	$\frac{1}{25}$

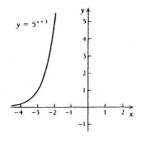

$y = 5^{x+3}$

12.

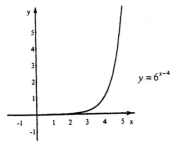

$y = 6^{x-4}$

13. Graph: $y = \left(\frac{1}{2}\right)^x$

We construct a table of values, thinking of y as f(x). Then we plot the points and connect them with a smooth curve.

$f(0) = \left(\frac{1}{2}\right)^0 = 1$

$f(1) = \left(\frac{1}{2}\right)^1 = \frac{1}{2}$

$f(2) = \left(\frac{1}{2}\right)^2 = \frac{1}{4}$

$f(3) = \left(\frac{1}{2}\right)^3 = \frac{1}{8}$

$f(-1) = \left(\frac{1}{2}\right)^{-1} = \frac{1}{\left(\frac{1}{2}\right)^1} = \frac{1}{\frac{1}{2}} = 2$

$f(-2) = \left(\frac{1}{2}\right)^{-2} = \frac{1}{\left(\frac{1}{2}\right)^2} = \frac{1}{\frac{1}{4}} = 4$

$f(-3) = \left(\frac{1}{2}\right)^{-3} = \frac{1}{\left(\frac{1}{2}\right)^3} = \frac{1}{\frac{1}{8}} = 8$

x	y, or f(x)
0	1
1	$\frac{1}{2}$
2	$\frac{1}{4}$
3	$\frac{1}{8}$
-1	2
-2	4
-3	8

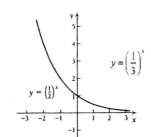

$y = \left(\frac{1}{3}\right)^x$

$y = \left(\frac{1}{2}\right)^x$

14.

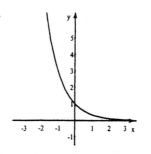

15. Graph: $y = \left(\frac{1}{5}\right)^x$

We construct a table of values, thinking of y as f(x). Then we plot the points and connect them with a smooth curve.

$f(0) = \left(\frac{1}{5}\right)^0 = 1$

$f(1) = \left(\frac{1}{5}\right)^1 = \frac{1}{5}$

$f(2) = \left(\frac{1}{5}\right)^2 = \frac{1}{25}$

$f(-1) = \left(\frac{1}{5}\right)^{-1} = \frac{1}{\frac{1}{5}} = 5$

$f(-2) = \left(\frac{1}{5}\right)^{-2} = \frac{1}{\frac{1}{25}} = 25$

x	y, or f(x)
0	1
1	$\frac{1}{5}$
2	$\frac{1}{25}$
-1	5
-2	25

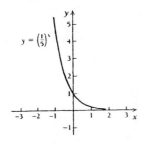

$y = \left(\frac{1}{5}\right)^x$

16.

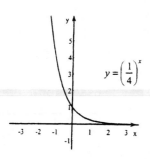

$y = \left(\frac{1}{4}\right)^x$

17. Graph: $y = 2^{2x-1}$

We construct a table of values, thinking of y as f(x). Then we plot the points and connect them with a smooth curve.

$f(0) = 2^{2\cdot0-1} = 2^{-1} = \frac{1}{2}$

$f(1) = 2^{2\cdot1-1} = 2^1 = 2$

$f(2) = 2^{2\cdot2-1} = 2^3 = 8$

$f(-1) = 2^{2(-1)-1} = 2^{-3} = \frac{1}{8}$

$f(-2) = 2^{2(-2)-1} = 2^{-5} = \frac{1}{32}$

x	y, or f(x)
0	$\frac{1}{2}$
1	2
2	8
-1	$\frac{1}{8}$
-2	$\frac{1}{32}$

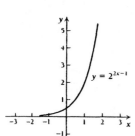

$y = 2^{2x-1}$

18.

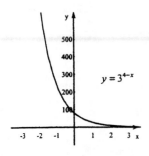

$y = 3^{4-x}$

19. Graph: $y = 2^{x-1} - 3$

We construct a table of values, thinking of y as f(x). Then we plot the points and connect them with a smooth curve.

$f(0) = 2^{0-1} - 3 = 2^{-1} - 3 = \frac{1}{2} - 3 = -\frac{5}{2}$

$f(1) = 2^{1-1} - 3 = 2^0 - 3 = 1 - 3 = -2$

$f(2) = 2^{2-1} - 3 = 2^1 - 3 = 2 - 3 = -1$

$f(3) = 2^{3-1} - 3 = 2^2 - 3 = 4 - 3 = 1$

$f(4) = 2^{4-1} - 3 = 2^3 - 3 = 8 - 3 = 5$

$f(-1) = 2^{-1-1} - 3 = 2^{-2} - 3 = \frac{1}{4} - 3 = -\frac{11}{4}$

$f(-2) = 2^{-2-1} - 3 = 2^{-3} - 3 = \frac{1}{8} - 3 = -\frac{23}{8}$

x	y, or f(x)
0	$-\frac{5}{2}$
1	-2
2	-1
3	1
4	5
-1	$-\frac{11}{4}$
-2	$-\frac{23}{8}$

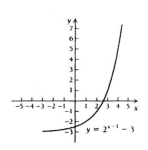

$y = 2^{x-1} - 3$

20.

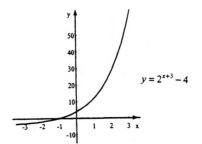

$y = 2^{x+3} - 4$

21. Graph: $x = 2^y$

We can find ordered pairs by choosing values for y and then computing values for x.

For y = 0, $x = 2^0 = 1$.

For y = 1, $x = 2^1 = 2$.

For y = 2, $x = 2^2 = 4$.

For y = 3, $x = 2^3 = 8$.

For y = -1, $x = 2^{-1} = \frac{1}{2^1} = \frac{1}{2}$.

For y = -2, $x = 2^{-2} = \frac{1}{2^2} = \frac{1}{4}$.

For y = -3, $x = 2^{-3} = \frac{1}{2^3} = \frac{1}{8}$.

x	y
1	0
2	1
4	2
8	3
$\frac{1}{2}$	-1
$\frac{1}{4}$	-2
$\frac{1}{8}$	-3

└── (1) Choose values for y.

└── (2) Compute values for x.

We plot these points and connect them with a smooth curve.

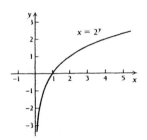

<u>22.</u>

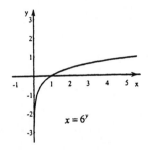

<u>23.</u>   Graph: $x = \left(\frac{1}{2}\right)^y$

We can find ordered pairs by choosing values for y and then computing values for x. Then we plot these points and connect them with a smooth curve.

For y = 0, $x = \left(\frac{1}{2}\right)^0 = 1.$

For y = 1, $x = \left(\frac{1}{2}\right)^1 = \frac{1}{2}.$

For y = 2, $x = \left(\frac{1}{2}\right)^2 = \frac{1}{4}.$

For y = 3, $x = \left(\frac{1}{2}\right)^3 = \frac{1}{8}.$

For y = -1, $x = \left(\frac{1}{2}\right)^{-1} = \frac{1}{\frac{1}{2}} = 2.$

For y = -2, $x = \left(\frac{1}{2}\right)^{-2} = \frac{1}{\frac{1}{4}} = 4.$

For y = -3, $x = \left(\frac{1}{2}\right)^{-3} = \frac{1}{\frac{1}{8}} = 8.$

x	y
1	0
$\frac{1}{2}$	1
$\frac{1}{4}$	2
$\frac{1}{8}$	3
2	-1
4	-2
8	-3

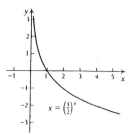

<u>24.</u>

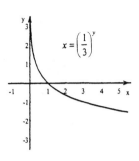

<u>25.</u>   Graph: $x = 5^y$

We can find ordered pairs by choosing values for y and then computing values for x. Then we plot these points and connect them with a smooth curve.

For y = 0, $x = 5^0 = 1.$

For y = 1, $x = 5^1 = 5.$

For y = 2, $x = 5^2 = 25.$

For y = -1, $x = 5^{-1} = \frac{1}{5}.$

For y = -2, $x = 5^{-2} = \frac{1}{25}.$

x	y
1	0
5	1
25	2
$\frac{1}{5}$	-1
$\frac{1}{25}$	-2

<u>26.</u>

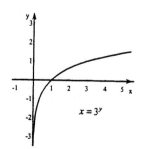

27. Graph: $x = \left(\frac{2}{3}\right)^y$

We can find ordered pairs by choosing values for y and then computing values for x. Then we plot these points and connect them with a smooth curve.

For y = 0, $x = \left(\frac{2}{3}\right)^0 = 1.$

For y = 1, $x = \left(\frac{2}{3}\right)^1 = \frac{2}{3}.$

For y = 2, $x = \left(\frac{2}{3}\right)^2 = \frac{4}{9}.$

For y = 3, $x = \left(\frac{2}{3}\right)^3 = \frac{8}{27}.$

For y = -1, $x = \left(\frac{2}{3}\right)^{-1} = \frac{1}{\frac{2}{3}} = \frac{3}{2}.$

For y = -2, $x = \left(\frac{2}{3}\right)^{-2} = \frac{1}{\frac{4}{9}} = \frac{9}{4}.$

For y = -3, $x = \left(\frac{2}{3}\right)^{-3} = \frac{1}{\frac{8}{27}} = \frac{27}{8}.$

x	y
1	0
$\frac{2}{3}$	1
$\frac{4}{9}$	2
$\frac{8}{27}$	3
$\frac{3}{2}$	-1
$\frac{9}{4}$	-2
$\frac{27}{8}$	-3

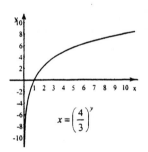

28.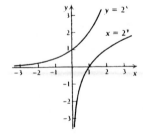

29. Graph $y = 2^x$ (see Exercise 1) and $x = 2^y$ (see Exercise 21) using the same set of axes.

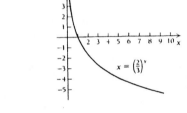

30.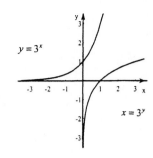

31. Graph $y = \left(\frac{1}{2}\right)^x$ (see Exercise 13) and $x = \left(\frac{1}{2}\right)^y$ (see Exercise 23) using the same set of axes.

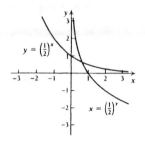

32.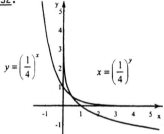

33. a) Keep in mind that t represents the number of years after 1985 and that N is given in millions.

For 1986, t = 1:
$N(1) = 7.5(6)^{0.5(1)} = 7.5(6)^{0.5} \approx 18.4$ million;

For 1987, t = 1987 - 1985, or 2:
$N(2) = 7.5(6)^{0.5(2)} = 7.5(6) = 45$ million;

For 1990, t = 1990 - 1985, or 5:
$N(5) = 7.5(6)^{0.5(5)} = 7.5(6)^{2.5} \approx 661.4$ million;

For 1995, t = 1995 - 1985, or 10:
$N(10) = 7.5(6)^{0.5(10)} = 7.5(6)^5 = 58,320$ million;

For 2000, t = 2000 - 1985, or 15:
$N(15) = 7.5(6)^{0.5(15)} = 7.5(6)^{7.5} \approx 5,142,752.7$ million

b) Use the function values computed in part (a) to draw the graph of the function.

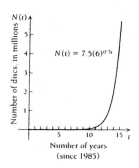

$N(t) = 7.5(6)^{0.5t}$

34. a) 4243; 6000; 8485; 12,000; 24,000

   b)

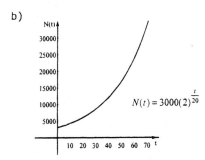

$N(t) = 3000(2)^{\frac{t}{20}}$

35. a) Substitute $50,000 for P and 9%, or 0.09, for i in the formula $A = P(1 + i)^t$:

   $A(t) = \$50,000(1 + 0.09)^t = \$50,000(1.09)^t$

   b) Substitute for t.

   $A(0) = \$50,000(1.09)^0 = \$50,000;$

   $A(4) = \$50,000(1.09)^4 \approx \$70,579.08;$

   $A(8) = \$50,000(1.09)^8 \approx \$99,628.13;$

   $A(10) = \$50,000(1.09)^{10} \approx \$118,368.18$

   c) We use the function values computed in part (b) to draw the graph. Note that the axes are scaled differently because of the large numbers.

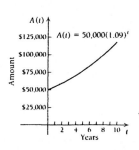

$A(t) = 50,000(1.09)^t$

36. a) 250,000; 62,500; 977; 0

b)

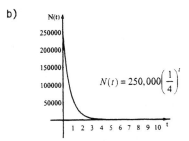

$N(t) = 250,000\left(\dfrac{1}{4}\right)^t$

37. a) Substitute for t.

   $V(0) = \$5200(0.75)^0 = \$5200;$

   $V(1) = \$5200(0.75)^1 = \$3900;$

   $V(2) = \$5200(0.75)^2 = \$2925;$

   $V(5) = \$5200(0.75)^5 \approx \$1233.98;$

   $V(10) = \$5200(0.75)^{10} \approx \$292.83$

   b) Use the function values computed in part (a) to draw the graph of the function.

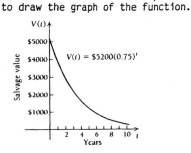

$V(t) = \$5200(0.75)^t$

38. a) 2.6 lb; 3.7 lb; 19.5 lb; 29.1 lb

   b)

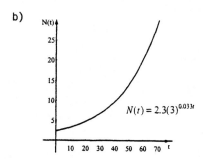

$N(t) = 2.3(3)^{0.033t}$

39. Use a calculator.

   a) $7^3 = 343$

   b) $7^{3.1} \approx 416.681217$

   c) $7^{3.14} \approx 450.409815$

   d) $7^{3.141} \approx 451.287125$

   e) $7^{3.1415} \approx 451.726421$

   f) $7^{3.14159} \approx 451.805540$

40. π^7

41. Since the bases are the same, the one with the larger exponent is the larger number. Thus $\pi^{3.2}$ is larger.

42.

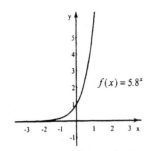

$f(x) = 5.8^x$

43. Graph: $f(x) = (2.7)^x$

Use a calculator with a power key to construct a table of values. (We will round values to the nearest hundredth.) Then plot these points and connect them with a smooth curve.

x	f(x)
0	1
1	2.7
2	7.29
3	19.68
-1	0.37
-2	0.14

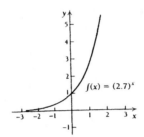

$f(x) = (2.7)^x$

44.

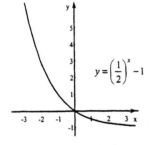

$y = \left(\dfrac{1}{2}\right)^x - 1$

45. Graph: $y = 2^x + 2^{-x}$

Construct a table of values, thinking of y as f(x). Then plot these points and connect them with a curve.

$f(0) = 2^0 + 2^{-0} = 1 + 1 = 2$

$f(1) = 2^1 + 2^{-1} = 2 + \dfrac{1}{2} = 2\dfrac{1}{2}$

$f(2) = 2^2 + 2^{-2} = 4 + \dfrac{1}{4} = 4\dfrac{1}{4}$

$f(3) = 2^3 + 2^{-3} = 8 + \dfrac{1}{8} = 8\dfrac{1}{8}$

$f(-1) = 2^{-1} + 2^{-(-1)} = \dfrac{1}{2} + 2 = 2\dfrac{1}{2}$

$f(-2) = 2^{-2} + 2^{-(-2)} = \dfrac{1}{4} + 4 = 4\dfrac{1}{4}$

$f(-3) = 2^{-3} + 2^{-(-3)} = \dfrac{1}{8} + 8 = 8\dfrac{1}{8}$

x	y, or f(x)
0	2
1	$2\dfrac{1}{2}$
2	$4\dfrac{1}{4}$
3	$8\dfrac{1}{8}$
-1	$2\dfrac{1}{2}$
-2	$4\dfrac{1}{4}$
-3	$8\dfrac{1}{8}$

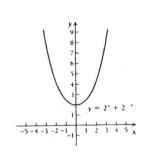

$y = 2^x + 2^{-x}$

46.

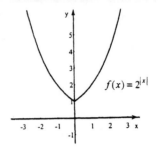

$f(x) = 2^{|x|}$

47. Graph: $y = 3^x + 3^{-x}$

We construct a table of values, thinking of y as f(x). Then plot these points and connect them with a curve.

$f(0) = 3^0 + 3^{-0} = 1 + 1 = 2$

$f(1) = 3^1 + 3^{-1} = 3 + \dfrac{1}{3} = 3\dfrac{1}{3}$

$f(2) = 3^2 + 3^{-2} = 9 + \dfrac{1}{9} = 9\dfrac{1}{9}$

$f(-1) = 3^{-1} + 3^{-(-1)} = \dfrac{1}{3} + 3 = 3\dfrac{1}{3}$

$f(-2) = 3^{-2} + 3^{-(-2)} = \dfrac{1}{9} + 9 = 9\dfrac{1}{9}$

x	y, or f(x)
0	2
1	$3\dfrac{1}{3}$
2	$9\dfrac{1}{9}$
-1	$3\dfrac{1}{3}$
-2	$9\dfrac{1}{9}$

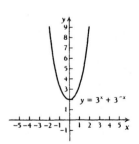

$y = 3^x + 3^{-x}$

48.

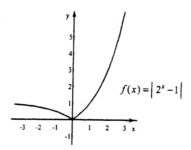

$$f(x) = \left| 2^x - 1 \right|$$

49. Graph: $y = 2^{-(x-1)}$

We construct a table of values, thinking of y as f(x). Then plot these points and connect them with a curve.

$f(0) = 2^{-(0-1)} = 2^1 = 2$

$f(1) = 2^{-(1-1)} = 2^0 = 1$

$f(2) = 2^{-(2-1)} = 2^{-1} = \frac{1}{2}$

$f(4) = 2^{-(4-1)} = 2^{-3} = \frac{1}{8}$

$f(-1) = 2^{-(-1-1)} = 2^2 = 4$

$f(-2) = 2^{-(-2-1)} = 2^3 = 8$

x	y, or f(x)
0	2
1	1
2	$\frac{1}{2}$
4	$\frac{1}{8}$
-1	4
-2	8

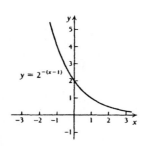

$$y = 2^{-(x-1)}$$

50.

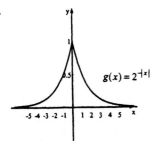

$$g(x) = 2^{-|x|}$$

51. Graph $y = |2^x - 2|$

We construct a table of values, thinking of y as f(x). Then plot these points and connect them with a curve.

$f(0) = |2^0 - 2| = |1 - 2| = 1$

$f(1) = |2^1 - 2| = |2 - 2| = 0$

$f(2) = |2^2 - 2| = |4 - 2| = 2$

$f(3) = |2^3 - 2| = |8 - 2| = 6$

$f(-1) = |2^{-1} - 2| = \left| \frac{1}{2} - 2 \right| = \frac{3}{2}$

$f(-2) = |2^{-2} - 2| = \left| \frac{1}{4} - 2 \right| = \frac{7}{4}$

$f(-4) = |2^{-4} - 2| = \left| \frac{1}{16} - 2 \right| = \frac{31}{16}$

x	y, or f(x)
0	1
1	0
2	2
3	6
-1	$\frac{3}{2}$
-2	$\frac{7}{4}$
-4	$\frac{31}{16}$

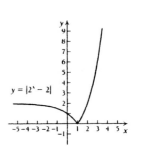

$$y = |2^x - 2|$$

52.

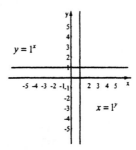

$$y = 1^x$$

$$x = 1^y$$

53. $y = 3^{-(x-1)}$ $x = 3^{-(y-1)}$

x	y
0	3
1	1
2	$\frac{1}{3}$
3	$\frac{1}{9}$
-1	9

x	y
3	0
1	1
$\frac{1}{3}$	2
$\frac{1}{9}$	3
9	-1

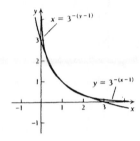

54. $\{x \mid x \leqslant 0\}$

55.   First we graph $y = 2^x$.

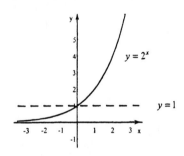

Then we study the graph to see which inputs give outputs greater than 1. (These are the inputs whose outputs are above the dashed line $y = 1$.) We see that all inputs greater than 0 give outputs greater than 1. The solution set is $\{x \mid x > 0\}$.

56.   a)   155.7 words per minute; 190.2 words per minute; 199.5 words per minute; 199.9999 words per minute

b)   Use the function values computed in part (a) to draw the graph of the function.

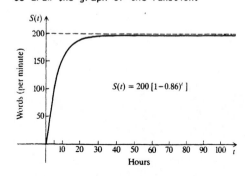

57.   Graph: $y = 2^{-x^2}$

$f(-3) = 2^{-3^2} = 2^{-9} = \frac{1}{512}$

$f(-2) = 2^{-2^2} = 2^{-4} = \frac{1}{16}$

$f(-1) = 2^{-1^2} = 2^{-1} = \frac{1}{2}$

$f(0) = 2^{-0^2} = 2^0 = 1$

$f(1) = 2^{-1^2} = 2^{-1} = \frac{1}{2}$

$f(2) = 2^{-2^2} = 2^{-4} = \frac{1}{16}$

x	y, or f(x)
-3	$\frac{1}{512}$
-2	$\frac{1}{16}$
-1	$\frac{1}{2}$
0	1
1	$\frac{1}{2}$
2	$\frac{1}{16}$

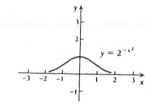

58.

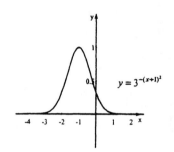

59.   Graph: $y = |2^{x^2} - 8|$

$f(-3) = |2^{(-3)^2} - 8| = |2^9 - 8| = 504$

$f(-2) = |2^{(-2)^2} - 8| = |2^4 - 8| = 8$

$f(-1) = |2^{(-1)^2} - 8| = |2^1 - 8| = 6$

$f(0) = |2^{0^2} - 8| = |2^0 - 8| = 7$

$f(1) = |2^{1^2} - 8| = |2^1 - 8| = 6$

$f(2) = |2^{2^2} - 8| = |2^4 - 8| = 8$

$f(3) = |2^{3^2} - 8| = |2^9 - 8| = 504$

x	y, or f(x)
-3	504
-2	8
-1	6
0	7
1	6
2	8
3	504

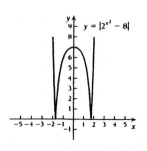

60. a) A(t) = $1,000,000(1.0885)^t$

b) 10.1 yr

c) 13.0 yr

d) $2,440,000

Exercise Set 5.3

1. Graph: $y = \log_7 x$

The equation $y = \log_7 x$ is equivalent to $7^y = x$. We can find ordered pairs by choosing values for y and computing the corresponding x-values.

For y = 0, $x = 7^0 = 1$.

For y = 1, $x = 7^1 = 7$.

For y = 2, $x = 7^2 = 49$.

For y = -1, $x = 7^{-1} = \frac{1}{7}$.

For y = -2, $x = 7^{-2} = \frac{1}{49}$.

x, or 7^y	y
1	0
7	1
49	2
$\frac{1}{7}$	-1
$\frac{1}{49}$	-2

(1) Select y.
(2) Compute x.

We plot the set of ordered pairs and connect the points with a smooth curve.

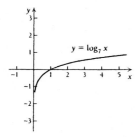

2.

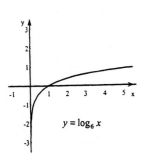

3. Graph: $y = \log_{10} x$

The equation $y = \log_{10} x$ is equivalent to $10^y = x$. We can find ordered pairs by choosing values for y and computing the corresponding x-values.

For y = 0, $x = 10^0 = 1$.

For y = 1, $x = 10^1 = 10$.

For y = 2, $x = 10^2 = 100$.

For y = -1, $x = 10^{-1} = \frac{1}{10}$.

For y = -2, $x = 10^{-2} = \frac{1}{100}$.

x, or 10^y	y
1	0
10	1
100	2
$\frac{1}{10}$	-1
$\frac{1}{100}$	-2

We plot the set of ordered pairs and connect the points with a smooth curve.

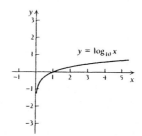

4.

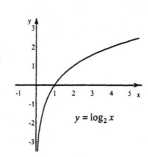

5. Graph: $f(x) = \log_4 x$

Think of f(x) as y. Then $y = \log_4 x$ is equivalent to $4^y = x$. We find ordered pairs by choosing values for y and computing the corresponding x-values. Then we plot the points and connect them with a smooth curve.

For y = 0, $x = 4^0 = 1$.

For y = 1, $x = 4^1 = 4$.

For y = 2, $x = 4^2 = 16$.

For y = -1, $x = 4^{-1} = \frac{1}{4}$.

For y = -2, $x = 4^{-2} = \frac{1}{16}$.

x, or 4^y	y
1	0
4	1
16	2
$\frac{1}{4}$	-1
$\frac{1}{16}$	-2

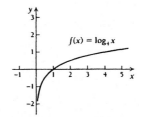

6.

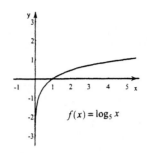

$f(x) = \log_5 x$

7. Graph: $f(x) = \log_{1/2} x$

Think of $f(x)$ as y. Then $y = \log_{1/2} x$ is equivalent to $\left(\frac{1}{2}\right)^y = x$. We construct a table of values, plot these points, and connect them with a smooth curve.

For $y = 0$, $x = \left(\frac{1}{2}\right)^0 = 1$.

For $y = 1$, $x = \left(\frac{1}{2}\right)^1 = \frac{1}{2}$.

For $y = 2$, $x = \left(\frac{1}{2}\right)^2 = \frac{1}{4}$.

For $y = -1$, $x = \left(\frac{1}{2}\right)^{-1} = 2$.

For $y = -2$, $x = \left(\frac{1}{2}\right)^{-2} = 4$.

For $y = -3$, $x = \left(\frac{1}{2}\right)^{-3} = 8$.

x, or $\left(\frac{1}{2}\right)^y$	y
1	0
$\frac{1}{2}$	1
$\frac{1}{4}$	2
2	-1
4	-2
8	-3

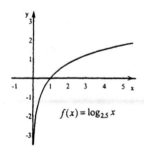

$f(x) = \log_{1/2} x$

8.

$f(x) = \log_{25} x$

9. Graph: $f(x) = \log_2 (x + 3)$

Think of $f(x)$ as y. Then $y = \log_2 (x + 3)$ is equivalent to $2^y = x + 3$, or $2^y - 3 = x$. We construct a table of values, plot these points, and connect them with a smooth curve.

For $y = 0$, $x = 2^0 - 3 = 1 - 3 = -2$

For $y = 1$, $x = 2^1 - 3 = 2 - 3 = -1$

For $y = 2$, $x = 2^2 - 3 = 4 - 3 = 1$

For $y = 3$, $x = 2^3 - 3 = 8 - 3 = 5$

For $y = -1$, $x = 2^{-1} - 3 = \frac{1}{2} - 3 = -\frac{5}{2}$

For $y = -2$, $x = 2^{-2} - 3 = \frac{1}{4} - 3 = -\frac{11}{4}$

x, or $2^y - 3$	y
-2	0
-1	1
1	2
5	3
$-\frac{5}{2}$	-1
$-\frac{11}{4}$	-2

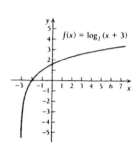

$f(x) = \log_2 (x + 3)$

Note that this is the graph of $f(x) = \log_2 x$ (see Exercise 4) shifted 3 units to the left.

10. $y = \log_3 (x - 2)$

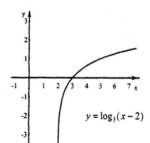

$y = \log_3 (x - 2)$

11. Graph: $f(x) = \log_2 x + 3$

Think of $f(x)$ as y. Then $y = \log_2 x + 3$ is equivalent to $y - 3 = \log_2 x$ or $2^{y-3} = x$. We construct a table of values, plot these points, and connect them with a smooth curve.

For $y = 0$, $x = 2^{0-3} = 2^{-3} = \frac{1}{8}$.

For $y = 1$, $x = 2^{1-3} = 2^{-2} = \frac{1}{4}$.

For $y = 2$, $x = 2^{2-3} = 2^{-1} = \frac{1}{2}$.

For $y = 3$, $x = 2^{3-3} = 2^0 = 1$.

For $y = 5$, $x = 2^{5-3} = 2^2 = 4$.

For $y = -1$, $x = 2^{-1-3} = 2^{-4} = \frac{1}{16}$.

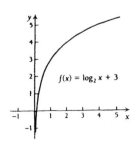

Note that this is the graph of $f(x) = \log_2 x$ (see Exercise 4) shifted upward 3 units.

12.

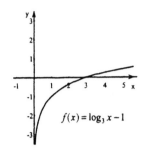

13. Graph $f(x) = 3^x$ (see Exercise Set 5.2, Exercise 2) and $f^{-1}(x) = \log_3 x$ (see Margin Exercise 1 in this section) on the same set of axes.

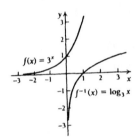

14.

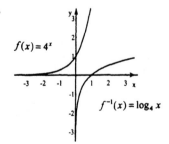

15. $10^3 = 1000 \longrightarrow 3 = \log_{10} 1000$ The exponent is the logarithm.
The base remains the same.

16. $2 = \log_{10} 100$

17. $5^{-3} = \dfrac{1}{125} \longrightarrow -3 = \log_5 \dfrac{1}{125}$ The exponent is the logarithm.
The base remains the same.

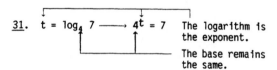

18. $-5 = \log_4 \dfrac{1}{1024}$

19. $8^{1/3} = 2 \longrightarrow \dfrac{1}{3} = \log_8 2$

20. $\dfrac{1}{4} = \log_{16} 2$

21. $10^{0.3010} = 2 \longrightarrow 0.3010 = \log_{10} 2$

22. $0.4771 = \log_{10} 3$

23. $e^3 = t \longrightarrow 3 = \log_e t$

24. $k = \log_p 3$

25. $Q^t = x \longrightarrow t = \log_Q x$

26. $m = \log_p V$

27. $e^3 = 20.0855 \longrightarrow 3 = \log_e 20.0855$

28. $2 = \log_e 7.3891$

29. $e^{-1} = 0.3679 \longrightarrow -1 = \log_e 0.3679$

30. $-6 = \log_e 0.00247$

31. $t = \log_4 7 \longrightarrow 4^t = 7$ The logarithm is the exponent.
The base remains the same.

32. $6^h = 29$

33. $\log_2 32 = 5 \longrightarrow 2^5 = 32$ The logarithm is the exponent.
The base remains the same.

34. $5^1 = 5$

35. $\log_{10} 0.1 = -1 \longrightarrow 10^{-1} = 0.1$

36. $10^{-2} = 0.01$

37. $\log_{10} 7 = 0.845 \longrightarrow 10^{0.845} = 7$

38. $10^{0.4771} = 3$

39. $\log_e 30 = 3.4012 \longrightarrow e^{3.4012} = 30$

40. $e^{2.3026} = 10$

41. $\log_t Q = k \longrightarrow t^k = Q$

42. $m^a = P$

43. $\log_e 0.38 = -0.9676 \longrightarrow e^{-0.9676} = 0.38$

44. $e^{-0.0987} = 0.906$

45. $\log_r M = -x \longrightarrow r^{-x} = M$

46. $c^{-w} = W$

47. $\log_{10} x = 3$

    $10^3 = x$ Converting to an exponential equation

    $1000 = x$ Computing 10^3

48. 4

49. $\log_3 3 = x$

    $3^x = 3$ Converting to an exponential equation

    $3^x = 3^1$

    $x = 1$ The exponents are the same.

50. -2

51. $\log_3 x = 2$

    $3^2 = x$ Converting to an exponential equation

    $9 = x$ Computing 3^2

52. 64

53. $\log_x 16 = 2$

    $x^2 = 16$ Converting to an exponential equation

    $x = 4$ or $x = -4$ Taking square roots

    $\log_4 16 = 2$ because $4^2 = 16$. Thus, 4 is a solution. Since all logarithm bases must be positive, $\log_{-4} 16$ is not defined and -4 is not a solution.

54. 4

55. $\log_2 x = -1$

    $2^{-1} = x$ Converting to an exponential equation

    $\frac{1}{2} = x$ Simplifying

56. $\frac{1}{9}$

57. $\log_8 x = \frac{1}{3}$

    $8^{1/3} = x$

    $2 = x$

58. 2

59. Let $\log_{10} 1000 = x$.

    Then $10^x = 1000$

    $10^x = 10^3$

    $x = 3$.

    Thus, $\log_{10} 1000 = 3$.

60. 7

61. Let $\log_{10} 0.1 = x$.

    Then $10^x = 0.1 = \frac{1}{10}$

    $10^x = 10^{-1}$

    $x = -1$.

    Thus, $\log_{10} 0.1 = -1$.

62. -3

63. Let $\log_{10} 1 = x$.

    Then $10^x = 1$

    $10^x = 10^0$ $(10^0 = 1)$

    $x = 0$.

    Thus, $\log_{10} 1 = 0$.

64. 1

65. Let $\log_5 625 = x$.

    Then $5^x = 625$

    $5^x = 5^4$

    $x = 4$.

    Thus, $\log_5 625 = 4$.

66. 6

67. Let $\log_5 \frac{1}{25} = x$.

    Then $5^x = \frac{1}{25}$

    $5^x = 5^{-2}$

    $x = -2$.

    Thus, $\log_5 \frac{1}{25} = -2$.

68. -4

69. Let $\log_3 1 = x$.

    Then $3^x = 1$

    $3^x = 3^0$ $(3^0 = 1)$

    $x = 0$.

    Thus, $\log_3 1 = 0$.

70. 1

71. Find $\log_e 1$.

    We know that $\log_e 1$ is the exponent to which e is raised to get 1. That exponent is 0. Therefore, $\log_e 1 = 0$.

72. 1

73. Find $\log_{81} 9$.

    We know that $\log_{81} 9$ is the exponent to which 81 is raised to get 9. Since $9 = \sqrt{81} = 81^{1/2}$, it follows that the exponent is $\frac{1}{2}$. Therefore, $\log_{81} 9 = \frac{1}{2}$.

74. $\frac{1}{3}$

75. Find $\log_e e^5$.

    We know that $\log_e e^5$ is the exponent to which e is raised to get e^5. That exponent is 5. Therefore, $\log_e e^5 = 5$.

76. -2

77. Graph: $y = \left(\frac{2}{3}\right)^x$ Graph: $y = \log_{2/3} x$, or $x = \left(\frac{2}{3}\right)^y$

x	y, or $\left(\frac{2}{3}\right)^x$	x, or $\left(\frac{2}{3}\right)^y$	y
0	1	1	0
1	$\frac{2}{3}$	$\frac{2}{3}$	1
2	$\frac{4}{9}$	$\frac{4}{9}$	2
3	$\frac{8}{27}$	$\frac{8}{27}$	3
-1	$\frac{3}{2}$	$\frac{3}{2}$	-1
-2	$\frac{9}{4}$	$\frac{9}{4}$	-2
-3	$\frac{27}{8}$	$\frac{27}{8}$	-3

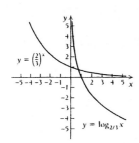

78.

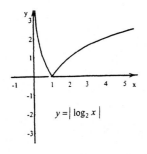

79. Graph: $y = \log_2 |x|$

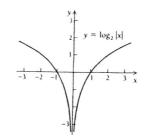

x	$\pm\frac{1}{4}$	$\pm\frac{1}{2}$	± 1	± 2	± 4
y	-2	-1	0	1	2

80. $27, \frac{1}{27}$

81. $\log_{125} x = \frac{2}{3}$

    $125^{2/3} = x$

    $(5^3)^{2/3} = x$

    $5^2 = x$

    $25 = x$

82. 4

83. $\log_{\sqrt{5}} x = -3$

    $(\sqrt{5})^{-3} = x$, or $5^{-3/2} = x$

84. $0, \frac{1}{2}$

85. $\log_4 (3x - 2) = 2$

    $4^2 = 3x - 2$

    $16 = 3x - 2$

    $18 = 3x$

    $6 = x$

86. $\frac{25}{16}$

87. $\log_x \sqrt[5]{36} = \frac{1}{10}$

    $x^{1/10} = \sqrt[5]{36}$

    $x^{1/10} = 36^{1/5}$

    $(x^{1/10})^{10} = (36^{1/5})^{10}$

    $x = 36^2$

    $x = 1296$

88. $-25, 4$

89. Let $\log_{1/4} \frac{1}{64} = x$.

Then $\left(\frac{1}{4}\right)^x = \frac{1}{64}$

$\left(\frac{1}{4}\right)^x = \left(\frac{1}{4}\right)^3$

$x = 3$.

Thus, $\log_{1/4} \frac{1}{64} = 3$.

90. 1

91. $\log_{10} (\log_4 (\log_3 81))$

$= \log_{10} (\log_4 4)$ $(\log_3 81 = 4)$

$= \log_{10} 1$ $(\log_4 4 = 1)$

$= 0$

92. 1

93. Let $\log_{\sqrt{3}} \frac{1}{81} = x$.

Then $(\sqrt{3})^x = \frac{1}{81}$

$(3^{1/2})^x = \frac{1}{3^4}$

$3^{x/2} = 3^{-4}$

$\frac{x}{2} = -4$

$x = -8$

Thus, $\log_{\sqrt{3}} \frac{1}{81} = -8$.

94. -2

95. $f(x) = 3^x$

3^x is defined for all real numbers. Thus the domain is the set of all real numbers.

96. $\{x \mid x > 0\}$

97. $f(x) = \log_a x^2$

The domain of a logarithmic function is the set of all positive real numbers. Since $x^2 > 0$ for all real numbers, except 0, the domain is $\{x \mid x \neq 0\}$.

98. $\{x \mid x > 0\}$

99. $f(x) = \log_{10} (3x - 4)$

The domain of a logarithmic function is the set of all positive real numbers. Thus, we solve

$3x - 4 > 0$

$3x > 4$

$x > \frac{4}{3}$

The domain is $\left\{x \mid x > \frac{4}{3}\right\}$.

100. $\{x \mid x \neq 0\}$

101. Let $y = \log_2 x$. The graph shows y is negative for $\{x \mid 0 < x < 1\}$.

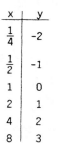

x	y
$\frac{1}{4}$	-2
$\frac{1}{2}$	-1
1	0
2	1
4	2
8	3

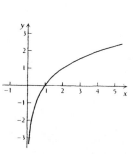

102.

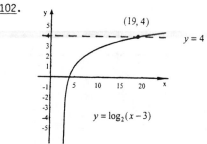

$\{x \mid x \geqslant 19\}$

Exercise Set 5.4

1. $\log_2 (64 \cdot 8) = \log_2 64 + \log_2 8$ Property 1

2. $\log_3 81 + \log_3 27$

3. $\log_4 (32 \cdot 64) = \log_4 32 + \log_4 64$ Property 1

4. $\log_5 125 + \log_5 25$

5. $\log_c QP = \log_c Q + \log_c P$ Property 1

6. $\log_t 9 + \log_t Y$

7. $\log_b 8 + \log_b 90 = \log_b (8 \cdot 90)$ Property 1
$= \log_b 720$

8. $\log_a (75 \cdot 2)$, or $\log_a 150$

9. $\log_c P + \log_c Q = \log_c PQ$ Property 1

10. $\log_e MT$

11. $\log_a x^4 = 4 \log_a x$ Property 2

12. $3 \log_b t$

13. $\log_c y^5 = 5 \log_c y$ Property 2

14. $8 \log_{10} y$

15. $\log_b Q^{-6} = -6 \log_b Q$ Property 2

16. $-6 \log_c K$

17. $\log_a \frac{76}{13} = \log_a 76 - \log_a 13$ Property 3

18. $\log_t M - \log_t 8$

19. $\log_b \frac{5}{4} = \log_b 5 - \log_b 4$ Property 3

20. $\log_a x - \log_a y$

21. $\log_a 18 - \log_a 5 = \log_a \frac{18}{5}$ Property 3

22. $\log_b \frac{54}{6}$, or $\log_b 9$

23. $\log_a x^3 y^2 z = \log_a x^3 + \log_a y^2 + \log_a z$ Property 1

       $= 3 \log_a x + 2 \log_a y + \log_a z$ Property 2

24. $\log_a 6 + \log_a x + 5 \log_a y + 4 \log_a z$

25. $\log_b \frac{x^2 y}{b^3} = \log_b x^2 y - \log_b b^3$ Property 3

       $= \log_b x^2 + \log_b y - \log_b b^3$ Property 1

       $= \log_b x^2 + \log_b y - 3$ Property 4

       $= 2 \log_b x + \log_b y - 3$ Property 2

26. $2 \log_b p + 5 \log_b q - 4 \log_b m - 9$

27. $\log_c \sqrt[3]{\frac{x^4}{y^3 z^2}}$

    $= \log_c \left[\frac{x^4}{y^3 z^2}\right]^{1/3}$

    $= \frac{1}{3} \log_c \frac{x^4}{y^3 z^2}$ Property 2

    $= \frac{1}{3}(\log_c x^4 - \log_c y^3 z^2)$ Property 3

    $= \frac{1}{3}[\log_c x^4 - (\log_c y^3 + \log_c z^2)]$ Property 1

    $= \frac{1}{3}(\log_c x^4 - \log_c y^3 - \log_c z^2)$ Removing parentheses

    $= \frac{1}{3}(4 \log_c x - 3 \log_c y - 2 \log_c z)$ Property 2

28. $\frac{1}{2}(6 \log_a x - 5 \log_a p - 8 \log_a q)$

29. $\log_a \sqrt[4]{\frac{m^8 n^{12}}{a^3 b^5}}$

    $= \frac{1}{4} \log_a \frac{m^8 n^{12}}{a^3 b^5}$ Property 2

    $= \frac{1}{4}(\log_a m^8 n^{12} - \log_a a^3 b^5)$ Property 3

    $= \frac{1}{4}[\log_a m^8 + \log_a n^{12} - (\log_a a^3 + \log_a b^5)]$ Property 1

    $= \frac{1}{4}(\log_a m^8 + \log_a n^{12} - \log_a a^3 - \log_a b^5)$ Removing parentheses

    $= \frac{1}{4}(\log_a m^8 + \log_a n^{12} - 3 - \log_a b^5)$ Property 4

    $= \frac{1}{4}(8 \log_a m + 12 \log_a n - 3 - 5 \log_a b)$ Property 2

30. $2 + \frac{3}{2} \log_a b$

31. $\frac{2}{5} \log_a x - \frac{1}{3} \log_a y = \log_a x^{\frac{2}{5}} - \log_a y^{\frac{1}{3}}$ Property 2

    $= \log_a \frac{x^{\frac{2}{5}}}{y^{\frac{1}{3}}}$ Property 3

32. $\log_a \frac{x^{\frac{1}{2}} y^4}{x^3}$, or $\log_a x^{-\frac{5}{2}} y^4$

33. $\log_a 2x + 3(\log_a x - \log_a y)$

    $= \log_a 2x + 3 \log_a x - 3 \log_a y$

    $= \log_a 2x + \log_a x^3 - \log_a y^3$ Property 2

    $= \log_a 2x^4 - \log_a y^3$ Property 1

    $= \log_a \frac{2x^4}{y^3}$ Property 3

34. $\log_a x$

35. $\log_a \frac{a}{\sqrt{x}} - \log_a \sqrt{ax}$

    $= \log_a ax^{-1/2} - \log_a a^{1/2} x^{1/2}$

    $= \log_a \frac{ax^{-1/2}}{a^{1/2} x^{1/2}}$ Property 3

    $= \log_a \frac{a^{1/2}}{x}$

    $= \log_a \frac{\sqrt{a}}{x}$

We can simplify this further as follows:

$\log_a \frac{\sqrt{a}}{x} = \log_a \sqrt{a} - \log_a x$ Property 3

       $= \log_a a^{1/2} - \log_a x$ Writing exponential notation

       $= \frac{1}{2} - \log_a x$ Property 4

36. $\log_a (x + 2)$

37. $\log_b 15 = \log_b (3 \cdot 5)$
$ = \log_b 3 + \log_b 5$ Property 1
$ = 0.5283 + 0.7740$
$ = 1.3023$

38. 0.2457

39. $\log_b \frac{3}{5} = \log_b 3 - \log_b 5$ Property 3
$\phantom{\log_b \frac{3}{5}} = 0.5283 - 0.7740$
$\phantom{\log_b \frac{3}{5}} = -0.2457$

40. -0.7740

41. $\log_b \frac{1}{3} = \log_b 1 - \log_b 3$ Property 3
$\phantom{\log_b \frac{1}{3}} = 0 - 0.5283$ $(\log_b 1 = 0)$
$\phantom{\log_b \frac{1}{3}} = -0.5283$

42. $\frac{1}{2}$

43. $\log_b \sqrt{b^3} = \log_b b^{3/2}$ Writing exponential notation
$\phantom{\log_b \sqrt{b^3}} = \frac{3}{2}$ Property 4

44. 1.7740

45. $\log_b 3b = \log_b 3 + \log_b b$ Property 1
$ = 0.5283 + 1$ $(\log_b b = 1)$
$ = 1.5283$

46. 1.0566

47. $\log_b 25 = \log_b 5^2$
$ = 2 \log_b 5$ Property 2
$ = 2(0.7740) = 1.548$

48. 2.0763

49. $\log_t t^{11} = 11$ Property 4

50. 3

51. $\log_e e^{|x-4|} = |x - 4|$ Property 4

52. $\sqrt{5}$

53. $3^{\log_3 4x} = 4x$ Property 5

54. $4x - 3$

55. $a^{\log_a Q} = Q$ Property 5

56. e^5

57. $a^{\log_a x} = 15$
$\phantom{a^{\log_a x}}x = 15$ Property 5

58. 4

59. $\log_e e^x = -7$
$x = -7$ Property 4

60. 2.7

61. $\frac{\log_a M}{\log_a N} = \log_a M - \log_a N$

Let $M = a^2$ and $N = a$. Then $\frac{\log_a a^2}{\log_a a} = \frac{2}{1} = 2$, but $\log_a a^2 - \log_a a = 2 - 1 = 1$. Thus the statement is false.

62. False

63. $\frac{\log_a M}{c} = \frac{1}{c} \log_a M = \log_a M^{1/c}$

The statement is true by Property 2.

64. True

65. $\log_a 2x = 2 \log_a x$

Let $x = a$. Then $\log_a 2a = \log_a 2 + \log_a a = \log_a 2 + 1$, but $2 \log_a a = 2 \cdot 1 = 2$. Thus the statement is false in general.

66. True

67. $\log_a (M + N) = \log_a M + \log_a N$

Let $M = N = a$. Then $\log_a (a + a) = \log_a 2a = \log_a 2 + 1$ (see Exercise 65), but $\log_a a + \log_a a = 1 + 1 = 2$. Thus the statement is false in general.

68. True

69. $\log_c a - \log_c b = \log_c \left(\frac{a}{b}\right)$ is true by Property 3.

70. False

71. $\log_a (x^8 - y^8) - \log_a (x^2 + y^2)$
$= \log_a \frac{x^8 - y^8}{x^2 + y^2}$ Property 3
$= \log_a \frac{(x^4 + y^4)(x^2 + y^2)(x + y)(x - y)}{x^2 + y^2}$
$= \log_a [(x^4 + y^4)(x^2 - y^2)]$ Simplifying
$= \log_a (x^6 - x^4y^2 + x^2y^4 - y^6)$ Multiplying

72. $\log_a (x^3 + y^3)$

73. $\log_a \sqrt{4 - x^2}$

$= \log_a (4 - x^2)^{1/2}$

$= \frac{1}{2} \log_a (4 - x^2)$

$= \frac{1}{2} \log_a [(2 + x)(2 - x)]$

$= \frac{1}{2} [\log_a (2 + x) + \log_a (2 - x)]$

74. $\frac{1}{2} \log_a (x - y) - \frac{1}{2} \log_a (x + y)$

75. $\log_a \dfrac{\sqrt[3]{x^2 z}}{\sqrt[3]{y^2 z^{-2}}}$

$= \log_a \left[\dfrac{x^2 z^3}{y^2}\right]^{1/3}$

$= \frac{1}{3} (\log_a x^2 z^3 - \log_a y^2)$

$= \frac{1}{3} (2 \log_a x + 3 \log_a z - 2 \log_a y)$

$= \frac{1}{3} [2 \cdot 2 + 3 \cdot 4 - 2 \cdot 3]$ Substituting

$= \frac{1}{3}(10)$

$= \frac{10}{3}$

76.   0, 5

77. $\log_a 3x = \log_a 3 + \log_a x$

$\log_a 3x = \log_a 3x$ Property 1

We get an equation that is true for all values of x for which the logarithm function is defined. Thus, the solution is $\{x \mid x > 0\}$, or $(0, \infty)$.

78. $\frac{1}{2}$

79. $3^{\log_3 (8x - 4)} = 5$

$\phantom{3^{\log_3}} 8x - 4 = 5$ Property 5

$\phantom{3^{\log_3} 8x -} 8x = 9$

$\phantom{3^{\log_3} 88x} x = \frac{9}{8}$

80. $\sqrt{7}$

81. $8^{2 \log_8 x + \log_8 x} = 27$

$\phantom{8^{2}} 8^{3 \log_8 x} = 27$

$\phantom{8^{2}} 8^{\log_8 x^3} = 27$

$\phantom{8^{2} 8^{\log_8}} x^3 = 27$ Property 5

$\phantom{8^{2} 8^{\log_8}} x = 3$

82. $\{x \mid x > 0\}$

83. $\log_b \dfrac{5}{x + 2} = \log_b 5 - \log_b (x + 2)$

$\log_b \dfrac{5}{x + 2} = \log_b \dfrac{5}{x + 2}$ Property 3

The equation is true for all values of x for which the logarithm function is defined. That is, for all x for which $\dfrac{5}{x + 2} > 0$, or

$ x + 2 > 0$

$ x > -2.$

The solution set is $\{x \mid x > -2\}$, or $(-2, \infty)$.

84.   -2

85. $\log_a x = 2$

$ a^2 = x$

Let $\log_{1/a} x = n$ and solve for n.

$ \log_{1/a} a^2 = n$ Substituting a^2 for x

$ \left[\dfrac{1}{a}\right]^n = a^2$

$ (a^{-1})^n = a^2$

$ a^{-n} = a^2$

$ -n = 2$

$ n = -2$

Thus, $\log_{1/a} x = -2$ when $\log_a x = 2$.

86. $\log_a \dfrac{1}{x} = \log_a 1 - \log_a x = -\log_a x$

87. $\log_a \dfrac{x + \sqrt{x^2 - 5}}{5} \cdot \dfrac{x - \sqrt{x^2 - 5}}{x - \sqrt{x^2 - 5}}$

$= \log_a \dfrac{5}{5(x - \sqrt{x^2 - 5})} = \log_a \dfrac{1}{x - \sqrt{x^2 - 5}}$

$= \log_a 1 - \log_a (x - \sqrt{x^2 - 5})$

$= -\log_a (x - \sqrt{x^2 - 5}).$

88.   Let $M = \log_a \left[\dfrac{1}{x}\right]$ or $\dfrac{1}{x} = a^M$ or $x = \left[\dfrac{1}{a}\right]^M$

or $\log_{\frac{1}{a}} x = M.$

89.   See the answer section in the text.

Exercise Set 5.5

1.   0.4771

2.   0.8451

3.   0.9031

4.   1.1139

5.   0.3692

6. 0.4871

7. 1.8129

8. 1.9243

9. 1.7952

10. 1.0334

11. 2.7259

12. 2.2923

13. 4.1271

14. 4.9689

15. -0.2441

16. -0.1612

17. -1.2840

18. -0.4123

19. -2.0084

20. -3.2905

21. 0.0011

22. 79,104.2833

23. Does not exist

24. Does not exist

25. 1.0986

26. 0.6931

27. 2.0794

28. 2.5649

29. 4.4067

30. 3.9120

31. 9.0318

32. 6.6962

33. -5.1328

34. -7.9020

35. 2.8044

36. 0.0017

37. 0.0000027

38. 54,515,738.62

39. Find $\log_4 100$. We will use common logarithms in the change-of-base formula:

$$\log_b M = \frac{\log_a M}{\log_a b}$$

Substitute 10 for a, 4 for b, and 100 for M.

$$\log_4 100 = \frac{\log_{10} 100}{\log_{10} 4}$$

$$\approx \frac{2.0000}{0.6021} \quad \text{(Using a calculator)}$$

$$\approx 3.3219$$

40. 2.7268

41. Find $\log_2 12$. We will use common logarithms in the change-of-base formula:

$$\log_b M = \frac{\log_a M}{\log_a b}$$

Substitute 10 for a, 2 for b, and 12 for M.

$$\log_2 12 = \frac{\log_{10} 12}{\log_{10} 2}$$

$$\approx \frac{1.0792}{0.3010} \quad \text{(Using a calculator)}$$

$$\approx 3.5850$$

42. 2.2920

43. Find $\log_{100} 0.3$. We will use common logarithms in the change-of-base formula:

$$\log_b M = \frac{\log_a M}{\log_a b}$$

Substitute 10 for a, 100 for b, and 0.3 for M.

$$\log_{100} 0.3 = \frac{\log_{10} 0.3}{\log_{10} 100}$$

$$\approx \frac{-0.52288}{2} \quad \text{(Using a calculator)}$$

$$= -0.26144$$

44. 0.7384

45. Find $\log_{0.5} 7$. We will use common logarithms in the change-of-base formula:

$$\log_b M = \frac{\log_a M}{\log_a b}$$

Substitute 10 for a, 0.5 for b and 7 for M.

$$\log_{0.5} 7 = \frac{\log_{10} 7}{\log_{10} 0.5}$$

$$\approx \frac{0.8451}{-0.3010} \quad \text{(Using a calculator)}$$

$$\approx -2.8074$$

46. -0.3010

47. Find $\log_3 0.3$. We will use common logarithms in the change-of-base formula:

$$\log_b M = \frac{\log_a M}{\log_b b}$$

Substitute 10 for a, 3 for b, and 0.3 for M.

$$\log_3 0.3 = \frac{\log_{10} 0.3}{\log_{10} 3}$$
$$\approx \frac{-0.5229}{0.4771} \quad \text{(Using a calculator)}$$
$$\approx -1.0959$$

48. -4.0589

49. Find $\log_\pi 100$. We will use common logarithms in the change-of-base formula:

$$\log_b M = \frac{\log_a M}{\log_a b}$$

Substitute 10 for a, π for b, and 100 for M.

$$\log_\pi 100 = \frac{\log_{10} 100}{\log_{10} \pi}$$
$$\approx \frac{2.0000}{0.4971} \quad \text{(Using a calculator)}$$
$$\approx 4.0229$$

50. 2.8119

51. See the answer section in the text.

52. $\log x = \frac{\ln x}{\ln 10} = \frac{\ln x}{\ln 10} \cdot \ln x = 0.4343 \ln x$

53. Graph $y = f(x) = 10^x$

We use a calculator to find function values.

x	y, or f(x)
-1	0.1
0	1
0.5	3.1623
1	10
1.5	31.6228
2	100
2.5	316.2278
3	1000
3.1	1258.9254

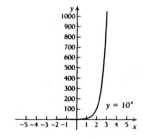

54.

55. Graph $y = f^{-1}(x) = \log x$.

We can interchange the first and second coordinates in each pair found in Exercise 53 to obtain ordered pairs on the graph of $y = \log x$.

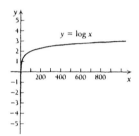

56.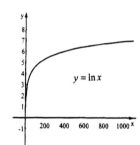

57. $\frac{\log_5 8}{\log_5 2} = \frac{\log_5 2^3}{\log_5 2} = \frac{3 \log_5 2}{\log_5 2} = 3$

58. $\frac{3}{2}$

59.
$$\log 872x = 5.3442$$
$$\log 872 + \log x = 5.3442$$
$$2.940516485 + \log x \approx 5.3442$$
$$\log x \approx 2.403683515$$
$$x = 253.3282 \quad \text{Finding the inverse}$$

60. ± 1554.4107

61.
$$\log 784 + \log x = \log 2322$$
$$2.894316063 + \log x \approx 3.365862215$$
$$\log x \approx 0.471546152$$
$$x \approx 2.9617 \quad \text{Finding the inverse}$$

62. 1.1799

63. Let a = e, b = 10, M = e. Substitute in the change-of-base formula.

$$\log e = \frac{\ln e}{\ln 10} = \frac{1}{\ln 10}$$

64. Let a = e, b = 10, M = M. Substitute in the change-of-base formula.

$$\log M = \frac{\ln M}{\ln 10}$$

65. Let a = M, b = b, M = M. Substitute in the change-of-base formula.

$$\log_b M = \frac{\log_M M}{\log_M b} = \frac{1}{\log_M b}$$

66. Let a = 10, b = e, M = M. Substitute in the change-of-base formula.

$$\ln M = \frac{\log M}{\log e}$$

67. See the answer section in the text.

68. 2, 2.25, 2.48832, 2.593742, 2.704814, 2.716924

69. $f(t) = t^{\frac{1}{t-1}}$

$f(0.5) = 0.5^{\frac{1}{0.5-1}} = 0.5^{\frac{1}{-0.5}} = 0.5^{-2} = 4$

$f(0.9) = 0.9^{\frac{1}{0.9-1}} = 0.9^{\frac{1}{-0.1}} = 0.9^{-10} \approx 2.867972$

$f(0.99) = 0.99^{\frac{1}{0.99-1}} = 0.99^{\frac{1}{-0.01}} = 0.99^{-100} \approx 2.731999$

$f(0.999) = 0.999^{\frac{1}{0.999-1}} = 0.999^{\frac{1}{-0.001}} = 0.999^{-1000} \approx 2.719642$

$f(0.9999) = 0.9999^{\frac{1}{0.9999-1}} = 0.9999^{\frac{1}{-0.0001}} = 0.9999^{-10,000} \approx 2.718418$

70. e^{π}

71. $e^{\sqrt{\pi}} \approx 5.8853$, $\sqrt{e^{\pi}} \approx 4.8105$

$e^{\sqrt{\pi}}$ is larger.

72. Use a calculator to find the inverse function in both base 10 and base e and compare with the value of x. Study graphs if possible.

73.

73. $y = 4.5^x$
$y = \log_{4.5} x$

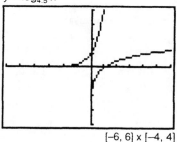

[−6, 6] x [−4, 4]

74. $y = 6.7^x$
$y = \log_{6.7} x$

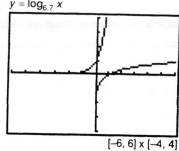

[−6, 6] x [−4, 4]

75. $y = 0.15^x$
$y = \log_{0.15} x$

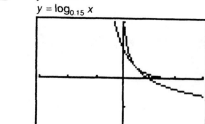

[−3, 3] x [−2, 2]

76. $y = 0.95^x$
$y = \log_{0.95} x$

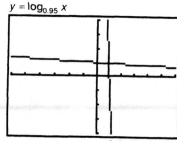

[−6, 6] x [−4, 4]

Exercise Set 5.6

1. Graph f(x) = 2e^X.

Use a calculator to find approximate values of
2e^X, and use these values to draw the graph.

x	2e^X
-2	0.3
-1	0.7
0	2
1	5.4
2	14.8

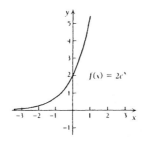

2.

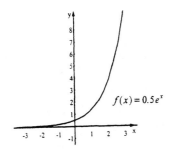

3. Graph f(x) = e^(1/2)X.

Use a calculator to find approximate values of
e^(1/2)X, and use these values to draw the graph.

x	e^(1/2)X
-2	0.4
-1	0.6
0	1
1	1.6
2	2.7

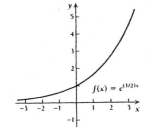

4.

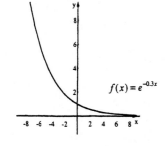

5. Graph f(x) = e^X+1.

Use a calculator to find approximate values of
e^X+1, and use these values to draw the graph.

x	e^X+1
-2	0.4
-1	1
0	2.7
1	7.4
2	20.1

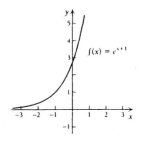

6.

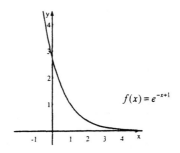

7. Graph f(x) = e^2X + 1.

Use a calculator to find approximate values of
e^2X + 1, and use these values to draw the graph.

x	e^2X + 1
-2	1.02
-1	1.1
0	2
1	8.4
2	55.6

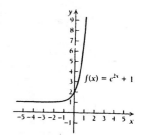

8.

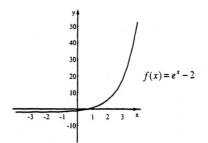

207

9. Graph $f(x) = 1 - e^{-0.01X}$, for nonnegative values of x. Use a calculator to find approximate values of $1 - e^{-0.01X}$ for nonnegative values of x, and use these values to draw the graph.

x	$1 - e^{-0.01X}$
0	0
100	0.63
200	0.86
300	0.95
400	0.98

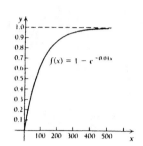

10.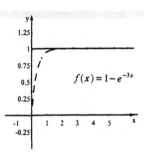

11. Graph $f(x) = 2(1 - e^{-X})$, for nonnegative values of x. Use a calculator to find approximate values of $2(1 - e^{-X})$ for nonnegative values of x, and use these values to draw the grpah.

x	$2(1 - e^{-X})$
0	0
1	1.26
2	1.73
3	1.90
4	1.96

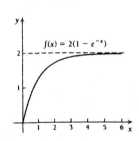

12.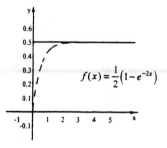

13. Graph $f(x) = 4 \ln x$.

Find some solutions with a calculator, plot them, and draw the graph.

x	0.25	0.5	1	1.5	2	3
4 ln x	-5.55	-2.77	0	1.62	2.77	4.39

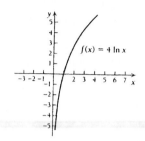

14.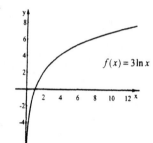

15. Graph $f(x) = \frac{1}{2} \ln x$.

Find some solutions with a calculator, plot them, and draw the graph.

x	0.5	1	2	4	7	8
$\frac{1}{2}$ ln x	-0.35	0	0.35	0.69	0.97	1.04

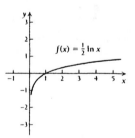

16.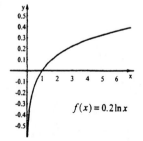

17. Graph f(x) = ln (x - 2).

   Find some solutions using a calculator, plot them, and draw the graph. When x = 3, f(x) = ln (3 - 2) = ln 1 = 0, and so on.

x	2.5	3	4	6	8	10
ln (x-2)	-0.69	0	0.69	1.39	1.79	2.08

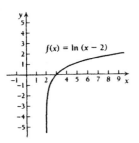

18.

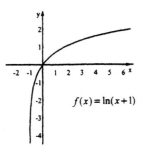

19. Graph f(x) = 2 - ln x.

   Find some solutions using a calculator, plot them, and draw the graph.

x	0.25	0.5	1	2	4
2 - ln x	3.39	2.69	2	1.31	0.61

x	7	8	10
2 - ln x	0.05	-0.08	-0.30

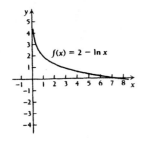

20.

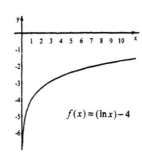

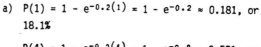

21. $P(t) = 1 - e^{-0.2t}$

   a) $P(1) = 1 - e^{-0.2(1)} = 1 - e^{-0.2} \approx 0.181$, or 18.1%

   $P(4) = 1 - e^{-0.2(4)} = 1 - e^{-0.8} \approx 0.551$, or 55.1%

   $P(6) = 1 - e^{-0.2(6)} = 1 - e^{-1.2} \approx 0.699$, or 69.9%

   $P(12) = 1 - e^{-0.2(12)} = 1 - e^{-2.4} \approx 0.909$, or 90.9%

   b) Plot the points (0, 0%), (1, 18.1%), (4, 55.1%), (6, 69.9%), and (12, 90.9%) and draw the graph.

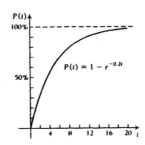

22. a) 26, 78, 91, 95

   b)

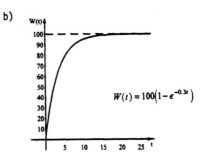

23. $V(t) = \$58(1 - e^{-1.1t}) + \$20$

   a) $V(1) = \$58(1 - e^{-1.1(1)}) + \$20$

   $= \$58(1 - e^{-1.1}) + \$20$

   $\approx \$58.69$

   $V(2) = \$58(1 - e^{-1.1(2)}) + \$20$

   $= \$58(1 - e^{-2.2}) + \$20$

   $\approx \$71.57$

   $V(4) = \$58(1 - e^{-1.1(4)}) + \$20$

   $= \$58(1 - e^{-4.4}) + \$20$

   $\approx \$77.29$

   $V(6) = \$58(1 - e^{-1.1(6)}) + \$20$

   $= \$58(1 - e^{-6.6}) + \$20$

   $\approx \$77.92$

   $V(12) = \$58(1 - e^{-1.1(12)}) + \$20$

   $= \$58(1 - e^{-13.2}) + \$20$

   $\approx \$77.99 +$

   b) Plot the points (0, $20), (1, $58.69), (2, $71.57), (4, $77.29), (6, $77.92), and (12, $77.99 +) and draw the graph.

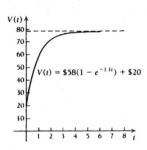

24. a) 0.259, or 25.9%; 0.593, or 59.3%
       0.835, or 83.5%; 0.973, or 97.3%

   b)

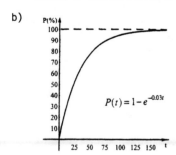

25. Graph $g(x) = e^{|x|}$.

   Use a calculator to find some values of $e^{|x|}$, and use these values to draw the graph.

x	$e^{|x|}$
-3	20.09
-2	7.39
-1	2.72
0	1
1	2.72
2	7.39

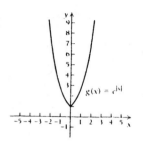

26.

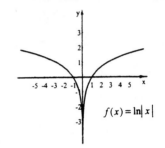

$f(x) = \ln|x|$

27. Graph $f(x) = |\ln x|$.

   Find some solutions using a calculator, plot them, and draw the graph.

x	0.25	0.5	1	3	6	10
$|\ln x|$	1.39	0.69	0	1.10	1.79	2.30

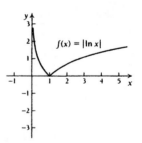

$f(x) = |\ln x|$

28.

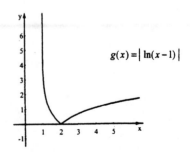

$g(x) = |\ln(x-1)|$

29. Graph $f(x) = \dfrac{e^x + e^{-x}}{2}$

Use a calculator to find some values of $\dfrac{e^x + e^{-x}}{2}$, and use these values to draw the graph.

x	$\dfrac{e^x + e^{-x}}{2}$
-3	13.76
-2	3.76
-1	1.54
0	1
1	1.54
2	3.76

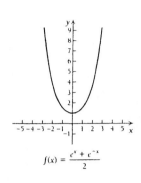

$f(x) = \dfrac{e^x + e^{-x}}{2}$

30.

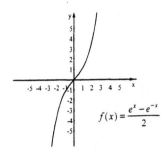

$f(x) = \dfrac{e^x - e^{-x}}{2}$

31. $N(t) = \dfrac{4800}{6 + 794e^{-0.4t}}$

a) $N(3) = \dfrac{4800}{6 + 794e^{-0.4(3)}} = \dfrac{4800}{6 + 794e^{-1.2}} \approx 20$

$N(5) = \dfrac{4800}{6 + 794e^{-0.4(5)}} = \dfrac{4800}{6 + 794e^{-2}} \approx 42$

$N(10) = \dfrac{4800}{6 + 794e^{-0.4(10)}} = \dfrac{4800}{6 + 794e^{-4}} \approx 234$

$N(15) = \dfrac{4800}{6 + 794e^{-0.4(15)}} = \dfrac{4800}{6 + 794e^{-6}} \approx 602$

b) Plot the points (0, 6), (3, 20), (5, 42), (10, 234), and (15, 602) and draw the graph.

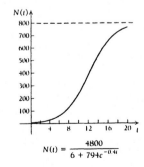

$N(t) = \dfrac{4800}{6 + 794e^{-0.4t}}$

32. a) 33, 183, 758, 1339, 1952, 1998

b)

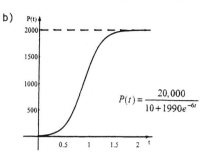

$P(t) = \dfrac{20,000}{10 + 1990e^{-6t}}$

33. a) $f(x) = x^2 e^{-x}$

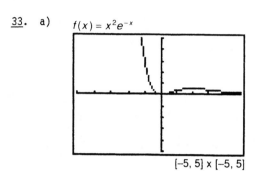

$[-5, 5] \times [-5, 5]$

b) From the graph we see that 0 is the only zero of the function.

c) The function has no maximum value. The minimum value is 0.

34. a) $f(x) = e^{-x^2}$

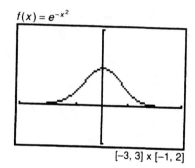

$[-3, 3] \times [-1, 2]$

b) None

c) Maximum: 1; no minimum

35. a) $f(x) = x^2 \ln x$

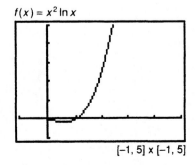

$[-1, 5] \times [-1, 5]$

b) From the graph we see that 1 is the only zero of the function.

c) The function has no maximum value. The minimum value is about -0.2.

36. a)

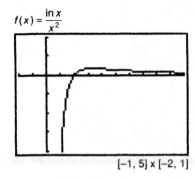

$f(x) = \dfrac{\ln x}{x^2}$

$[-1, 5] \times [-2, 1]$

b) 1

c) Maximum: about 0.18; no minimum

37.-42. See Exercises 25 - 30.

Exercise Set 5.7

1. $2^x = 32$
   $2^x = 2^5$
   $x = 5$ The exponents are the same.

2. 4

3. $3^x = 81$
   $3^x = 3^4$
   $x = 4$ The exponents are the same.

4. 4

5. $2^{2x} = 8$
   $2^{2x} = 2^3$
   $2x = 3$ The exponents are the same.
   $x = \dfrac{3}{2}$

6. $\dfrac{1}{2}$

7. $3^{7x} = 27$
   $3^{7x} = 3^3$
   $7x = 3$
   $x = \dfrac{3}{7}$

8. 1

9. $2^x = 33$
   $\log 2^x = \log 33$ Taking the common logarithm on both sides
   $x \log 2 = \log 33$ Property 2
   $x = \dfrac{\log 33}{\log 2}$
   $x \approx \dfrac{1.5185}{0.3010}$
   $x \approx 5.0444$

10. 4.3219

11. $2^x = 40$
    $\log 2^x = \log 40$
    $x \log 2 = \log 40$
    $x = \dfrac{\log 40}{\log 2}$
    $x \approx \dfrac{1.6021}{0.3010}$
    $x \approx 5.3219$

12. 4.2479

13. $5^{4x-7} = 125$
    $5^{4x-7} = 5^3$
    $4x - 7 = 3$
    $4x = 10$
    $x = \dfrac{10}{4} = \dfrac{5}{2}$

14. -1

15. $3^{x^2+4x} = \dfrac{1}{27}$
    $3^{x^2+4x} = 3^{-3}$
    $x^2 + 4x = -3$
    $x^2 + 4x + 3 = 0$
    $(x + 3)(x + 1) = 0$
    $x = -3$ or $x = -1$

16. $\dfrac{1}{2}$, -3

17. $84^x = 70$
    $\log 84^x = \log 70$
    $x \log 84 = \log 70$
    $x = \dfrac{\log 70}{\log 84}$
    $x \approx \dfrac{1.8451}{1.9243}$
    $x \approx 0.9589$

18. 0.6910

19. $e^t = 1000$

$\ln e^t = \ln 1000$

$t = \ln 1000$ Property 4

$t \approx 6.9078$

20. 4.6052

21. $e^{-t} = 0.3$

$\ln e^{-t} = \ln 0.3$

$-t = \ln 0.3$

$t = -\ln 0.3$

$t \approx 1.2040$

22. 3.2189

23. $e^{-0.03t} = 0.08$

$\ln e^{-0.03t} = \ln 0.08$

$-0.03t = \ln 0.08$

$t = \dfrac{\ln 0.08}{-0.03}$

$t \approx \dfrac{-2.5257}{-0.03}$

$t \approx 84.1910$

24. 15.4033

25. $3^x = 2^{x-1}$

$\ln 3^x = \ln 2^{x-1}$

$x \ln 3 = (x - 1) \ln 2$

$x \ln 3 = x \ln 2 - \ln 2$

$\ln 2 = x \ln 2 - x \ln 3$

$\ln 2 = x(\ln 2 - \ln 3)$

$\dfrac{\ln 2}{\ln 2 - \ln 3} = x$

$\dfrac{0.6931}{0.6931 - 1.0986} \approx x$

$-1.7095 \approx x$

26. -20.6377

27. $(3.9)^x = 48$

$\log (3.9)^x = \log 48$

$x \log 3.9 = \log 48$

$x = \dfrac{\log 48}{\log 3.9}$

$x \approx \dfrac{1.6812}{0.5911}$

$x \approx 2.8444$

28. 2.6731

29. $250 - (1.87)^x = 0$

$250 = (1.87)^x$

$\log 250 = \log (1.87)^x$

$\log 250 = x \log 1.87$

$\dfrac{\log 250}{\log 1.87} = x$

$\dfrac{2.3979}{0.2718} \approx x$

$8.8211 \approx x$

30. 2.7716

31. $4^{2x} = 8^{3x-4}$

$(2^2)^{2x} = (2^3)^{3x-4}$

$2^{4x} = 2^{9x-12}$ Multiplying exponents

$4x = 9x - 12$

$-5x = -12$

$x = \dfrac{12}{5}$

32. -16

33. $\dfrac{e^x - e^{-x}}{t} = 5$

$e^x - e^{-x} = 5t$ Multiplying by t

$e^{2x} - 1 = 5te^x$ Multiplying by e^x

$(e^x)^2 - 5t \cdot e^x - 1 = 0$

This equation is reducible to quadratic with $u = e^x$. We use the quadratic formula with $a = 1$, $b = -5t$ and $c = -1$.

$e^x = \dfrac{-(-5t) \pm \sqrt{(-5t)^2 - 4 \cdot 1 \cdot (-1)}}{2 \cdot 1}$

$= \dfrac{5t \pm \sqrt{25t^2 + 4}}{2}$

$\ln e^x = \ln \left(\dfrac{5t \pm \sqrt{25t^2 + 4}}{2}\right)$ Taking the natural logarithm on both sides

$x = \ln \left(\dfrac{5t \pm \sqrt{25t^2 + 4}}{2}\right)$ Property 4

34. ± 1.5668

35. $\dfrac{e^x + e^{-x}}{e^x - e^{-x}} = t$

$e^x + e^{-x} = te^x - te^{-x}$ Multiplying by $e^x - e^{-x}$

$e^{2x} + 1 = te^{2x} - t$ Multiplying by e^x

$t + 1 = te^{2x} - e^{2x}$

$t + 1 = e^{2x}(t - 1)$

$\dfrac{t + 1}{t - 1} = e^{2x}$

$\ln \left(\dfrac{t + 1}{t - 1}\right) = \ln e^{2x}$

$\ln \left(\dfrac{t + 1}{t - 1}\right) = 2x$

$\dfrac{1}{2} \ln \left(\dfrac{t + 1}{t - 1}\right) = x$

36. $\frac{1}{2} \log_5 \left(\frac{t + 1}{1 - t}\right)$

37. $\log_5 x = 4$
    $\quad x = 5^4$ Writing an equivalent exponential equation
    $\quad x = 625$

38. 4

39. $\log_5 x = -3$
    $\quad x = 5^{-3}$ Writing an equivalent exponential equation
    $\quad x = \frac{1}{125}$

40. 5

41. $\log x = 2$ The base is 10.
    $\quad x = 10^2$
    $\quad x = 100$

42. 10

43. $\log x = -3$ The base is 10.
    $\quad x = 10^{-3}$
    $\quad x = \frac{1}{1000}$, or 0.001

44. 10^{-4}, or 0.0001

45. $\ln x = 1$
    $\quad x = e^1 = e$

46. e^3

47. $\ln x = -2$
    $\quad x = e^{-2}$

48. e^{-1}

49. $\log_5 (8 - 7x) = 3$
    $\quad 5^3 = 8 - 7x$ Writing an equivalent exponential equation
    $\quad 125 = 8 - 7x$
    $\quad 117 = -7x$
    $\quad -\frac{117}{7} = x$

    The answer checks. The solution is $-\frac{117}{7}$.

50. $\frac{22}{3}$

51. $\log x + \log (x - 9) = 1$ The base is 10.
    $\quad \log_{10} [x(x - 9)] = 1$ Property 1
    $\quad\quad x(x - 9) = 10^1$
    $\quad\quad x^2 - 9x = 10$
    $\quad\quad x^2 - 9x - 10 = 0$
    $\quad\quad (x - 10)(x + 1) = 0$

    $x = 10$ or $x = -1$

    Check: For 10:
    $$\frac{\log x + \log (x - 9) = 1}{\log 10 + \log (10 - 9) \;\big|\; 1}$$
    $$\log 10 + \log 1$$
    $$1 + 0$$
    $$1$$

    For -1:
    $$\frac{\log x + \log (x - 9) = 1}{\log (-1) + \log (-1 - 9) \;\big|\; 1}$$

    The number -1 does not check, because negative numbers do not have logarithms. The solution is 10.

52. 1

53. $\log x - \log (x + 3) = -1$ The base is 10.
    $\quad \log_{10} \frac{x}{x + 3} = -1$ Property 3
    $\quad\quad \frac{x}{x + 3} = 10^{-1}$
    $\quad\quad \frac{x}{x + 3} = \frac{1}{10}$
    $\quad\quad 10x = x + 3$
    $\quad\quad 9x = 3$
    $\quad\quad x = \frac{1}{3}$

    The answer checks. The solution is $\frac{1}{3}$.

54. 1

55. $\log_2 (x + 1) + \log_2 (x - 1) = 3$
    $\quad \log_2 [(x + 1)(x - 1)] = 3$ Property 1
    $\quad\quad (x + 1)(x - 1) = 2^3$
    $\quad\quad\quad x^2 - 1 = 8$
    $\quad\quad\quad x^2 = 9$
    $\quad\quad\quad x = \pm 3$

    The number 3 checks, but -3 does not. The solution is 3.

56. $\frac{1}{63}$

57. $\log_8 (x + 1) - \log_8 x = \log_8 4$

$\log_8 \left[\dfrac{x + 1}{x}\right] = \log_8 4$ Property 3

$\dfrac{x + 1}{x} = 4$ Taking antilogarithms

$x + 1 = 4x$

$1 = 3x$

$\dfrac{1}{3} = x$

The answer checks. The solution is $\dfrac{1}{3}$.

58. $\dfrac{21}{8}$

59. $\log_4 (x + 3) + \log_4 (x - 3) = 2$

$\log_4 [(x + 3)(x - 3)] = 2$ Property 1

$(x + 3)(x - 3) = 4^2$

$x^2 - 9 = 16$

$x^2 = 25$

$x = \pm 5$

The number 5 checks, but -5 does not. The solution is 5.

60. $\sqrt{41}$

61.

$\log \sqrt[4]{x} = \sqrt{\log x}$

$\log x^{1/4} = \sqrt{\log x}$ Writing exponential notation

$\dfrac{1}{4} \log x = \sqrt{\log x}$ Property 2

$\dfrac{1}{16} (\log x)^2 = \log x$ Squaring both sides

$\dfrac{1}{16} (\log x)^2 - \log x = 0$

Let $u = \log x$, substitute and solve for u.

$\dfrac{1}{16}u^2 - u = 0$

$u\left[\dfrac{1}{16}u - 1\right] = 0$

$u = 0$ or $u = 16$

$\log x = 0$ or $\log x = 16$

$x = 1$ or $x = 10^{16}$

Both numbers check. The solutions are 1 and 10^{16}.

62. 1, 10^9

63. $\log_5 \sqrt{x^2 + 1} = 1$

$\sqrt{x^2 + 1} = 5^1$

$x^2 + 1 = 25$ Squaring both sides

$x^2 = 24$

$x = \pm\sqrt{24}$, or $\pm 2\sqrt{6}$

Both numbers check. The solutions are $\pm 2\sqrt{6}$.

64. 1, 10,000

65. $\log x^2 = (\log x)^2$

$2 \log x = (\log x)^2$

$0 = (\log x)^2 - 2 \log x$

Let $u = \log x$.

$0 = u^2 - 2u$

$0 = u(u - 2)$

$u = 0$ or $u = 2$

$\log x = 0$ or $\log x = 2$

$x = 10^0$ or $x = 10^2$

$x = 1$ or $x = 100$

Both numbers check. The solutions are 1 and 100.

66. 27, $\dfrac{1}{3}$

67. $\log_3 (\log_4 x) = 0$

$\log_4 x = 3^0$ Writing an equivalent exponential equation

$\log_4 x = 1$

$x = 4^1$

$x = 4$

The answer checks. The solution is 4.

68. 10^{100}

69. $(\log_a x)^{-1} = \log_a x^{-1}$

$\dfrac{1}{\log_a x} = \log_a x^{-1}$ Writing $(\log_a x)^{-1}$ with a positive exponent

$\dfrac{1}{\log_a x} = -\log_a x$ Property 2

$1 = -(\log_a x)^2$ Multiplying by $\log_a x$

$-1 = (\log_a x)^2$

Since $(\log_a x)^2 \geqslant 0$ for all values of x, the equation has no real-number solution. The solution set is $\emptyset$.

70. 25, $\dfrac{1}{25}$

71. $\log_7 \sqrt{x^2 - 9} = 1$

    $\sqrt{x^2 - 9} = 7^1$

    $x^2 - 9 = 49$ Squaring both sides

    $x^2 = 58$

    $x = \pm\sqrt{58}$

Both numbers check. The solutions are $\pm\sqrt{58}$.

72. -1

73. $\log\,(\log x) = 3$ The base is 10.

    $\log x = 10^3 = 1000$

    $x = 10^{1000}$

74. $-3, -1$

75. $\log_3 |x| = 2$

    $|x| = 3^2 = 9$

    $x = 9$ or $x = -9$

76. $125, -125$

77. $\log x^{\log x} = 4$

    $\log x\,(\log x) = 4$ Property 2

      $(\log x)^2 = 4$

    $\log x = 2$ or $\log x = -2$

    $x = 10^2$ or $x = 10^{-2}$

    $x = 100$ or $x = \dfrac{1}{100}$

Both answers check. The solutions are 100 and $\dfrac{1}{100}$.

78. $\dfrac{1}{2}$, 5000

79. $\log_a a^{x^2 + 5x} = 24$

    $x^2 + 5x = 24$ Property 4

    $x^2 + 5x - 24 = 0$

    $(x + 8)(x - 3) = 0$

    $x = -8$ or $x = 3$

80. 100, 10

81. $x^{\log_{10} x} = \dfrac{x^{-4}}{1000}$

    $x^{\log_{10} x} \cdot x^4 = \dfrac{1}{1000}$ Multiplying by x^4

    $x^{\log_{10} x + 4} = \dfrac{1}{1000}$ Adding exponents

    $\log_{10} x^{\log_{10} x + 4} = \log_{10} \dfrac{1}{1000}$

    $(\log_{10} x + 4)\log_{10} x = -3$

    $(\log_{10} x)^2 + 4 \log_{10} x + 3 = 0$

Let $u = \log_{10} x$.

    $u^2 + 4u + 3 = 0$

    $(u + 3)(u + 1) = 0$

    $u = -3$ or $u = -1$

    $\log_{10} x = -3$ or $\log_{10} x = -1$

    $x = 10^{-3}$ or $x = 10^{-1}$

    $x = \dfrac{1}{1000}$ or $x = \dfrac{1}{10}$

Both numbers check. The solutions are $\dfrac{1}{1000}$ and $\dfrac{1}{10}$.

82. 1.2091, 0.4307

83. $(32^{x-2})(64^{x+1}) = 16^{2x-3}$

    $(2^5)^{x-2}(2^6)^{x+1} = (2^4)^{2x-3}$

    $(2^{5x-10})(2^{6x+6}) = 2^{8x-12}$ Multiplying exponents

    $2^{11x-4} = 2^{8x-12}$ Simplifying

    $11x - 4 = 8x - 12$

    $3x = -8$

    $x = -\dfrac{8}{3}$

84. 0.1957

85. $4^{3x} - 4^{3x-1} = 48$

    $4^{3x-1}(4 - 1) = 48$ Factoring

    $(3)4^{3x-1} = 48$

    $4^{3x-1} = 16$

    $4^{3x-1} = 4^2$

    $3x - 1 = 2$

    $3x = 3$

    $x = 1$

86. 100, $\dfrac{1}{10}$

87. $\dfrac{(e^{3x+1})^2}{e^4} = e^{10x}$

    $\dfrac{e^{6x+2}}{e^4} = e^{10x}$

    $e^{6x-2} = e^{10x}$

    $6x - 2 = 10x$

    $-2 = 4x$

    $-\dfrac{1}{2} = x$

88. $\dfrac{7}{4}$

89.
$$P = P_0\, e^{kt}$$
$$\ln P = \ln (P_0\, e^{kt})$$
$$\ln P = \ln P_0 + \ln e^{kt}$$
$$\ln P = \ln P_0 + kt \qquad \text{Property 4}$$
$$\ln P - \ln P_0 = kt$$
$$\frac{\ln P - \ln P_0}{k} = t$$

90. $\dfrac{\ln P_0 - \ln P}{k}$

91.
$$T = T_0 + (T_1 - T_0)\, e^{-kt}$$
$$T - T_0 = (T_1 - T_0)\, e^{-kt}$$
$$\frac{T - T_0}{T_1 - T_0} = e^{-kt}$$
$$\ln \left(\frac{T - T_0}{T_1 - T_0}\right) = \ln e^{-kt}$$
$$\ln \left(\frac{T - T_0}{T_1 - T_0}\right) = -kt \qquad \text{Property 4}$$
$$-\frac{1}{k} \ln \left(\frac{T - T_0}{T_1 - T_0}\right) = t$$

92. $Q = a^b y^{\frac{1}{3}}$, or $a^b \sqrt[3]{y}$

93. $\log_5 125 = 3$ and $\log_{125} 5 = \frac{1}{3}$, so
$x = (\log_{125} 5)^{\log_5 125}$ is equivalent to
$x = \left(\frac{1}{3}\right)^3 = \frac{1}{27}$. Then $\log_3 x = \log_3 \frac{1}{27} = -3$.

94. 38

95. $|\log_a x| = \log_a |x|$

Case I: Assume $0 < a < 1$.

    A. Assume $0 < x \leqslant 1$. Then $|\log_a x| = \log_a x$ and $|x| = x$. We have:
$$\log_a x = \log_a x$$
This is true for all values of x in the interval under consideration, $0 < x \leqslant 1$.

    B. Assume $x > 1$. Then $|\log_a x| = -\log_a x$ and $|x| = x$. We have:
$$-\log_a x = \log_a x$$
$$0 = 2 \log_a x$$
$$0 = \log_a x^2$$
$$a^0 = x^2$$
$$1 = x^2$$
$$\pm 1 = x$$
This yields no solution, since we have assumed $x > 1$.

Case II. Assume $a > 1$.

    A. Assume $0 < x < 1$. Then $|\log_a x| = -\log_a x$ and $|x| = x$. We have:
$$-\log_a x = \log_a x$$
$$0 = 2 \log_a x$$
$$\pm 1 = x \qquad \text{(See Case IB above)}$$
This yields no solution, since we have assumed $0 < x < 1$.

    B. Assume $x \geqslant 1$. Then $|\log_a x| = \log_a x$ and $|x| = x$. We have:
$$\log_a x = \log_a x$$
This is true for all values of x in the interval under consideration, $x \geqslant 1$.

The solution is as follows:
If $0 < a < 1$, $0 < x \leqslant 1$, and if $a > 1$, $x \geqslant 1$.

96. 5

97.
$$(0.5)^x < \frac{4}{5}$$
$$\log (0.5)^x < \log \frac{4}{5}$$
$$x \log 0.5 < \log 0.8 \quad \text{Property 2; writing } \frac{4}{5} \text{ as } 0.8$$
$$x > \frac{\log 0.8}{\log 0.5} \quad \text{Reversing the inequality symbol } (\log 0.5 < 0)$$
$$\text{or} \quad x > 0.3219$$

98. 2^{10}

99.
$$2 \log_3 (x - 2y) = \log_3 x + \log_3 y$$
$$\log_3 (x - 2y)^2 = \log_3 (xy) \quad \text{Properties 2 and 1}$$
$$(x - 2y)^2 = xy$$
$$x^2 - 4xy + 4y^2 = xy$$
$$x^2 - 5xy + 4y^2 = 0$$
$$(x - 4y)(x - y) = 0$$
$$x - 4y = 0 \quad \text{or} \quad x - y = 0$$
$$x = 4y \quad \text{or} \quad x = y$$
$$\frac{x}{y} = 4 \qquad\qquad \frac{x}{y} = 1$$

100. 88

101. $a = \log_8 225$, so $8^a = 225 = 15^2$.

$b = \log_2 15$, so $2^b = 15$.

Then
$$8^a = (2^b)^2$$
$$(2^3)^a = 2^{2b}$$
$$2^{3a} = 2^{2b}$$
$$3a = 2b$$
$$a = \frac{2}{3}b.$$

102. The general solution is $\left[x, \dfrac{\frac{3}{2}\log 3 - x \log 2}{\log 3}\right]$.

When $x = 0$, we have $\left(0, \dfrac{3}{2}\right)$.

103.-122. See 69.-88.

Exercise Set 5.8

1. a) One billion is 1000 million, so we set $N(t) = 1000$ and solve for t:

$$1000 = 7.5(6)^{0.5t}$$
$$\frac{1000}{7.5} = (6)^{0.5t}$$
$$\log \frac{1000}{7.5} = \log(6)^{0.5t}$$
$$\log 1000 - \log 7.5 = 0.5t \log 6$$
$$t = \frac{\log 1000 - \log 7.5}{0.5 \log 6}$$
$$t \approx 5.5$$

After about 5.5 years, one billion compact discs will be sold in a year.

b) When $t = 0$, $N(t) = 7.5(6)^{0.5(0)} = 7.5(6)^0 = 7.5(1) = 7.5$. Twice this initial number is 15, so we set $N(t) = 15$ and solve for t:

$$15 = 7.5(6)^{0.5t}$$
$$2 = (6)^{0.5t}$$
$$\log 2 = \log(6)^{0.5t}$$
$$\log 2 = 0.5t \log 6$$
$$t = \frac{\log 2}{0.5 \log 6} \approx 0.8$$

The doubling time is about 0.8 year.

2. a) 86.4 minutes

b) 300.5 minutes

c) 20 minutes

3. a) We have
$$A(t) = \$50,000(1.09)^t.$$

b) We set $A(t) = \$450,000$ and solve for t:

$$450,000 = 50,000(1.09)^t$$
$$\frac{450,000}{50,000} = (1.09)^t$$
$$9 = (1.09)^t$$
$$\log 9 = \log(1.09)^t \quad \text{Taking the common logarithm on both sides}$$
$$\log 9 = t \log 1.09 \quad \text{Property 2}$$
$$t = \frac{\log 9}{\log 1.09} \approx 25.5$$

It will take about 25.5 years for the \$50,000 to grow to \$450,000.

c) We set $A(t) = \$100,000$ and solve for t:

$$100,000 = 50,000(1.09)^t$$
$$2 = (1.09)^t$$
$$\log 2 = \log(1.09)^t$$
$$\qquad\qquad \text{Taking the common logarithm on both sides}$$
$$\log 2 = t \log 1.09 \quad \text{Property 2}$$
$$t = \frac{\log 2}{\log 1.09} \approx 8.04$$

The doubling time is about 8.04 years.

4. a) 1.0 year

b) 7.3 years

5. a) We set $V(t) = \$1200$ and solve for t.

$$1200 = 5200(0.8)^t$$
$$\frac{3}{13} = (0.8)^t$$
$$\log \frac{3}{13} = \log(0.8)^t$$
$$\log \frac{3}{13} = t \log 0.8$$
$$t = \frac{\log \left(\frac{3}{13}\right)}{\log 0.8} \approx 6.6$$

After about 6.6 years the salvage value will be \$1200.

b) We set $V(t) = \frac{1}{2}(\$5200)$, or \$2600, and solve for t.

$$2600 = 5200(0.8)^t$$
$$0.5 = (0.8)^t$$
$$\log 0.5 = \log(0.8)^t$$
$$\log 0.5 = t \log 0.8$$
$$t = \frac{\log 0.5}{\log 0.8} \approx 3.1$$

After about 3.1 years the salvage value will be half of its original value.

6. a) 59.7 years

b) 19.1 years

7. a) $S(0) = 68 - 20 \log(0 + 1) = 68 - 20 \log 1 = 68 - 20(0) = 68\%$

b) $S(4) = 68 - 20 \log(4 + 1) = 68 - 20 \log 5 \approx 54\%$

$S(24) = 68 - 20 \log(24 + 1) = 68 - 20 \log 25 \approx 40\%$

c) Using the values we computed in parts (a) and (b) and any others we wish to calculate, we sketch the graph:

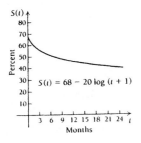

$S(t) = 68 - 20 \log (t + 1)$

Months

d) We set $S(t) = 50$ and solve for t:

$$50 = 68 - 20 \log (t + 1)$$
$$-18 = -20 \log (t + 1)$$
$$0.9 = \log (t + 1)$$
$$10^{0.9} = t + 1 \qquad \text{Using the definintion of logarithms}$$
$$7.9 \approx t + 1$$
$$6.9 \approx t$$

After about 6.9 months, the average score was 50.

8. a) 78%

b) 68%; 57%

c)

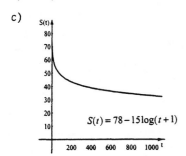

$S(t) = 78 - 15\log(t + 1)$

d) 1584 months

9. a) $N(1) = 1000 + 200 \log 1 = 1000 + 200(0) = 1000$ units

b) $N(5) = 1000 + 200 \log 5 \approx 1140$ units

c) Using the values computed in parts (a) and (b) and any others we wish to calculate, we can sketch the graph:

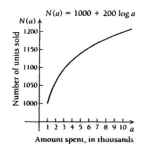

$N(a) = 1000 + 200 \log a$

Amount spent, in thousands

d) Set $N(a) = 1276$ and solve for a.

$$1276 = 1000 + 200 \log a$$
$$276 = 200 \log a$$
$$1.38 = \log a$$
$$10^{1.38} = a \qquad \text{Using the definition of logarithms}$$
$$23.98833 \approx a$$

About \$23,988.33 would have to be spent.

10. a) 2000 units

b) 2452 units

c)

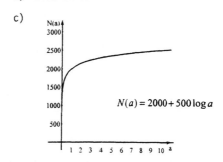

$N(a) = 2000 + 500\log a$

d) \$1,000,000 thousand, or \$1,000,000,000

11. We substitute 1.6×10^{-4} for H^+ in the formula for pH.

$$pH = -\log [H^+] = -\log [1.6 \times 10^{-4}] =$$
$$-[\log 1.6 + \log 10^{-4}] \approx -[0.2041 - 4] =$$
$$-[-3.7959] = 3.7959 \approx 3.8$$

The pH of pineapple juice is about 3.8.

12. 7.4

13. We substitute 6.3×10^{-5} for H^+ in the formula for pH.

$$pH = -\log [H^+] = -\log [6.3 \times 10^{-5}] =$$
$$-[\log 6.3 + \log 10^{-5}] \approx -[0.7993 - 5] =$$
$$-[-4.2007] = 4.2007 \approx 4.2$$

The pH of a tomato is about 4.2.

14. 7.8

15. We substitute 5.4 for pH in the formula and solve for H^+.

$$5.4 = -\log [H^+]$$
$$-5.4 = \log [H^+]$$
$$10^{-5.4} = H^+ \qquad \text{Using the definition of logarithm}$$
$$0.0000040 \approx H^+$$
$$4.0 \times 10^{-6} \approx H^+$$

The hydrogen ion concentration of rainwater is about 4.0×10^{-6} moles/liter.

16. 10^{-7} moles/liter

17. We substitute 4.8 for pH in the formula and solve for H^+.

$$4.8 = -\log [H^+]$$
$$-4.8 = \log [H^+]$$
$$10^{-4.8} = H^+ \quad \text{Using the definition of logarithm}$$
$$0.000016 \approx H^+$$
$$1.6 \times 10^{-5} \approx H^+$$

The hydrogen ion concentration of wine is about 1.6×10^{-5} moles/liter.

18. 6.3×10^{-4} moles/liter

19. We substitute into the formula.

$$R = \log \frac{10^{8.25} I_0}{I_0} = \log 10^{8.25} = 8.25$$

20. 5

21. We substitute into the formula.

$$R = \log \frac{10^{9.6} \cdot I_0}{I_0} = \log 10^{9.6} = 9.6$$

22. $10^{7.2} \cdot I_0$

23. We substitute into the formula.

$$L = 10 \log \frac{2510 \, I_0}{I_0} = 10 \log 2510 \approx 10(3.399674) \approx$$
34 decibels

24. 64 decibels

25. We substitute into the formula.

$$L = 10 \log \frac{10^6 \, I_0}{I_0} = 10 \log 10^6 = 10(6) =$$
60 decibels

26. 90 decibels

27. a) Substitute 100 for S(t) and solve for t.

$$100 = 200[1 - (0.86)^t]$$
$$0.5 = 1 - (0.86)^t$$
$$(0.86)^t = 0.5$$
$$\log(0.86)^t = \log 0.5$$
$$t \log 0.86 = \log 0.5$$
$$t = \frac{\log 0.5}{\log 0.86} \approx 4.6$$

The typist's speed will be 100 words per minute after she has studied typing for about 4.6 hours.

b) Substitute 150 for S(t) and solve for t.

$$150 = 200[1 - (0.86)^t]$$
$$0.75 = 1 - (0.86)^t$$
$$(0.86)^t = 0.25$$
$$\log(0.86)^t = \log 0.25$$
$$t \log 0.86 = \log 0.25$$
$$t = \frac{\log 0.25}{\log 0.86} \approx 9.2$$

About 9.2 hours of studying occurred in the course.

28. a) $A = \$100,000(1.02)^{4t}$

b) 29.1 years

c) 8.8 years

29. $$R = \log \frac{I}{I_0}$$
$$\frac{I}{I_0} = 10^R$$
$$I = 10^R I_0$$

30. $I = 10^{0.1L} I_0$

31. $$pH = -\log [H^+]$$
$$-pH = \log [H^+]$$
$$[H^+] = 10^{-pH}$$

32. a) $$L_1 = 10 \log \frac{I_1}{I_0}, \quad L_2 = 10 \log \frac{I_2}{I_0}$$
$$L_2 - L_1 = 10 \log \frac{I_2}{I_0} - 10 \log \frac{I_1}{I_0}$$
$$= 10 \log \frac{\frac{I_2}{I_0}}{\frac{I_1}{I_0}}$$
$$= 10 \log \frac{I_2}{I_1}$$

b) $L_1 + L_1 = 10 \log \left(\frac{I_1 I_2}{I_0^2} \right)$

33. a) We substitute -1.5 for M_1 and -0.3 for M_2.

$$-0.3 - (-1.5) = 2.5 \log \frac{I_1}{I_2}$$
$$1.2 = 2.5 \log \frac{I_1}{I_2}$$
$$0.48 = \log \frac{I_1}{I_2}$$
$$10^{0.48} = \frac{I_1}{I_2}$$
$$3 \approx \frac{I_1}{I_2}$$

b) We substitute -1.5 for M_1 and 12.6 for M_2.

$$12.6 - (-1.5) = 2.5 \log \frac{I_1}{I_2}$$

$$14.1 = 2.5 \log \frac{I_1}{I_2}$$

$$5.64 = \log \frac{I_1}{I_2}$$

$$10^{5.64} = \frac{I_1}{I_2}$$

$$436{,}515.8 \approx \frac{I_1}{I_2}$$

c) We substitute 5 for $M_2 - M_1$.

$$5 = 2.5 \log \frac{I_1}{I_2}$$

$$2 = \log \frac{I_1}{I_2}$$

$$10^2 = \frac{I_1}{I_2}$$

$$100 = \frac{I_1}{I_2}$$

$$100 I_2 = I_1$$

One star is 100 times brighter than the other.

34. a) 3.4×10^{22} watts

b) 8.5×10^{27} watts

c) $L = 3.9 \times 10^{\frac{69.75 - M}{2.5}}$

Exercise Set 5.9

1. We substitute 7900 for P, since P is in thousands.

   $R(P) = 0.37 \ln P + 0.05$

   $R(7900) = 0.37 \ln 7900 + 0.05$

   ≈ 3.4 ft/sec   Finding $\ln 7900$ on a calculator

2. 2.5 ft/sec

3. We substitute 50.4 for P in the function, since P is in thousands.

   $R(50.4) = 0.37 \ln 50.4 + 0.05$

   ≈ 1.5 ft/sec

4. 1.8 ft/sec

5. a) We can use the function $C(t) = C_0 e^{kt}$ as a model. At $t = 0$ (1962), the cost was 5¢, or $0.05. We substitute 0.05 for C_0 and 9.7%, or 0.097, for k:

   $$C(t) = 0.05 e^{0.097t}$$

   b) In 1990, t = 1990 - 1962, or 28. We substitute 28 for t:

   $$C(28) = 0.05 e^{0.097(28)}$$
   $$= 0.05 e^{2.716}$$
   $$\approx 0.76$$

In 1993 a Hershey bar will cost about $0.76.

In 2000, t = 2000 - 1962, or 38. We substitute 38 for t:

$$C(38) = 0.05 e^{0.097(38)}$$
$$= 0.05 e^{3.686}$$
$$\approx 1.99$$

In 2000 a Hershey bar will cost about $1.99.

c) We set $C(t) = 5$ and solve for t:

$$5 = 0.05 e^{0.097t}$$
$$100 = e^{0.097t}$$
$$\ln 100 = \ln e^{0.097t}$$
$$\ln 100 = 0.097t$$
$$\frac{\ln 100}{0.097} = t$$
$$47.5 \approx t$$

A Hershey bar will cost $5 about 47.5 years after 1962.

d) Using the ordered pairs found in parts (b) and (c) and any others we wish to compute, we draw the graph:

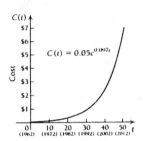

e) To find the doubling time, we set $C(t) = 2(\$0.05)$, or $0.10 and solve for t:

$$0.1 = 0.05 e^{0.097t}$$
$$2 = e^{0.097t}$$
$$\ln 2 = \ln e^{0.097t}$$
$$\ln 2 = 0.097t$$
$$\frac{\ln 2}{0.097} = t$$

(Note: We could also have used the expression relating the growth rate k and doubling time T: $T = \frac{\ln 2}{k}$.)

$$7.1 \approx t$$

The doubling time is about 7.1 years.

6. a) $P(t) = 5e^{0.028t}$, where t is the number of years after 1987 and P is in billions.

b) 6.4 billion; 7.2 billion

c) About 6.5 years after 1987

d)

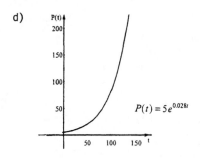

$P(t) = 5e^{0.028t}$

<u>7</u>. a) We can use the function $P(t) = P_0\,e^{kt}$ as a model. We substitute 6%, or 0.06, for k.

$P(t) = P_0\,e^{0.06t}$

b) In 1995, t = 1995 – 1967, or 28. We substitute 28 for t and $100 for P_0.

$C(28) = \$100\,e^{0.06(28)}$

$= \$100\,e^{1.68}$

$\approx \$536.56$

Goods and services that cost $100 in 1967 will cost about $536.56 in 1995.

c) In 2000, t = 2000 – 1967, or 33. We substitute 33 for t and $100 for P_0.

$C(33) = \$100\,e^{0.06(33)}$

$= \$100\,e^{1.98}$

$\approx \$724.27$

Goods and services that cost $100 in 1967 will cost about $724.27 in 2000.

d) Using the ordered pairs found in parts (b) and (c) and any others we wish to compute, we draw the graph:

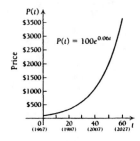

$P(t) = 100e^{0.06t}$

<u>8</u>. a) $P(t) = 100\,e^{0.117t}$

b) 227

c)

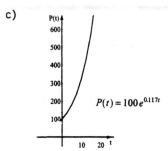

$P(t) = 100\,e^{0.117t}$

d) 5.9 days

<u>9</u>. a) We plot the ordered pairs given in the table and sketch the following graph. We draw a curve as close to the data points as possible. It is very close to the graph of an exponential function.

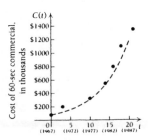

b) The exponential growth function is $C(t) = C_0\,e^{kt}$. Substituting 80 for C_0, we have

$C(t) = 80\,e^{kt}$,

where t is the number of years after 1967.

To find k using the data point C(21) = $1350 thousand, we substitute $1350 for C(t) and 21 for t and solve for k:

$1350 = 80\,e^{k(21)}$

$16.875 = e^{21k}$

$\ln 16.875 = \ln e^{21k}$

$\ln 16.875 = 21k$

$\dfrac{\ln 16.875}{21} = k$

$0.135 \approx k$

The exponential growth function for the cost of a Super Bowl commercial is

$C(t) = 80\,e^{0.135t}$.

c) The year 1995 is 28 years from 1967. We let t = 28 and find C(28):

$C(28) = 80\,e^{0.135(28)} \approx 3505$

The cost of a 60-second commercial will be about $3505 thousand, or $3,505,000 in 1995.

d) We set C(t) = 3000 ($3000 thousand is $3,000,000) and solve for t:

$3000 = 80\,e^{0.135t}$

$37.5 = e^{0.135t}$

$\ln 37.5 = \ln e^{0.135t}$

$\ln 37.5 = 0.135t$

$\dfrac{\ln 37.5}{0.135} = t$

$26.8 \approx t$

The cost of a commercial will be $3,000,000 about 26.8 years after 1967.

e) We can set $C(t) = 160$ and solve for t, or we can use the expression relating growth rate k and doubling time T, $T = \frac{\ln 2}{k}$. We will use the latter.

$$T = \frac{\ln 2}{0.135} \approx 5.1 \text{ years}$$

10. a) $k \approx 0.03$; $P(t) = 52\, e^{0.03t}$

b) 107¢, or $1.07

c) 23 years

d) In about 58 years from 1970

11. a) Substitute 0.09 for k:
$$P(t) = P_0\, e^{0.09t}$$

b) To find the balance after one year, set $P_0 = 5000$ and $t = 1$. We find $P(1)$:

$$P(1) = 5000\, e^{0.09(1)} = 5000\, e^{0.09} \approx \$5470.87$$

To find the balance after 2 years, set $P_0 = 5000$ and $t = 2$. We find $P(2)$:

$$P(2) = 5000\, e^{0.09(2)} = 5000\, e^{0.18} \approx \$5986.09$$

c) We will use the expression relating growth rate k and doubling time T. (We could also set $P(t) = 10,000$ and solve for t.)

$$T = \frac{\ln 2}{0.09} \approx 7.7 \text{ years}$$

12. a) $P(t) = P_0\, e^{0.1t}$

b) $38,680.98; $42,749.10

c) 6.9 years

13. We will use the expression relating growth rate k and doubling time T:
$$T = \frac{\ln 2}{k}.$$

Substitute 3.5%, or 0.035, for k:
$$T = \frac{\ln 2}{0.035} \approx 19.8 \text{ years}$$

14. 69.3 years

15. We will use the expression relating growth rate k and doubling time T:

$$k = \frac{\ln 2}{T}$$

$$k = \frac{\ln 2}{7} \approx 0.099$$

The annual interest rate is about 9.9%.

16. 0.128, or 12.8%

17. a) The exponential growth function is $V(t) = V_0\, e^{kt}$, where t is the number of years after 1947. Since $V_0 = 84,000$ we have

$$V(t) = 84,000\, e^{kt}.$$

In 1987 ($t = 40$), we know that $V(t) = 53,900,000$. We substitute and solve for k.

$$53,900,000 = 84,000\, e^{k(40)}$$

$$\frac{1925}{3} = e^{40k}$$

$$\ln \frac{1925}{3} = \ln e^{40k}$$

$$\ln \frac{1925}{3} = 40k$$

$$\frac{\ln \frac{1925}{3}}{40} = k$$

$$0.16 \approx k$$

The exponential growth rate is about 0.16, or 16%. The exponential growth function is $V(t) = 84,000\, e^{0.16t}$, where t is the number of years after 1947.

b) In 2007, $t = 60$. We find $V(60)$.
$$V(60) = 84,000\, e^{0.16(60)} = 84,000\, e^{9.6} \approx \$1,240,241,652$$

c) We will use the expression relating growth rate k and doubling time T. (We could also set $V(t) = 168,000$ and solve for t.)

$$T = \frac{\ln 2}{k} = \frac{\ln 2}{0.16} \approx 4.3 \text{ years}$$

d) We set $V(t) = 1,000,000,000$ and solve for t.
$$1,000,000,000 = 84,000\, e^{0.16t}$$

$$\frac{250,000}{21} = e^{0.16t}$$

$$\ln \frac{250,000}{21} = \ln e^{0.16t}$$

$$\ln \frac{250,000}{21} = 0.16t$$

$$\frac{\ln \frac{250,000}{21}}{0.16} = t$$

$$58.7 \approx t$$

The value of the painting will be $1 billion about 58.7 years after 1947.

18. a) $k \approx 0.31$; $V(t) = 7.75\, e^{0.31t}$, where t is the number of years after 1983.

b) $319.80; $1506.72

c) 2.2 years

d) 17.9 years

19. Using the expression relating growth rate k and doubling time T, we substitute 4%, or 0.04, for k:

$$T = \frac{\ln 2}{0.04} \approx 17.3 \text{ years from 1990}$$

20. About 6.9 years from 1990

21. a) The exponential growth function with $P_0 = 2,812,000$ is $P(t) = 2,812,000\, e^{kt}$, where t is the number of years after 1970. To find k we substitue 14 for t and 3,097,000 for P and solve for k.

$$3,097,000 = 2,812,000\, e^{k(14)}$$

$$\frac{3,097,000}{2,812,000} = e^{14k}$$

$$\ln \frac{3,097,000}{2,812,000} = \ln e^{14k}$$

$$\ln \frac{3,097,000}{2,812,000} = 14k$$

$$\frac{\ln \frac{3,097,000}{2,812,000}}{14} = k$$

$$0.007 \approx k$$

The value of k is about 0.007, or 0.7%. The exponential growth function is

$$P(t) = 2,812,000\, e^{0.007t},$$

where t is the number of years after 1970.

b) In 1996, t = 1996 - 1970, or 26. We find P(26).

$$P(26) = 2,812,000\, e^{0.007(26)} =$$
$$2,812,000\, e^{0.182} \approx 3,373,315$$

In 1996 the population of Los Angeles will be about 3,373,315.

22. a) $k \approx 0.018$, or 1.8%; $P(t) = 786,000\, e^{0.018t}$, where t is the number of years after 1980.

b) 1,126,597

23. We will use the function derived in Example 7:
$$P(t) = P_0\, e^{-0.00012t}$$

If the mummy has lost 46% of its carbon-14, then 54% (P_0) is the amount present. To find the age of the mummy, we substitute 54% (P_0), or $0.54\, P_0$, for P(t) and solve for t.

$$0.54\, P_0 = P_0\, e^{-0.00012t}$$
$$0.54 = e^{-0.00012t}$$
$$\ln 0.54 = \ln e^{-0.00012t}$$
$$\ln 0.54 \approx -0.00012t$$
$$t \approx \frac{\ln 0.54}{-0.00012} \approx 5135 \text{ years}$$

24. 3590 years

25. We use the function $P(t) = P_0\, e^{-kt}$, k > 0. When t = 3, $P(t) = \frac{1}{2} P_0$. We substitute and solve for k.

$$\frac{1}{2} P_0 = P_0\, e^{-k(3)}, \text{ or } \frac{1}{2} = e^{-3k}$$
$$\ln \frac{1}{2} = \ln e^{-3k} = -3k$$
$$k = \frac{\ln 0.5}{-3} \approx 0.231$$

The decay rate is 0.231, or 23.1% per minute.

26. 0.032, or 3.2% per year

27. We use the function $P(t) = P_0\, e^{-kt}$, k > 0. For iodine-131, k = 9.6%, or 0.096. To find the half-life we substitute 0.096 for k and $\frac{1}{2} P_0$ for P(t), and solve for t.

$$\frac{1}{2} P_0 = P_0\, e^{-0.096t}, \text{ or } \frac{1}{2} = e^{-0.096t}$$
$$\ln \frac{1}{2} = \ln e^{-0.096t} = -0.096t$$
$$t = \frac{\ln 0.5}{-0.096} \approx 7.2 \text{ days}$$

28. 11 years

29. a) The value of k in the function is 0.007, so the animal loses 0.7% of its weight each day.

b) Find W(30).
$$W(30) = W_0\, e^{-0.007(30)} = W_0\, e^{-0.21} \approx 0.811\, W_0$$
After 30 days, about 0.811, or 81.1%, of its initial weight W_0 remains.

30. a) 6.7 watts

b) 115.5 days

c) 298.6 days

d) 60 watts

31. a) Substitute 14.7 for P_0 and 2000 for a.
$$P = 14.7\, e^{-0.00005(2000)} = 14.7\, e^{-0.1} \approx 13.3 \text{ lb/in}^2$$

b) Substitute 14.7 for P_0 and 14,162 for a.
$$P = 14.7\, e^{-0.00005(14,162)} = 14.7\, e^{-0.7081} \approx 7.24 \text{ lb/in}^2$$

c) Substitute 1.47 for P and 14.7 for P_0, and solve for a.
$$1.47 = 14.7\, e^{-0.00005a}$$
$$0.1 = e^{-0.00005a}$$
$$\ln 0.1 = \ln e^{-0.00005a}$$
$$\ln 0.1 = -0.00005a$$
$$a = \frac{\ln 0.1}{-0.00005} \approx 46,052 \text{ ft}$$

d) Substitute 14.7 for P_0 and 0.39 for P and solve for a.
$$0.39 = 14.7\, e^{-0.00005a}$$
$$\frac{0.39}{14.7} = e^{-0.00005a}$$
$$\ln \frac{0.39}{14.7} = \ln e^{-0.00005a}$$
$$\ln \frac{0.39}{14.7} = -0.00005a$$
$$\frac{\ln \frac{0.39}{14.7}}{-0.00005} = a$$
$$a \approx 72,589 \text{ ft}$$

32. a) $46,000

    b) $6225.42

33. a) At 1 m: $I = I_0 e^{-1.4(1)} \approx 0.247 I_0$
          24.7% of I_0 remains.

       At 3 m: $I = I_0 e^{-1.4(3)} \approx 0.015 I_0$
          1.5% of I_0 remains.

       At 5 m: $I = I_0 e^{-1.4(5)} \approx 0.0009 I_0$
          0.09% of I_0 remains.

       At 50 m: $I = I_0 e^{-1.4(50)} \approx (3.98 \times 10^{-31}) I_0$
          $3.98 \times 10^{-31} = (3.98 \times 10^{-29}) \times 10^{-2}$, so
          (3.98×10^{-29})% remains.

    b) $I = I_0 e^{-1.4(10)} \approx 0.0000008 I_0$
       0.00008% remains

34. $11,846.39

35. $v = c \ln R$

    $\dfrac{v}{c} = \ln R$

    $R = e^{\frac{v}{c}}$

36. $t = -\dfrac{L}{R}\left[\ln\left(1 - \dfrac{iR}{V}\right)\right]$

37. Measure the atmospheric pressure P at the top of building. Substitute that value in the equation $P = 14.7 e^{-0.00005a}$, and solve for the height, or altitude, a, of the top of the building. Also measure the atmospheric pressure at the base of the building and solve for the altitude of the base. Then subtract to find the height of the building.

38. 51.9° F

39. To find k we substitute 94.6 for T_1, 70 for T_0, 60 for t (1 hr = 60 min), and 93.4 for $T(t)$.

    $93.4 = 70 + |94.6 - 70| e^{-k(60)}$

    $23.4 = 24.6 e^{-60k}$

    $\dfrac{23.4}{24.6} = e^{-60k}$

    $\ln \dfrac{23.4}{24.6} = \ln e^{-60k}$

    $\ln \dfrac{23.4}{24.6} = -60k$

    $k = \dfrac{\ln \dfrac{23.4}{24.6}}{-60} \approx 0.0008$

    The function is

    $T(t) = 70 + 24.6 e^{-0.0008t}.$

    We substitute 98.6 for $T(t)$ and solve for t.

$98.6 = 70 + 24.6 e^{-0.0008t}$

$28.6 = 24.6 e^{-0.0008t}$

$\dfrac{28.6}{24.6} = e^{-0.0008t}$

$\ln \dfrac{28.6}{24.6} = \ln e^{-0.0008t}$

$\ln \dfrac{28.6}{24.6} = -0.0008t$

$t = \dfrac{\ln \dfrac{28.6}{24.6}}{-0.0008} \approx -188$

The murder was committed at approximately 188 minutes, or about 3 hours, before 12:00 PM, or at about 9:00 AM.

Exercise Set 6.1

1. We replace x by $\frac{1}{2}$ and y by 1.

$3x + y = \frac{5}{2}$	
$3 \cdot \frac{1}{2} + 1$	$\frac{5}{2}$
$\frac{3}{2} + \frac{2}{2}$	
$\frac{5}{2}$	

$2x - y = \frac{1}{4}$	
$2 \cdot \frac{1}{2} - 1$	$\frac{1}{4}$
$1 - 1$	
0	

The ordered pair $\left(\frac{1}{2}, 1\right)$ is not a solution of $2x - y = \frac{1}{4}$. Therefore it is not a solution of the system of equations.

2. Yes

3. Graph both lines on the same set of axes. The x and y-intercepts of the line x + y = 2 are (2, 0) and (0, 2). Plot these points and draw the line they determine. Next graph the line 3x + y = 0. Three of the points on the line are (0, 0), (-1, 3), and (2, -6). Plot these points and draw the line they determine.

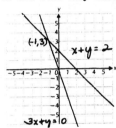

The solution (point of intersection) seems to be the point (-1, 3).

Check:

x + y = 2	
-1 + 3	2
2	

3x + y = 0	
3(-1) + 3	0
-3 + 3	
0	

The solution is (-1, 3).

4. (3, -2)

5. Graph both lines on the same set of axes. The intercepts of y + 1 = 2x are $\left(\frac{1}{2}, 0\right)$ and (0, -1). The intercepts of y - 1 = 2x are $\left(-\frac{1}{2}, 0\right)$ and (0, 1).

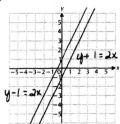

The lines have no point in common. The system has no solution.

6. Infinitely many solutions

7. x - 5y = 4,   (1)
   y = 7 - 2x   (2)

   Substitute 7 - 2x for y in Equation (1) and solve for x.

   $x - 5(7 - 2x) = 4$
   $x - 35 + 10x = 4$
   $11x = 39$
   $x = \frac{39}{11}$

   Now substitute $\frac{39}{11}$ for x in either Equation (1) or or (2) and solve for y. We use Equation (2).

   $y = 7 - 2 \cdot \frac{39}{11}$
   $y = 7 - \frac{78}{11}$
   $y = -\frac{1}{11}$

   Check:

x - 5y = 4	
$\frac{39}{11} - 5\left(-\frac{1}{11}\right)$	4
$\frac{39}{11} + \frac{5}{11}$	
$\frac{44}{11}$	
4	

y = 7 - 2x	
$-\frac{1}{11}$	$7 - 2 \cdot \frac{39}{11}$
$-\frac{1}{11}$	$7 - \frac{78}{11}$
	$-\frac{1}{11}$

   The solution is $\left(\frac{39}{11}, -\frac{1}{11}\right)$.

8. $\left(\frac{11}{8}, -\frac{7}{8}\right)$

9. x - 3y = 2,   (1)
   6x + 5y = -34   (2)

   We multiply equation (1) by -6 and add the result to equation (2).

   x - 3y = 2,
   23y = -46 ← $\begin{array}{r} -6x + 18y = -12 \\ 6x + 5y = -34 \\ \hline 23y = -46 \end{array}$

   Now we solve the second equation for y. Then we substitute the result in the first equation to find x.

   $23y = -46$ $x - 3(-2) = 2$
   $y = -2$ $x + 6 = 2$
                        $x = -4$

   The solution is (-4, -2).

10. (3, -1)

11. $0.3x + 0.2y = -0.9,$
    $0.2x - 0.3y = -0.6$

    Multiply each equation by 10 to eliminate decimals.

    $3x + 2y = -9,$ (1)
    $2x - 3y = -6$ (2)

    Multiply equation (2) by 3 to make the x-coefficient a multiple of 3.

    $3x + 2y = -9,$
    $6x - 9y = -18$

    Then multiply equation (1) by -2 and add the result to equation (2).

    $$\begin{array}{ll} 3x + 2y = -9 & \\ -13y = 0 \longleftarrow & \begin{array}{r} -6x - 4y = 18 \\ \underline{6x - 9y = -18} \\ -13y = 0 \end{array} \end{array}$$

    Now solve the second equation for y and substitute in the first equation to find x.

    $$\begin{array}{ll} -13y = 0 & 3x + 2\cdot0 = -9 \\ y = 0 & 3x = -9 \\ & x = -3 \end{array}$$

    The solution is (-3, 0).

12. $\left(\dfrac{1}{2}, -\dfrac{2}{3}\right)$

13. $\dfrac{1}{5}x + \dfrac{1}{2}y = 6,$

    $\dfrac{3}{5}x - \dfrac{1}{2}y = 2$

    Multiply each equation by 10 to clear of fractions.

    $2x + 5y = 60,$ (1)
    $6x - 5y = 20$ (2)

    Multiply equation (1) by -3 and add the result to equation (2).

    $$\begin{array}{ll} 2x + 5y = 60, & \\ -20y = -160 \longleftarrow & \begin{array}{r} -6x - 15y = -180 \\ \underline{6x - 5y = 20} \\ -20y = -160 \end{array} \end{array}$$

    Solve the second equation for y and substitute in the first equation to find x.

    $$\begin{array}{ll} -20y = -160 & 2x + 5\cdot8 = 60 \\ y = 8 & 2x + 40 = 60 \\ & 2x = 20 \\ & x = 10 \end{array}$$

    The solution is (10, 8).

14. (-12, -15)

15. $2a = 5 - 3b,$
    $4a = 11 - 7b$

    First write each equation in the form Ax + By = C.

    $2a + 3b = 5,$ (1)
    $4a + 7b = 11$ (2)

    Multiply equation (1) by -2 and add the result to equation (2).

    $$\begin{array}{ll} 2a + 3b = 5, & \\ b = 1 \longleftarrow & \begin{array}{r} -4a - 6b = -10 \\ \underline{4a + 7b = 11} \\ b = 1 \end{array} \end{array}$$

    Substitute in the first equation to find a.

    $$\begin{array}{l} 2a + 3\cdot1 = 5 \\ 2a + 3 = 5 \\ 2a = 2 \\ a = 1 \end{array}$$

    The solution is (1, 1).

16. (3, 1)

17. Familiarize and Translate:

    Let x and y represent the numbers. A system of equations results directly from the statements in the problem.

    $x + y = -10,$ (The sum is -10.)
    $x - y = 1$ (The difference is 1.)

    Carry out: We solve the system.

    Multiply the first equation by -1 and add the result to the second equation.

    $$\begin{array}{l} x + y = -10, \\ -2y = 11 \end{array}$$

    Solve the second equation for y and then substitute into the first equation to find x.

    $$\begin{array}{ll} -2y = 11 & x + \left(-\dfrac{11}{2}\right) = -10 \\ y = -\dfrac{11}{2} & x = -\dfrac{20}{2} + \dfrac{11}{2} = -\dfrac{9}{2} \end{array}$$

    Check:

    The sum of $-\dfrac{9}{2}$ and $-\dfrac{11}{2}$ is $-\dfrac{20}{2}$, or -10. The difference is $-\dfrac{9}{2} - \left(-\dfrac{11}{2}\right)$, or $\dfrac{2}{2}$ which is 1.

    State:

    The numbers are $-\dfrac{9}{2}$ and $-\dfrac{11}{2}$.

18. $\dfrac{9}{2}, -\dfrac{11}{2}$

19. Familiarize:

It helps to make a drawing. Then organize the information in a table. Let b = the speed of the boat and s = the speed of the stream. The speed upstream is b - s. The speed downstream is b + s.

$$d_1 = 46 \quad t_1 = 2 \quad r_1 = b + s$$

Downstream

$$d_2 = 51 \quad t_2 = 3 \quad r_2 = b - s$$

Upstream

	Distance	Speed	Time
Downstream	46	b + s	2
Upstream	51	b - s	3

Translate:

Using d = rt in each row of the table, we get a system of equations.

$$46 = (b + s)2 = 2b + 2s, \quad \text{or} \quad b + s = 23,$$
$$51 = (b - s)3 = 3b - 3s \qquad b - s = 17$$

Carry out:

Multiply the first equation by -1 and add the result to the second equation.

$$b + s = 23,$$
$$-2s = -6$$

Solve the second equation for s and then substitute in the first equation to find b.

$$-2s = -6 \qquad b + 3 = 23$$
$$s = 3 \qquad b = 20$$

Check:

The speed downstream is 20 + 3, or 23 km/h. The distance downstream is 23·2, or 46 km. The speed upstream is 20 - 3, or 17 km/h. The distance upstream is 17·3, or 51 km. The values check.

State:

The speed of the boat is 20 km/h. The speed of the stream is 3 km/h.

20. Plane: 875 km/h, wind: 125 km/h

21. Familiarize:

We organize the information in a table. We let x = the number of liters of antifreeze A and y = the number of liters of antifreeze B.

	A	B	Mixture
Amount of antifreeze	x	y	20 liters
Percent of alcohol	18%	10%	15%
Amount of alcohol in antifreeze	18%x	10%y	0.15 × 20, or 3 liters

Translate:

If we add x and y in the first row, we get 20.

$$x + y = 20$$

If we add the amounts of alcohol in the third row, we get 3.

$$18\%x + 10\%y = 3$$

After changing percents to decimals and clearing, we have the following system:

$$x + y = 20,$$
$$18x + 10y = 300$$

Carry out:

Multiply -18 times the first equation and add the result to the second equation.

$$x + y = 20,$$
$$-8y = -60$$

Solve the second equation for y and substitute in the first equation to find x.

$$-8y = -60 \qquad x + 7.5 = 20$$
$$y = 7.5 \qquad x = 12.5$$

Check:

We add the amounts of antifreeze: 12.5 L + 7.5 L = 20 L. Thus the amount of antifreeze checks. Next, we check the amount of alcohol: 18%(12.5) + 10%(7.5) = 2.25 + 0.75, or 3 L. The amount of alcohol also checks.

State:

The solution of the problem is 12.5 L of A and 7.5 L of B.

22. A: 15 L, B: 35 L

23. Familiarize:

We first make a drawing and organize the information in a table. Each car travels the same amount of time. Let t = the time and d = the distance traveled by the slower car. Then 528 - d = the distance traveled by the faster car.

t hours   80 km/h   96 km/h   t hours

d kilometers   528 - d kilometers

├───────── 528 km ─────────┤

	Distance	Speed	Time
Slow car	d	80	t
Fast car	528 - d	96	t

Translate:

Using d = rt in each row of the table, we get

$$d = 80t, \qquad d - 80t = 0,$$
$$\text{or}$$
$$528 - d = 96t \qquad d + 96t = 528.$$

Carry out:

Multiply the first equation by -1 and add the result to the second equation.

$$d - 80t = 0,$$
$$176t = 528$$

Solve the second equation for t. (We do not need to find d.

$$176t = 528$$
$$t = 3$$

Check:

In 3 hours, one car travels 80·3, or 240 km, and the other car travels 96·3, or 288 km. The sum of the distances is 240 + 288, or 528 km. The value checks.

State:

In 3 hours the cars will be 528 km apart.

24. 375 km

25. Familiarize:

We first make a drawing and then organize the information in a table. The time is the same for each plane. Let t = the time and d = the distance traveled by the slower plane. Then 780 - d = the distance traveled by the faster plane.

t hours   190 km/h   200 km/h   t hours

d kilometers   780 - d kilometers

————— 780 km —————

	Distance	Speed	Time
Slow plane	d	190	t
Fast plane	780 - d	200	t

Translate:

Using d = rt in each row of the table, we get

$$d = 190t,$$ $$d - 190t = 0,$$
         or
$$780 - d = 200t.$$ $$d + 200t = 780.$$

Carry out:

Multiply the first equation by -1 and add the result to the second equation.

$$d - 190t = 0,$$
$$390t = 780$$

Solve the second equation for t. (We do not need to find d.)

$$390t = 780$$
$$t = 2$$

Check:

In 2 hours, one plane travels 2·190, or 380 km, and the other plane travels 2·200, or 400 km. The total distance is 380 + 400, or 780 km. The value checks.

State:

The planes will meet in 2 hours.

26. $1\frac{3}{4}$ hr

27. Familiarize and Translate:

Let x = the number of white scarves and y = the number of printed scarves. Thus, $4.95x worth of white ones and $7.95y worth of printed ones were sold. The total number of scarves was 40.

$$x + y = 40$$

The total sales were $282.

$$4.95x + 7.95y = 282, \text{ or } 495x + 795y = 28,200$$

We solve the following system of equations:

$$x + y = 40,$$
$$495x + 795y = 28,200$$

Carry out:

Multiply -495 times the first equation and add the result to the second equation.

$$x + y = 40,$$
$$300y = 8400$$

Solve the second equation for y and then substitute in the first equation to find x.

$$300y = 8400$$ $$x + 28 = 40$$
$$y = 28$$ $$x = 12$$

Check:

The total number of scarves was 12 + 28, or 40. Total sales were 4.95(12) + 7.95(28), or 59.40 + 222.60, or $282. The values check.

State:

12 white scarves and 28 printed ones were sold.

28. 8 white, 22 yellow

29. Familiarize and Translate:

Let x = Paula's age now and y = Bob's age now. It helps to organize the information in a table.

	Age now	Age four years from now
Paula	x	x + 4
Bob	y	y + 4

Paula is 12 years older than Bob.
$$x = y + 12, \text{ or } x - y = 12$$

Four years from now, Bob will be $\frac{2}{3}$ as old as Paula.

$$y + 4 = \frac{2}{3}(x + 4)$$

or

$$3(y + 4) = 2(x + 4)$$
$$3y + 12 = 2x + 8$$
$$4 = 2x - 3y$$

We now have a system of equations.

$$x - y = 12,$$
$$2x - 3y = 4$$

Carry out:

Multiply the first equation by -2 and add the result to the second equation.

$$x - y = 12,$$
$$-y = -20$$

Complete the solution.

$$-y = -20 \qquad x - 20 = 12$$
$$y = 20 \qquad x = 32$$

Check:

If Paula is 32 and Bob 20, Paula is 12 years older than Bob. In four years, Paula will be 36 and Bob 24. Thus Bob will be $\frac{2}{3}$ as old as Paula $\left(24 = \frac{2}{3} \cdot 36\right)$. The ages check.

State:

Paula is 32, and Bob is 20.

30. Carlos: 28, Maria: 20

31. Familiarize:

If helps to make a drawing. We let ℓ = the length and w = the width.

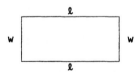

The formula for the perimeter is $P = 2\ell + 2w$.

Translate:

The perimeter is 190 m.

$$2\ell + 2w = 190, \quad \text{or} \quad \ell + w = 95$$

The width is one-fourth the length.

$$w = \frac{1}{4}\ell$$

The resulting system is

$$\ell + w = 95,$$
$$w = \frac{1}{4}\ell.$$

Carry out:

Substitute $\frac{1}{4}\ell$ for w in the first equation and solve for ℓ.

$$\ell + w = 95$$
$$\ell + \frac{1}{4}\ell = 95$$
$$\frac{5}{4}\ell = 95$$
$$\frac{4}{5} \cdot \frac{5}{4}\ell = \frac{4}{5} \cdot 95$$
$$\ell = 76$$

Then substitute 76 for ℓ in one of the equations of the system and solve for w.

$$\ell + w = 95$$
$$76 + w = 95$$
$$w = 19$$

Check:

If ℓ = 76 m and w = 19 m, then the width is one-fourth the length. The perimeter is $2 \cdot 76 + 2 \cdot 19$, or 190 m. The values check.

State:

The length is 76 m; the width is 19 m.

32. 160 m, 154 m

33. Familiarize:

We first make a drawing.

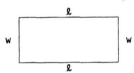

We let ℓ = the length and w = the width. The formula for the perimeter is $P = 2\ell + 2w$.

Translate:

The perimeter is 384 m.

$$2\ell + 2w = 384, \quad \text{or} \quad \ell + w = 192$$

The length is 82 m greater than the width.

$$\ell = w + 82$$

The resulting system is

$$\ell + w = 192,$$
$$\ell = w + 82.$$

Carry out:

Substitute w + 82 for ℓ in the first equation and solve for w.

$$\ell + w = 192$$
$$w + 82 + w = 192$$
$$2w + 82 = 192$$
$$2w = 110$$
$$w = 55$$

Then substitute 55 for w in one of the original equations and solve for ℓ.

ℓ = w + 82

ℓ = 55 + 82

ℓ = 137

Check:

The length is 82 greater than the width (137 = 55 + 82). The perimeter is 2·137 + 2·55, or 384 m. The values check.

State:

The length is 137 m; the width is 55 m.

34. 372 cm²

35. Familiarize:

Let x = the amount invested at 9%, and let y = the amount invested at 10%. The interest earned by these investments is 9%x, or 0.09x, and 10%y, or 0.1y, respectively.

Translate:

The total investment is $15,000.

x + y = 15,000

The total interest is $1432.

0.09x + 0.1y = 1432, or 9x + 10y = 143,200

The resulting system is

x + y = 15,000,

9x + 10y = 143,200.

Carry out:

Multiply the first equation by -9 and add.

x + y = 15,000,

y = 8200

Complete the solution.

x + 8200 = 15,000

x = 6800

Check:

The total investment is $8200 + $6800, or $15,000. The total interest is 0.09($6800) + 0.1($8200), or $612 + $820, or $1432.

State:

$6800 is invested at 9%, and $8200 is invested at 10%.

36. $30,000 at 5%, $40,000 at 6%

37. Familiarize:

Let d = the number of dimes and n = the number of nickels. The value of the dimes is $0.10d, and the value of the nickels is $0.05n.

Translate:

The total number of coins is 34.

d + n = 34

The total value is $1.90.

0.10d + 0.05n = 1.90, or 10d + 5n = 190

The resulting system is

d + n = 34,

10d + 5n = 190.

Carry out:

Multiply the first equation by -10 and add.

d + n = 34,

-5n = -150

Complete the solution.

-5n = -150   d + 30 = 34

n = 30   d = 4

Check:

The total number of coins is 4 + 30, or 34. The total value is $0.10(4) + $0.05(30), or $0.40 + $1.50, or $1.90.

State:

There are 4 dimes and 30 nickels.

38. 21 dimes, 22 quarters

39. Familiarize:

We organize the information in a table. Let x = the number of pounds of $4.05 tobacco and y = the number of pounds of $2.70 tobacco.

	$4.05 tobacco	$2.70 tobacco	Mixture
Amount of tobacco	x	y	15 lb
Price per pound	$4.05	$2.70	$3.15
Total cost	$4.05x	$2.70y	$3.15 × 15, or $47.25

Translate:

If we add x and y in the first row, we get 15.

x + y = 15

If we add the amounts in the third row, we get $47.25.

4.05x + 2.70y = 47.25, or 405x + 270y = 4725

The resulting system is

x + y = 15,

405x + 270y = 4725.

Carry out:

Multiply the first equation by -405 and add.

x + y = 15,

-135y = -1350

Complete the solution.

-135y = -1350   x + 10 = 15

y = 10   x = 5

Check:

The total amount of tobacco is 5 + 10, or 15 lb.

The total value of the mixture is $4.05(5) + $2.70(10), or $20.25 + $27.00, or $47.25.

State:

The dealer should use 5 lb of the $4.05 tobacco and 10 lb of the $2.70 tobacco.

40. Candy: 14 lb, nuts: 6 lb

41. $\frac{x + y}{4} - \frac{x - y}{3} = 1$

$\frac{x - y}{2} + \frac{x + y}{4} = -9$

First multiply each equation by the LCM to clear fractions.

$3(x + y) - 4(x - y) = 12$ (Multiplying by 12)

$2(x - y) + (x + y) = -36$ (Multiplying by 4)

After simplifying, the resulting system is

$-x + 7y = 12,$

$3x - y = -36.$

Multiply the first equation by 3 and add the result to the second equation.

$-x + 7y = 12$

$20y = 0$

Complete the solution.

$20y = 0$ $-x + 7 \cdot 0 = 12$

$y = 0$ $-x = 12$

$x = -12$

The solution is $(-12, 0)$.

42. $(0, 0)$

43. Familiarize:

We first make a drawing and then organize the information in a table. Using $d = rt$, we let $8t_1$ = the distance jogging and $4t_2$ = the distance walking.

Jogging 8 km/h t_1 hr   Walking 4 km/h t_2 hr

Home   University

| 6 km |

	Distance	Speed	Time
Jogging	$8t_1$	8	t_1
Walking	$4t_2$	4	t_2
Total	6		1

Translate:

The total distance is 6 km. The total time is 1 hour. A system of equations results from the columns in the table.

$t_1 + t_2 = 1,$

$8t_1 + 4t_2 = 6$

Carry out:

Multiply the first equation by -8 and add the result to the second equation.

$t_1 + t_2 = 1$

$-4t_2 = -2$

Complete the solution.

$-4t_2 = -2$ $t_1 + t_2 = 1$

$t_2 = \frac{1}{2}$ $t_1 = \frac{1}{2}$

The jogging distance, $8t_1$, is $8 \cdot \frac{1}{2}$, or 4 km.

The walking distance, $4t_2$, is $4 \cdot \frac{1}{2}$, or 2 km.

Check:

The total distance is 4 + 2, or 6 km. The total time is $\frac{1}{2} + \frac{1}{2}$, or 1 hour. The values check.

State:

She jogs 4 km on each trip.

44. James: 32 yr, Joan: 14 yr

45. Familiarize:

Let x = the number of one-book orders and y = the number of two-book orders. Then $12x$ = the amount taken in from the one-book orders and $20y$ = the amount taken in from the two-book orders. It helps to organize the information in a table.

	One-book orders	Two-book orders	Total
Number of orders	x	y	
Number of books	x	2y	880
Amount taken in	$12x	$20y	$9840

Translate:

The total number of books sold was 880.

$x + 2y = 880$ (See Row 2 of the table.)

The total amount of money taken in was $9840.

$12x + 20y = 9840$, or $3x + 5y = 2460$ (See Row 3 of the table)

The resulting system of equations is

$x + 2y = 880,$

$3x + 5y = 2460.$

Carry out:

Multiply the first equation by -3 and add the result to the second equation.

$$x + 2y = 880,$$
$$-y = -180$$

Complete the solution.

$$-y = -180 \qquad x + 2 \cdot 180 = 880$$
$$y = 180 \qquad x + 360 = 880$$
$$x = 520$$

Check:

The total number of books sold was $520 + 2 \cdot 180$, or 880. The total amount taken in was $12 \cdot 520 + 20 \cdot 180$, or \$9840. The values check.

State:

180 members ordered two books.

46. 82

47. Familiarize and Translate:

Let x = the numerator and y = the denominator. The numerator is 12 more than the denominator.

$x = y + 12$, or $x - y = 12$

The sum of the numerator and the denominator is 5 more than three times the denominator.

$x + y = 3y + 5$, or $x - 2y = 5$

The resulting system is

$$x - y = 12,$$
$$x - 2y = 5.$$

Carry out:

Multiply the first equation by -1 and add the result to the second equation.

$$x - y = 12,$$
$$-y = -7$$

Complete the solution.

$$-y = -7 \qquad x - 7 = 12$$
$$y = 7 \qquad x = 19$$

Check:

If $x = 19$ and $y = 7$, the fraction is $\frac{19}{7}$. The numerator is 12 more than the denominator. The sum of the numerator and the denominator, $19 + 7$ (or 26), is 5 more than three times the denominator ($26 = 5 + 3 \cdot 7$). The numbers check.

State:

The reciprocal of $\frac{19}{7}$ is $\frac{7}{19}$.

48. 137°

49. Familiarize and Translate:

We let r_1 and r_2 represent the speeds of the trains. Organize the information in a table. Using $d = rt$, we let $3r_1$, $2r_2$, $1.5r_1$, and $3r_2$ represent the distances the trains travel.

First situation:

3 hours	r_1 km/h		r_2 km/h	2 hours
Union				Central

|———— 216 km ————|

Second situation:

1.5 hours	r_1 km/h		r_2 km/h	3 hours
Union				Central

|———— 216 km ————|

	Distance traveled in first situation	Distance traveled in second situation
Train$_1$ (from Union to Central)	$3r_1$	$1.5r_1$
Train$_2$ (from Central to Union)	$2r_2$	$3r_2$
Total	216	216

The total distance in each situation is 216 km. Thus, we have a system of equations.

$$3r_1 + 2r_2 = 216,$$
$$1.5r_1 + 3r_2 = 216.$$

Carry out:

Multiply the second equation by 2 to make the x-coefficient a multiple of 3.

$$3r_1 + 2r_2 = 216,$$
$$3r_1 + 6r_2 = 432$$

Now multiply the first equation by -1 and add.

$$3r_1 + 2r_2 = 216,$$
$$4r_2 = 216$$

Complete the solution.

$$4r_2 = 216 \qquad 3r_1 + 2 \cdot 54 = 216$$
$$r_2 = 54 \qquad 3r_1 + 108 = 216$$
$$3r_1 = 108$$
$$r_1 = 36$$

Check:

If $r_1 = 36$ and $r_2 = 54$, the total distance the trains travel in the first situation is $3 \cdot 36 + 2 \cdot 54$, or 216 km. The total distance they travel in the second situation is $1.5 \cdot 36 + 3 \cdot 54$, or 216 km. The values check.

State:

The speed of the first train is 36 km/h. The speed of the second train is 54 km/h.

50. $4\frac{4}{7}$ L

51. Familiarize:

Let x = the value of the horse.

Let y = the value of the compensation the stablehand was to receive after one year.

Translate:

If the stablehand had worked for one year, he would have received $240 and one horse.

$y = 240 + x$

$100 and one horse is $\frac{7}{12}$ of one year's compensation.

$100 + x = \frac{7}{12}y$

We have a system of equations:

$y = 240 + x,$

$100 + x = \frac{7}{12}y$

Carry out:

We substitute $240 + x$ for y in the second equation.

$$100 + x = \frac{7}{12}(240 + x)$$

$$100 + x = 140 + \frac{7}{12}x$$

$$\frac{5}{12}x = 40$$

$$x = 96$$

Check:

If the value of the horse is $96, the compensation after one year would have been $240 + $96, or $336. Since $\frac{7}{12} \cdot$ $336 = $196, or $100 + $96, the value checks.

State:

The value of the horse is $96.

52. 5

53. Familiarize and Translate:

Let x = the number of girls in the family and y = the number of boys. Then Phil has x sisters and y - 1 brothers, and Phyllis has x - 1 sisters and y brothers.

	Brothers	Sisters
Phil	y - 1	x
Phyllis	y	x - 1

Phil has the same number of brothers and sisters.

$y - 1 = x,$ or $x - y = -1$

Phyllis has twice as many brothers as sisters.

$y = 2(x - 1),$ or $2x - y = 2$

We now have a system of equations.

$x - y = -1,$

$2x - y = 2$

Carry out:

Multiply the first equation by -2 and add the result to the second equation.

$x - y = -1,$

$\quad\quad y = 4$

Complete the solution.

$x - 4 = -1$

$\quad x = 3$

Check:

If there are 3 girls and 4 boys in the family, Phil has 3 brothers and 3 sisters and Phyllis has 2 sisters and 4 brothers. Thus, Phil has the same number of brothers and sisters, and Phyllis has twice as many brothers as sisters.

State:

There are 3 girls and 4 boys in the family.

54. City: 261 mi, highway: 204 mi

55. Substituting the given solutions in the equation $y = mx + b$, we get a system of equations.

$$\begin{array}{ccc} 3 = -2m + b, & & -2m + b = 3, \\ & \text{or} & \\ -5 = 4m + b & & 4m + b = -5 \end{array}$$

Multiply the first equation by 2 and add.

$-2m + b = 3,$

$\quad\quad 3b = 1$

Complete the solution.

$$\begin{array}{cc} 3b = 1 & -2m + \frac{1}{3} = 3 \\[4pt] b = \frac{1}{3} & -2m = \frac{8}{3} \\[4pt] & m = -\frac{4}{3} \end{array}$$

56. $A = \frac{1}{10}, B = -\frac{7}{10}$

57. $\frac{1}{x} - \frac{3}{y} = 2$

$\frac{6}{x} + \frac{5}{y} = -34$

Substitute u for $\frac{1}{x}$ and v for $\frac{1}{y}$ to obtain a linear system.

$u - 3v = \quad 2,$

$6u + 5v = -34$

Multiply the first equation by -6 and add the result to the second equation.

$u - 3v = 2,$

$\quad\quad 23v = -46$

Complete the solution for u and v.

$$23v = -46 \qquad u - 3(-2) = 2$$
$$v = -2 \qquad u + 6 = 2$$
$$u = -4$$

If $v = -2$, then $\frac{1}{y} = -2$, or $y = -\frac{1}{2}$.

If $u = -4$, then $\frac{1}{x} = -4$, or $x = -\frac{1}{4}$.

The values check. The solution of the original system is $\left(-\frac{1}{4}, -\frac{1}{2}\right)$.

58. (−125, 100)

59. $3|x| + 5|y| = 30$
$5|x| + 3|y| = 34$

Substitute u for |x| and v for |y| to obtain a linear system.

$3u + 5v = 30,$
$5u + 3v = 34$

Multiply the second equation by 3 to make the x-coefficient a multiple of 3.

$3u + 5v = 30,$
$15u + 9v = 102$

Now multiply the first equation by −5 and add.

$3u + 5v = 30,$
$-16v = -48$

Complete the solution for u and v.

$$-16v = -48 \qquad 3u + 5\cdot 3 = 30$$
$$v = 3 \qquad 3u + 15 = 30$$
$$3u = 15$$
$$u = 5$$

If $v = 3$, then $|y| = 3$ and $y = 3$ or -3.
If $u = 5$, then $|x| = 5$ and $x = 5$ or -5.

Thus, the solution set for the original system is {(5, 3), (5, −3), (−5, 3), (−5, −3)}.

60. $(0, \sqrt[3]{3})$

61. Familiarize:

First we convert the given distances to miles:

$300 \text{ ft} = \frac{300}{5280} \text{ mi} = \frac{5}{88} \text{ mi},$

$500 \text{ ft} = \frac{500}{5280} \text{ mi} = \frac{25}{264} \text{ mi}$

Then at 10 mph, the student can run to point P in $\frac{\frac{5}{88}}{10}$, or $\frac{1}{176}$ hr, and she can run to point Q in $\frac{\frac{25}{264}}{10}$, or $\frac{5}{528}$ hr $\left(\text{using } d = rt, \text{ or } \frac{d}{r} = t\right)$.

Let d represent the distance, in miles, from the train to point P in the drawing in the text, and let r represent the speed of the train, in miles per hour.

We organize the information in a table.

	Distance	Speed	Time
Going to P	d	r	$\frac{1}{176}$
Going to Q	$d + \frac{5}{88} + \frac{25}{264}$	r	$\frac{5}{528}$

Translate:

Using d = rt in each row of the table, we get a system of equations.

$$d = r\left(\frac{1}{176}\right),$$
$$d + \frac{5}{88} + \frac{25}{264} = r\left(\frac{5}{528}\right)$$

Carry out:

We substitute $\frac{r}{176}$ for d in the second equation.

$$\frac{r}{176} + \frac{5}{88} + \frac{25}{264} = \frac{5r}{528}$$
$$3r + 30 + 50 = 5r \quad \text{(Multiplying by 528)}$$
$$80 = 2r$$
$$40 = r$$

Check:
The check is left to the student.

State:
The train is traveling 40 mph.

62. (−0.187, 2.493)

63. (−4.699, 5.071)

64. (2.309, −2.457)

65. (−0.210, 0.505)

Exercise Set 6.2

1. We substitute (−1, 1, 0) in each of the three equations.

$2x + 3y - 5z = 1$	
$2(-1) + 3\cdot 1 - 5\cdot 0$	1
$-2 + 3 - 0$	
1	

$6x - 6y + 10z = 3$	
$6(-1) - 6\cdot 1 + 10\cdot 0$	3
$-6 - 6 + 0$	
-12	

We do not need to proceed. Since (−1, 1, 0) is not a solution of one of the equations, it is not a solution of the system.

2. Yes

3. $\begin{aligned} x + y + z &= 2, \quad &\text{(P1)} \\ 6x - 4y + 5z &= 31, \quad &\text{(P2)} \\ 5x + 2y + 2z &= 13 \quad &\text{(P3)} \end{aligned}$

Multiply (P1) by -6 and add the result to (P2). We also multiply (P1) by -5 and add the result to (P3).

$\begin{aligned} x + y + z &= 2 \quad &\text{(P1)} \\ -10y - z &= 19 \quad &\text{(P2)} \\ -3y - 3z &= 3 \quad &\text{(P3)} \end{aligned}$

Multiply (P3) by 10 to make the y-coefficient a multiple of the y-coefficient in (P2).

$\begin{aligned} x + y + z &= 2 \quad &\text{(P1)} \\ -10y - z &= 19 \quad &\text{(P2)} \\ -30y - 30z &= 30 \quad &\text{(P3)} \end{aligned}$

Multiply (P2) by -3 and add the result to (P3).

$\begin{aligned} x + y + z &= 2 \quad &\text{(P1)} \\ -10y - z &= 19 \quad &\text{(P2)} \\ -27z &= -27 \quad &\text{(P3)} \end{aligned}$

Solve (P3) for z.

$\begin{aligned} -27z &= -27 \\ z &= 1 \end{aligned}$

Back-substitute 1 for z in (P2) and solve for y.

$\begin{aligned} -10y - z &= 19 \\ -10y - 1 &= 19 \\ -10y &= 20 \\ y &= -2 \end{aligned}$

Back-substitute 1 for z and -2 for y in (P1) and solve for x.

$\begin{aligned} x + y + z &= 2 \\ x + (-2) + 1 &= 2 \\ x - 1 &= 2 \\ x &= 3 \end{aligned}$

The solution is (3, -2, 1).

4. (-2, -1, 4)

5. $\begin{aligned} x - y + 2z &= -3 \quad &\text{(P1)} \\ x + 2y + 3z &= 4 \quad &\text{(P2)} \\ 2x + y + z &= -3 \quad &\text{(P3)} \end{aligned}$

Multiply (P1) by -1 and add the result to (P2). Also multiply (P1) by -2 and add the result to (P3).

$\begin{aligned} x - y + 2z &= -3 \quad &\text{(P1)} \\ 3y + z &= 7 \quad &\text{(P2)} \\ 3y - 3z &= 3 \quad &\text{(P3)} \end{aligned}$

Multiply (P2) by -1 and add the result to (P3).

$\begin{aligned} x - y + 2z &= -3 \quad &\text{(P1)} \\ 3y + z &= 7 \quad &\text{(P2)} \\ -4z &= -4 \quad &\text{(P3)} \end{aligned}$

Solve (P3) for z.

$\begin{aligned} -4z &= -4 \\ z &= 1 \end{aligned}$

Back-substitute 1 for z in (P2) and solve for y.

$\begin{aligned} 3y + z &= 7 \\ 3y + 1 &= 7 \\ 3y &= 6 \\ y &= 2 \end{aligned}$

Back-substitute 1 for z and 2 for y in (P1) and solve for x.

$\begin{aligned} x - y + 2z &= -3 \\ x - 2 + 2 \cdot 1 &= -3 \\ x &= -3 \end{aligned}$

The solution is (-3, 2, 1).

6. (1, 2, 3)

7. $\begin{aligned} 4a + 9b \quad\;\; &= 8 \quad &\text{(P1)} \\ 8a \quad\;\;\; + 6c &= -1 \quad &\text{(P2)} \\ 6b + 6c &= -1 \quad &\text{(P3)} \end{aligned}$

Multiply (P1) by -2 and add the result to (P2).

$\begin{aligned} 4a + 9b \quad\;\; &= 8 \quad &\text{(P1)} \\ -18b + 6c &= -17 \quad &\text{(P2)} \\ 6b + 6c &= -1 \quad &\text{(P3)} \end{aligned}$

Multiply (P3) by 3 to make the b-coefficient a multiple of the b-coefficient in (P2).

$\begin{aligned} 4a + 9b \quad\;\; &= 8 \quad &\text{(P1)} \\ -18b + 6c &= -17 \quad &\text{(P2)} \\ 18b + 18c &= -3 \quad &\text{(P3)} \end{aligned}$

Add (P2) to (P3).

$\begin{aligned} 4a + 9b \quad\;\; &= 8 \quad &\text{(P1)} \\ -18b + 6c &= -17 \quad &\text{(P2)} \\ 24c &= -20 \quad &\text{(P3)} \end{aligned}$

Solve (P3) for c.

$\begin{aligned} 24c &= -20 \\ c &= -\frac{20}{24} = -\frac{5}{6} \end{aligned}$

Back-substitute $-\frac{5}{6}$ for c in (P2) and solve for b.

$\begin{aligned} -18b + 6c &= -17 \\ -18b + 6\left(-\frac{5}{6}\right) &= -17 \\ -18b - 5 &= -17 \\ -18b &= -12 \\ b &= \frac{12}{18} = \frac{2}{3} \end{aligned}$

Back-substitute $\frac{2}{3}$ for b in (P1) and solve for a.

$$4a + 9b = 8$$
$$4a + 9 \cdot \frac{2}{3} = 8$$
$$4a + 6 = 8$$
$$4a = 2$$
$$a = \frac{1}{2}$$

The solution is $\left(\frac{1}{2}, \frac{2}{3}, -\frac{5}{6}\right)$.

<u>8.</u> $\left(4, \frac{1}{2}, -\frac{1}{2}\right)$

<u>9.</u>

$w + x + y + z = 2$		(P1)	
$w + 2x + 2y + 4z = 1$		(P2)	
$-w + x - y - z = -6$		(P3)	
$-w + 3x + y - z = -2$		(P4)	

Multiply (P1) by -1 and add to (P2). Add (P1) to (P3) and to (P4).

$$w + x + y + z = 2 \qquad (P1)$$
$$x + y + 3z = -1 \qquad (P2)$$
$$2x = -4 \qquad (P3)$$
$$4x + 2y = 0 \qquad (P4)$$

Solve (P3) for x.

$$2x = -4$$
$$x = -2$$

Back-substitute -2 for x in (P4) and solve for y.

$$4(-2) + 2y = 0$$
$$-8 + 2y = 0$$
$$2y = 8$$
$$y = 4$$

Back-substitute -2 for x and 4 for y in (P2) and solve for z.

$$-2 + 4 + 3z = -1$$
$$3z = -3$$
$$z = -1$$

Back-substitute -2 for x, 4 for y, and -1 for z in (P1) and solve for w.

$$w - 2 + 4 - 1 = 2$$
$$w = 1$$

The solution is (1, -2, 4, -1).

<u>10.</u> (-3, -1, 0, 4)

<u>11.</u> Familiarize and Translate:

We let x, y, and z represent the first, second, and third numbers, respectively.

The sum of the three numbers is 26.

$$x + y + z = 26$$

Twice the first minus the second is 2 less than the third.

$$2x - y = z - 2, \quad \text{or} \quad 2x - y - z = -2$$

The third is the second minus three times the first.

$$z = y - 3x, \quad \text{or} \quad 3x - y + z = 0$$

We now have a system of three equations.

$$x + y + z = 26,$$
$$2x - y - z = -2,$$
$$3x - y + z = 0$$

Carry out:

We solve the system. The solution is (8, 21, -3).

Check:

The sum of the numbers is 8 + 21 + (-3), or 26. Twice the first minus the second (2·8 - 21 = -5) is two less than the third (-3 - 2 = -5). The third (-3) is the second minus three times the first (21 - 3·8 = -3). The numbers check.

State:

The numbers are 8, 21, and -3.

<u>12.</u> 4, 2, -1

<u>13.</u> Familiarize:

We make a drawing and use A, B, and C for the measures of the angles.

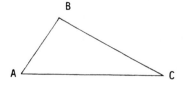

We must use the fact that the measures of the angles of a triangle add up to 180°.

Translate:

The sum of the angle measures is 180°.

$$A + B + C = 180$$

The measure of angle B is three times the measure of angle A.

$$B = 3A$$

The measure of angle C is 30° greater than the measure of angle A.

$$C = A + 30$$

We now have a system of equations.

$$A + B + C = 180,$$
$$B = 3A,$$
$$C = A + 30$$

Carry out:

We solve the system. The solution is (30, 90, 60).

Check:

The sum of the angle measures is 30 + 90 + 60, or 180°. The measure of angle B, 90°, is three times 30°, the measure of angle A. The measure of angle C, 60°, is 30° greater than the measure of angle A, 30°. The values check.

State:

The measures of angles A, B, and C are 30°, 90°, and 60°, respectively.

14. A = 34°, B = 104°, C = 42°

15. Familiarize and Translate:

Let x, y, and z represent the number of quarts picked on Monday, Tuesday, and Wednesday, respectively.

She picked a total of 87 quarts.

$x + y + z = 87$

On Tuesday she picked 15 quarts more than on Monday.

$y = x + 15,$ or $x - y = -15$

On Wednesday she picked 3 quarts fewer than on Tuesday.

$z = y - 3,$ or $y - z = 3$

We now have a system of equations.

$x + y + z = 87,$
$x - y \quad\;\; = -15,$
$\quad\; y - z = 3$

Carry out:

We solve the system. The solution is (20, 35, 32).

Check:

The total quarts picked in three days was 20 + 35 + 32, or 87 quarts. On Tuesday she picked 35 quarts which is 15 more quarts than on Monday (20 + 15 = 35). On Wednesday she picked 32 quarts which is 3 quarts less than on Tuesday (35 - 3 = 32). The amounts check.

State:

The farmer picked 20 quarts on Monday, 35 quarts on Tuesday, and 32 quarts on Wednesday.

16. Thursday: $21, Friday: $18, Saturday: $27

17. Familiarize and Translate:

Let A, B, and C represent the number of board-feet of lumber produced per day by sawmills A, B, and C, respectively.

All three together can produce 7400 board-feet in a day.

$A + B + C = 7400$

A and B together can produce 4700 board-feet in a day.

$A + B = 4700$

B and C together can produce 5200 board-feet in a day.

$B + C = 5200$

We now have a system of equations.

$A + B + C = 7400$
$A + B \quad\;\; = 4700$
$\quad\; B + C = 5200$

Carry out:

We solve the system. The solution is (2200, 2500, 2700).

Check:

All three can produce 2200 + 2500 + 2700, or 7400 board-feet per day. A and B together can produce 2200 + 2500, or 4700 board-feet. B and C together can produce 2500 + 2700, or 5200 board-feet. All values check.

State:

In a day, sawmill A can produce 2200 board-feet, sawmill B can produce 2500 board-feet, and sawmill C can produce 2700 board-feet.

18. A: 1500, B: 1900, C: 2300

19. Familiarize and Translate:

Let A, B, and C represent the number of linear feet per hour welders A, B, and C can weld, respectively.

Together A, B, and C can weld 37 linear feet.

$A + B + C = 37$

A and B together can weld 22 linear feet.

$A + B = 22$

A and C together can weld 25 linear feet.

$A + C = 25$

We now have a system of equations.

$A + B + C = 37,$
$A + B \quad\;\; = 22,$
$A \quad\; + C = 25$

Carry out:

We solve the system. The solution is (10, 12, 15).

Check:

Together all three can weld 10 + 12 + 15, or 37 linear feet per hour. A and B together can weld 10 + 12, or 22 linear feet. A and C together can weld 10 + 15, or 25 linear feet. The values check.

State:

Welder A can weld 10 linear feet per hour.
Welder B can weld 12 linear feet per hour.
Welder C can weld 15 linear feet per hour.

20. A: 900 gal/hr, B: 1300 gal/hr, C: 1500 gal/hr

21. Familiarize:

Let x, y, and z represent the number of par-3, par-4, and par-5 holes, respectively. A golfer who shoots par on every hole has 3x from the par-3 holes, 4y from the par-4 holes, and 5z from the par-5 holes.

Translate:

The total number of holes is 18.

$x + y + z = 18$

A golfer who shoots par on every hole has a total of 72.

$3x + 4y + 5z = 72$

The sum of the number of par-3 holes and the number of par-5 holes is 8.

$x + z = 8$

We have a system of equations.

$$x + y + z = 18,$$
$$3x + 4y + 5z = 72,$$
$$x + z = 8$$

Carry out:

We solve the system. The solution is (4, 10, 4).

Check:

The total number of holes is 4 + 10 + 4, or 18. A golfer who shoots par on every hole has $3 \cdot 4 + 4 \cdot 10 + 5 \cdot 4$, or 72. The sum of the number of par-3 holes and the number of par-5 holes is 4 + 4, or 8. The values check.

State:

There are 4 par-3 holes, 10 par-4 holes, and 4 par-5 holes.

22. Par-3: 6, par-4: 8, par-5: 4

23. Familiarize:

Let x, y, and z represent the amounts invested at 7%, 8%, and 9%, respectively. Then the interest from each investment is 7%x or 0.07x, 8%y or 0.08y, and 9%z or 0.09z.

Translate:

The total investment is $2500.

$x + y + z = 2500$

The total interest is $212.

$0.07x + 0.08y + 0.09z = 212$, or

$7x + 8y + 9z = 21,200$

The amount invested at 9% is $1100 more than the amount invested at 8%.

$z = y + 1100$

We have a system of equations.

$$x + y + z = 2500,$$
$$7x + 8y + 9z = 21,200,$$
$$ -y + z = 1100$$

Carry out:

We solve the system. The solution is (400, 500, 1600).

Check:

The total investment is $400 + $500 + $1600, or $2500. The total interest is 0.07($400) + 0.08($500) + 0.09($1600), or $28 + $40 + $144, or $212. The amount invested at 9%, $1600, is $1100 more than $500, the amount invested at 8%. The values check.
State:

$400 is invested at 7%, $500 is invested at 8%, and $1600 is invested at 9%.

24. $300 at 8%, $300 at 9%, $2900 at 10%

25. We wish to find a quadratic function

$$f(x) = ax^2 + bx + c$$

containing the three given points.

When we substitute, we get

for (1, 4) $4 = a \cdot 1^2 + b \cdot 1 + c$
for (-1, -2) $-2 = a \cdot (-1)^2 + b \cdot (-1) + c$
for (2, 13) $13 = a \cdot 2^2 + b \cdot 2 + c$

We now have a system of equations in three unknowns, a, b, and c:

$$a + b + c = 4,$$
$$a - b + c = -2,$$
$$4a + 2b + c = 13$$

We solve this system, obtaining (2, 3, -1). Thus the function we are looking for is

$$f(x) = 2x^2 + 3x - 1.$$

26. $f(x) = 3x^2 - x + 2$

27. a) We wish to find a quadratic function

$$E(t) = at^2 + bt + c$$

containing the three given points.

When we substitute, we get

for (1, 38) $38 = a \cdot 1^2 + b \cdot 1 + c$
for (2, 66) $66 = a \cdot 2^2 + b \cdot 2 + c$
for (3, 86) $86 = a \cdot 3^2 + b \cdot 3 + c$

We now have a system of equations in three unknowns, a, b, and c.

$$a + b + c = 38,$$
$$4a + 2b + c = 66,$$
$$9a + 3b + c = 86$$

We solve the system, obtaining (-4, 40, 2). Thus, the function we are looking for is $E(t) = -4t^2 + 40t + 2$.

b) $E(t) = -4t^2 + 40t + 2$

$E(4) = -4 \cdot 4^2 + 40 \cdot 4 + 2$

$= -64 + 160 + 2$

$= \$98$

28. a) $E(t) = 2500t^2 - 6500t + 5000$

    b) $19,000

29. a) We wish to find a quadratic function
    $$f(x) = ax^2 + bx + c$$
    containing the points (5, 1121), (7, 626), and (9, 967).

    When we substitute, we get
    $$1121 = 25a + 5b + c,$$
    $$626 = 49a + 7b + c,$$
    $$967 = 81a + 9b + c.$$

    We solve this system of equations, obtaining (104.5, -1501.5, 6016). Thus,
    $$f(x) = 104.5x^2 - 1501.5x + 6016.$$

    b) $f(4) = 104.5(4)^2 - 1501.5(4) + 6016 = 1682$
    $f(6) = 104.5(6)^2 - 1501.5(6) + 6016 = 769$
    $f(10) = 104.5(10)^2 - 1501.5(10) + 6016 = 1451$

30. a) $f(x) = \frac{1}{300,000}x^2 + \frac{7}{3000}x$

    b) 2.925 hr

    c) 407

31. $$\frac{2}{x} - \frac{1}{y} - \frac{3}{z} = -1,$$
    $$\frac{2}{x} - \frac{1}{y} + \frac{1}{z} = -9,$$
    $$\frac{1}{x} + \frac{2}{y} - \frac{4}{z} = 17$$

    First substitute u for $\frac{1}{x}$, v for $\frac{1}{y}$, and w for $\frac{1}{z}$ and solve for u, v, and w.
    $$2u - v - 3w = -1,$$
    $$2u - v + w = -9,$$
    $$u + 2v - 4w = 17$$

    Solving this system we get (-1, 5, -2).

    If $u = -1$ and $u = \frac{1}{x}$, then $\frac{1}{x} = -1$, or $x = -1$.

    If $v = 5$ and $v = \frac{1}{y}$, then $\frac{1}{y} = 5$, or $y = \frac{1}{5}$.

    If $w = -2$ and $w = \frac{1}{z}$, then $\frac{1}{z} = -2$, or $z = -\frac{1}{2}$.

    The solution of the original system is $\left(-1, \frac{1}{5}, -\frac{1}{2}\right)$.

32. $\left(-\frac{1}{2}, -1, -\frac{1}{3}\right)$

33. **Familiarize:**
    Let a, b, and c represent the time it would take A, B, and C, working alone, to do the job, respectively. Then A can do $\frac{1}{a}$ of the job in 1 hr, $\frac{2}{a}$ of the job in 2 hr and so on.

Translate:
Working together, A, B, and C can do a job in 2 hr.
$$\frac{2}{a} + \frac{2}{b} + \frac{2}{c} = 1$$

Working together, B and C can do the job in 4 hr.
$$\frac{4}{b} + \frac{4}{c} = 1$$

Working together, A and B can do the job in $\frac{12}{5}$ hr.
$$\frac{\frac{12}{5}}{a} + \frac{\frac{12}{5}}{b} = 1, \text{ or } \frac{12}{a} + \frac{12}{b} = 5$$

We have a system of equations.
$$\frac{2}{a} + \frac{2}{b} + \frac{2}{c} = 1,$$
$$\frac{4}{b} + \frac{4}{c} = 1,$$
$$\frac{12}{a} + \frac{12}{b} = 5$$

Carry out:
Substitute x for $\frac{1}{a}$, y for $\frac{1}{b}$, and z for $\frac{1}{c}$ and solve for x, y, and z.
$$2x + 2y + 2z = 1,$$
$$4y + 4z = 1,$$
$$12x + 12y = 5$$

Solving this system we get $\left(\frac{1}{4}, \frac{1}{6}, \frac{1}{12}\right)$.

If $x = \frac{1}{4}$ and $x = \frac{1}{a}$, then $\frac{1}{a} = \frac{1}{4}$ or $a = 4$.

If $y = \frac{1}{6}$ and $y = \frac{1}{b}$, then $\frac{1}{b} = \frac{1}{6}$ or $b = 6$.

If $z = \frac{1}{12}$ and $z = \frac{1}{c}$, then $\frac{1}{c} = \frac{1}{12}$ or $c = 12$.

The solution of the original system is (4, 6, 12).

Check:
The check is left to the student.

State:
Working alone, A can do the job in 4 hr, B can do the job in 6 hr, and C can do the job in 12 hr.

34. A: 24 hr, B: 12 hr, C: $4\frac{4}{5}$ hr

35. Label the angle measures at the tips of the stars a, b, c, d, and e. Also label the angles of the pentagon p, q, r, s, and t.

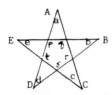

Using the geometric fact that the sum of the angle measures of a triangle is 180°, we get 5 equations.

$p + b + d = 180$

$q + c + e = 180$

$r + a + d = 180$

$s + b + e = 180$

$t + a + c = 180$

Adding these equations, we get

$(p + q + r + s + t) + 2a + 2b + 2c + 2d + 2e =$
$$5(180)$$

The sum of the angle measures of any convex polygon with n sides is given by the formula $S = (n - 2)180$. Thus $p + q + r + s + t = (5 - 2)180$, or 540. We substitute and solve for $a + b + c + d + e$.

$540 + 2(a + b + c + d + e) = 900$

$2(a + b + c + d + e) = 360$

$a + b + c + d + e = 180$

The sum of the angle measures at the tips of the star is 180°.

36. 1869

37. Substituting, we get

$A + \frac{3}{4}B + 3C = 12,$

$\frac{4}{3}A + B + 2C = 12,$

$2A + B + C = 12,$ or

$4A + 3B + 12C = 48,$

$4A + 3B + 6C = 36,$ (Clearing fractions)

$2A + B + C = 12.$

Solving the system, we get $A = 3$, $B = 4$, and $C = 2$. The equation is $3x + 4y + 2z = 12$.

38. $y = 2 - \frac{1}{2}x - \frac{1}{4}z$

39. Familiarize and Translate:

Let a, s, and c represent the number of adults, students, and children in attendance, respectively.

The total attendance was 100.

$a + s + c = 100$

The total amount of money taken in was $100. [Express 50 cents as $\frac{1}{2}$ dollar.]

$10a + 3s + \frac{1}{2}c = 100$

The resulting system is

$a + s + c = 100,$

$10a + 3s + \frac{1}{2}c = 100.$

Carry out:

Multiply the first equation by 3 and then subtract it from the second equation to obtain $7a - \frac{5}{2}c = -200$ or $a = \frac{5}{14}(c - 80)$ where $(c - 80)$ is a positive multiple of 14 (because a must be a positive integer). That is $(c - 80) = k \cdot 14$ or $c = 80 + k \cdot 14$, where k is a positive integer. If $k > 1$, then $c > 100$. This is impossible since the total attendance is 100. Thus $k = 1$, so $c = 80 + 1 \cdot 14 = 94$. Then $a = \frac{5}{14}(94 - 80) = \frac{5}{14} \cdot 14 = 5$, and $5 + s + 94 = 10$, or $s = 1$.

Check:

The total attendance is $5 + 1 + 94$, or 100.

The total amount of money taken in was

$\$10 \cdot 5 + \$3 \cdot 1 + \$\frac{1}{2} \cdot 94 = \$100$. The result checks.

State:

There were 5 adults, 1 student, and 94 children in attendance.

40. Art: 181, Bob: 176, Carl: 158, Denny: 190, Emmett: 180

Exercise Set 6.3

Note: The answers to Exercise 19 are included in the answers to Exercises 1 - 17.

1. $9x - 3y = 15,$ (1)

   $6x - 2y = 10$ (2)

   Multiply equation (2) by 3 to make the x-coefficient a multiple of 9.

   $9x - 3y = 15,$

   $18x - 6y = 30$

   Multiply equation (1) by -2 and add the result to equation (2).

   Now we have

   $9x - 3y = 15,$

   $0 = 0.$

   which illustrates that the original system of equations is a _dependent_ system of equations. For this particular system, there is an infinite solution set. The system is _consistent_.

These solutions can be described by expressing one variable in terms of the other. Since $0 = 0$ contributes nothing, we solve $9x - 3y = 15$ for x and obtain

$$x = \frac{3y + 15}{9}, \quad \text{or} \quad \frac{y + 5}{3}$$

The ordered pairs in the solution set can be described in terms of y only.

$$\left(\frac{y + 5}{3}, \ y \right)$$

Any value chosen for y gives a value for x, and thus an ordered pair in the solution set.

We could have solved $9x - 3y = 15$ for y obtaining $y = \frac{9x - 15}{3}$ or $3x - 5$ and described the solution set in terms of x only.

$(x, 3x - 5)$

Then any value chosen for x would give a value for y, and thus an ordered pair in the solution set.

A few of the solutions are $(0, -5)$, $(1, -2)$, and $(-1, -8)$.

2. $\emptyset$

3.
$$5c + 2d = 24, \qquad (1)$$
$$30c + 12d = 10 \qquad (2)$$

Multiply equation (1) by -6 and add the result to equation (2).

This gives us
$$5c + 2d = 24,$$
$$0 = -134.$$

The second equation says that $0 \cdot x + 0 \cdot y = -134$. There are no numbers x and y for which this is true. Thus, the system has no solutions. The solution set is $\emptyset$. The system is <u>inconsistent</u>. This system is not equivalent to a system of fewer than 2 linear equations. The system is <u>independent</u>.

4. $\emptyset$

5.
$$3x + 2y = 5,$$
$$4y = 10 - 6x$$

Rewrite each equation in the form $Ax + By = C$.
$$3x + 2y = 5, \qquad (1)$$
$$6x + 4y = 10 \qquad (2)$$

Multiply equation (1) by -2 and add.
$$3x + 2y = 5,$$
$$0 = 0$$

The system is <u>dependent</u>. The equation $3x + 2y = 5$ has infinitely many solutions, so the system is <u>consistent</u>.

If we solve $3x + 2y = 5$ for x obtaining
$$x = \frac{5 - 2y}{3},$$

the ordered pairs in the solution set can be described in terms of y only.

$$\left(\frac{5 - 2y}{3}, \ y \right)$$

If we solve $3x + 2y = 5$ for y obtaining
$$y = \frac{5 - 3x}{2},$$

the ordered pairs in the solution set can be described in terms of x only.

$$\left(x, \ \frac{5 - 3x}{2} \right)$$

A few of the solutions are $\left(0, \frac{5}{2} \right)$, $(3, -2)$, and $(-1, 4)$.

6. $\left(\frac{7y - 2}{5}, \ y \right)$ or $\left(x, \ \frac{5x + 2}{7} \right)$; $(1, 1)$, $\left(0, \frac{2}{7} \right)$, $(8, 6)$, etc.

7.
$$12y - 8x = 6,$$
$$4x + 3 = 6y$$

Rewrite each equation in the form $Ax + By = C$.
$$-8x + 12y = 6, \qquad (1)$$
$$4x - 6y = -3 \qquad (2)$$

Multiply equation (2) by 2 to make the x-coefficient a multiple of -8.
$$-8x + 12y = 6,$$
$$8x - 12y = -6$$

Now add equation (1) to equation (2).
$$-8x + 12y = 6,$$
$$0 = 0$$

The system is <u>dependent</u>. The equation $-8x + 12y = 6$ has infinitely many solutions, so the system is <u>consistent</u>.

If we solve $-8x + 12y = 6$ for x obtaining
$$x = \frac{6y - 3}{4},$$

the ordered pairs in the solution set can be described in terms of y only.

$$\left(\frac{6y - 3}{4}, \ y \right)$$

If we solve $-8x + 12y = 6$ for y obtaining
$$y = \frac{4x + 3}{6},$$

the ordered pairs in the solution set can be described in terms of x only.

$$\left(x, \ \frac{4x + 3}{6} \right)$$

A few of the solutions are $\left(-\frac{3}{4}, 0 \right)$, $\left(\frac{9}{4}, 2 \right)$ and $\left(-\frac{9}{4}, -1 \right)$.

8. $\left(\frac{6y + 5}{8}, \ y \right)$ or $\left(x, \ \frac{8x - 5}{6} \right)$; $\left(0, -\frac{5}{6} \right)$, $\left(1, \frac{1}{2} \right)$, $\left(-1, -\frac{13}{6} \right)$, etc.

9. $x + 2y - z = -8$, (P1)

$2x - y + z = 4$, (P2)

$8x + y + z = 2$ (P3)

Multiply (P1) by -2 and add the result to (P2).
Multiply (P1) by -8 and add the result to (P3).

$x + 2y - z = -8$, (P1)

$-5y + 3z = 20$, (P2)

$-15y + 9z = 66$ (P3)

Multiply (P2) by -3 and add the result to (P3).

$x + 2y - z = -8$, (P1)

$-5y + 3z = 20$, (P2)

$0 = 6$ (P3)

Since in (P3) we obtain a false equation $0 = 6$, the system has no solution. The solution set is $\emptyset$. The system is <u>inconsistent</u>. This system is not equivalent to a system of fewer than 3 equations. The system is <u>independent</u>.

10. $\left(\dfrac{z + 28}{11}, \dfrac{5z + 8}{11}, z\right)$; $\left(\dfrac{28}{11}, \dfrac{8}{11}, 0\right)$, $\left(\dfrac{29}{11}, \dfrac{13}{11}, 1\right)$,

$\left(\dfrac{27}{11}, \dfrac{3}{11}, -1\right)$, etc.

11. $2x + y - 3z = 1$,

$x - 4y + z = 6$,

$4x - 16y + 4z = 24$

First interchange the first two equations so all the x-coefficients are multiples of the first.

$x - 4y + z = 6$, (P1)

$2x + y - 3z = 1$, (P2)

$4x - 16y + 4z = 24$ (P3)

Multiply (P1) by -2 and add the result to (P2).
Multiply (P1) by -4 and add the result to (P3).

$x - 4y + z = 6$, (P1)

$9y - 5z = -11$, (P2)

$0 = 0$ (P3)

Now we know the system is <u>dependent</u>.

Solve (P2) for y.

$9y - 5z = -11$

$9y = 5z - 11$

$y = \dfrac{5z - 11}{9}$

Substitute $\dfrac{5z - 11}{9}$ for y in (P1) and solve for x.

$x - 4y + z = 6$

$x - 4\left(\dfrac{5z - 11}{9}\right) + z = 6$

$9x - 4(5z - 11) + 9z = 54$

$9x - 20z + 44 + 9z = 54$

$9x = 11z + 10$

$x = \dfrac{11z + 10}{9}$

The ordered triples in the solution set can be described as follows:

$\left(\dfrac{11z + 10}{9}, \dfrac{5z - 11}{9}, z\right)$

There is an infinite number of solutions. The system is <u>consistent</u>.

A few of the solutions are $\left(\dfrac{10}{9}, -\dfrac{11}{9}, 0\right)$,

$\left(\dfrac{7}{3}, -\dfrac{2}{3}, 1\right)$, and $\left(\dfrac{32}{9}, -\dfrac{1}{9}, 2\right)$.

12. $\left(\dfrac{z + 5}{5}, -\dfrac{7}{5}z, z\right)$; $(1,0,0)$, $(2,-7,5)$ $\left(\dfrac{6}{5}, -\dfrac{7}{5}, 1\right)$, etc.

13. $2x + y - 3z = 0$,

$x - 4y + z = 0$,

$4x - 16y + 4z = 0$

Note that this is a system of homogeneous equations. The trivial solution is $(0, 0, 0)$. There may or may not be other solutions.

First we interchange the first two equations so all the x-coefficients are multiples of the first:

$x - 4y + z = 0$, (P1)

$2x + y - 3z = 0$, (P2)

$4x - 16y + 4z = 0$ (P3)

Multiply (P1) by -2 and add to (P2).
Multiply (P1) by -4 and add to (P3).

$x - 4y + z = 0$, (P1)

$9y - 5z = 0$, (P2)

$0 = 0$ (P3)

Now we know that the system is <u>dependent</u>.

Solve (P2) for y.

$9y - 5z = 0$

$9y = 5z$

$y = \dfrac{5}{9}z$

Substitute $\dfrac{5}{9}z$ for y in (P1) and solve for x.

$x - 4y + z = 0$

$x - 4\left(\dfrac{5}{9}z\right) + z = 0$

$x - \dfrac{20}{9}z + \dfrac{9}{9}z = 0$

$x - \dfrac{11}{9}z = 0$

$x = \dfrac{11}{9}z$

The ordered triples in the solution set can be described as follows:

$\left(\dfrac{11}{9}z, \dfrac{5}{9}z, z\right)$

There is an infinite number of solutions. The system is <u>consistent</u>.

A few of the solutions are (0, 0, 0), $\left(\frac{11}{8}, \frac{5}{18}, \frac{1}{2}\right)$ and $\left(-\frac{11}{9}, -\frac{5}{9}, -1\right)$.

14. $\left(\frac{1}{5}z, -\frac{7}{5}z, z\right)$; (0,0,0), (1,-7,5), $\left(\frac{1}{5}, -\frac{7}{5}, 1\right)$, etc.

15. $x + y - z = -3$, (P1)
    $x + 2y + 2z = -1$ (P2)

    Multiply (P1) by -1 and add to (P2).
    $x + y - z = -3$,
    $y + 3z = 2$

    Solve (P2) for y.
    $y + 3z = 2$
    $y = -3z + 2$

    Substitute $-3z + 2$ for y in (P1) and solve for x.
    $x + y - z = -3$
    $x + (-3z + 2) - z = -3$
    $x - 4z + 2 = -3$
    $x = 4z - 5$

    The system has an infinite set of solutions. The system is consistent. The ordered triples in the solutions can be described as follows:
    $(4z - 5, -3z + 2, z)$

    A few of the solutions are (-5, 2, 0), (-9, 5, -1), and (3, -4, 2).

    The system is dependent.

16. $\left(-\frac{7}{2}z, -\frac{19}{2}z, z\right)$; (0,0,0), (-7,-19,2), $\left(-\frac{7}{2}, -\frac{19}{2}, 1\right)$, etc.

17. $2x + y + z = 0$,
    $x + y - z = 0$,
    $x + 2y + 2z = 0$

    Any homogeneous system like this always has a solution, because (0, 0, 0) is a solution. There may or may not be other solutions.

    Interchange the first two equations.
    $x + y - z = 0$, (P1)
    $2x + y + z = 0$, (P2)
    $x + 2y + 2z = 0$ (P3)

    Multiply (P1) by -2 and add to (P2). Multiply (P1) by -1 and add to (P3).
    $x + y - z = 0$, (P1)
    $-y + 3z = 0$, (P2)
    $y + 3z = 0$ (P3)

    Add (P2) to (P3).
    $x + y - z = 0$, (P1)
    $-y + 3z = 0$, (P2)
    $6z = 0$ (P3)

Solve (P3) for z.
$6z = 0$
$z = 0$

Substitute 0 for z in (P2) and solve for y.
$-y + 3z = 0$
$-y + 3 \cdot 0 = 0$
$-y = 0$
$y = 0$

Substitute 0 for z and 0 for y in (P1) and solve for x.
$x + y - z = 0$
$x + 0 - 0 = 0$
$x = 0$

The only solution is (0, 0, 0). The system is consistent. The system is not equivalent to a system of fewer than 3 equations. Thus, the system is independent.

18. (0, 0, 0)

19. Answers in Exercises 1 - 17.

20. Consistent: 6, 8, 10, 12, 14, 16, 18, the others are inconsistent; dependent: 6, 8, 10, 12, 14, 16, the others are independent

21. $4.026x - 1.448y = 18.32$,
    $0.724y = -9.16 + 2.013x$

    Multiply each equation by 1000 and rewrite in the form $Ax + By = C$.
    $4026x - 1448y = 18,320$,
    $2013x - 724y = 9160$

    Interchange the equations.
    $2013x - 724y = 9160$, (1)
    $4026x - 1448y = 18,320$ (2)

    Multiply equation (1) by -2 and add.
    $2013x - 724y = 9160$,
    $0 = 0$

    The system is dependent.

    If we solve $2013x - 724y = 9160$ for x obtaining
    $x = \frac{9160 + 724y}{2013}$,
    the ordered pairs in the solution set can be described in terms of y only.
    $\left(\frac{9160 + 724y}{2013}, y\right)$

    If we solve $2013x - 724y = 9160$ for y obtaining
    $y = \frac{2013x - 9160}{724}$,
    the ordered pairs in the solution set can be described in terms of x only.
    $\left(x, \frac{2013x - 9160}{724}\right)$

22. $\left[\dfrac{5260 - 142y}{40,570}, y\right]$ or $\left[x, \dfrac{5260 - 40,570x}{142}\right]$

23. a) $w + x + y + z = 4,$ (P1)

    $w + x + y + z = 3,$ (P2)

    $w + x + y + z = 3$ (P3)

    Multiply (P1) by -1 and add the result to (P2) and to (P3).

    $w + x + y + z = 4,$ (P1)

    $\quad\quad\quad\quad 0 = -1,$ (P2)

    $\quad\quad\quad\quad 0 = -1$ (P3)

    Since we obtain the false equation 0 = -1, the solution set is ∅.

    b) Since there is no solution, the system is inconsistent.

    c) Since (P2) and (P3) are identical, the system is equivalent to a system of fewer than 3 equations. It is dependent.

24. a) $(2x + 5y, x, y, 3x - 4y)$

    b) consistent

    c) dependent

25. $6x - 9y = -3,$ (1)

    $-4x + 6y = k$ (2)

    Multiply equation (2) by 3 to make the x-coefficient a multiple of 6.

    $6x - 9y = -3,$ (1)

    $-12x + 18y = 3k$ (2)

    Multiply equation (1) by 2 and add the result to Equation (2).

    $6x - 9y = -3,$

    $\quad\quad\quad 0 = -6 + 3k$

If the system is to be dependent, it must be true that

$$-6 + 3k = 0 \quad \text{or}$$
$$k = 2.$$

26. 25

27. Let a = the number of par 3 holes,

    b = the number of par 4 holes, and

    c = the number of par 5 holes.

    Then $a + b + c = 18,$

    $3a + 4b + 5c = 72,$

    $\quad\quad\quad a = c$

    or, equivalently,

    $a + b + c = 18,$

    $\quad\quad b + 2c = 18,$

    $\quad\quad\quad\quad 0 = 0$

    The solutions are of the form {(a, 18 - 2a, a)} where $1 \leqslant a \leqslant 8$, and a is an integer.

    All permissible solutions are: (1, 16, 1),

    (2, 14, 2),

    (3, 12, 3),

    (4, 10, 4),

    (5, 8, 5),

    (6, 6, 6),

    (7, 4, 7),

    (8, 2, 8).

28. 12, 24, 36, 48

Exercise Set 6.4

---

1. $4x + 2y = 11,$

    $3x - \ y = 2$

    Write a matrix using only the constants.

    $$\begin{bmatrix} 4 & 2 & 11 \\ 3 & -1 & 2 \end{bmatrix}$$

    Multiply row 2 by 4 to make the first number in row 2 a multiple of 4.

    $$\begin{bmatrix} 4 & 2 & 11 \\ 12 & -4 & 8 \end{bmatrix}$$

    Multiply row 1 by -3 and add it to row 2.

    $$\begin{bmatrix} 4 & 2 & 11 \\ 0 & -10 & -25 \end{bmatrix}$$

Putting the variables back in, we have

$4x + 2y = 11$,   (1)

$-10y = -25$ (2)

Solve (2) for y.

$-10y = -25$

$y = \frac{5}{2}$

Substitute $\frac{5}{2}$ for y in (1) and solve for x.

$4x + 2y = 11$

$4x + 2 \cdot \frac{5}{2} = 11$

$4x + 5 = 11$

$4x = 6$

$x = \frac{3}{2}$

The solution is $\left(\frac{3}{2}, \frac{5}{2}\right)$.

2. $\left(-\frac{1}{3}, -4\right)$

3. $x + 2y - 3z = 9$,
   $2x - y + 2z = -8$,
   $3x - y - 4z = 3$

Write a matrix using only the constants.

$$\begin{bmatrix} 1 & 2 & -3 & 9 \\ 2 & -1 & 2 & -8 \\ 3 & -1 & -4 & 3 \end{bmatrix}$$

Multiply row 1 by -2 and add it to row 2.
Multiply row 1 by -3 and add it to row 3.

$$\begin{bmatrix} 1 & 2 & -3 & 9 \\ 0 & -5 & 8 & -26 \\ 0 & -7 & 5 & -24 \end{bmatrix}$$

Multiply row 3 by 5.

$$\begin{bmatrix} 1 & 2 & -3 & 9 \\ 0 & -5 & 8 & -26 \\ 0 & -35 & 25 & -120 \end{bmatrix}$$

Multiply row 2 by -7 and add to row 3.

$$\begin{bmatrix} 1 & 2 & -3 & 9 \\ 0 & -5 & 8 & -26 \\ 0 & 0 & -31 & 62 \end{bmatrix}$$

Putting the variables back in, we have

$x + 2y - 3z = 9$,   (1)

$-5y + 8z = -26$,   (2)

$-31z = 62$ (3)

Solve (3) for z.

$-31z = 62$

$z = -2$

Substitute -2 for z in (2) and solve for y.

$-5y + 8z = -26$

$-5y + 8(-2) = -26$

$-5y - 16 = -26$

$-5y = -10$

$y = 2$

Substitute -2 for z and 2 for y in (1) and solve for x.

$x + 2y - 3z = 9$

$x + 2(2) - 3(-2) = 9$

$x + 10 = 9$

$x = -1$

The solution is (-1, 2, -2).

4. (0, 2, 1)

5. $5x - 3y = -2$,
   $4x + 2y = 5$

Write a matrix using only the constants.

$$\begin{bmatrix} 5 & -3 & -2 \\ 4 & 2 & 5 \end{bmatrix}$$

Multiply row 2 by 5.

$$\begin{bmatrix} 5 & -3 & -2 \\ 20 & 10 & 25 \end{bmatrix}$$

Multiply row 1 by -4 and add it to row 2.

$$\begin{bmatrix} 5 & -3 & -2 \\ 0 & 22 & 33 \end{bmatrix}$$

Putting the variables back in, we have

$5x - 3y = -2$,   (1)

$22y = 33$ (2)

Solve (2) for y.

$22y = 33$

$y = \frac{3}{2}$

Substitute $\frac{3}{2}$ for y in (1) and solve for x.

$5x - 3y = -2$

$5x - 3 \cdot \frac{3}{2} = -2$

$5x - \frac{9}{2} = -\frac{4}{2}$

$5x = \frac{5}{2}$

$x = \frac{1}{2}$

The solution is $\left(\frac{1}{2}, \frac{3}{2}\right)$.

6. $\left[-1, \frac{5}{2}\right]$

7. $4x - y - 3z = 1,$
    $8x + y - z = 5,$
    $2x + y + 2z = 5$

Write a matrix using only the constants.

$$\begin{bmatrix} 4 & -1 & -3 & 1 \\ 8 & 1 & -1 & 5 \\ 2 & 1 & 2 & 5 \end{bmatrix}$$

First interchange rows 1 and 3 so that each number below the first number in the first row is a multiple of that number.

$$\begin{bmatrix} 2 & 1 & 2 & 5 \\ 8 & 1 & -1 & 5 \\ 4 & -1 & -3 & 1 \end{bmatrix}$$

Multiply row 1 by −4 and add it to row 2.
Multiply row 1 by −2 and add it to row 3.

$$\begin{bmatrix} 2 & 1 & 2 & 5 \\ 0 & -3 & -9 & -15 \\ 0 & -3 & -7 & -9 \end{bmatrix}$$

Multiply row 2 by −1 and add it to row 3.

$$\begin{bmatrix} 2 & 1 & 2 & 5 \\ 0 & -3 & -9 & -15 \\ 0 & 0 & 2 & 6 \end{bmatrix}$$

Putting the variables back in, we have
$2x + y + 2z = 5,$ (1)
$-3y - 9z = -15,$ (2)
$2z = 6$ (3)

Solve (3) for z.
$2z = 6$
$z = 3$

Substitute 3 for z in (2) and solve for y.
$-3y - 9z = -15$
$-3y - 9(3) = -15$
$-3y - 27 = -15$
$-3y = 12$
$y = -4$

Substitute 3 for z and −4 for y in (1) and solve for x.
$2x + y + 2z = 5$
$2x + (-4) + 2(3) = 5$
$2x - 4 + 6 = 5$
$2x = 3$
$x = \frac{3}{2}$

The solution is $\left[\frac{3}{2}, -4, 3\right]$.

8. $\left(2, \frac{1}{2}, -2\right)$

9. $p + q + r = 1,$
   $p + 2q + 3r = 4,$
   $4p + 5q + 6r = 7$

   Write a matrix using only the constants.

   $$\begin{bmatrix} 1 & 1 & 1 & 1 \\ 1 & 2 & 3 & 4 \\ 4 & 5 & 6 & 7 \end{bmatrix}$$

   Multiply row 1 by -1 and add it to row 2.
   Multiply row 1 by -4 and add it to row 3.

   $$\begin{bmatrix} 1 & 1 & 1 & 1 \\ 0 & 1 & 2 & 3 \\ 0 & 1 & 2 & 3 \end{bmatrix}$$

   Multiply row 2 by -1 and add it to row 3.

   $$\begin{bmatrix} 1 & 1 & 1 & 1 \\ 0 & 1 & 2 & 3 \\ 0 & 0 & 0 & 0 \end{bmatrix}$$

   Putting the variables back in, we have
   $p + q + r = 1,$
   $q + 2r = 3,$
   $0 = 0$

   This system is dependent. In this case it has an infinite set of solutions.

   Solve $q + 2r = 3$ for q.
   $q = -2r + 3$

   Substitute $-2r + 3$ for q in the first equation and solve for p.
   $$p + q + r = 1$$
   $$p + (-2r + 3) + r = 1$$
   $$p - r + 3 = 1$$
   $$p - r = -2$$
   $$p = r - 2$$

   The solutions can be described as follows:
   $(r - 2, -2r + 3, r)$

10. No solution

11. $-2w + 2x + 2y - 2z = -10,$
    $w + x + y + z = -5,$
    $3w + x - y + 4z = -2,$
    $w + 3x - 2y + 2z = -6$

    Write a matrix using only the constants.

    $$\begin{bmatrix} -2 & 2 & 2 & -2 & -10 \\ 1 & 1 & 1 & 1 & -5 \\ 3 & 1 & -1 & 4 & -2 \\ 1 & 3 & -2 & 2 & -6 \end{bmatrix}$$

Interchange rows 1 and 2.

$$\begin{bmatrix} 1 & 1 & 1 & 1 & -5 \\ -2 & 2 & 2 & -2 & -10 \\ 3 & 1 & -1 & 4 & -2 \\ 1 & 3 & -2 & 2 & -6 \end{bmatrix}$$

Multiply row 1 by 2 and add it to row 2.
Multiply row 1 by -3 and add it to row 3.
Multiply row 1 by -1 and add it to row 4.

$$\begin{bmatrix} 1 & 1 & 1 & 1 & -5 \\ 0 & 4 & 4 & 0 & -20 \\ 0 & -2 & -4 & 1 & 13 \\ 0 & 2 & -3 & 1 & -1 \end{bmatrix}$$

Interchange rows 2 and 3.

$$\begin{bmatrix} 1 & 1 & 1 & 1 & -5 \\ 0 & -2 & -4 & 1 & 13 \\ 0 & 4 & 4 & 0 & -20 \\ 0 & 2 & -3 & 1 & -1 \end{bmatrix}$$

Multiply row 2 by 2 and add it to row 3.
Add row 2 to row 4.

$$\begin{bmatrix} 1 & 1 & 1 & 1 & -5 \\ 0 & -2 & -4 & 1 & 13 \\ 0 & 0 & -4 & 2 & 6 \\ 0 & 0 & -7 & 2 & 12 \end{bmatrix}$$

Multiply row 4 by 4.

$$\begin{bmatrix} 1 & 1 & 1 & 1 & -5 \\ 0 & -2 & -4 & 1 & 13 \\ 0 & 0 & -4 & 2 & 6 \\ 0 & 0 & -28 & 8 & 48 \end{bmatrix}$$

Multiply row 3 by -7 and add it to row 4.

$$\begin{bmatrix} 1 & 1 & 1 & 1 & -5 \\ 0 & -2 & -4 & 1 & 13 \\ 0 & 0 & -4 & 2 & 6 \\ 0 & 0 & 0 & -6 & 6 \end{bmatrix}$$

Putting the variables back in, we get

$$w + x + y + z = -5, \qquad (1)$$
$$-2x - 4y + z = 13, \qquad (2)$$
$$-4y + 2z = 6, \qquad (3)$$
$$-6z = 6 \qquad (4)$$

Solve (4) for z.

$$-6z = 6$$
$$z = -1$$

Substitute -1 for z in (3) and solve for y.

$$-4y + 2(-1) = 6$$
$$-4y = 8$$
$$y = -2$$

      Substitute -2 for y and -1 for z in (2) and solve for x.

$$-2x - 4(-2) + (-1) = 13$$
$$-2x + 8 - 1 = 13$$
$$-2x = 6$$
$$x = -3$$

      Substitute -3 for x, -2 for y, and -1 for z in (1) and solve for w.

$$w + (-3) + (-2) + (-1) = -5$$
$$w - 6 = -5$$
$$w = 1$$

      The solution is (1, -3, -2, -1).

<u>12.</u> (7, 4, 5, 6)

<u>13.</u> $A + B = \begin{bmatrix} 1 & 2 \\ 4 & 3 \end{bmatrix} + \begin{bmatrix} -3 & 5 \\ 2 & -1 \end{bmatrix} = \begin{bmatrix} 1 + (-3) & 2 + 5 \\ 4 + 2 & 3 + (-1) \end{bmatrix} = \begin{bmatrix} -2 & 7 \\ 6 & 2 \end{bmatrix}$

<u>14.</u> $\begin{bmatrix} -2 & 7 \\ 6 & 2 \end{bmatrix}$

<u>15.</u> $E + 0 = \begin{bmatrix} 1 & 3 \\ 2 & 6 \end{bmatrix} + \begin{bmatrix} 0 & 0 \\ 0 & 0 \end{bmatrix} = \begin{bmatrix} 1 + 0 & 3 + 0 \\ 2 + 0 & 6 + 0 \end{bmatrix} = \begin{bmatrix} 1 & 3 \\ 2 & 6 \end{bmatrix}$

<u>16.</u> $\begin{bmatrix} 2 & 4 \\ 8 & 6 \end{bmatrix}$

<u>17.</u> $3F = 3\begin{bmatrix} 3 & 3 \\ -1 & -1 \end{bmatrix} = \begin{bmatrix} 3 \cdot 3 & 3 \cdot 3 \\ 3 \cdot (-1) & 3 \cdot (-1) \end{bmatrix} = \begin{bmatrix} 9 & 9 \\ -3 & -3 \end{bmatrix}$

<u>18.</u> $\begin{bmatrix} -1 & -1 \\ -1 & -1 \end{bmatrix}$

<u>19.</u> $3F = 3\begin{bmatrix} 3 & 3 \\ -1 & -1 \end{bmatrix} = \begin{bmatrix} 9 & 9 \\ -3 & -3 \end{bmatrix}, \quad 2A = 2\begin{bmatrix} 1 & 2 \\ 4 & 3 \end{bmatrix} = \begin{bmatrix} 2 & 4 \\ 8 & 6 \end{bmatrix}$

      Then
$3F + 2A = \begin{bmatrix} 9 & 9 \\ -3 & -3 \end{bmatrix} + \begin{bmatrix} 2 & 4 \\ 8 & 6 \end{bmatrix} = \begin{bmatrix} 9 + 2 & 9 + 4 \\ -3 + 8 & -3 + 6 \end{bmatrix} = \begin{bmatrix} 11 & 13 \\ 5 & 3 \end{bmatrix}$

<u>20.</u> $\begin{bmatrix} 4 & -3 \\ 2 & 4 \end{bmatrix}$

<u>21.</u> $B - A = \begin{bmatrix} -3 & 5 \\ 2 & -1 \end{bmatrix} - \begin{bmatrix} 1 & 2 \\ 4 & 3 \end{bmatrix} = \begin{bmatrix} -3 & 5 \\ 2 & -1 \end{bmatrix} + \begin{bmatrix} -1 & -2 \\ -4 & -3 \end{bmatrix}$ [B - A = B + (-A)]

      $= \begin{bmatrix} -3 + (-1) & 5 + (-2) \\ 2 + (-4) & -1 + (-3) \end{bmatrix} = \begin{bmatrix} -4 & 3 \\ -2 & -4 \end{bmatrix}$

22. $\begin{bmatrix} 1 & 3 \\ -6 & 17 \end{bmatrix}$

23. $BA = \begin{bmatrix} -3 & 5 \\ 2 & -1 \end{bmatrix}\begin{bmatrix} 1 & 2 \\ 4 & 3 \end{bmatrix} = \begin{bmatrix} -3\cdot1 + 5\cdot4 & -3\cdot2 + 5\cdot3 \\ 2\cdot1 + (-1)4 & 2\cdot2 + (-1)3 \end{bmatrix} = \begin{bmatrix} 17 & 9 \\ -2 & 1 \end{bmatrix}$

24. $\begin{bmatrix} 0 & 0 \\ 0 & 0 \end{bmatrix}$

25. $CD = \begin{bmatrix} 1 & -1 \\ -1 & 1 \end{bmatrix}\begin{bmatrix} 1 & 1 \\ 1 & 1 \end{bmatrix} = \begin{bmatrix} 1\cdot1 + (-1)\cdot1 & 1\cdot1 + (-1)\cdot1 \\ -1\cdot1 + 1\cdot1 & -1\cdot1 + 1\cdot1 \end{bmatrix} = \begin{bmatrix} 0 & 0 \\ 0 & 0 \end{bmatrix}$

26. $\begin{bmatrix} 0 & 0 \\ 0 & 0 \end{bmatrix}$

27. $AI = \begin{bmatrix} 1 & 2 \\ 4 & 3 \end{bmatrix}\begin{bmatrix} 1 & 0 \\ 0 & 1 \end{bmatrix} = \begin{bmatrix} 1\cdot1 + 2\cdot0 & 1\cdot0 + 2\cdot1 \\ 4\cdot1 + 3\cdot0 & 4\cdot0 + 3\cdot1 \end{bmatrix} = \begin{bmatrix} 1 & 2 \\ 4 & 3 \end{bmatrix}$ (Note: $AI = A$)

28. $\begin{bmatrix} 1 & 2 \\ 4 & 3 \end{bmatrix}$

29. $AB = \begin{bmatrix} 1 & 0 & -2 \\ 0 & -1 & 3 \\ 3 & 2 & 4 \end{bmatrix}\begin{bmatrix} -1 & -2 & 5 \\ 1 & 0 & -1 \\ 2 & -3 & 1 \end{bmatrix}$

$= \begin{bmatrix} 1(-1) + 0\cdot1 + (-2)2 & 1(-2) + 0\cdot0 + (-2)(-3) & 1\cdot5 + 0(-1) + (-2)1 \\ 0(-1) + (-1)1 + 3\cdot2 & 0(-2) + (-1)0 + 3(-3) & 0\cdot5 + (-1)(-1) + 3\cdot1 \\ 3(-1) + 2\cdot1 + 4\cdot2 & 3(-2) + 2\cdot0 + 4(-3) & 3\cdot5 + 2(-1) + 4\cdot1 \end{bmatrix}$

$= \begin{bmatrix} -5 & 4 & 3 \\ 5 & -9 & 4 \\ 7 & -18 & 17 \end{bmatrix}$

30. $\begin{bmatrix} 14 & 12 & 16 \\ -2 & -2 & -6 \\ 5 & 5 & -9 \end{bmatrix}$

31. $CI = \begin{bmatrix} -2 & 9 & 6 \\ -3 & 3 & 4 \\ 2 & -2 & 1 \end{bmatrix} \begin{bmatrix} 1 & 0 & 0 \\ 0 & 1 & 0 \\ 0 & 0 & 1 \end{bmatrix}$

$= \begin{bmatrix} -2\cdot1 + 9\cdot0 + 6\cdot0 & -2\cdot0 + 9\cdot1 + 6\cdot0 & -2\cdot0 + 9\cdot0 + 6\cdot1 \\ -3\cdot1 + 3\cdot0 + 4\cdot0 & -3\cdot0 + 3\cdot1 + 4\cdot0 & -3\cdot0 + 3\cdot0 + 4\cdot1 \\ 2\cdot1 + (-2)0 + 1\cdot0 & 2\cdot0 + (-2)1 + 1\cdot0 & 2\cdot0 + (-2)0 + 1\cdot1 \end{bmatrix}$

$= \begin{bmatrix} -2 & 9 & 6 \\ -3 & 3 & 4 \\ 2 & -2 & 1 \end{bmatrix}$ (Note: CI = C)

32. $\begin{bmatrix} -2 & 9 & 6 \\ -3 & 3 & 4 \\ 2 & -2 & 1 \end{bmatrix}$

33. $\begin{bmatrix} -3 & 2 \end{bmatrix} \begin{bmatrix} 4 \\ -2 \end{bmatrix} = \begin{bmatrix} -3\cdot4 + 2(-2) \end{bmatrix} = \begin{bmatrix} -16 \end{bmatrix}$

34. $\begin{bmatrix} -14 \end{bmatrix}$

35. $\begin{bmatrix} -5 & 1 & 2 \end{bmatrix} \begin{bmatrix} 1 & 3 \\ -1 & 0 \\ 4 & -2 \end{bmatrix} = \begin{bmatrix} -5\cdot1 + 1(-1) + 2\cdot4 & -5\cdot3 + 1\cdot0 + 2(-2) \end{bmatrix} = \begin{bmatrix} 2 & -19 \end{bmatrix}$

36. $\begin{bmatrix} -8 \\ 2 \\ -6 \end{bmatrix}$

37. $\begin{bmatrix} 3 & -2 & 4 \\ 2 & 1 & -5 \end{bmatrix} \begin{bmatrix} x \\ y \\ z \end{bmatrix} = \begin{bmatrix} 17 \\ 13 \end{bmatrix}$

38. $\begin{bmatrix} 3 & 2 & 5 \\ 4 & -3 & 2 \end{bmatrix} \begin{bmatrix} x \\ y \\ z \end{bmatrix} = \begin{bmatrix} 9 \\ 10 \end{bmatrix}$

39. $\begin{bmatrix} 1 & -1 & 2 & -4 \\ 2 & -1 & -1 & 1 \\ 1 & 4 & -3 & -1 \\ 3 & 5 & -7 & 2 \end{bmatrix} \begin{bmatrix} x \\ y \\ z \\ w \end{bmatrix} = \begin{bmatrix} 12 \\ 0 \\ 1 \\ 9 \end{bmatrix}$

$\underline{40.}$ $\begin{bmatrix} 2 & 4 & -5 & 12 \\ 4 & -1 & 12 & -1 \\ -1 & 4 & 0 & 2 \\ 2 & 10 & 1 & 0 \end{bmatrix} \begin{bmatrix} x \\ y \\ z \\ w \end{bmatrix} = \begin{bmatrix} 2 \\ 5 \\ 13 \\ 5 \end{bmatrix}$

$\underline{41.}$ $A = \begin{bmatrix} -1 & 0 \\ 2 & 1 \end{bmatrix}$, $B = \begin{bmatrix} 1 & -1 \\ 0 & 2 \end{bmatrix}$

$(A + B)(A - B) = \begin{bmatrix} 0 & -1 \\ 2 & 3 \end{bmatrix} \begin{bmatrix} -2 & 1 \\ 2 & -1 \end{bmatrix} = \begin{bmatrix} -2 & 1 \\ 2 & -1 \end{bmatrix}$

$A^2 - B^2 = \begin{bmatrix} -1 & 0 \\ 2 & 1 \end{bmatrix} \begin{bmatrix} -1 & 0 \\ 2 & 1 \end{bmatrix} - \begin{bmatrix} 1 & -1 \\ 0 & 2 \end{bmatrix} \begin{bmatrix} 1 & -1 \\ 0 & 2 \end{bmatrix}$

$= \begin{bmatrix} 1 & 0 \\ 0 & 1 \end{bmatrix} - \begin{bmatrix} 1 & -3 \\ 0 & 4 \end{bmatrix} = \begin{bmatrix} 0 & 3 \\ 0 & -3 \end{bmatrix}$

Thus, $(A + B)(A - B) \neq A^2 - B^2$.

$\underline{42.}$ $(A + B)(A + B) = \begin{bmatrix} -2 & -3 \\ 6 & 7 \end{bmatrix}$, $A^2 + 2AB + B^2 = \begin{bmatrix} 0 & -1 \\ 4 & 5 \end{bmatrix}$

Thus, $(A + B)(A + B) \neq A^2 + 2AB + B^2$.

$\underline{43.}$ $(A + B)(A - B) = \begin{bmatrix} -2 & 1 \\ 2 & -1 \end{bmatrix}$ (See Exercise 41.)

$A^2 = \begin{bmatrix} 1 & 0 \\ 0 & 1 \end{bmatrix}$ (See Exercise 41.)

$BA = \begin{bmatrix} 1 & -1 \\ 0 & 2 \end{bmatrix} \begin{bmatrix} -1 & 0 \\ 2 & 1 \end{bmatrix} = \begin{bmatrix} -3 & -1 \\ 4 & 2 \end{bmatrix}$

$AB = \begin{bmatrix} -1 & 0 \\ 2 & 1 \end{bmatrix} \begin{bmatrix} 1 & -1 \\ 0 & 2 \end{bmatrix} = \begin{bmatrix} -1 & 1 \\ 2 & 0 \end{bmatrix}$

$B^2 = \begin{bmatrix} 1 & -3 \\ 0 & 4 \end{bmatrix}$ (See Exercise 41.)

$A^2 + BA - AB - B^2 = \begin{bmatrix} 1 & 0 \\ 0 & 1 \end{bmatrix} + \begin{bmatrix} -3 & -1 \\ 4 & 2 \end{bmatrix} - \begin{bmatrix} -1 & 1 \\ 2 & 0 \end{bmatrix} - \begin{bmatrix} 1 & -3 \\ 0 & 4 \end{bmatrix}$

$= \begin{bmatrix} -2 & 1 \\ 2 & -1 \end{bmatrix}$

Therefore, $(A + B)(A - B) = A^2 + BA - AB - B^2$.

44. $(A + B)(A + B) = \begin{bmatrix} -2 & -3 \\ 6 & 7 \end{bmatrix}$ (See Exercise 42.)

$A^2 + BA + AB + B^2$

$= \begin{bmatrix} 1 & 0 \\ 0 & 1 \end{bmatrix} + \begin{bmatrix} -3 & -1 \\ 4 & 2 \end{bmatrix} + \begin{bmatrix} -1 & 1 \\ 2 & 0 \end{bmatrix} + \begin{bmatrix} 1 & -3 \\ 0 & 4 \end{bmatrix}$ (See Exercise 43.)

$= \begin{bmatrix} -2 & -3 \\ 6 & 7 \end{bmatrix}$

Therefore, $(A + B)(A + B) = A^2 + BA + AB + B^2$.

45. $A + B = \begin{bmatrix} a_{11} + b_{11} & a_{12} + b_{12} \\ a_{21} + b_{21} & a_{22} + b_{22} \end{bmatrix}$

$= \begin{bmatrix} b_{11} + a_{11} & b_{12} + a_{12} \\ b_{21} + a_{21} & b_{22} + a_{22} \end{bmatrix}$ (By commutativity of addition of real numbers)

$= B + A$

46. $A + (B + C) = \begin{bmatrix} a_{11} + (b_{11} + c_{11}) & a_{12} + (b_{12} + c_{12}) \\ a_{21} + (b_{21} + c_{21}) & a_{22} + (b_{22} + c_{22}) \end{bmatrix}$

$= \begin{bmatrix} (a_{11} + b_{11}) + c_{11} & (a_{12} + b_{12}) + c_{12} \\ (a_{21} + b_{21}) + c_{21} & (a_{22} + b_{22}) + c_{22} \end{bmatrix}$ (By associativity of addition of real numbers)

$= (A + B) + C$

47. See the answer section in the text.

48. $k(A + B) = \begin{bmatrix} k(a_{11} + b_{11}) & k(a_{12} + b_{12}) \\ k(a_{21} + b_{21}) & k(a_{22} + b_{22}) \end{bmatrix}$

$= \begin{bmatrix} ka_{11} + kb_{11} & ka_{12} + kb_{12} \\ ka_{21} + kb_{21} & ka_{22} + kb_{22} \end{bmatrix}$ (By the distributive law of real numbers)

$= \begin{bmatrix} ka_{11} & ka_{12} \\ ka_{21} & ka_{22} \end{bmatrix} + \begin{bmatrix} kb_{11} & kb_{12} \\ kb_{21} & kb_{22} \end{bmatrix}$

$= kA + kB$

49. $AI = \begin{bmatrix} a_{11} & a_{12} \\ a_{21} & a_{22} \end{bmatrix} \begin{bmatrix} 1 & 0 \\ 0 & 1 \end{bmatrix} = \begin{bmatrix} a_{11} \cdot 1 + a_{12} \cdot 0 & a_{11} \cdot 0 + a_{12} \cdot 1 \\ a_{21} \cdot 1 + a_{22} \cdot 0 & a_{21} \cdot 0 + a_{22} \cdot 1 \end{bmatrix}$

$= \begin{bmatrix} a_{11} & a_{12} \\ a_{21} & a_{22} \end{bmatrix} = A$

$IA = \begin{bmatrix} 1 & 0 \\ 0 & 1 \end{bmatrix} \begin{bmatrix} a_{11} & a_{12} \\ a_{21} & a_{22} \end{bmatrix} = \begin{bmatrix} 1 \cdot a_{11} + 0 \cdot a_{21} & 1 \cdot a_{12} + 0 \cdot a_{22} \\ 0 \cdot a_{11} + 1 \cdot a_{21} & 0 \cdot a_{12} + 1 \cdot a_{22} \end{bmatrix}$

$= \begin{bmatrix} a_{11} & a_{12} \\ a_{21} & a_{22} \end{bmatrix} = A$

Thus, $AI = IA = A$.

<u>50</u>. Find BC. Then A(BC). Find AB. Then (AB)C. Then compare A(BC) and (AB)C.

Exercise Set 6.5

<u>1</u>. $\begin{vmatrix} -2 & -\sqrt{5} \\ -\sqrt{5} & 3 \end{vmatrix} = -2 \cdot 3 - (-\sqrt{5})(-\sqrt{5}) = -6 - 5 = -11$

<u>2</u>. $2\sqrt{5} + 12$

<u>3</u>. $\begin{vmatrix} x & 4 \\ x & x^2 \end{vmatrix} = x \cdot x^2 - x \cdot 4 = x^3 - 4x$

<u>4</u>. $3y^2 + 2y$

<u>5</u>. $\begin{vmatrix} 3 & 1 & 2 \\ -2 & 3 & 1 \\ 3 & 4 & -6 \end{vmatrix} = 3\begin{vmatrix} 3 & 1 \\ 4 & -6 \end{vmatrix} - (-2)\begin{vmatrix} 1 & 2 \\ 4 & -6 \end{vmatrix} + 3\begin{vmatrix} 1 & 2 \\ 3 & 1 \end{vmatrix}$

$= 3(-22) + 2(-14) + 3(-5) = -66 - 28 - 15 = -109$

<u>6</u>. -9

<u>7</u>. $\begin{vmatrix} x & 0 & -1 \\ 2 & x & x^2 \\ -3 & x & 1 \end{vmatrix} = x\begin{vmatrix} x & x^2 \\ x & 1 \end{vmatrix} - 2\begin{vmatrix} 0 & -1 \\ x & 1 \end{vmatrix} + (-3)\begin{vmatrix} 0 & -1 \\ x & x^2 \end{vmatrix}$

$= x(x - x^3) - 2(x) - 3(x) = x^2 - x^4 - 2x - 3x$

$= -x^4 + x^2 - 5x$

<u>8</u>. $-2x^3$

<u>9</u>. a_{11} is the element in the 1st row and 1st column. $a_{11} = 7$
a_{32} is the element in the 3rd row and 2nd column. $a_{32} = 2$
a_{22} is the element in the 2nd row and 2nd column. $a_{22} = 0$

<u>10</u>. $a_{13} = -6$, $a_{31} = 1$, $a_{23} = -3$

<u>11</u>. To find M_{11} we delete the 1st row and the 1st column.

$$A = \begin{bmatrix} 7 & -4 & -6 \\ 2 & 0 & -3 \\ 1 & 2 & -5 \end{bmatrix}$$

We calculate the determinant of the matrix formed by the remaining elements.

$M_{11} = \begin{vmatrix} 0 & -3 \\ 2 & -5 \end{vmatrix} = 0(-5) - 2(-3) = 0 + 6 = 6$

To find M_{32} we delete the 3rd row and the 2nd column.

$$A = \begin{bmatrix} 7 & -4 & -6 \\ 2 & 0 & -3 \\ 1 & 2 & -5 \end{bmatrix}$$

We calculate the determinant of the matrix formed by the remaining elements.

$$M_{32} = \begin{vmatrix} 7 & -6 \\ 2 & -3 \end{vmatrix} = 7(-3) - 2(-6) = -21 + 12 = -9$$

To find M_{22} we delete the 2nd row and the 2nd column.

$$A = \begin{bmatrix} 7 & -4 & -6 \\ 2 & 0 & -3 \\ 1 & 2 & -5 \end{bmatrix}$$

We calculate the determinant of the matrix formed by the remaining elements.

$$M_{22} = \begin{vmatrix} 7 & -6 \\ 1 & -5 \end{vmatrix} = 7(-5) - 1(-6) = -35 + 6 = -29.$$

<u>12.</u> $M_{13} = 4$, $M_{31} = 12$, $M_{23} = 18$

<u>13.</u> From Exercise 11, we know that $M_{11} = 6$, $M_{32} = -9$, and
$M_{22} = -29$. We use this information to find the desired cofactors.

$A_{11} = (-1)^{1+1} M_{11} = (-1)^2 \cdot 6 = 1 \cdot 6 = 6$

$A_{32} = (-1)^{3+2} M_{32} = (-1)^5 \cdot (-9) = -1 \cdot (-9) = 9$

$A_{22} = (-1)^{2+2} M_{22} = (-1)^4 \cdot (-29) = 1 \cdot 29 \; -29$

<u>14.</u> $A_{13} = 4$, $A_{31} = 12$, $A_{23} = -18$

<u>15.</u>

$$A = \begin{bmatrix} 7 & -4 & -6 \\ 2 & 0 & -3 \\ 1 & 2 & -5 \end{bmatrix}$$

Find $|A|$ by expanding about the second row.

$|A| = (2)(-1)^{2+1} \cdot \begin{vmatrix} -4 & -6 \\ 2 & -5 \end{vmatrix} + (0)(-1)^{2+2} \cdot \begin{vmatrix} 7 & -6 \\ 1 & -5 \end{vmatrix} + (-3)(-1)^{2+3} \cdot \begin{vmatrix} 7 & -4 \\ 1 & 2 \end{vmatrix}$

$= 2(-1)(32) + 0 + (-3)(-1)(18) = -64 + 54 = -10$

<u>16.</u> -10

<u>17.</u>

$$A = \begin{bmatrix} 7 & -4 & -6 \\ 2 & 0 & -3 \\ 1 & 2 & -5 \end{bmatrix}$$

Find $|A|$ by expanding about the third column.

$|A| = (-6)(-1)^{1+3} \cdot \begin{vmatrix} 2 & 0 \\ 1 & 2 \end{vmatrix} + (-3)(-1)^{2+3} \cdot \begin{vmatrix} 7 & -4 \\ 1 & 2 \end{vmatrix} + (-5)(-1)^{3+3} \cdot \begin{vmatrix} 7 & -4 \\ 2 & 0 \end{vmatrix}$

$= -6(1)(4) + (-3)(-1)(18) + (-5)(1)(8) = -24 + 54 - 40 = -10$

18. -10

19. To find M_{41} we delete the 4th row and the 1st column.

$$A = \begin{bmatrix} 1 & 0 & 0 & -2 \\ 4 & 1 & 0 & 0 \\ 5 & 6 & 7 & 8 \\ -2 & -3 & -1 & 0 \end{bmatrix}$$

We calculate the determinant of the matrix formed by the remaining elements.

$$M_{41} = \begin{vmatrix} 0 & 0 & -2 \\ 1 & 0 & 0 \\ 6 & 7 & 8 \end{vmatrix} = 0 + 0 + (-2)(-1)^{1+3} \cdot \begin{vmatrix} 1 & 0 \\ 6 & 7 \end{vmatrix} = -2 \cdot 1 \cdot 7 = -14$$

To find M_{33} we delete the 3rd row and the 3rd column.

$$A = \begin{bmatrix} 1 & 0 & 0 & -2 \\ 4 & 1 & 0 & 0 \\ 5 & 6 & 7 & 8 \\ -2 & -3 & -2 & 0 \end{bmatrix}$$

We calculate the determinant of the matrix formed by the remaining elements.

$$M_{33} = \begin{vmatrix} 1 & 0 & -2 \\ 4 & 1 & 0 \\ -2 & -3 & 0 \end{vmatrix} = -2(-1)^{1+3} \cdot \begin{vmatrix} 4 & 1 \\ -2 & -3 \end{vmatrix} + 0 + 0 = -2 \cdot 1 \cdot (-10) = 20$$

20. $M_{12} = 32$, $M_{44} = 7$

21.

$$A = \begin{bmatrix} 1 & 0 & 0 & -2 \\ 4 & 1 & 0 & 0 \\ 5 & 6 & 7 & 8 \\ -2 & -3 & -1 & 0 \end{bmatrix} \qquad (a_{24} \text{ is } 0; \quad a_{43} \text{ is } -1)$$

A_{24} is the cofactor of the element a_{24}.

$A_{24} = (-1)^{2+4} M_{24}$ (M_{24} is the minor of a_{24})

$$= (-1)^6 \begin{vmatrix} 1 & 0 & 0 \\ 5 & 6 & 7 \\ -2 & -3 & -1 \end{vmatrix} \qquad \text{Think:} \quad A = \begin{bmatrix} 1 & 0 & 0 & -2 \\ 4 & 1 & 0 & 0 \\ 5 & 6 & 7 & 8 \\ -2 & -3 & -1 & 0 \end{bmatrix}$$

$$= (1)\left[1(-1)^{1+1} \cdot \begin{vmatrix} 6 & 7 \\ -3 & -1 \end{vmatrix} + 0(-1)^{1+2} \cdot \begin{vmatrix} 5 & 7 \\ -2 & -1 \end{vmatrix} + 0(-1)^{1+3} \cdot \begin{vmatrix} 5 & 6 \\ -2 & -3 \end{vmatrix} \right]$$

(Here we expanded about the first row.)

$$= 1(1)[6(-1) - (-3)7] + 0 + 0 = -6 + 21 = 15$$

A_{43} is the cofactor of the element a_{43}.

$A_{43} = (-1)^{4+3} M_{43}$ (M_{43} is the minor of a_{43})

$$= (-1)^7 \begin{vmatrix} 1 & 0 & -2 \\ 4 & 1 & 0 \\ 5 & 6 & 8 \end{vmatrix} \qquad \text{Think:} \quad A = \begin{bmatrix} 1 & 0 & 0 & -2 \\ 4 & 1 & 0 & 0 \\ 5 & 6 & 7 & 8 \\ -2 & -3 & -1 & 0 \end{bmatrix}$$

$$= (-1)\left[4(-1)^{2+1} \cdot \begin{vmatrix} 0 & -2 \\ 6 & 8 \end{vmatrix} + 1(-1)^{2+2} \cdot \begin{vmatrix} 1 & -2 \\ 5 & 8 \end{vmatrix} + 0(-1)^{2+3} \cdot \begin{vmatrix} 1 & 0 \\ 5 & 6 \end{vmatrix} \right]$$

(Here we expanded about the second row.)

$$= (-1)[4(-1)(0 \cdot 8 - 6(-2)) + 1(1 \cdot 8 - 5(-2)) + 0]$$
$$= (-1)[-4(12) + 18] = (-1)(-30) = 30$$

<u>22.</u> $A_{22} = -10$, $A_{34} = 1$

<u>23.</u>

$$A = \begin{bmatrix} \boxed{1 \quad 0 \quad 0 \quad -2} \\ 4 & 1 & 0 & 0 \\ 5 & 6 & 7 & 8 \\ -2 & -3 & -1 & 0 \end{bmatrix}$$

Find |A| by expanding about the first row.

$$|A| = (1)(-1)^{1+1} \cdot \begin{vmatrix} \boxed{1 \quad 0 \quad 0} \\ 6 & 7 & 8 \\ -3 & -1 & 0 \end{vmatrix} + 0 + 0 + (-2)(-1)^{1+4} \cdot \begin{vmatrix} \boxed{4 \quad 1 \quad 0} \\ 5 & 6 & 7 \\ -2 & -3 & -1 \end{vmatrix}$$

$$= 1 \cdot \left[1(-1)^{1+1} \cdot \begin{vmatrix} 7 & 8 \\ -1 & 0 \end{vmatrix} + 0 + 0 \right] + 2 \cdot \left[4(-1)^{1+1} \cdot \begin{vmatrix} 6 & 7 \\ -3 & -1 \end{vmatrix} \right] +$$

$$1(-1)^{1+2} \cdot \begin{vmatrix} 5 & 7 \\ -2 & -1 \end{vmatrix} + 0]$$

$$= 1[1(0 - (-8))] + 2[4(-6 - (-21)) + (-1)(-5 - (-14))]$$
$$= 1 \cdot 8 + 2 \cdot (60 - 9) = 8 + 102 = 110$$

<u>24.</u> 110

<u>25.</u>

$$\begin{vmatrix} \boxed{5} & -4 & 2 & -2 \\ \boxed{3} & -3 & -4 & 7 \\ \boxed{-2} & 3 & 2 & 4 \\ \boxed{-8} & 9 & 5 & -5 \end{vmatrix} \qquad \text{(Here we will expand about the first column.)}$$

$$= 5(-1)^{1+1} \cdot \begin{vmatrix} -3 & -4 & 7 \\ 3 & 2 & 4 \\ 9 & 5 & -5 \end{vmatrix} + 3(-1)^{2+1} \cdot \begin{vmatrix} -4 & 2 & -2 \\ 3 & 2 & 4 \\ 9 & 5 & -5 \end{vmatrix} +$$

$$(-2)(-1)^{3+1} \cdot \begin{vmatrix} -4 & 2 & -2 \\ -3 & -4 & 7 \\ 9 & 5 & -5 \end{vmatrix} + (-8)(-1)^{4+1} \cdot \begin{vmatrix} -4 & 2 & -2 \\ -3 & -4 & 7 \\ 3 & 2 & 4 \end{vmatrix}$$

$$= 5\left[-3(-1)^{1+1} \cdot \begin{vmatrix} 2 & 4 \\ 5 & -5 \end{vmatrix} + 3(-1)^{2+1} \cdot \begin{vmatrix} -4 & 7 \\ 5 & -5 \end{vmatrix} + 9(-1)^{3+1} \cdot \begin{vmatrix} -4 & 7 \\ 2 & 4 \end{vmatrix}\right] +$$

$$(-3)\left[-4(-1)^{1+1} \cdot \begin{vmatrix} 2 & 4 \\ 5 & -5 \end{vmatrix} + 3(-1)^{2+1} \cdot \begin{vmatrix} 2 & -2 \\ 5 & -5 \end{vmatrix} + 9(-1)^{3+1} \cdot \begin{vmatrix} 2 & -2 \\ 2 & 4 \end{vmatrix}\right] +$$

$$(-2)\left[-4(-1)^{1+1} \cdot \begin{vmatrix} -4 & 7 \\ 5 & -5 \end{vmatrix} + (-3)(-1)^{2+1} \cdot \begin{vmatrix} 2 & -2 \\ 5 & -5 \end{vmatrix} + 9(-1)^{3+1} \cdot \begin{vmatrix} 2 & -2 \\ -4 & 7 \end{vmatrix}\right] +$$

$$8\left[-4(-1)^{1+1} \cdot \begin{vmatrix} -4 & 7 \\ 2 & 4 \end{vmatrix} + (-3)(-1)^{2+1} \cdot \begin{vmatrix} 2 & -2 \\ 2 & 4 \end{vmatrix} + 3(-1)^{3+1} \cdot \begin{vmatrix} 2 & -2 \\ -4 & 7 \end{vmatrix}\right]$$

$= 5[-3(-30) + (-3)(-15) + 9(-30)] + (-3)[-4(-30) + (-3)(0) + 9(12)] +$
$(-2)[-4(-15) + 3(0) + 9(6)] + 8[-4(-30) + 3(12) + 3(6)]$

$= 5(90 + 45 - 270) + (-3)(120 + 0 + 108) + (-2)(60 + 0 + 54) +$
$8(120 + 36 + 18)$

$= 5(-135) + (-3)(228) + (-2)(114) + 8(174) = -675 - 684 - 228 + 1392 = -195$

26. xyzw

27. $\begin{vmatrix} -4 & 5 \\ 6 & 10 \end{vmatrix}$

$= 5\begin{vmatrix} -4 & 1 \\ 6 & 2 \end{vmatrix}$ (Theorem 7, factoring 5 out of the second column)

$= 5 \cdot 2\begin{vmatrix} -4 & 1 \\ 3 & 1 \end{vmatrix}$ (Theorem 7, factoring 2 out of the second row)

$= 5 \cdot 2\begin{vmatrix} -4 & 1 \\ 7 & 0 \end{vmatrix}$ (Theorem 8, multiplying each element of the first row by -1 and adding the products to the corresponding elements of the second row)

$= 5 \cdot 2(0 - 7)$

$= 10(-7) = -70$

28. -6

29. $\begin{vmatrix} 2 & 1 & 1 \\ 2 & -3 & -1 \\ -4 & 5 & 2 \end{vmatrix}$

$= \begin{vmatrix} 2 & 1 & 1 \\ 4 & -2 & 0 \\ -8 & 3 & 0 \end{vmatrix}$ (Theorem 8, adding the first row to the second row; also multiplying each element of the first row by -2 and adding the products to the corresponding elements of the third row)

$= 1(-1)^{1+3}\begin{vmatrix} 4 & -2 \\ -8 & 3 \end{vmatrix} + 0 + 0$ (Expanding the determinant about the third column)

$= 1(12 - 16) = -4$

30. -9

31.
$$\begin{vmatrix} 11 & -15 & 20 \\ 16 & 24 & -8 \\ 6 & 9 & 15 \end{vmatrix}$$

$$= 8 \cdot \begin{vmatrix} 11 & -15 & 20 \\ 2 & 3 & -1 \\ 6 & 9 & 15 \end{vmatrix} = 8 \cdot 3 \begin{vmatrix} 11 & -15 & 20 \\ 2 & 3 & -1 \\ 2 & 3 & 5 \end{vmatrix}$$ (Theorem 7, factoring 8 out of the second row and 3 out of the third row)

$$= 8 \cdot 3 \cdot 3 \begin{vmatrix} 11 & -5 & 20 \\ 2 & 1 & -1 \\ 2 & 1 & 5 \end{vmatrix}$$ (Theorem 7, factoring 3 out of the second column)

$$= 8 \cdot 3 \cdot 3 \begin{vmatrix} 11 & -5 & 20 \\ 2 & 1 & -1 \\ 0 & 0 & 6 \end{vmatrix}$$ (Theorem 8, multiplying each element of the second row by -1 and adding the products to the corresponding elements of the third row)

$$= 8 \cdot 3 \cdot 3 \cdot 6 \begin{vmatrix} 11 & -5 & 20 \\ 2 & 1 & -1 \\ 0 & 0 & 1 \end{vmatrix}$$ (Theorem 7 factoring 6 out of the third row)

$$= 8 \cdot 3 \cdot 3 \cdot 6 \left[0 + 0 + 1(-1)^{3+3} \begin{vmatrix} 11 & -5 \\ 2 & 1 \end{vmatrix} \right]$$ (Expanding the determinant about the third row)

$$= 8 \cdot 3 \cdot 3 \cdot 6 \cdot 21 = 9072$$

32. -546

33.
$$\begin{vmatrix} -3 & 0 & 2 & 6 \\ 2 & 4 & 0 & -1 \\ -1 & 0 & -5 & 2 \\ 0 & -1 & -2 & -3 \end{vmatrix}$$

$$= \begin{vmatrix} 0 & 0 & 17 & 0 \\ 2 & 4 & 0 & -1 \\ -1 & 0 & -5 & 2 \\ 0 & -1 & -2 & -3 \end{vmatrix}$$ (Theorem 8, multiplying each element of the third row by -3 and adding the products to the corresponding elements of the first row)

$$= 17 \begin{vmatrix} 0 & 0 & 1 & 0 \\ 2 & 4 & 0 & -1 \\ -1 & 0 & -5 & 2 \\ 0 & -1 & -2 & -3 \end{vmatrix}$$ (Theorem 7, factoring 17 out of the first row)

$$= 17 \left[0 + 0 + 1(-1)^{1+3} \begin{vmatrix} 2 & 4 & -1 \\ -1 & 0 & 2 \\ 0 & -1 & -3 \end{vmatrix} + 0 \right]$$ (Expanding the determinant about the first row)

$$= -17 \begin{vmatrix} 2 & 4 & -1 \\ 1 & 0 & -2 \\ 0 & -1 & -3 \end{vmatrix}$$

(Theorem 7, factoring -1 out of the second row)

$$= -17 \begin{vmatrix} 0 & 4 & 3 \\ 1 & 0 & -2 \\ 0 & -1 & -3 \end{vmatrix}$$

(Theorem 8, multiplying each element of the second row by -2 and adding the products to the corresponding elements in the first row)

$$= -17 \left[0 + 1(-1)^{2+1} \begin{vmatrix} 4 & 3 \\ -1 & -3 \end{vmatrix} + 0 \right]$$

(Expanding the determinant about the first column)

$$= -17 \cdot (-1) \cdot (-12 + 3) = 17(-9) = -153$$

34. 746

35.
$$\begin{vmatrix} x & y & z \\ 0 & 0 & 0 \\ p & q & r \end{vmatrix} = 0$$

(Theorem 4)

36. 0

37.
$$\begin{vmatrix} 2a & t & -7a \\ 2b & u & -7b \\ 2c & v & -7c \end{vmatrix}$$

$$= 2(-7) \begin{vmatrix} a & t & a \\ b & u & b \\ c & v & c \end{vmatrix}$$

(Theorem 7, factoring 2 out of the first column and -7 out of the third column)

$$= 2 \cdot (-7) \cdot 0 = 0$$

(Theorem 6; columns 1 and 3 are the same)

38. 0

39.
$$\begin{vmatrix} x^2 & x & 1 \\ y^2 & y & 1 \\ z^2 & z & 1 \end{vmatrix}$$

$$= \begin{vmatrix} x^2 - y^2 & x - y & 0 \\ y^2 & y & 1 \\ z^2 - y^2 & z - y & 0 \end{vmatrix}$$

(Theorem 8, multiplying each element of the second row by -1 and adding the products to the corresponding elements of the first and third rows)

$$= (x - y)(z - y) \begin{vmatrix} x + y & 1 & 0 \\ y^2 & y & 1 \\ z + y & 1 & 0 \end{vmatrix}$$

(Theorem 7, factoring x - y out of the first row and z - y out of the third row)

$$= (x - y)(z - y) \begin{vmatrix} x - z & 0 & 0 \\ y^2 & y & 1 \\ z + y & 1 & 0 \end{vmatrix}$$

(Theorem 8, multiplying each element of the third row by -1 and adding the products to the corresponding elements of the first row)

$= (x - y)(z - y)(x - z)(-1)$ (Expanding the determinant about the first row)

$= (x - y)(z - y)(z - x)$, or $(x - y)(y - z)(x - z)$, or $(y - x)(z - y)(x - z)$

40. $(c - a)(b - c)(a - b)$, or $(a - c)(c - b)(a - b)$, or
     $(a - c)(b - c)(b - a)$

41. $\begin{vmatrix} x & x^2 & x^3 \\ y & y^2 & y^3 \\ z & z^2 & z^3 \end{vmatrix}$

$= xyz \cdot \begin{vmatrix} 1 & x & x^2 \\ 1 & y & y^2 \\ 1 & z & z^2 \end{vmatrix}$ (Theorem 7, factoring x out of the first row, y out of the second row, and z out of the third row)

$= xyz \cdot \begin{vmatrix} 0 & x - z & x^2 - z^2 \\ 0 & y - z & y^2 - z^2 \\ 1 & z & z^2 \end{vmatrix}$ (Theorem 8, multiplying each element of the third row by -1 and adding the products to the corresponding elements of of the first and second rows)

$= xyz(x - z)(y - z) \cdot \begin{vmatrix} 0 & 1 & x + z \\ 0 & 1 & y + z \\ 1 & z & z^2 \end{vmatrix}$ (Theorem 7, factoring x - z out of the first row and y - z out of the second row)

$= xyz(x - z)(y - z) \cdot \begin{vmatrix} 0 & 0 & x - y \\ 0 & 1 & y + z \\ 1 & z & z^2 \end{vmatrix}$ (Theorem 8, multiplying each element of the second row by -1 and adding the products to the corresponding elements of the first row)

$= xyz(x - z)(y - z)(x - y) \cdot \begin{vmatrix} 0 & 0 & 1 \\ 0 & 1 & y + z \\ 1 & z & z^2 \end{vmatrix}$ (Theorem 7, factoring x - y out of the first row)

$= xyz(x - z)(y - z)(x - y) \cdot \left[0 + 0 + 1(-1)^{1+3} \begin{vmatrix} 0 & 1 \\ 1 & z \end{vmatrix} \right]$ (Expanding the determinant about the first row)

$= xyz(x - z)(y - z)(x - y)(-1)$ $\left[\begin{vmatrix} 0 & 1 \\ 1 & z \end{vmatrix} = 0 - 1 = -1 \right]$

$= -xyz(x - z)(y - z)(x - y)$, or $xyz(z - x)(y - z)(x - y)$,
  or $xyz(x - z)(z - y)(x - y)$, or $xyz(x - z)(y - z)(y - x)$

42. $(c - a)(b - c)(a - b)(a + b + c)$, or $(a - c)(c - b)(a - b)(a + b + c)$, or
     $(a - c)(b - c)(b - a)(a + b + c)$

43. $-2x + 4y = 3$
     $3x - 7y = 1$

$$x = \frac{\begin{vmatrix} 3 & 4 \\ 1 & -7 \end{vmatrix}}{\begin{vmatrix} -2 & 4 \\ 3 & -7 \end{vmatrix}} = \frac{3(-7) - 1(4)}{-2(-7) - 3(4)} = \frac{-25}{2}$$

$$y = \frac{\begin{vmatrix} -2 & 3 \\ 3 & 1 \end{vmatrix}}{\begin{vmatrix} -2 & 4 \\ 3 & -7 \end{vmatrix}} = \frac{-2(1) - 3(3)}{2} = \frac{-11}{2}$$

The solution is. $\left(-\frac{25}{2}, -\frac{11}{2}\right)$.

44. $\left(\frac{9}{19}, \frac{51}{38}\right)$

45. $\sqrt{3}x + \pi y = -5$
     $\pi x - \sqrt{3}y = 4$

$$x = \frac{\begin{vmatrix} -5 & \pi \\ 4 & -\sqrt{3} \end{vmatrix}}{\begin{vmatrix} \sqrt{3} & \pi \\ \pi & -\sqrt{3} \end{vmatrix}} = \frac{5\sqrt{3} - 4\pi}{-3 - \pi^2} = \frac{-5\sqrt{3} + 4\pi}{3 + \pi^2}$$

$$y = \frac{\begin{vmatrix} \sqrt{3} & -5 \\ \pi & 4 \end{vmatrix}}{\begin{vmatrix} \sqrt{3} & \pi \\ \pi & -\sqrt{3} \end{vmatrix}} = \frac{4\sqrt{3} + 5\pi}{-3 - \pi^2} = \frac{-4\sqrt{3} - 5\pi}{3 + \pi^2}$$

The solution is $\left(\frac{-5\sqrt{3} + 4\pi}{3 + \pi^2}, \frac{-4\sqrt{3} - 5\pi}{3 + \pi^2}\right)$, or

$\left(\frac{-5\sqrt{3} + 4\pi}{3 + \pi^2}, \frac{4\sqrt{3} + 5\pi}{-3 - \pi^2}\right)$.

46. $\left(\frac{2\pi - 3\sqrt{5}}{\pi^2 + 5}, \frac{-3\pi - 2\sqrt{5}}{\pi^2 + 5}\right)$

47. $3x + 2y - z = 4$
     $3x - 2y + z = 5$
     $4x - 5y - z = -1$

$$x = \frac{\begin{vmatrix} 4 & 2 & -1 \\ 5 & -2 & 1 \\ -1 & -5 & -1 \end{vmatrix}}{\begin{vmatrix} 3 & 2 & -1 \\ 3 & -2 & 1 \\ 4 & -5 & -1 \end{vmatrix}}$$

$$= \frac{4\begin{vmatrix} -2 & 1 \\ -5 & -1 \end{vmatrix} - 5\begin{vmatrix} 2 & -1 \\ -5 & -1 \end{vmatrix} + (-1)\begin{vmatrix} 2 & -1 \\ -2 & 1 \end{vmatrix}}{3\begin{vmatrix} -2 & 1 \\ -5 & -1 \end{vmatrix} - 3\begin{vmatrix} 2 & -1 \\ -5 & -1 \end{vmatrix} + 4\begin{vmatrix} 2 & -1 \\ -2 & 1 \end{vmatrix}}$$

$$= \frac{4(7) - 5(-7) - 1(0)}{3(7) - 3(-7) + 4(0)}$$

$$= \frac{28 + 35}{21 + 21} = \frac{63}{42} = \frac{3}{2}$$

$$y = \frac{\begin{vmatrix} 3 & 4 & -1 \\ 3 & 5 & 1 \\ 4 & -1 & -1 \end{vmatrix}}{42}$$

$$= \frac{3\begin{vmatrix} 5 & 1 \\ -1 & -1 \end{vmatrix} - 3\begin{vmatrix} 4 & -1 \\ -1 & -1 \end{vmatrix} + 4\begin{vmatrix} 4 & -1 \\ 5 & 1 \end{vmatrix}}{42}$$

$$= \frac{3(-4) - 3(-5) + 4(9)}{42}$$

$$= \frac{-12 + 15 + 36}{42} = \frac{39}{42} = \frac{13}{14}$$

$$z = \frac{\begin{vmatrix} 3 & 2 & 4 \\ 3 & -2 & 5 \\ 4 & -5 & -1 \end{vmatrix}}{42}$$

$$= \frac{3\begin{vmatrix} -2 & 5 \\ -5 & -1 \end{vmatrix} - 3\begin{vmatrix} 2 & 4 \\ -5 & -1 \end{vmatrix} + 4\begin{vmatrix} 2 & 4 \\ -2 & 5 \end{vmatrix}}{42}$$

$$= \frac{3(27) - 3(18) + 4(18)}{42}$$

$$= \frac{81 - 54 + 72}{42} = \frac{99}{42} = \frac{33}{14}$$

The solution is $\left(\frac{3}{2}, \frac{13}{14}, \frac{33}{14}\right)$.

48. $\left(-1, -\frac{6}{7}, \frac{11}{7}\right)$

49. $0x + 6y + 6z = -1$
    $8x + 0y + 6z = -1$
    $4x + 9y + 0z = 8$

$$x = \frac{\begin{vmatrix} -1 & 6 & 6 \\ -1 & 0 & 6 \\ 8 & 9 & 0 \end{vmatrix}}{\begin{vmatrix} 0 & 6 & 6 \\ 8 & 0 & 6 \\ 4 & 9 & 0 \end{vmatrix}}$$

$$= \frac{-1\begin{vmatrix} 0 & 6 \\ 9 & 0 \end{vmatrix} - (-1)\begin{vmatrix} 6 & 6 \\ 9 & 0 \end{vmatrix} + 8\begin{vmatrix} 6 & 6 \\ 0 & 6 \end{vmatrix}}{0\begin{vmatrix} 0 & 6 \\ 9 & 0 \end{vmatrix} - 8\begin{vmatrix} 6 & 6 \\ 9 & 0 \end{vmatrix} + 4\begin{vmatrix} 6 & 6 \\ 0 & 6 \end{vmatrix}}$$

$$= \frac{-1(-54) + 1(-54) + 8(36)}{0(-54) - 8(-54) + 4(36)} = \frac{54 - 54 + 288}{0 + 432 + 144}$$

$$= \frac{288}{576} = \frac{1}{2}$$

$$y = \frac{\begin{vmatrix} 0 & -1 & 6 \\ 8 & -1 & 6 \\ 4 & 8 & 0 \end{vmatrix}}{576} = \frac{384}{576} = \frac{2}{3}$$

$$z = \frac{\begin{vmatrix} 0 & 6 & -1 \\ 8 & 0 & -1 \\ 4 & 9 & 8 \end{vmatrix}}{576} = \frac{-480}{576} = -\frac{5}{6}$$

The solution is $\left(\frac{1}{2}, \frac{2}{3}, -\frac{5}{6}\right)$.

50. $\left(-\frac{31}{16}, \frac{25}{16}, -\frac{29}{8}\right)$

51. $\begin{vmatrix} x & 5 \\ -4 & x \end{vmatrix} = 24$

    $x^2 + 20 = 24$

    $x^2 = 4$

    $x = \pm 2$

52. $3, -2$

53. $\begin{vmatrix} x & -3 \\ -1 & x \end{vmatrix} \geqslant 0$

    $x^2 - 3 \geqslant 0$

    The solutions of $f(x) = x^2 - 3 = 0$ are $-\sqrt{3}$ and $\sqrt{3}$. They divide the real-number line as shown.

    | A | B | C |

    $-\sqrt{3} \qquad \sqrt{3}$

    We try test numbers in each interval.

    $f(-2) = (-2)^2 - 3 = 4 - 3 = 1$
    $f(0) = 0^2 - 3 = -3$
    $f(2) = 2^2 - 3 = 4 - 3 = 1$

    Since the inequality is $\geqslant$, we include $\pm\sqrt{3}$ in the solution set. The solution set is $\{x \mid x \leqslant -\sqrt{3} \text{ or } x \geqslant \sqrt{3}\}$.

54. $\{y \mid -\sqrt{10} < y < \sqrt{10}\}$

55. $\begin{vmatrix} x+3 & 4 \\ x-3 & 5 \end{vmatrix} = -7$

    $(x + 3)(5) - (x - 3)4 = -7$

    $5x + 15 - 4x + 12 = -7$

    $x + 27 = -7$

    $x = -34$

56. 3

57. $\begin{vmatrix} 2 & x & 1 \\ 1 & 2 & -1 \\ 3 & 4 & -2 \end{vmatrix} = -6$

    $2\begin{vmatrix} 2 & -1 \\ 4 & -2 \end{vmatrix} - 1\begin{vmatrix} x & 1 \\ 4 & -2 \end{vmatrix} + 3\begin{vmatrix} x & 1 \\ 2 & -1 \end{vmatrix} = -6$

    $2(-4 + 4) - 1(-2x - 4) + 3(-x - 2) = -6$

    $2x + 4 - 3x - 6 = -6$

    $-x - 2 = -6$

    $-x = -4$

    $x = 4$

58. 0

59. – 64. Answers may vary

59. $2L + 2W = \begin{vmatrix} L & -W \\ 2 & 2 \end{vmatrix}$

60. $\begin{vmatrix} x & -h \\ x & r \end{vmatrix}$

61.

$$a^2 + b^2 = \begin{vmatrix} a & b \\ -b & a \end{vmatrix}$$

62.

$$\begin{vmatrix} \frac{1}{2}h & -b \\ \frac{1}{2}h & a \end{vmatrix}$$

63.

$$2\pi r^2 + 2\pi rh = \begin{vmatrix} 2\pi r & 2\pi r \\ -h & r \end{vmatrix}$$

64.

$$\begin{vmatrix} x^2 & 1 \\ Q^2 & y^2 \end{vmatrix}$$

65.

$$\begin{vmatrix} x & y & 1 \\ x_1 & y_1 & 1 \\ x_2 & y_2 & 1 \end{vmatrix} = 0$$

We first evaluate the determinant. Here we evaluate by expanding about the 1st row.

$$x(-1)^{1+1} \cdot \begin{vmatrix} y_1 & 1 \\ y_2 & 1 \end{vmatrix} + y(-1)^{1+2} \cdot \begin{vmatrix} x_1 & 1 \\ x_2 & 1 \end{vmatrix} + 1(-1)^{1+3} \cdot \begin{vmatrix} x_1 & y_1 \\ x_2 & y_2 \end{vmatrix} = 0$$

$$x(y_1 - y_2) - y(x_1 - x_2) + 1(x_1 y_2 - x_2 y_1) = 0$$

$$xy_1 - xy_2 - yx_1 + yx_2 + x_1 y_2 - x_2 y_1 = 0$$

The two-point equation of the line that contains the points (x_1, y_1) and (x_2, y_2) is

$$y - y_1 = \frac{y_2 - y_1}{x_2 - x_1}(x - x_1)$$

$$(y - y_1)(x_2 - x_1) = (y_2 - y_1)(x - x_1) \qquad \text{(Multiplying by } x_2 - x_1\text{)}$$

$$yx_2 - yx_1 - y_1 x_2 + y_1 x_1 = y_2 x - y_2 x_1 - y_1 x + y_1 x_1 \qquad \text{(Using FOIL)}$$

$$xy_1 - xy_2 - yx_1 + yx_2 + x_1 y_2 - x_2 y_1 = 0$$

Thus $\begin{vmatrix} x & y & 1 \\ x_1 & y_1 & 1 \\ x_2 & y_2 & 1 \end{vmatrix} = 0$ and $y - y_1 = \frac{y_2 - y_1}{x_2 - x_1}(x - x_1)$ are equivalent.

66. From Exercise 65, we know that the equation of line ℓ through (x_2, y_2), (x_3, y_3) is $\begin{vmatrix} x & y & 1 \\ x_2 & y_2 & 1 \\ x_3 & y_3 & 1 \end{vmatrix} = 0.$

Then (x_1, y_1) also lies on ℓ if and only if

$$\begin{vmatrix} x_1 & y_1 & 1 \\ x_2 & y_2 & 1 \\ x_3 & y_3 & 1 \end{vmatrix} = 0.$$

67. See the answer section in the text.

68. NOTICE: Parallel lines, by definition, are not coincident;

    E_1, E_2 are the given equations;

    m_1, m_2 are slopes;

    D, D_x, D_y are the given determinants.

(I) If D = 0, then $a_1b_2 - a_2b_1 = 0$. If b_1, $b_2 \neq 0$ then
$\dfrac{-a_1}{b_1} = \dfrac{-a_2}{b_2}$ and $m_1 = m_2$. Thus, ℓ_1 and ℓ_2 are either parallel
or coincident. If either of b_1, b_2 is 0, so is the other, since
$a_1b_2 - a_2b_1 = 0$; and both lines are vertical and either parallel or
coincident.

(II) If the lines coincide, $E_2 = k \cdot E_1$, $k \neq 0$; and

    ℓ_1: $a_1x + b_1y = c_1$,

    ℓ_2: $ka_1x + kb_1y = kc_1$.

    Then $D_x = 0$ and $D_y = 0$. Thus a sufficient condition for
    distinct lines is $D_x \neq 0$ or $D_y \neq 0$.

    In conclusion, sufficient conditions for two lines to be parallel
    are: D = 0 and $D_x \neq 0$ or $D_y \neq 0$.

Exercise Set 6.6

1.
$$A = \begin{bmatrix} 3 & 2 \\ 5 & 3 \end{bmatrix}$$

$$\begin{bmatrix} 3 & 2 & 1 & 0 \\ 5 & 3 & 0 & 1 \end{bmatrix} = \begin{bmatrix} 3 & 2 & 1 & 0 \\ 15 & 9 & 0 & 3 \end{bmatrix} \quad R_2' = 3R_2$$

$$= \begin{bmatrix} 3 & 2 & 1 & 0 \\ 0 & -1 & -5 & 3 \end{bmatrix} \quad R_2' = -5R_1 + R_2$$

$$= \begin{bmatrix} 3 & 2 & 1 & 0 \\ 0 & 1 & 5 & -3 \end{bmatrix} \quad R_2' = -1 \cdot R_2$$

$$= \begin{bmatrix} 3 & 0 & -9 & 6 \\ 0 & 1 & 5 & -3 \end{bmatrix} \quad R_1' = -2R_2 + R_1$$

$$= \begin{bmatrix} 1 & 0 & -3 & 2 \\ 0 & 1 & 5 & -3 \end{bmatrix} \quad R_1' = \tfrac{1}{3}R_1$$

$$A^{-1} = \begin{bmatrix} -3 & 2 \\ 5 & -3 \end{bmatrix}$$

2.
$$\begin{bmatrix} 2 & -5 \\ -1 & 3 \end{bmatrix}$$

<u>3.</u>

$$A = \begin{bmatrix} 11 & 3 \\ 7 & 2 \end{bmatrix}$$

$$\begin{bmatrix} 11 & 3 & 1 & 0 \\ 7 & 2 & 0 & 1 \end{bmatrix} = \begin{bmatrix} 11 & 3 & 1 & 0 \\ 77 & 22 & 0 & 11 \end{bmatrix} \quad R_2' = 11R_2$$

$$= \begin{bmatrix} 11 & 3 & 1 & 0 \\ 0 & 1 & -7 & 11 \end{bmatrix} \quad R_2' = -7R_1 + R_2$$

$$= \begin{bmatrix} 11 & 0 & 22 & -33 \\ 0 & 1 & -7 & 11 \end{bmatrix} \quad R_1' = -3R_2 + R_1$$

$$= \begin{bmatrix} 1 & 0 & 2 & -3 \\ 0 & 1 & -7 & 11 \end{bmatrix} \quad R_1' = \frac{1}{11} R_1$$

$$A^{-1} = \begin{bmatrix} 2 & -3 \\ -7 & 11 \end{bmatrix}$$

<u>4.</u>

$$\begin{bmatrix} -3 & 5 \\ 5 & -8 \end{bmatrix}$$

<u>5.</u>

$$A = \begin{bmatrix} 4 & -3 \\ 1 & 2 \end{bmatrix}$$

$$\begin{bmatrix} 4 & -3 & 1 & 0 \\ 1 & 2 & 0 & 1 \end{bmatrix} = \begin{bmatrix} 1 & 2 & 0 & 1 \\ 4 & -3 & 1 & 0 \end{bmatrix} \quad \begin{matrix} R_1' = R_2 \\ R_2' = R_1 \end{matrix}$$

$$= \begin{bmatrix} 1 & 2 & 0 & 1 \\ 0 & -11 & 1 & -4 \end{bmatrix} \quad R_2' = -4R_1 + R_2$$

$$= \begin{bmatrix} 11 & 22 & 0 & 11 \\ 0 & -11 & 1 & -4 \end{bmatrix} \quad R_1' = 11R_1$$

$$= \begin{bmatrix} 11 & 0 & 2 & 3 \\ 0 & -11 & 1 & -4 \end{bmatrix} \quad R_1' = 2R_2 + R_1$$

$$= \begin{bmatrix} 1 & 0 & \frac{2}{11} & \frac{3}{11} \\ 0 & 1 & -\frac{1}{11} & \frac{4}{11} \end{bmatrix} \quad \begin{matrix} R_1' = \frac{1}{11} R_1 \\ R_2' = -\frac{1}{11} R_2 \end{matrix}$$

$$A^{-1} = \begin{bmatrix} \frac{2}{11} & \frac{3}{11} \\ -\frac{1}{11} & \frac{4}{11} \end{bmatrix}$$

<u>6.</u>

$$\begin{bmatrix} 0 & 1 \\ -1 & 0 \end{bmatrix}$$

7.

$$A = \begin{bmatrix} 3 & 1 & 0 \\ 1 & 1 & 1 \\ 1 & -1 & 2 \end{bmatrix}$$

$$\begin{bmatrix} 3 & 1 & 0 & 1 & 0 & 0 \\ 1 & 1 & 1 & 0 & 1 & 0 \\ 1 & -1 & 2 & 0 & 0 & 1 \end{bmatrix}$$

$$= \begin{bmatrix} 1 & 1 & 1 & 0 & 1 & 0 \\ 3 & 1 & 0 & 1 & 0 & 0 \\ 1 & -1 & 2 & 0 & 0 & 1 \end{bmatrix} \begin{array}{l} R_1' = R_2 \\ R_2' = R_1 \end{array}$$

$$= \begin{bmatrix} 1 & 1 & 1 & 0 & 1 & 0 \\ 0 & -2 & -3 & 1 & -3 & 0 \\ 0 & -2 & 1 & 0 & -1 & 1 \end{bmatrix} \begin{array}{l} R_2' = -3R_1 + R_2 \\ R_3' = -1 \cdot R_1 + R_3 \end{array}$$

$$= \begin{bmatrix} 1 & 1 & 1 & 0 & 1 & 0 \\ 0 & -2 & -3 & 1 & -3 & 0 \\ 0 & 0 & 4 & -1 & 2 & 1 \end{bmatrix} \begin{array}{l} \\ \\ R_3' = -1 \cdot R_2 + R_3 \end{array}$$

$$= \begin{bmatrix} 4 & 4 & 4 & 0 & 4 & 0 \\ 0 & -8 & -12 & 4 & -12 & 0 \\ 0 & 0 & 4 & -1 & 2 & 1 \end{bmatrix} \begin{array}{l} R_1' = 4R_1 \\ R_2' = 4R_2 \end{array}$$

$$= \begin{bmatrix} 4 & 4 & 0 & 1 & 2 & -1 \\ 0 & -8 & 0 & 1 & -6 & 3 \\ 0 & 0 & 4 & -1 & 2 & 1 \end{bmatrix} \begin{array}{l} R_1' = -1 \cdot R_3 + R_1 \\ R_2' = 3R_3 + R_2 \end{array}$$

$$= \begin{bmatrix} 8 & 8 & 0 & 2 & 4 & -2 \\ 0 & -8 & 0 & 1 & -6 & 3 \\ 0 & 0 & 4 & -1 & 2 & 1 \end{bmatrix} \begin{array}{l} R_1' = 2R_1 \end{array}$$

$$= \begin{bmatrix} 8 & 0 & 0 & 3 & -2 & 1 \\ 0 & -8 & 0 & 1 & -6 & 3 \\ 0 & 0 & 4 & -1 & 2 & 1 \end{bmatrix} \begin{array}{l} R_1' = R_2 + R_1 \end{array}$$

$$= \begin{bmatrix} 1 & 0 & 0 & \frac{3}{8} & -\frac{1}{4} & \frac{1}{8} \\ 0 & 1 & 0 & -\frac{1}{8} & \frac{3}{4} & -\frac{3}{8} \\ 0 & 0 & 1 & -\frac{1}{4} & \frac{1}{2} & \frac{1}{4} \end{bmatrix} \begin{array}{l} R_1' = \frac{1}{8} R_1 \\ R_2' = -\frac{1}{8} R_2 \\ R_3' = \frac{1}{4} R_3 \end{array}$$

$$A^{-1} = \begin{bmatrix} \frac{3}{8} & -\frac{1}{4} & \frac{1}{8} \\ -\frac{1}{8} & \frac{3}{4} & -\frac{3}{8} \\ -\frac{1}{4} & \frac{1}{2} & \frac{1}{4} \end{bmatrix}$$

8.
$$\begin{bmatrix} -\frac{1}{2} & \frac{1}{2} & \frac{1}{2} \\ 1 & 0 & -1 \\ \frac{3}{2} & -\frac{1}{2} & -\frac{1}{2} \end{bmatrix}$$

9.
$$A = \begin{bmatrix} 1 & -1 & 2 \\ 0 & 1 & 3 \\ 2 & 1 & -2 \end{bmatrix}$$

$$\begin{bmatrix} 1 & -1 & 2 & 1 & 0 & 0 \\ 0 & 1 & 3 & 0 & 1 & 0 \\ 2 & 1 & -2 & 0 & 0 & 1 \end{bmatrix}$$

$$= \begin{bmatrix} 1 & -1 & 2 & 1 & 0 & 0 \\ 0 & 1 & 3 & 0 & 1 & 0 \\ 0 & 3 & -6 & -2 & 0 & 1 \end{bmatrix} \quad R_3' = -2R_1 + R_3$$

$$= \begin{bmatrix} 1 & -1 & 2 & 1 & 0 & 0 \\ 0 & 1 & 3 & 0 & 1 & 0 \\ 0 & 0 & -15 & -2 & -3 & 1 \end{bmatrix} \quad R_3' = -3R_2 + R_3$$

$$= \begin{bmatrix} 15 & -15 & 30 & 15 & 0 & 0 \\ 0 & 5 & 15 & 0 & 5 & 0 \\ 0 & 0 & -15 & -2 & -3 & 1 \end{bmatrix} \quad \begin{matrix} R_1' = 15R_1 \\ R_2' = 5R_2 \end{matrix}$$

$$= \begin{bmatrix} 15 & -15 & 0 & 11 & -6 & 2 \\ 0 & 5 & 0 & -2 & 2 & 1 \\ 0 & 0 & -15 & -2 & -3 & -1 \end{bmatrix} \quad \begin{matrix} R_1' = 2R_3 + R_1 \\ R_2' = R_3 + R_2 \end{matrix}$$

$$= \begin{bmatrix} 15 & 0 & 0 & 5 & 0 & 5 \\ 0 & 5 & 0 & -2 & 2 & 1 \\ 0 & 0 & -15 & -2 & -3 & -1 \end{bmatrix} \quad R_1' = 3R_2 + R_1$$

$$= \begin{bmatrix} 1 & 0 & 0 & \frac{1}{3} & 0 & \frac{1}{3} \\ 0 & 1 & 0 & -\frac{2}{5} & \frac{2}{5} & \frac{1}{5} \\ 0 & 0 & 1 & \frac{2}{15} & \frac{1}{5} & -\frac{1}{15} \end{bmatrix} \quad \begin{matrix} R_1' = \frac{1}{15}R_1 \\ R_2' = \frac{1}{5}R_2 \\ R_3' = -\frac{1}{15}R_3 \end{matrix}$$

$$A^{-1} = \begin{bmatrix} \frac{1}{3} & 0 & \frac{1}{3} \\ -\frac{2}{5} & \frac{2}{5} & \frac{1}{5} \\ \frac{2}{15} & \frac{1}{5} & -\frac{1}{15} \end{bmatrix}$$

<u>10.</u> $\begin{bmatrix} -1 & 5 & 2 \\ -1 & 3 & 1 \\ \frac{1}{2} & -1 & -\frac{1}{2} \end{bmatrix}$

<u>11.</u> We cannot obtain the identity matrix on the left using the Gauss-Jordan reduction method. Thus, A^{-1} does not exist.

<u>12.</u> A^{-1} does not exist.

<u>13.</u>

$A = \begin{bmatrix} 1 & 2 & 3 & 4 \\ 0 & 1 & 3 & -5 \\ 0 & 0 & 1 & -2 \\ 0 & 0 & 0 & -1 \end{bmatrix}$

$\left[\begin{array}{cccc|cccc} 1 & 2 & 3 & 4 & 1 & 0 & 0 & 0 \\ 0 & 1 & 3 & -5 & 0 & 1 & 0 & 0 \\ 0 & 0 & 1 & -2 & 0 & 0 & 1 & 0 \\ 0 & 0 & 0 & -1 & 0 & 0 & 0 & 1 \end{array}\right]$

$= \left[\begin{array}{cccc|cccc} 1 & 2 & 3 & 0 & 1 & 0 & 0 & 4 \\ 0 & 1 & 3 & 0 & 0 & 1 & 0 & -5 \\ 0 & 0 & 1 & 0 & 0 & 0 & 1 & -2 \\ 0 & 0 & 0 & -1 & 0 & 0 & 0 & 1 \end{array}\right]$
$\begin{array}{l} R_1' = 4R_4 + R_1 \\ R_2' = -5R_4 + R_2 \\ R_3' = -2R_4 + R_3 \end{array}$

$= \left[\begin{array}{cccc|cccc} 1 & 2 & 0 & 0 & 1 & 0 & -3 & 10 \\ 0 & 1 & 0 & 0 & 0 & 1 & -3 & 1 \\ 0 & 0 & 1 & 0 & 0 & 0 & 1 & -2 \\ 0 & 0 & 0 & -1 & 0 & 0 & 0 & 1 \end{array}\right]$
$\begin{array}{l} R_1' = -3R_3 + R_1 \\ R_2' = -3R_3 + R_2 \end{array}$

$= \left[\begin{array}{cccc|cccc} 1 & 0 & 0 & 0 & 1 & -2 & 3 & 8 \\ 0 & 1 & 0 & 0 & 0 & 1 & -3 & 1 \\ 0 & 0 & 1 & 0 & 0 & 0 & 1 & -2 \\ 0 & 0 & 0 & -1 & 0 & 0 & 0 & 1 \end{array}\right]$
$R_1' = -2R_2 + R_1$

$= \left[\begin{array}{cccc|cccc} 1 & 0 & 0 & 0 & 1 & -2 & 3 & 8 \\ 0 & 1 & 0 & 0 & 0 & 1 & -3 & 1 \\ 0 & 0 & 1 & 0 & 0 & 0 & 1 & -2 \\ 0 & 0 & 0 & 1 & 0 & 0 & 0 & -1 \end{array}\right]$
$R_4' = -1 \cdot R_4$

$A^{-1} = \begin{bmatrix} 1 & -2 & 3 & 8 \\ 0 & 1 & -3 & 1 \\ 0 & 0 & 1 & -2 \\ 0 & 0 & 0 & -1 \end{bmatrix}$

14. A^{-1} does not exist.

15. We cannot obtain the identity matrix on the left using the Gauss-Jordan reduction method. Then, A^{-1} does not exist.

16. A^{-1} does not exist.

17. $11x + 3y = -4,$
    $7x + 2y = 5$

    We write a matrix equation equivalent to this system.

    $$\begin{bmatrix} 11 & 3 \\ 7 & 2 \end{bmatrix}\begin{bmatrix} x \\ y \end{bmatrix} = \begin{bmatrix} -4 \\ 5 \end{bmatrix}$$

    We solve the matrix equation by multiplying by A^{-1} on the left on each side.

    $$\begin{bmatrix} 2 & -3 \\ -7 & 11 \end{bmatrix}\begin{bmatrix} 11 & 3 \\ 7 & 2 \end{bmatrix}\begin{bmatrix} x \\ y \end{bmatrix} = \begin{bmatrix} 2 & -3 \\ -7 & 11 \end{bmatrix}\begin{bmatrix} -4 \\ 5 \end{bmatrix}$$

    $$\begin{bmatrix} 1 & 0 \\ 0 & 1 \end{bmatrix}\begin{bmatrix} x \\ y \end{bmatrix} = \begin{bmatrix} -23 \\ 83 \end{bmatrix}$$

    $$\begin{bmatrix} x \\ y \end{bmatrix} = \begin{bmatrix} -23 \\ 83 \end{bmatrix}$$

    The solution of the system of equations is $(-23, 83)$.

18. $(28, -46)$

19. $3x + y \quad\ = 2,$
    $2x - y + 2z = -5,$
    $x + y + z = 5$

    We write a matrix equation equivalent to this system.

    $$\begin{bmatrix} 3 & 1 & 0 \\ 2 & -1 & 2 \\ 1 & 1 & 1 \end{bmatrix}\begin{bmatrix} x \\ y \\ z \end{bmatrix} = \begin{bmatrix} 2 \\ -5 \\ 5 \end{bmatrix}$$

    We solve the matrix equation by multiplying by A^{-1} on the left on each side.

    $$\frac{1}{9}\begin{bmatrix} 3 & 1 & -2 \\ 0 & -3 & 6 \\ -3 & 2 & 5 \end{bmatrix}\begin{bmatrix} 3 & 1 & 0 \\ 2 & -1 & 2 \\ 1 & 1 & 1 \end{bmatrix}\begin{bmatrix} x \\ y \\ z \end{bmatrix} = \frac{1}{9}\begin{bmatrix} 3 & 1 & -2 \\ 0 & -3 & 6 \\ -3 & 2 & 5 \end{bmatrix}\begin{bmatrix} 2 \\ -5 \\ 5 \end{bmatrix}$$

    $$\begin{bmatrix} 1 & 0 & 0 \\ 0 & 1 & 0 \\ 0 & 0 & 1 \end{bmatrix}\begin{bmatrix} x \\ y \\ z \end{bmatrix} = \frac{1}{9}\begin{bmatrix} -9 \\ 45 \\ 9 \end{bmatrix}$$

    $$\begin{bmatrix} x \\ y \\ z \end{bmatrix} = \begin{bmatrix} -1 \\ 5 \\ 1 \end{bmatrix}$$

    The solution of the system of equations is $(-1, 5, 1)$.

20. $(1, -7, -3)$

<u>21.</u> $4x - 3y = 2,$ $\begin{bmatrix} 4 & -3 \\ 1 & 2 \end{bmatrix} \begin{bmatrix} x \\ y \end{bmatrix} = \begin{bmatrix} 2 \\ -1 \end{bmatrix}$
    $x + 2y = -1$

The coefficient matrix is the matrix in Exercise 5. Its inverse is

$\begin{bmatrix} \frac{2}{11} & \frac{3}{11} \\ -\frac{1}{11} & \frac{4}{11} \end{bmatrix}$, or $\frac{1}{11}\begin{bmatrix} 2 & 3 \\ -1 & 4 \end{bmatrix}$.

$\begin{bmatrix} x \\ y \end{bmatrix} = \begin{bmatrix} \frac{2}{11} & \frac{3}{11} \\ -\frac{1}{11} & \frac{4}{11} \end{bmatrix} \begin{bmatrix} 2 \\ -1 \end{bmatrix} = \begin{bmatrix} \frac{1}{11} \\ -\frac{6}{11} \end{bmatrix}$

The solution of the system is $\left(\frac{1}{11}, -\frac{6}{11}\right)$.

<u>22.</u>
$\begin{bmatrix} 3 & 5 \\ 2 & 4 \end{bmatrix} \begin{bmatrix} x \\ y \end{bmatrix} = \begin{bmatrix} -4 \\ -2 \end{bmatrix}$, $A^{-1} = \begin{bmatrix} 2 & -\frac{5}{2} \\ -1 & \frac{3}{2} \end{bmatrix}$, $(-3, 1)$

<u>23.</u> $7x - 2y = -3,$ $\begin{bmatrix} 7 & -2 \\ 9 & 3 \end{bmatrix} \begin{bmatrix} x \\ y \end{bmatrix} = \begin{bmatrix} -3 \\ 4 \end{bmatrix}$
    $9x + 3y = 4$

The inverse of the coefficient matrix is $\begin{bmatrix} \frac{1}{13} & \frac{2}{39} \\ -\frac{3}{13} & \frac{7}{39} \end{bmatrix}$, or $\frac{1}{39}\begin{bmatrix} 3 & 2 \\ -9 & 7 \end{bmatrix}$.

$\begin{bmatrix} x \\ y \end{bmatrix} = \begin{bmatrix} \frac{1}{13} & \frac{2}{39} \\ -\frac{3}{13} & \frac{7}{39} \end{bmatrix} \begin{bmatrix} -3 \\ 4 \end{bmatrix} = \begin{bmatrix} -\frac{1}{39} \\ \frac{55}{39} \end{bmatrix}$

The solution of the system of equations is $\left(-\frac{1}{39}, \frac{55}{39}\right)$.

<u>24.</u> $\begin{bmatrix} 5 & 3 \\ 4 & -1 \end{bmatrix} \begin{bmatrix} x \\ y \end{bmatrix} = \begin{bmatrix} -2 \\ 1 \end{bmatrix}$, $A^{-1} = \frac{1}{17}\begin{bmatrix} 1 & 3 \\ 4 & -5 \end{bmatrix}$, $\left(\frac{1}{17}, -\frac{13}{17}\right)$

<u>25.</u> $x \quad\ + z = 1,$ $\begin{bmatrix} 1 & 0 & 1 \\ 2 & 1 & 0 \\ 1 & -1 & 1 \end{bmatrix} \begin{bmatrix} x \\ y \\ z \end{bmatrix} = \begin{bmatrix} 1 \\ 3 \\ 4 \end{bmatrix}$
    $2x + y \quad\ = 3,$
    $x - y + z = 4$

The inverse of the coefficient matrix is

$\frac{1}{2}\begin{bmatrix} -1 & 1 & 1 \\ 2 & 0 & -2 \\ 3 & -1 & -1 \end{bmatrix}$.

$\begin{bmatrix} x \\ y \\ z \end{bmatrix} = \frac{1}{2}\begin{bmatrix} -1 & 1 & 1 \\ 2 & 0 & -2 \\ 3 & -1 & -1 \end{bmatrix} \begin{bmatrix} 1 \\ 3 \\ 4 \end{bmatrix} = \begin{bmatrix} 3 \\ -3 \\ -2 \end{bmatrix}$

The solution of the system of equations is $(3, -3, -2)$.

26.
$$\begin{bmatrix} 1 & 2 & 3 \\ 2 & -3 & 4 \\ -3 & 5 & -6 \end{bmatrix} \begin{bmatrix} x \\ y \\ z \end{bmatrix} = \begin{bmatrix} -1 \\ 2 \\ 4 \end{bmatrix}, \ A^{-1} = \begin{bmatrix} -2 & 27 & 17 \\ 0 & 3 & 2 \\ 1 & -11 & -7 \end{bmatrix}, \ (124, \ 14, \ -51)$$

27.
$$\begin{array}{l} 2w - 3x + 4y - 5z = 0, \\ 3w - 2x + 7y - 3z = 2, \\ w + x - y + z = 1, \\ -w - 3x - 6y + 4z = 6 \end{array} \qquad \begin{bmatrix} 2 & -3 & 4 & -5 \\ 3 & -2 & 7 & -3 \\ 1 & 1 & -1 & 1 \\ -1 & -3 & -6 & 4 \end{bmatrix} \begin{bmatrix} w \\ x \\ y \\ z \end{bmatrix} = \begin{bmatrix} 0 \\ 2 \\ 1 \\ 6 \end{bmatrix}$$

The inverse of the coefficient matrix is

$$\frac{1}{11,165} \begin{bmatrix} 1430 & 605 & 6985 & 495 \\ 440 & -1045 & 2145 & -1870 \\ -2035 & 2145 & -2640 & -275 \\ -3025 & 2585 & -605 & 1100 \end{bmatrix}.$$

$$\begin{bmatrix} w \\ x \\ y \\ z \end{bmatrix} = \frac{1}{11,165} \begin{bmatrix} 1430 & 605 & 6985 & 495 \\ 440 & -1045 & 2145 & -1870 \\ -2035 & 2145 & -2640 & -275 \\ -3025 & 2585 & -605 & 1100 \end{bmatrix} \begin{bmatrix} 0 \\ 2 \\ 1 \\ 6 \end{bmatrix}$$

$$= \begin{bmatrix} 1 \\ -1 \\ 0 \\ 1 \end{bmatrix}$$

The solution of the system of equations is $(1, -1, 0, 1)$.

28.
$$\begin{bmatrix} 5 & -4 & 3 & -2 \\ 1 & 4 & -2 & 3 \\ 2 & -3 & 6 & -9 \\ 3 & -5 & 2 & -4 \end{bmatrix} \begin{bmatrix} w \\ x \\ y \\ z \end{bmatrix} = \begin{bmatrix} -6 \\ -5 \\ 14 \\ -3 \end{bmatrix},$$

$$A^{-1} = \frac{1}{302} \begin{bmatrix} 18 & 75 & 1 & 45 \\ -10 & 59 & 33 & -25 \\ 112 & -87 & 23 & -173 \\ 82 & -61 & -29 & -97 \end{bmatrix}, \ (-2, \ 1, \ 2, \ -1)$$

29.
$$\begin{bmatrix} a & b & c \\ d & e & f \\ g & h & i \end{bmatrix} \begin{bmatrix} 1 & 0 & 0 \\ 0 & 1 & 0 \\ 0 & 0 & 1 \end{bmatrix} = \begin{bmatrix} a & b & c \\ d & e & f \\ g & h & i \end{bmatrix}, \qquad AI = A$$

$$\begin{bmatrix} 1 & 0 & 0 \\ 0 & 1 & 0 \\ 0 & 0 & 1 \end{bmatrix} \begin{bmatrix} a & b & c \\ d & e & f \\ g & h & i \end{bmatrix} = \begin{bmatrix} a & b & c \\ d & e & f \\ g & h & i \end{bmatrix} \qquad IA = A$$

30. A^{-1} exists if and only if $x \neq 0$. $A^{-1} = \begin{bmatrix} \frac{1}{x} \end{bmatrix}$

31.
$$A = \begin{bmatrix} x & 0 \\ 0 & y \end{bmatrix}$$

A^{-1} exists if and only if $xy \neq 0$.

$$\begin{bmatrix} x & 0 & 1 & 0 \\ 0 & y & 0 & 1 \end{bmatrix} = \begin{bmatrix} 1 & 0 & \frac{1}{x} & 0 \\ 0 & 1 & 0 & \frac{1}{y} \end{bmatrix} \quad \begin{matrix} R_1' = \frac{1}{x} R_1 \\ R_2' = \frac{1}{y} R_2 \end{matrix}$$

$$A^{-1} = \begin{bmatrix} \frac{1}{x} & 0 \\ 0 & \frac{1}{y} \end{bmatrix}$$

32. A^{-1} exists if and only if $xyz \neq 0$.

$$\begin{bmatrix} 0 & 0 & \frac{1}{z} \\ 0 & \frac{1}{y} & 0 \\ \frac{1}{x} & 0 & 0 \end{bmatrix}$$

33.
$$A = \begin{bmatrix} x & 1 & 1 & 1 \\ 0 & y & 0 & 0 \\ 0 & 0 & z & 0 \\ 0 & 0 & 0 & w \end{bmatrix}$$

A^{-1} exists if and only if $xyzw \neq 0$.

$$\begin{bmatrix} x & 1 & 1 & 1 & 1 & 0 & 0 & 0 \\ 0 & y & 0 & 0 & 0 & 1 & 0 & 0 \\ 0 & 0 & z & 0 & 0 & 0 & 1 & 0 \\ 0 & 0 & 0 & w & 0 & 0 & 0 & 1 \end{bmatrix}$$

$$= \begin{bmatrix} -xw & -w & -w & -w & -w & 0 & 0 & 0 \\ 0 & y & 0 & 0 & 0 & 1 & 0 & 0 \\ 0 & 0 & z & 0 & 0 & 0 & 1 & 0 \\ 0 & 0 & 0 & w & 0 & 0 & 0 & 1 \end{bmatrix} \quad R_1' = -wR_1$$

$$= \begin{bmatrix} -xw & -w & -w & 0 & -w & 0 & 0 & 1 \\ 0 & y & 0 & 0 & 0 & 1 & 0 & 0 \\ 0 & 0 & z & 0 & 0 & 0 & 1 & 0 \\ 0 & 0 & 0 & w & 0 & 0 & 0 & 1 \end{bmatrix} \quad R_1' = R_4 + R_1$$

$$= \begin{bmatrix} -xzw & -zw & -zw & 0 & -zw & 0 & 0 & z \\ 0 & y & 0 & 0 & 0 & 1 & 0 & 0 \\ 0 & 0 & z & 0 & 0 & 0 & 1 & 0 \\ 0 & 0 & 0 & w & 0 & 0 & 0 & 1 \end{bmatrix} \quad R_1' = zR_1$$

$$= \begin{bmatrix} -xzw & -zw & 0 & 0 & -zw & 0 & w & z \\ 0 & y & 0 & 0 & 0 & 1 & 0 & 0 \\ 0 & 0 & z & 0 & 0 & 0 & 1 & 0 \\ 0 & 0 & 0 & w & 0 & 0 & 0 & 1 \end{bmatrix} \quad R_1' = wR_3 + R_1$$

$$= \begin{bmatrix} -xzyw & -zyw & 0 & 0 & -zyw & 0 & wy & yz \\ 0 & y & 0 & 0 & 0 & 1 & 0 & 0 \\ 0 & 0 & z & 0 & 0 & 0 & 1 & 0 \\ 0 & 0 & 0 & w & 0 & 0 & 0 & 1 \end{bmatrix} \quad R_1' = yR_1$$

$$= \begin{bmatrix} -xzyw & 0 & 0 & 0 & -zyw & zw & wy & yz \\ 0 & y & 0 & 0 & 0 & 1 & 0 & 0 \\ 0 & 0 & z & 0 & 0 & 0 & 1 & 0 \\ 0 & 0 & 0 & w & 0 & 0 & 0 & 1 \end{bmatrix} \quad R_1' = zwR_2 + R_1$$

$$= \begin{bmatrix} 1 & 0 & 0 & 0 & \frac{1}{x} & -\frac{1}{xy} & -\frac{1}{xz} & -\frac{1}{xw} \\ 0 & 1 & 0 & 0 & 0 & \frac{1}{y} & 0 & 0 \\ 0 & 0 & 1 & 0 & 0 & 0 & \frac{1}{z} & 0 \\ 0 & 0 & 0 & 1 & 0 & 0 & 0 & \frac{1}{w} \end{bmatrix} \quad \begin{array}{l} R_1' = -\frac{1}{xzyw}R_1 \\[4pt] R_2' = \frac{1}{y}R_2 \\[4pt] R_3' = \frac{1}{z}R_3 \\[4pt] R_4' = \frac{1}{w}R_4 \end{array}$$

$$A^{-1} = \begin{bmatrix} \frac{1}{x} & -\frac{1}{xy} & -\frac{1}{xz} & -\frac{1}{xw} \\ 0 & \frac{1}{y} & 0 & 0 \\ 0 & 0 & \frac{1}{z} & 0 \\ 0 & 0 & 0 & \frac{1}{w} \end{bmatrix}$$

<u>34.</u> $\begin{bmatrix} a_1 & b_1 \\ a_2 & b_2 \end{bmatrix} \begin{bmatrix} x \\ y \end{bmatrix} = \begin{bmatrix} c_1 \\ c_2 \end{bmatrix}$ or $A \cdot X = C.$

Then $X = A^{-1}C$ or

$$X = \begin{bmatrix} \dfrac{b_2}{a_1b_2 - b_1a_2} & -\dfrac{b_1}{a_1b_2 - b_1a_2} \\[10pt] -\dfrac{a_2}{a_1b_2 - b_1a_2} & \dfrac{a_1}{a_1b_2 - b_1a_2} \end{bmatrix} \begin{bmatrix} c_1 \\[10pt] c_2 \end{bmatrix}.$$

Now $a_1b_2 - b_1a_2 = |A|$, so

$$X = \begin{vmatrix} \dfrac{b_2c_1 - b_1c_2}{|A|} \\[10pt] \dfrac{-a_2c_1 + a_1c_2}{|A|} \end{vmatrix}$$

Thus $x = \dfrac{b_2c_1 - b_1c_2}{|A|} = \dfrac{\begin{vmatrix} c_1 & b_1 \\ c_2 & b_2 \end{vmatrix}}{|A|} = \dfrac{D_x}{D}$; and

$y = \dfrac{a_1c_1 - a_2c_1}{|A|} = \dfrac{\begin{vmatrix} a_1 & c_1 \\ a_2 & c_2 \end{vmatrix}}{|A|} = \dfrac{D_y}{D}$.

Exercise Set 6.7

1. We replace x by -4 and y by 2.

$$\begin{array}{c|c} 2x + y > -5 \\ \hline 2(-4) + 2 & -5 \\ -8 + 2 & \\ -6 & \text{FALSE} \end{array}$$

Since -6 > -5 is false, (-4,2) is not a solution.

2. Yes

3. We replace x by 8 and y by 14.

$$\begin{array}{c|c} 2y - 3x \geqslant 5 \\ \hline 2\cdot14 - 3\cdot8 & 5 \\ 28 - 24 & \\ 4 & \text{FALSE} \end{array}$$

Since 4 ⩾ 5 is false, (8,14) is not a solution.

4. No

5. Graph: y > 2x

We first graph the line y = 2x. We draw the line dashed since the inequality symbol is >. To determine which half-plane to shade, test a point not on the line. We try (1,1) and substitute:

$$\begin{array}{c|c} y > 2x \\ \hline 1 & 2\cdot1 \\ & 2 \quad \text{FALSE} \end{array}$$

Since 1 > 2 is false, (1,1) is not a solution, nor are any points in the half-plane containing (1,1). The points in the opposite half-plane are solutions, so we shade that half-plane and obtain the graph.

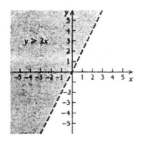

6. 2y < x

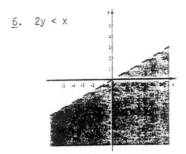

7. Graph: y + x ⩾ 0

First graph the equation y + x = 0. A few solutions of this equation are (-2, 2), (0, 0), and (3, -3). Plot these points and draw the line solid since the inequality is ⩾.

Next determine which half-plane to shade by trying some point off the line. Here we use (2, 2) as a check.

$$\begin{array}{c|c} y + x \geqslant 0 \\ \hline 2 + 2 & 0 \\ 4 & \text{TRUE} \end{array}$$

Since 4 ⩾ 0 is true, (2, 2) is a solution. Thus shade the half-plane containing (2, 2).

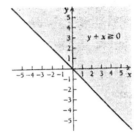

8. y + x < 0

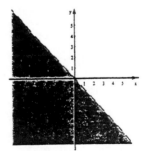

9. Graph: y > x - 3

First graph the equation y = x - 3. The intercepts are (0,-3) and (3,0). Draw the line dashed since the inequality is >. To determine which half-plane to shade, test a point not on the line. We try (0, 0).

$$\begin{array}{c|c} y > x - 3 \\ \hline 0 & 0 - 3 \\ & -3 \quad \text{TRUE} \end{array}$$

Since 0 > -3 is true, (0,0) is a solution. Thus we shade the half-plane containing (0,0).

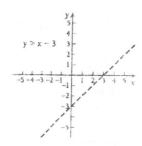

<u>10.</u> y ≤ x + 4

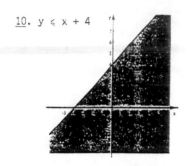

<u>11.</u> Graph: x + y < 4

First graph the equation x + y = 4. The intercepts are (0,4) and (4,0). Draw the line dashed since the inequality is <. To determine which half-plane to shade, test a point not on the line. We try (0,0).

$$\frac{x + y < 4}{\begin{array}{c|c} 0 + 0 & 4 \\ 0 & \text{TRUE} \end{array}}$$

Since 0 < 4 is true, (0,0) is a solution. Thus we shade the half-plane containing (0,0).

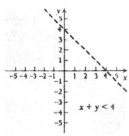

<u>12.</u> x - y ≥ 5

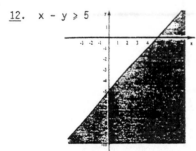

<u>13.</u> Graph: 3x - 2y ≤ 6

First graph the equation 3x - 2y = 6. The y-intercept is (0, -3), and the x-intercept is (2, 0). Plot these points and draw the line solid since the inequality is ≤.

Next complete the graph by shading the correct half-plane. Since the line does not contain the origin, check the point (0, 0) in the inequality to see if we get a true sentence.

$$\frac{3x - 2y \le 6}{\begin{array}{c|c} 3(0) - 2(0) & 6 \\ 0 & 6 \ \text{TRUE} \end{array}}$$

Since 0 ≤ 6 is a true sentence, (0, 0) is a solution. Thus shade the half-plane containing (0, 0).

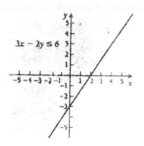

<u>14.</u> 2x - 5y < 10

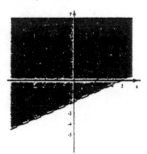

<u>15.</u> Graph: 3y + 2x ≥ 6

First graph the equation 3y + 2x = 6. The y-intercept is (0, 2), and the x-intercept is (3, 0). Plot these points and draw the line solid since the inequality is ≥.

Next complete the graph by shading the correct half-plane. Since the line does not contain the origin, check the point (0, 0) in the inequality to see if we get a true sentence.

$$\frac{3y + 2x \ge 6}{\begin{array}{c|c} 3\cdot0 + 2\cdot0 & 6 \\ 0 & \text{FALSE} \end{array}}$$

Since 0 ≥ 6 is false, (0, 0) is not a solution. We shade the half-plane which does not contain (0, 0).

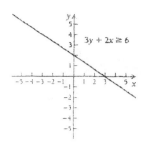

$3y + 2x \geq 6$

18. $2x - 6y \geqslant 8 + 4y$, or $2x - 10y \geqslant 8$

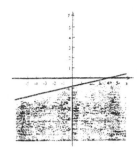

16. $2y + x \leqslant 4$

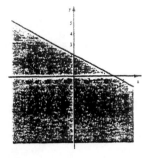

19. Graph: $x < -4$

We first graph the equation $x = -4$. The line is parallel to the y-axis with x-intercept $(-4, 0)$. We draw the line dashed since the inequality is $<$.

Next complete the graph by shading the correct half-plane. Since the line does not contain the origin, check the point $(0, 0)$.

$x < -4$	
0	-4 FALSE

Since $0 < -4$ is false, $(0, 0)$ is not a solution. Thus, we shade the half-plane which does not contain the origin.

17. Graph: $3x - 2 \leqslant 5x + y$

$$-2 \leqslant 2x + y \qquad \text{(Adding } -3x\text{)}$$

First graph $2x + y = -2$. The y-intercept is $(0, -2)$, and the x-intercept is $(-1, 0)$. Plot these points and draw the line solid since the inequality is $\leqslant$.

Next complete the graph by shading the correct half-plane. Since the line does not contain the origin, check the point $(0, 0)$.

$2x + y \geqslant -2$	
$2(0) + 0$	-2
0	-2 TRUE

Since $0 \geqslant -2$ is a true sentence, $(0, 0)$ is a solution. Thus shade the half-plane containing the origin.

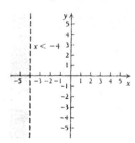

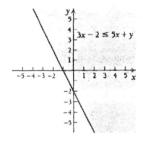

$3x - 2 \leq 5x + y$

20. $y \geqslant 5$

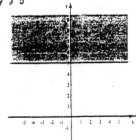

21. Graph: y > -3

First we graph the equation y = -3. We draw the line dashed since the inequality is >. To determine which half-plane to shade we test a point not on the line. We try (0,0).

y > -3
0 | -3 TRUE

Since 0 > -3 is true, (0,0) is a solution. We shade the half-plane containing (0,0).

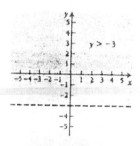

22. x ≤ 5

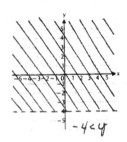

23. Graph: -4 < y < -1

This is a conjunction of two inequalities

-4 < y and y < -1.

We can graph -4 < y and y < -1 separately and then graph the intersection.

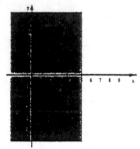

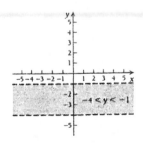

24. -1 < y < 4

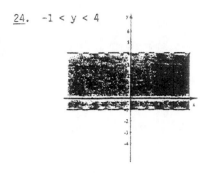

25. Graph: -4 ≤ x ≤ 4

This is a conjunction of inequalities

-4 ≤ x and x ≤ 4.

We can graph -4 ≤ x and x ≤ 4 separately and then graph the intersection.

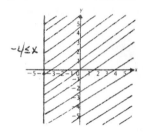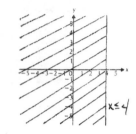

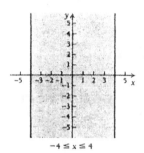

-4 ≤ x ≤ 4

26. -3 ≤ x ≤ 3

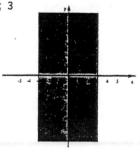

27. Graph: $y \geqslant |x|$

First graph $y = |x|$. A few solutions of this
equation are (0, 0), (1, 1), (-1, 1), (4, 4),
and (-4, 4). Plot these points and draw the
graph solid since the inequality is $\geqslant$.

$y = |x|$

Note that in $y \geqslant |x|$, y is by itself. We
interpret the graph of the inequality as the set
of all ordered pairs (x, y) where the second
coordinate y is greater than or equal to the
absolute value of the first. Decide below which
pairs satisfy the inequality and which do not.

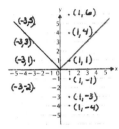

We see that any pair above the graph of $y = |x|$
is a solution as well as those in the graph of
$y = |x|$. The graph is as follows.

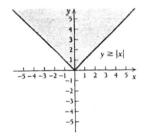

28. $y \leqslant |x|$

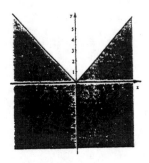

29. Graph: $y \leqslant x$,

$y \geqslant 3 - x$

We graph the lines $y = x$ and $y = 3 - x$ using
solid lines. The arrows at the ends of the lines
indicate the region for each inequality. Note
where the regions overlap and shade the region of
solutions.

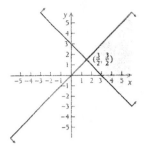

We find the vertex $\left(\frac{3}{2}, \frac{3}{2}\right)$ by solving the system

$y = x,$

$y = 3 - x.$

30. Graph: $y \geqslant x$,

$y \leqslant x - 5$

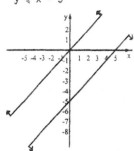

31. Graph $y \geqslant x$,

$y \leqslant x - 4$

We graph the lines $y = x$ and $y = x - 4$ using
solid lines. The arrows at the ends of the lines
indicate the region for each inequality. The
regions do not overlap, so there are no vertices.

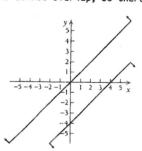

32. $y \geqslant x$,

   $y \leqslant 2 - x$

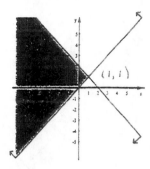

33. Graph: $y \geqslant -3$,

   $x \geqslant 1$

We graph the lines $y = -3$ and $x = 1$ using solid lines. The arrows at the ends of the lines indicate the region for each inequality. Note where the regions overlap and shade the region of solutions.

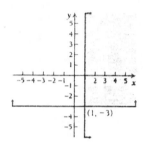

We find the vertex $(1, -3)$ by solving the system

$$y = -3,$$
$$x = 1.$$

34. $y \leqslant -2$,

   $x \geqslant 2$

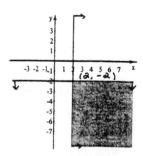

35. Graph: $x \leqslant 3$,

   $y \geqslant 2 - 3x$

We graph the lines $x = 3$ and $y = 2 - 3x$ using solid lines. The arrows at the ends of the lines indicate the region for each inequality. Note where the regions overlap and shade the region of solutions.

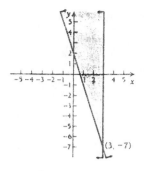

We find the vertex $(3,-7)$ by solving the system

$$x = 3,$$
$$y = 2 - 3x.$$

36. $x \geqslant -2$,

   $y \leqslant 3 - 2x$

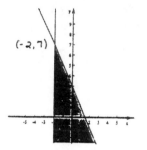

37. Graph: $x + y \leqslant 1$,

   $x - y \leqslant 2$

We graph the lines $x + y = 1$ and $x - y = 2$ using solid lines. The arrows at the ends of the lines indicate the region for each inequality. Note where the regions overlap and shade the region of solutions.

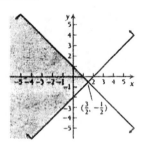

We find the vertex $\left(\frac{3}{2}, -\frac{1}{2}\right)$ by solving the system

$$x + y = 1,$$
$$x - y = 2.$$

38. y + 3x ⩾ 0,
     y + 3x ⩽ 2

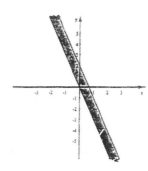

39. Graph: 2y - x ⩽ 2,
              y + 3x ⩾ -1

We graph the lines 2y - x = 2 and y + 3x = -1
using solid lines. The arrows at the ends of the
lines indicate the region for each inequality.
Note where the regions overlap and shade the
region of solutions.

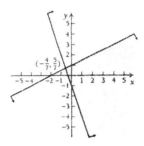

We find the vertex $\left(-\frac{4}{7}, \frac{5}{7}\right)$ by solving the system

        2y - x = 2,
        y + 3x = -1.

40. y ⩽ 2x + 1,
     y ⩾ -2x + 1,
     x ⩽ 2

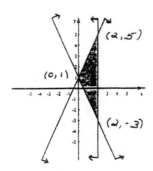

41. Graph: x - y ⩽ 2,
              x + 2y ⩾ 8,
                  y ⩽ 4

We graph the lines x - y = 2, x + 2y = 8, and
y = 4 using solid lines. The arrows at the ends
of the lines indicate the region for each
inequality. Note where the regions overlap and
shade the region of solutions.

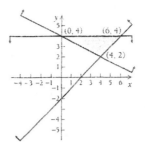

We find the vertex (0,4) by solving the system
        x + 2y = 8,
             y = 4.

We find the vertex (6,4) by solving the system
        x - y = 2,
            y = 4.

We find the vertex (4,2) by solving the system
        x - y = 2,
        x + 2y = 8.

42. x + 2y ⩽ 12,
     2x + y ⩽ 12,
         x ⩾ 0,
         y ⩾ 0

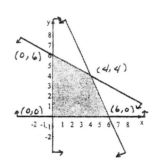

43. Graph: 4y - 3x ⩾ -12,
              4y + 3x ⩾ -36,
                   y ⩽ 0,
                   x ⩽ 0

Shade the intersection of the graphs of the four
inequalities.

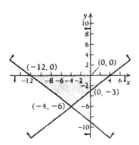

We find the vertex (-12, 0) by solving the system
        4y + 3x = -36,
             y = 0.

We find the vertex (0,0) by solving the system
        y = 0,
        x = 0.

We find the vertex (0,-3) by solving the system

$$4y - 3x = -12$$
$$x = 0.$$

We find the vertex (-4,-6) by solving the system

$$4y - 3x = -12,$$
$$4y + 3x = -36.$$

44. $8x + 5y \leqslant 40,$
    $x + 2y \leqslant 8,$
    $x \geqslant 0,$
    $y \geqslant 0$

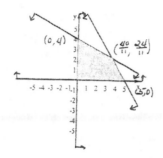

5. Graph: $3x + 4y \geqslant 12,$
    $5x + 6y \leqslant 30,$
    $1 \leqslant x \leqslant 3$

Shade the intersection of the graphs of the given inequalities.

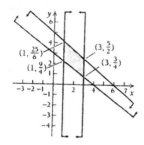

We find the vertex $\left[1, \dfrac{25}{6}\right]$ by solving the system

$$5x + 6y = 30,$$
$$x = 1.$$

We find the vertex $\left[3, \dfrac{5}{2}\right]$ by solving the system

$$5x + 6y = 30,$$
$$x = 3.$$

We find the vertex $\left[3, \dfrac{3}{4}\right]$ by solving the system

$$3x + 4y = 12,$$
$$x = 3.$$

We find the vertex $\left[1, \dfrac{9}{4}\right]$ by solving the system

$$3x + 4y = 12,$$
$$x = 1.$$

46. Graph: $y - x \geqslant 1,$
    $y - x \leqslant 3,$
    $2 \leqslant x \leqslant 5$

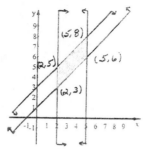

47. The graph of the domain is as follows:

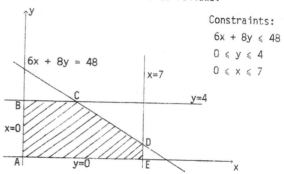

Constraints:
$6x + 8y \leqslant 48$
$0 \leqslant y \leqslant 4$
$0 \leqslant x \leqslant 7$

We need to find the coordinate of each vertex.

Vertex A: (0, 0)

Vertex B:
  We solve the system $x = 0$ and $y = 4$.
  The coordinates of point B are (0, 4).

Vertex C:
  We solve the system $6x + 8y = 48$ and $y = 4$.
  The coordinates of point C are $\left[\dfrac{8}{3}, 4\right]$.

Vertex D:
  We solve the system $6x + 8y = 48$ and $x = 7$.
  The coordinates of point D are $\left[7, \dfrac{3}{4}\right]$.

Vertex E:
  We solve the system $x = 7$ and $y = 0$.
  The coordinates of point E are (7, 0).

We compute the value of P for each vertex.

Vertex	$P = 17x - 3y + 60$
A(0, 0)	$17 \cdot 0 - 3 \cdot 0 + 60 = 60$
B(0, 4)	$17 \cdot 0 - 3 \cdot 4 + 60 = 48$
C$\left[\dfrac{8}{3}, 4\right]$	$17 \cdot \dfrac{8}{3} - 3 \cdot 4 + 60 = 66\dfrac{2}{3}$
D$\left[7, \dfrac{3}{4}\right]$	$17 \cdot 7 - 3 \cdot \dfrac{3}{4} + 60 = 176\dfrac{3}{4}$
E(7, 0)	$17 \cdot 7 - 3 \cdot 0 + 60 = 179$

The maximum value of P is 179 when $x = 7$ and $y = 0$.

The minimum value of P is 48 when $x = 0$ and $y = 4$.

48. The maximum value of Q is 151 when $x = 3$ and $y = \frac{5}{4}$. The minimum value of Q is $78\frac{2}{5}$ when $x = \frac{4}{5}$ and $y = 4$.

49. The graph of the domain is as follows:

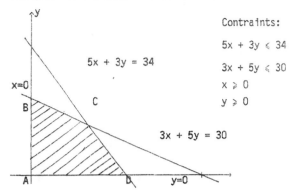

Contraints:

$5x + 3y \leqslant 34$

$3x + 5y \leqslant 30$

$x \geqslant 0$

$y \geqslant 0$

We need to find the coordinates of each vertex.

Vertex A: (0, 0)

Vertex B:

We solve the system $3x + 5y = 30$ and $x = 0$. The coordinates of point B are (0, 6).

Vertex C:

We solve the system $5x + 3y = 34$ and $3x + 5y = 30$. The coordinates of point C are (5, 3).

Vertex D:

We solve the system $5x + 3y = 34$ and $y = 0$. The coordinates of point D are $\left(\frac{34}{5}, 0\right)$.

We compute the value of F for each vertex.

Vertex	$F = 5x + 36y$
A(0, 0)	$5\cdot 0 + 36\cdot 0 = 0$
B(0, 6)	$5\cdot 0 + 36\cdot 6 = 216$
C(5, 3)	$5\cdot 5 + 36\cdot 3 = 133$
$D\left(\frac{34}{5}, 0\right)$	$5 \cdot \frac{34}{5} + 36\cdot 0 = 34$

The maximum value of F is 216 when $x = 0$ and $y = 6$.

The minimum value of F is 0 when $x = 0$ and $y = 0$.

50. The maximum value of G is 81.2 when $x = 0$ and $y = 5.8$. The minimum value of G is 0 when $x = 0$ and $y = 0$.

51. Let x = the number of Biscuit Jumbos and y = the number of Mitimite Biscuits to be made per day. The income I is given by

$$I = \$0.10x + \$0.08y$$

subject to the constraints

$x + y \leqslant 200,$

$2x + y \leqslant 300$

$x \geqslant 0,$

$y \geqslant 0.$

We graph the system of inequalities, determine the vertices, and find the value of I at each vertex.

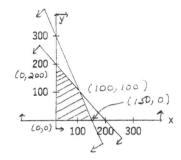

Vertex	$I = \$0.10x + \$0.08y$
(0, 0)	$\$0.10(0) + \$0.08(0) = \$0$
(0, 200)	$\$0.10(0) + \$0.08(200) = \$16$
(100, 100)	$\$0.10(100) + \$0.08(100) = \$18$
(150, 0)	$\$0.10(150) + \$0.08(0) = \$15$

The company will have a maximum income of $18 when 100 of each type of biscuit is made.

52. The maximum number of miles is 480 when the car uses 9 gal and the moped uses 3 gal.

53. Let x = the number of type A questions and y = the number of type B questions you answer. The score S is given by

$$S = 10x + 25y$$

subject to the constraints

$3 \leqslant x \leqslant 12,$

$4 \leqslant y \leqslant 15,$

$x + y \leqslant 20.$

We graph the system of inequalities, determine the vertices, and find the value of S at each vertex.

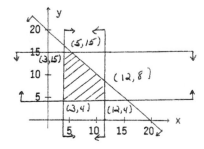

Vertex	S = 10x + 25y
(3, 4)	10·3 + 25·4 = 130
(3, 15)	10·3 + 25·15 = 405
(5, 15)	10·5 + 25·15 = 425
(12, 8)	10·12 + 25·8 = 320
(12, 4)	10·12 + 25·4 = 220

The maximum score is 425 points when 5 type A questions and 15 type B questions are answered.

54. The maximum test score is 102 when 8 questions of type A and 10 questions of type B are answered.

55. Let x = the number of units of lumber and y = the number of units of plywood produced per week. The profit P is given by

$$P = \$20x + \$30y$$

subject to the constraints

$$x + y \leqslant 400,$$
$$x \geqslant 100,$$
$$y \geqslant 150.$$

We graph the system of inequalities, determine the vertices and find the value of P at each vertex.

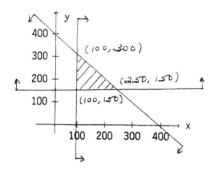

Vertex	P = \$20x + \$30y
(100, 150)	\$20·100 + \$30·150 = \$6500
(100, 300)	\$20·100 + \$30·300 = \$11,000
(250, 150)	\$20·250 + \$30·150 = \$9500

The maximum profit of \$11,000 is achieved by producing 100 units of lumber and 300 units of plywood.

56. The maximum profit of \$8000 occurs by planting 80 acres of corn and 160 acres of oats.

57. Let x = the amount invested in corporate bonds and y = the amount invested in municipal bonds. The income I is given by

$$I = 0.08x + 0.075y$$

subject to the constraints

$$x + y \leqslant \$40,000,$$
$$\$6000 \leqslant x \leqslant \$22,000,$$
$$y \leqslant \$30,000.$$

We graph the system of inequalities, determine the vertices, and find the value of I at each vertex.

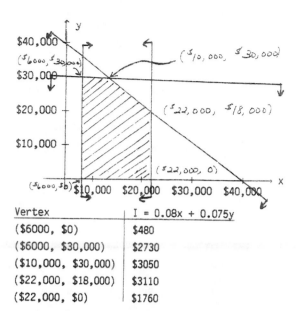

Vertex	I = 0.08x + 0.075y
(\$6000, \$0)	\$480
(\$6000, \$30,000)	\$2730
(\$10,000, \$30,000)	\$3050
(\$22,000, \$18,000)	\$3110
(\$22,000, \$0)	\$1760

The maximum income of \$3110 occurs when \$22,000 is invested in corporate bonds and \$18,000 is invested in municipal bonds.

58. The maximum interest income is \$1395 when \$7000 is invested in bank X and \$15,000 is invested in bank Y.

59. Let x = the number of batches of Smello and y = the number of batches of Roppo to be made. We organize the information in a table.

	Composition		Number of lbs Available
	Smello	Roppo	
Number of Batches	x	y	
English tobacco	12	8	3000
Virginia tobacco	0	8	2000
Latakia tobacco	4	0	500
Profit per Batch	\$10.56	\$6.40	

The profit P is given by

$$P = \$10.56x + \$6.40y$$

subject to the constraints

$$12x + 8y \leqslant 3000,$$
$$8y \leqslant 2000,$$
$$4x \leqslant 500,$$
$$x \geqslant 0,$$
$$y \geqslant 0.$$

We graph the system of inequalities, determine the vertices, and find the value of P at each vertex.

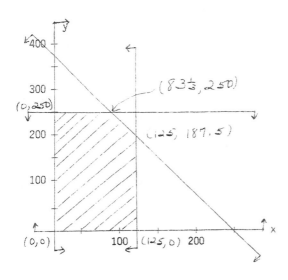

Vertex	P = $10.56x + $6.40y
(0, 0)	$0
(125, 0)	$1320
(125, 187.5)	$2520
$\left(83\frac{1}{3}, 250\right)$	$2480
(0, 250)	$1600

The maximum profit of $2520 occurs when 125 batches of Smello and 187.5 batches of Roppo are made.

60. The maximum profit per day is $192 when 2 knit suits and 4 worsted suits are made.

61. Graph: $y \geqslant x^2 - 2$
            $y \leqslant 2 - x^2$

Graph the equation $y = x^2 - 2$. A few solutions are (0, -2), (1, -1), (-1, -1), (2, 2), and (-2, 2). Plot these points and draw the graph solid since the inequality is $\geqslant$. Since for any point above the graph y is greater than $x^2 - 2$, we shade above $y = x^2 - 2$.

$y \geqslant x^2 - 2$

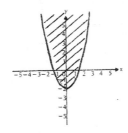

Graph the equation $y = 2 - x^2$. A few solutions are (0, 2), (1, 1), (-1, 1), (2, -2), and (-2, -2). Plot these points and draw the graph solid since the inequality is $\leqslant$. Since for any point below the graph y is less than $2 - x^2$, we shade below $y = 2 - x^2$.

$y < 2 - x^2$

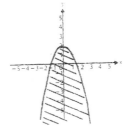

Now graph the intersection of the graphs.

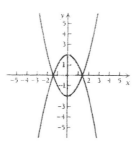

62. $y < x + 1$
     $y \geqslant x^2$

63. The team gets 2w points from w wins and t points from t ties. The number of wins and ties must each be nonnegative.

We have
    $2w + t \geqslant 60,$
        $w \geqslant 0,$
        $t \geqslant 0.$

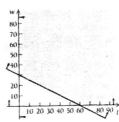

64. $75a + 35c > 1000,$
          $a \geqslant 0,$
          $c \geqslant 0$

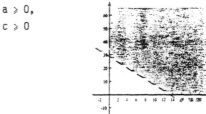

65. Graph: 2L + 2W ≤ 248,
          0 ≤ L ≤ 74,
          0 ≤ W ≤ 50

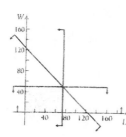

66. |x| + |y| ≤ 1

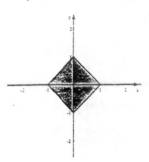

67.

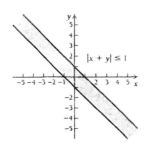

$|x + y| \le 1$

68. |x - y| > 0

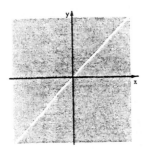

69.

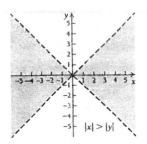

$|x| > |y|$

70. The maximum income of $3350 occurs when 25 chairs
     and 9.5 sofas are made. (A more practical answer
     is that the maximum income of $3200 is achieved
     when 25 chairs and 9 sofas are made.)

71. $3.5x + 5.6y \le 10$
       $2.1x - 12.3y \ge 9$
       $x \ge -2.3$

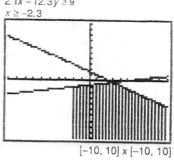

$[-10, 10] \times [-10, 10]$

Vertices: (-2.30, 3.11), (-2.30, -1.12),
           (3.17, -0.19)

72. $20x - 45y > 105$
       $12.5x + 13y < 10$
       $0.7x - 7y > -35$

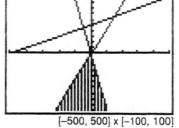

$[-500, 500] \times [-100, 100]$

Vertices: (21.29, 7.13), (-3.99, 4.60),
           (2.21, 1.35)

Exercise Set 6.8

1. $\dfrac{x + 7}{(x - 3)(x + 2)} = \dfrac{A}{x - 3} + \dfrac{B}{x + 2}$

    $\dfrac{x + 7}{(x - 3)(x + 2)} = \dfrac{A(x + 2) + B(x - 3)}{(x - 3)(x + 2)}$ (Adding)

Equate the numerators:

$x + 7 = A(x + 2) + B(x - 3)$

Let $x + 2 = 0$, or $x = -2$. Then we get

$-2 + 7 = 0 + B(-2 - 3)$

     $5 = -5B$

    $-1 = B$

Next let $x - 3 = 0$, or $x = 3$. Then we get

$3 + 7 = A(3 + 2) + 0$

    $10 = 5A$

     $2 = A$

The decomposition is as follows:

$$\frac{2}{x-3} + \frac{-1}{x+2}$$

2. $\dfrac{1}{x+1} + \dfrac{1}{x-1}$

3. $\dfrac{7x-1}{6x^2 - 5x + 1}$

$= \dfrac{7x-1}{(3x-1)(2x-1)}$ (Factoring the denominator)

$= \dfrac{A}{3x-1} + \dfrac{B}{2x-1}$

$= \dfrac{A(2x-1) + B(3x-1)}{(3x-1)(2x-1)}$ (Adding)

Equate the numerators:

$7x - 1 = A(2x - 1) + B(3x - 1)$

Let $2x - 1 = 0$, or $x = \frac{1}{2}$. Then we get

$7\left(\frac{1}{2}\right) - 1 = 0 + B\left[3 \cdot \frac{1}{2} - 1\right]$

$\qquad \frac{5}{2} = \frac{1}{2} B$

$\qquad 5 = B$

Next let $3x - 1 = 0$, or $x = \frac{1}{3}$. We get

$7\left(\frac{1}{3}\right) - 1 = A\left[2 \cdot \frac{1}{3} - 1\right] + 0$

$\qquad \frac{7}{3} - 1 = A\left[\frac{2}{3} - 1\right]$

$\qquad \frac{4}{3} = -\frac{1}{3} A$

$\qquad -4 = A$

The decomposition is as follows:

$$\frac{-4}{3x-1} + \frac{5}{2x-1}$$

4. $\dfrac{-5}{4x+3} + \dfrac{7}{3x-5}$

5. $\dfrac{3x^2 - 11x - 26}{(x^2 - 4)(x+1)}$

$= \dfrac{3x^2 - 11x - 26}{(x+2)(x-2)(x+1)}$ (Factoring the denominator)

$= \dfrac{A}{x+2} + \dfrac{B}{x-2} + \dfrac{C}{x+1}$

$= \dfrac{A(x-2)(x+1) + B(x+2)(x+1) + C(x+2)(x-2)}{(x+2)(x-2)(x+1)}$

(Adding)

Equate the numerators:

$3x^2 - 11x - 26 = A(x-2)(x+1) +$
$\qquad\qquad B(x+2)(x+1) + C(x+2)(x-2)$

Let $x + 2 = 0$, or $x = -2$. Then, we get

$3(-2)^2 - 11(-2) - 26 = A(-2-2)(-2+1) + 0 + 0$

$\qquad 12 + 22 - 26 = A(-4)(-1)$

$\qquad\qquad 8 = 4A$

$\qquad\qquad 2 = A$

Next let $x - 2 = 0$, or $x = 2$. Then, we get

$3 \cdot 2^2 - 11 \cdot 2 - 26 = 0 + B(2+2)(2+1) + 0$

$\qquad 12 - 22 - 26 = B \cdot 4 \cdot 3$

$\qquad\qquad -36 = 12B$

$\qquad\qquad -3 = B$

Finally let $x + 1 = 0$, or $x = -1$. We get

$3(-1)^2 - 11(-1) - 26 = 0 + 0 + C(-1+2)(-1-2)$

$\qquad 3 + 11 - 26 = C(1)(-3)$

$\qquad\qquad -12 = -3C$

$\qquad\qquad 4 = C$

The decomposition is as follows:

$$\frac{2}{x+2} + \frac{-3}{x-2} + \frac{4}{x+1}$$

6. $\dfrac{6}{x-4} + \dfrac{3}{x-2} + \dfrac{-4}{x+1}$

7. $\dfrac{9}{(x+2)^2(x-1)}$

$= \dfrac{A}{x+2} + \dfrac{B}{(x+2)^2} + \dfrac{C}{x-1}$

$= \dfrac{A(x+2)(x-1) + B(x-1) + C(x+2)^2}{(x+2)^2(x-1)}$ (Adding)

Equate the numerators:

$9 = A(x+2)(x-1) + B(x-1) + C(x+2)^2$ (1)

Let $x - 1 = 0$, or $x = 1$. Then, we get

$9 = 0 + 0 + C(1+2)^2$

$9 = 9C$

$1 = C$

Next let $x + 2 = 0$, or $x = -2$. Then, we get

$\quad 9 = 0 + B(-2-1) + 0$

$\quad 9 = -3B$

$-3 = B$

To find A we first simplify Eq. (1).

$9 = A(x^2 + x - 2) + B(x-1) + C(x^2 + 4x + 4)$

$\quad = Ax^2 + Ax - 2A + Bx - B + Cx^2 + 4Cx + 4C$

$\quad = (A+C)x^2 + (A+B+4C)x + (-2A - B + 4C)$

Then we equate the coefficients of x^2.

$0 = A + C$

$0 = A + 1$ (Substituting 1 for C)

$-1 = A$

The decomposition is as follows:

$$\frac{-1}{x+2} + \frac{-3}{(x+2)^2} + \frac{1}{x-1}$$

8. $\dfrac{1}{x-2} + \dfrac{3}{(x-2)^2} + \dfrac{-2}{(x-2)^3}$

9. $$\frac{2x^2 + 3x + 1}{(x^2 - 1)(2x - 1)}$$

$$= \frac{2x^2 + 3x + 1}{(x + 1)(x - 1)(2x - 1)} \quad \text{(Factoring the denominator)}$$

$$= \frac{A}{x + 1} + \frac{B}{x - 1} + \frac{C}{2x - 1}$$

$$= \frac{A(x-1)(2x-1) + B(x+1)(2x-1) + C(x+1)(x-1)}{(x + 1)(x - 1)(2x - 1)}$$

(Adding)

Equate the numerators:

$$2x^2 + 3x + 1 = A(x - 1)(2x - 1) +$$
$$B(x + 1)(2x - 1) + C(x + 1)(x - 1)$$

Let $x + 1 = 0$, or $x = -1$. Then, we get

$$2(-1)^2 + 3(-1) + 1 = A(-1 - 1)[2(-1) - 1] + 0 + 0$$
$$2 - 3 + 1 = A(-2)(-3)$$
$$0 = 6A$$
$$0 = A$$

Next let $x - 1 = 0$, or $x = 1$. Then, we get

$$2 \cdot 1^2 + 3 \cdot 1 + 1 = 0 + B(1 + 1)(2 \cdot 1 - 1) + 0$$
$$2 + 3 + 1 = B \cdot 2 \cdot 1$$
$$6 = 2B$$
$$3 = B$$

Finally we let $2x - 1 = 0$, or $x = \frac{1}{2}$. We get

$$2\left[\frac{1}{2}\right]^2 + 3\left[\frac{1}{2}\right] + 1 = 0 + 0 + C\left[\frac{1}{2} + 1\right]\left[\frac{1}{2} - 1\right]$$
$$\frac{1}{2} + \frac{3}{2} + 1 = C \cdot \frac{3}{2} \cdot \left[-\frac{1}{2}\right]$$
$$3 = -\frac{3}{4}C$$
$$-4 = C$$

The decomposition is as follows:

$$\frac{3}{x - 1} + \frac{-4}{2x - 1}$$

10. $$\frac{-4}{x - 3} + \frac{3}{x - 2} + \frac{2}{x - 1}$$

11. $$\frac{x^4 - 3x^3 - 3x^2 + 10}{(x + 1)^2(x - 3)}$$

$$= \frac{x^4 - 3x^3 - 3x^2 + 10}{x^3 - x^2 - 5x - 3} \quad \text{(Multiplying the denominator)}$$

Since the degree of the numerator is greater than the degree of the denominator, we divide.

$$
\begin{array}{r}
x - 2 \\
x^3 - x^2 - 5x - 3 \overline{\smash{\big)}\ x^4 - 3x^3 - 3x^2 + 0x + 10} \\
\underline{x^4 - x^3 - 5x^2 - 3x} \\
-2x^3 + 2x^2 + 3x + 10 \\
\underline{-2x^3 + 2x^2 + 10x + 6} \\
-7x + 4
\end{array}
$$

The original expression is thus equivalent to the following:

$$x - 2 + \frac{-7x + 4}{x^3 - x^2 - 5x - 3}$$

We proceed to decompose the fraction.

$$\frac{-7x + 4}{(x + 1)^2(x - 3)}$$

$$= \frac{A}{x + 1} + \frac{B}{(x + 1)^2} + \frac{C}{x - 3}$$

$$= \frac{A(x + 1)(x - 3) + B(x - 3) + C(x + 1)^2}{(x + 1)^2(x - 3)} \quad \text{(Adding)}$$

Equate the numerators:

$$-7x + 4 = A(x + 1)(x - 3) + B(x - 3) + C(x + 1)^2 \quad (1)$$

Let $x - 3 = 0$, or $x = 3$. Then, we get

$$-7 \cdot 3 + 4 = 0 + 0 + C(3 + 1)^2$$
$$-17 = 16C$$
$$-\frac{17}{16} = C$$

Let $x + 1 = 0$, or $x = -1$. Then, we get

$$-7(-1) + 4 = 0 + B(-1 - 3) + 0$$
$$11 = -4B$$
$$-\frac{11}{4} = B$$

To find A we first simplify Eq. (1).

$$-7x + 4 = A(x^2 - 2x - 3) + B(x - 3) + C(x^2 + 2x + 1)$$
$$= Ax^2 - 2Ax - 3A + Bx - 3B + Cx^2 - 2Cx + C$$
$$= (A + C)x^2 + (-2A + B - 2C)x + (-3A - 3B + C)$$

Then equate the coefficients of x^2.

$$0 = A + C$$

Substituting $-\frac{17}{16}$ for C, we get $A = \frac{17}{16}$.

The decomposition is as follows:

$$\frac{17/16}{x + 1} + \frac{-11/4}{(x + 1)^2} + \frac{-17/16}{x - 3}$$

The original expression is equivalent to the following:

$$x - 2 + \frac{17/16}{x + 1} + \frac{-11/4}{(x + 1)^2} + \frac{-17/16}{x - 3}$$

12. $$10x - 5 + \frac{6}{x - 3} + \frac{14}{x + 2}$$

13. $$\frac{-x^2 + 2x - 13}{(x^2 + 2)(x - 1)}$$

$$= \frac{Ax + B}{x^2 + 2} + \frac{C}{x - 1}$$

$$= \frac{(Ax + B)(x - 1) + C(x^2 + 2)}{(x^2 + 2)(x - 1)} \quad \text{(Adding)}$$

Equate the numerators:

$$-x^2 + 2x - 13 = (Ax + B)(x - 1) + C(x^2 + 2) \quad (1)$$

Let $x - 1 = 0$, or $x = 1$. Then we get

$$-1^2 + 2 \cdot 1 - 13 = 0 + C(1^2 + 2)$$
$$-1 + 2 - 13 = C(1 + 2)$$
$$-12 = 3C$$
$$-4 = C$$

To find A and B we first simplify Eq. (1).

$-x^2 + 2x - 13 = Ax^2 - Ax + Bx - B + Cx^2 + 2C$

$\qquad\qquad\qquad = (A+C)x^2 + (-A + B)x + (-B + 2C)$

Equate the coefficients of x^2:

$-1 = A + C$

Substituting -4 for C, we get $A = 3$.

Equate the constant terms:

$-13 = -B + 2C$

Substituting -4 for C, we get $B = 5$.

The decomposition is as follows:

$\dfrac{3x + 5}{x^2 + 2} + \dfrac{-4}{x - 1}$

14. $\dfrac{41x + 3}{x^2 + 1} + \dfrac{-15}{x + 5}$

15. $\qquad \dfrac{6 + 26x - x^2}{(2x - 1)(x + 2)^2}$

$= \dfrac{A}{2x - 1} + \dfrac{B}{x + 2} + \dfrac{C}{(x + 2)^2}$

$= \dfrac{A(x + 2)^2 + B(2x - 1)(x + 2) + C(2x - 1)}{(2x - 1)(x + 2)^2}$

$\qquad\qquad\qquad\qquad\qquad\qquad$ (Adding)

Equate the numerators:

$6 + 26x - x^2 = A(x + 2)^2 + B(2x - 1)(x + 2) +$
$\qquad\qquad\qquad C(2x - 1) \qquad (1)$

Let $2x - 1 = 0$, or $x = \frac{1}{2}$. Then, we get

$6 + 26 \cdot \frac{1}{2} - \left[\frac{1}{2}\right]^2 = A\left[\frac{1}{2} + 2\right]^2 + 0 + 0$

$\qquad 6 + 13 - \frac{1}{4} = A\left[\frac{5}{2}\right]^2$

$\qquad\qquad \frac{75}{4} = \frac{25}{4} A$

$\qquad\qquad\quad 3 = A$

Let $x + 2 = 0$, or $x = -2$. We get

$6 + 26(-2) - (-2)^2 = 0 + 0 + C[2(-2) - 1]$

$\qquad 6 - 52 - 4 = -5C$

$\qquad\qquad -50 = -5C$

$\qquad\qquad\quad 10 = C$

To find B we first simplify Eq. (1).

$6 + 26x - x^2 = A(x^2 + 4x + 4) + B(2x^2 + 3x - 2) +$
$\qquad\qquad\qquad C(2x - 1)$

$\qquad\qquad = Ax^2 + 4Ax + 4A + 2Bx^2 + 3Bx - 2B +$
$\qquad\qquad\quad 2Cx - C$

$\qquad\qquad = (A + 2B)x^2 + (4A + 3B + 2C)x +$
$\qquad\qquad\quad (4A - 2B - C)$

Equate the coefficients of x^2:

$-1 = A + 2B$

Substituting 3 for A, we obtain $B = -2$.

The decomposition is as follows:

$\dfrac{3}{2x - 1} + \dfrac{-2}{x + 2} + \dfrac{10}{(x + 2)^2}$

16. $\dfrac{1}{x - 1} + \dfrac{-1}{x + 1} + \dfrac{3}{(x + 1)^2} + \dfrac{-3}{(x + 1)^3} + \dfrac{2}{(x + 1)^4}$

17. $\dfrac{6x^3 + 5x^2 + 6x - 2}{2x^2 + x - 1}$

Since the degree of the numerator is greater than the degree of the denominator, we divide.

$$
\begin{array}{r}
3x + 1 \\
2x^2 + x - 1 \overline{\smash{)}\, 6x^3 + 5x^2 + 6x - 2} \\
\underline{6x^3 + 3x^2 - 3x} \\
2x^2 + 9x - 2 \\
\underline{2x^2 + x - 1} \\
8x - 1
\end{array}
$$

The original expression is equivalent to

$3x + 1 + \dfrac{8x - 1}{2x^2 + x - 1}$

We proceed to decompose the fraction.

$\dfrac{8x - 1}{2x^2 + x - 1} = \dfrac{8x - 1}{(2x - 1)(x + 1)} \qquad$ (Factoring the denominator)

$= \dfrac{A}{2x - 1} + \dfrac{B}{x + 1}$

$= \dfrac{A(x + 1) + B(2x - 1)}{(2x - 1)(x + 1)} \qquad$ (Adding)

Equate the numerators:

$8x - 1 = A(x + 1) + B(2x - 1)$

Let $x + 1 = 0$, or $x = -1$. Then we get

$8(-1) - 1 = 0 + B[2(-1) - 1]$

$\qquad -8 - 1 = B(-2 - 1)$

$\qquad\qquad -9 = -3B$

$\qquad\qquad\quad 3 = B$

Next let $2x - 1 = 0$, or $x = \frac{1}{2}$. We get

$8\left[\frac{1}{2}\right] - 1 = A\left[\frac{1}{2} + 1\right] + 0$

$\qquad 4 - 1 = A\left[\frac{3}{2}\right]$

$\qquad\qquad 3 = \frac{3}{2} A$

$\qquad\qquad 2 = A$

The decomposition is

$\dfrac{2}{2x - 1} + \dfrac{3}{x + 1}.$

The original expression is equivalent to

$3x + 1 + \dfrac{2}{2x - 1} + \dfrac{3}{x + 1}.$

18. $2x - 1 + \dfrac{1}{x + 3} + \dfrac{-4}{x - 1}$

19. $$\frac{2x^2 - 11x + 5}{(x - 3)(x^2 + 2x - 5)}$$

$$= \frac{A}{x - 3} + \frac{Bx + C}{x^2 + 2x - 5}$$

$$= \frac{A(x^2 + 2x - 5) + (Bx + C)(x - 3)}{(x - 3)(x^2 + 2x - 5)} \quad \text{(Adding)}$$

Equate the numerators:

$$2x^2 - 11x + 5 = A(x^2 + 2x - 5) +$$
$$(Bx + C)(x - 3) \quad (1)$$

Let $x - 3 = 0$, or $x = 3$. Then, we get

$$2 \cdot 3^2 - 11 \cdot 3 + 5 = A(3^2 + 2 \cdot 3 - 5) + 0$$
$$18 - 33 + 5 = A(9 + 6 - 5)$$
$$-10 = 10A$$
$$-1 = A$$

To find B and C, we first simplify Eq. (1).

$$2x^2 - 11x + 5 = Ax^2 + 2Ax - 5A + Bx^2 - 3Bx + Cx - 3C$$
$$= (A + B)x^2 + (2A - 3B + C)x + (-5A - 3C)$$

Equate the coefficients of x^2:

$$2 = A + B$$

Substituting -1 for A, we get B = 3.

Equate the constant terms:

$$5 = -5A - 3C$$

Substituting -1 for A, we get C = 0.

The decomposition is as follows:

$$\frac{-1}{x - 3} + \frac{3x}{x^2 + 2x - 5}$$

20. $$\frac{2}{x - 5} + \frac{x}{x^2 + x - 4}$$

21. $$\frac{-4x^2 - 2x + 10}{(3x + 5)(x + 1)^2}$$

The decomposition looks like

$$\frac{A}{3x + 5} + \frac{B}{x + 1} + \frac{C}{(x + 1)^2}.$$

Add and equate the numerators.

$$-4x^2 - 2x + 10 = A(x + 1)^2 + B(3x + 5)(x + 1) + C(3x + 5)$$
$$= A(x^2 + 2x + 1) + B(3x^2 + 8x + 5) + C(3x + 5)$$

or

$$-4x^2 - 2x + 10 = (A + 3B)x^2 + (2A + 8B + 3C)x + (A + 5B + 5C)$$

Then equate corresponding coefficients.

$$-4 = A + 3B \qquad \text{(Coefficients of } x^2\text{-terms)}$$
$$-2 = 2A + 8B + 3C \qquad \text{(Coefficients of } x\text{-terms)}$$
$$10 = A + 5B + 5C \qquad \text{(Constant terms)}$$

We solve this system of three equations and find A = 5, B = -3, C = 4.

The decomposition is

$$\frac{5}{3x + 5} + \frac{-3}{x + 1} + \frac{4}{(x + 1)^2}.$$

22. $$\frac{6}{x - 4} + \frac{1}{2x - 1} + \frac{-3}{(2x - 1)^2}$$

23. $$\frac{36x + 1}{12x^2 - 7x - 10} = \frac{36x + 1}{(4x - 5)(3x + 2)}$$

The decomposition looks like

$$\frac{A}{4x - 5} + \frac{B}{3x + 2}.$$

Add and equate the numerators.

$$36x + 1 = A(3x + 2) + B(4x - 5)$$

or $36x + 1 = (3A + 4B)x + (2A - 5B)$

Then equate corresponding coefficients.

$$36 = 3A + 4B \qquad \text{(Coefficients of } x\text{-terms)}$$
$$1 = 2A - 5B \qquad \text{(Constant terms)}$$

We solve this system of equations and find A = 8 and B = 3.

The decomposition is

$$\frac{8}{4x - 5} + \frac{3}{3x + 2}.$$

24. $$\frac{7}{6x - 3} + \frac{-4}{x + 7}$$

25. $$\frac{-4x^2 - 9x + 8}{(3x^2 + 1)(x - 2)}$$

The decomposition looks like

$$\frac{Ax + B}{3x^2 + 1} + \frac{C}{x - 2}.$$

Add and equate the numerators.

$$-4x^2 - 9x + 8 = (Ax + B)(x - 2) + C(3x^2 + 1)$$
$$= Ax^2 - 2Ax + Bx - 2B + 3Cx^2 + C$$

or $-4x^2 - 9x + 8 = (A + 3C)x^2 + (-2A + B)x + (-2B + C)$

Then equate corresponding coefficients.

$$-4 = A + 3C \qquad \text{(Coefficients of } x^2\text{-terms)}$$
$$-9 = -2A + B \qquad \text{(Coefficients of } x\text{-terms)}$$
$$8 = -2B + C \qquad \text{(Constant terms)}$$

We solve this system of equations and find A = 2, B = -5, C = -2.

The decomposition is

$$\frac{2x - 5}{3x^2 + 1} + \frac{-2}{x - 2}.$$

26. $$\frac{5x + 1}{x^2 + 4} + \frac{6}{x - 8}$$

27. $\dfrac{x}{x^4 - a^4}$

$= \dfrac{x}{(x^2 + a^2)(x + a)(x - a)}$ (Factoring the denominator)

$= \dfrac{Ax + B}{x^2 + a^2} + \dfrac{C}{x + a} + \dfrac{D}{x - a}$

$= \dfrac{(Ax+B)(x+a)(x-a) + C(x^2+a^2)(x-a) + D(x^2+a^2)(x+a)}{(x^2 + a^2)(x + a)(x - a)}$

Equate the numerators:

$x = (Ax + B)(x + a)(x - a) + C(x^2 + a^2)(x - a) +$
$\qquad D(x^2 + a^2)(x + a)$

Let $x - a = 0$, or $x = a$. Then, we get

$a = 0 + 0 + D(a^2 + a^2)(a + a)$

$a = D(2a^2)(2a)$

$a = 4a^3 D$

$\dfrac{1}{4a^2} = D$

Let $x + a = 0$, or $x = -a$. We get

$-a = 0 + C[(-a)^2 + a^2](-a - a) + 0$

$-a = C(2a^2)(-2a)$

$-a = -4a^3 C$

$\dfrac{1}{4a^2} = C$

Equate the coefficients of x^3:

$0 = A + C + D$

Substituting $\dfrac{1}{4a^2}$ for C and for D, we get

$A = -\dfrac{1}{2a^2}.$

Equate the constant terms:

$0 = -Ba^2 - Ca^3 + Da^3$

Substitute $\dfrac{1}{4a^2}$ for C and for D. Then solve for B.

$0 = -Ba^2 - \dfrac{1}{4a^2} \cdot a^3 + \dfrac{1}{4a^2} \cdot a^3$

$0 = -Ba^2$

$0 = B$

The decomposition is as follows:

$\dfrac{-\dfrac{1}{2a^2} x}{x^2 + a^2} + \dfrac{\dfrac{1}{4a^2}}{x + a} + \dfrac{\dfrac{1}{4a^2}}{x - a}$

28. $\dfrac{-1}{x + 1} + \dfrac{1}{x - 2} + \dfrac{6}{(x - 2)^2} + \dfrac{12}{(x - 2)^3} + \dfrac{24}{(x - 2)^4}$

29. $f(x) = \dfrac{3x}{x^2 + 5x + 4} = \dfrac{3x}{(x + 4)(x + 1)}$

$\dfrac{3x}{(x + 4)(x + 1)} = \dfrac{A}{x + 4} + \dfrac{B}{x + 1}$

$= \dfrac{A(x + 1) + B(x + 4)}{(x + 4)(x + 1)}$

Equate the numerators:

$3x = A(x + 1) + B(x + 4)$

Let $x + 1 = 0$, or $x = -1$. We get

$3(-1) = 0 + B(-1 + 4)$

$-3 = 3B$

$-1 = B$

Next let $x + 4 = 0$, or $x = -4$. Then we get

$3(-4) = A(-4 + 1) + 0$

$-12 = -3A$

$4 = A$

The decomposition is $f(x) = \dfrac{4}{x + 4} + \dfrac{-1}{x + 1}$, or

$\dfrac{4}{x + 4} - \dfrac{1}{x + 1}.$

We graph by addition of ordinates (adding respective fraction values.)

If $x = 4$, $y = \dfrac{4}{4 + 4} - \dfrac{1}{4 + 1} = \dfrac{1}{2} - \dfrac{1}{5} = \dfrac{3}{10}.$

If $x = 2$, $y = \dfrac{4}{2 + 4} - \dfrac{1}{2 + 1} = \dfrac{2}{3} - \dfrac{1}{3} = \dfrac{1}{3}.$

If $x = 0$, $y = \dfrac{4}{0 + 4} - \dfrac{1}{0 + 1} = 1 - 1 = 0.$

If $x = -2$, $y = \dfrac{4}{-2 + 4} - \dfrac{1}{-2 + 1} = 2 + 1 = 3.$

Continue adding fraction values as above. Then plot these points and draw the graph. Note that $x = -4$ and $x = -1$ are vertical asymptotes.

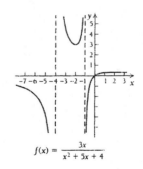

$f(x) = \dfrac{3x}{x^2 + 5x + 4}$

30. $f(x) = \dfrac{\frac{1}{2}}{x - 3} + \dfrac{\frac{1}{2}}{x + 1}$

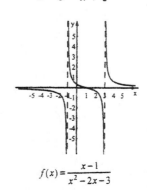

$f(x) = \dfrac{x - 1}{x^2 - 2x - 3}$

31. $\dfrac{1 + \ln x^2}{(\ln x + 2)(\ln x - 3)^2} = \dfrac{1 + 2\ln x}{(\ln x + 2)(\ln x - 3)^2}$

Let $u = \ln x$. Then we have:

$$\dfrac{1 + 2u}{(u + 2)(u - 3)^2} = \dfrac{A}{u + 2} + \dfrac{B}{(u - 3)} + \dfrac{C}{(u - 3)^2}$$

$$= \dfrac{A(u-3)^2 + B(u+2)(u-3) + C(u+2)}{(u + 2)(u - 3)^2}$$

Equate the numerators:

$1 + 2u = A(u - 3)^2 + B(u + 2)(u - 3) + C(u + 2)$

Let $u - 3 = 0$, or $u = 3$.

$1 + 2\cdot3 = 0 + 0 + C(5)$

$7 = 5C$

$\dfrac{7}{5} = C$

Let $u + 2 = 0$, or $u = -2$.

$1 + 2(-2) = A(-2 - 3)^2 + 0 + 0$

$-3 = 25A$

$-\dfrac{3}{25} = A$

To find B, we equate the coefficients of u^2:

$0 = A + B$

Substituting $-\dfrac{3}{25}$ for A and solving for B, we get $B = \dfrac{3}{25}$.

The decomposition of $\dfrac{1 + 2u}{(u + 2)(u - 3)^2}$ is as follows:

$$-\dfrac{3}{25(u + 2)} + \dfrac{3}{25(u - 3)} + \dfrac{7}{5(u - 3)^2}$$

Substituting $\ln x$ for u we get

$$-\dfrac{3}{25(\ln x + 2)} + \dfrac{3}{25(\ln x - 3)} + \dfrac{7}{5(\ln x - 3)^2}.$$

32. $\dfrac{1}{e^x + 1} - \dfrac{1}{2e^x + 1}$

Exercise Set 7.1

1.
$$x^2 - y^2 = 0$$
$$(x + y)(x - y) = 0$$

$x + y = 0$ or $x - y = 0$

$y = -x$ or $y = x$

The graph consists of two intersecting lines.

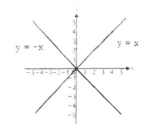

2.

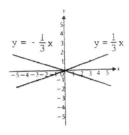

3.
$$3x^2 + xy - 2y^2 = 0$$
$$(3x - 2y)(x + y) = 0$$

$3x - 2y = 0$ or $x + y = 0$

$y = \frac{3}{2}x$ or $y = -x$

The graph consists of two intersecting lines.

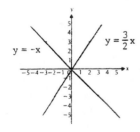

4.

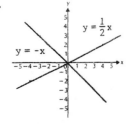

5. $2x^2 + y^2 = 0$

The expression $2x^2 + y^2$ is not factorable in the real-number system. The only real-number solution of the equation is $(0, 0)$. The graph is the point $(0, 0)$.

6. $5x^2 + y^2 = -3$

Since squares of numbers are never negative, the left side of the equation can never be negative. The equation has no real-number solution, hence has no graph.

7. $\dfrac{x^2}{4} + \dfrac{y^2}{1} = 1$

$$\dfrac{x^2}{2^2} + \dfrac{y^2}{1^2} = 1$$

The center of the ellipse is $(0, 0)$; $a = 2$ and $b = 1$.

Two of the vertices are $(-2, 0)$ and $(2, 0)$. These are also the x-intercepts. The other two vertices are $(0, -1)$ and $(0, 1)$. These are also the y-intercepts.

Since $a > b$, we find c using $c^2 = a^2 - b^2$.

$c^2 = a^2 - b^2 = 4 - 1 = 3$

Thus $c = \sqrt{3}$.

The foci are on the x-axis. They are $(-\sqrt{3}, 0)$ and $(\sqrt{3}, 0)$.

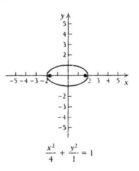

$$\frac{x^2}{4} + \frac{y^2}{1} = 1$$

8.   V: $(-1, 0)$, $(1, 0)$,
        $(0, -2)$, $(0, 2)$

     F: $(0, -\sqrt{3})$, $(0, \sqrt{3})$

9. $16x^2 + 9y^2 = 144$

$\dfrac{x^2}{9} + \dfrac{y^2}{16} = 1$ $\left[\text{Multiplying by } \dfrac{1}{144}\right]$

$\dfrac{x^2}{3^2} + \dfrac{y^2}{4^2} = 1$

The center of the ellipse is (0, 0); a = 3 and b = 4.

Two of the vertices are (-3, 0) and (3, 0). These are also the x-intercepts. The other two vertices are (0, -4) and (0, 4). These are also the y-intercepts.

Since b > a, we find c using $c^2 = b^2 - a^2$.

$c^2 = b^2 - a^2 = 16 - 9 = 7$

Thus $c = \sqrt{7}$.

The foci are on the y-axis. They are (0, $-\sqrt{7}$) and (0, $\sqrt{7}$).

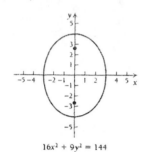

$16x^2 + 9y^2 = 144$

10.   V: (-4, 0),   (4, 0),

(0, -3),   (0, 3)

F: $(-\sqrt{7}, 0)$, $(\sqrt{7}, 0)$

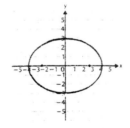

11. $2x^2 + 3y^2 = 6$

$\dfrac{x^2}{3} + \dfrac{y^2}{2} = 1$ $\left[\text{Multiplying by } \dfrac{1}{6}\right]$

$\dfrac{x^2}{(\sqrt{3})^2} + \dfrac{y^2}{(\sqrt{2})^2} = 1$

The center of the ellipse is (0, 0); a = $\sqrt{3}$ and b = $\sqrt{2}$.

Two of the vertices are ($-\sqrt{3}$, 0) and ($\sqrt{3}$, 0). These are also the x-intercepts. The other two vertices are (0, $-\sqrt{2}$) and (0, $\sqrt{2}$). These are also the y-intercepts.

Since a > b, we find c using $c^2 = a^2 - b^2$.

$c^2 = a^2 - b^2 = 3 - 2 = 1$

Thus c = 1.

The foci are on the x-axis. They are (-1, 0) and (1, 0).

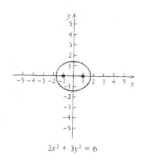

$2x^2 + 3y^2 = 6$

12.   V: $(-\sqrt{7}, 0)$, $(\sqrt{7}, 0)$

(0, $-\sqrt{5}$),   (0, $\sqrt{5}$)

F: $(-\sqrt{2}, 0)$, $(\sqrt{2}, 0)$

13. $4x^2 + 9y^2 = 1$

$\dfrac{x^2}{\dfrac{1}{4}} + \dfrac{y^2}{\dfrac{1}{9}} = 1$

$\dfrac{x^2}{\left(\dfrac{1}{2}\right)^2} + \dfrac{y^2}{\left(\dfrac{1}{3}\right)^2} = 1$

The center of the ellipse is (0, 0); a = $\dfrac{1}{2}$ and b = $\dfrac{1}{3}$.

Two of the vertices are $\left[-\dfrac{1}{2}, 0\right]$ and $\left[\dfrac{1}{2}, 0\right]$. These are also the x-intercepts. The other two vertices are $\left[0, -\dfrac{1}{3}\right]$ and $\left[0, \dfrac{1}{3}\right]$. These are also the y-intercepts.

Since a > b, we find c using $c^2 = a^2 - b^2$.

$c^2 = a^2 - b^2 = \dfrac{1}{4} - \dfrac{1}{9} = \dfrac{5}{36}$

Thus $c = \dfrac{\sqrt{5}}{6}$.

The foci are on the x-axis. They are $\left[-\dfrac{\sqrt{5}}{6}, 0\right]$ and $\left[\dfrac{\sqrt{5}}{6}, 0\right]$.

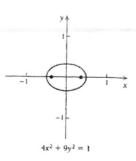

$4x^2 + 9y^2 = 1$

14. V: $\left(-\frac{1}{5}, 0\right)$, $\left(\frac{1}{5}, 0\right)$,

    $\left(0, -\frac{1}{4}\right)$, $\left(0, \frac{1}{4}\right)$

    F: $\left(0, -\frac{3}{20}\right)$, $\left(0, \frac{3}{20}\right)$

15. $\frac{(x-1)^2}{4} + \frac{(y-2)^2}{1} = 1$

    $\frac{(x-1)^2}{2^2} + \frac{(y-2)^2}{1^2} = 1$

    The center is (1, 2); a = 2 and b = 1. Since a > b, we find c using $c^2 = a^2 - b^2$.

    $c^2 = a^2 - b^2 = 4 - 1 = 3$

    Thus $c = \sqrt{3}$.

    Consider the ellipse $\frac{x^2}{2^2} + \frac{y^2}{1^2} = 1$.

    The center is (0, 0); a = 2, b = 1, and $c = \sqrt{3}$.
    The vertices are (-2, 0), (2, 0), (0, -1), and (0, 1).
    The foci are $(-\sqrt{3}, 0)$ and $(\sqrt{3}, 0)$.

    The vertices and foci of the translated ellipse are found by translation in the same way in which the center has been translated.

    The vertices are
    (-2 + 1, 0 + 2), (2 + 1, 0 + 2), (0 + 1, -1 + 2), and (0 + 1, 1 + 2).
    or
    (-1, 2), (3, 2), (1, 1), and (1, 3).

    The foci are
    $(-\sqrt{3} + 1, 0 + 2)$ and $(\sqrt{3} + 1, 0 + 2)$
    or
    $(1 - \sqrt{3}, 2)$ and $(1 + \sqrt{3}, 2)$.

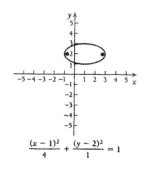

$\frac{(x-1)^2}{4} + \frac{(y-2)^2}{1} = 1$

16. C: (1, 2)

    V: (0, 2),   (2, 2),
       (1, 0),   (1, 4)

    F: $(1, 2 - \sqrt{3})$,
       $(1, 2 + \sqrt{3})$

17. $\frac{(x+3)^2}{25} + \frac{(y-2)^2}{16} = 1$

    $\frac{[x-(-3)]^2}{5^2} + \frac{(y-2)^2}{4^2} = 1$

    The center is (-3, 2); a = 5 and b = 4. Since a > b, we find c using $c^2 = a^2 - b^2$.

    $c^2 = a^2 - b^2 = 25 - 16 = 9$

    Thus c = 3.

    Consider the ellipse $\frac{x^2}{5^2} + \frac{y^2}{4^2} = 1$.

    The center is (0, 0); a = 5, b = 4, and c = 3.
    The vertices are (-5, 0), (5, 0), (0, -4) and (0, 4).
    The foci are (-3, 0) and (3, 0).

    The vertices and foci of the translated ellipse are found by translation in the same way in which the center has been translated.

    The vertices are
    (-5 - 3, 0 + 2), (5 - 3, 0 + 2), (0 - 3, -4 + 2), and (0 - 3, 4 + 2)
    or
    (-8, 2), (2, 2), (-3, -2), and (-3, 6).

    The foci are
    (-3 - 3, 0 + 2) and (3 - 3, 0 + 2)
    or
    (-6, 2) and (0, 2).

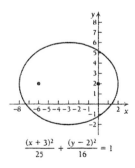

$\frac{(x+3)^2}{25} + \frac{(y-2)^2}{16} = 1$

18. C: (2, -3)

    V: (-3, -3),   (7, -3),
       (2, -7),   (2, 1)

    F: (-1, -3),   (5, -3)

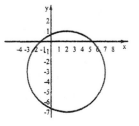

19. $3(x + 2)^2 + 4(y - 1)^2 = 192$

$$\frac{(x + 2)^2}{64} + \frac{(y - 1)^2}{48} = 1 \qquad \left[\text{Multiplying by } \frac{1}{192}\right]$$

$$\frac{(x + 2)^2}{8^2} + \frac{(y - 1)^2}{(4\sqrt{3})^2} = 1$$

The center is $(-2, 1)$; $a = 8$ and $b = 4\sqrt{3}$. Since $a > b$, we find c using $c^2 = a^2 - b^2$.

$c^2 = a^2 - b^2 = 64 - 48 = 16$

Thus $c = 4$.

Consider the ellipse $\frac{x^2}{8^2} + \frac{y^2}{(4\sqrt{3})^2} = 1$

The center is $(0, 0)$; $a = 8$, $b = 4\sqrt{3}$, and $c = 4$.

The vertices are $(-8, 0)$, $(8, 0)$, $(0, -4\sqrt{3})$ and $(0, 4\sqrt{3})$.

The foci are $(-4, 0)$ and $(4, 0)$.

The vertices and foci of the translated ellipse are found by translation in the same way in which the center has been translated.

The vertices are

$(-8 - 2, 0 + 1)$, $(8 - 2, 0 + 1)$, $(0 - 2, -4\sqrt{3} + 1)$ and $(0 - 2, 4\sqrt{3} + 1)$

or

$(-10, 1)$, $(6, 1)$, $(-2, 1 - 4\sqrt{3})$, and $(-2, 1 + 4\sqrt{3})$.

The foci are

$(-4 - 2, 0 + 1)$ and $(4 - 2, 0 + 1)$

or

$(-6, 1)$ and $(2, 1)$.

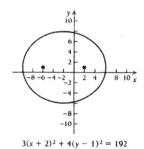

$3(x + 2)^2 + 4(y - 1)^2 = 192$

20. C: (5, 5)

V: $(5 - 4\sqrt{3}, 5)$, $(5 + 4\sqrt{3}, 5)$,
    $(5, -3)$, $(5, 13)$

F: $(5, 1)$, $(5, 9)$

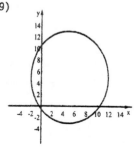

21. $4x^2 + 9y^2 - 16x + 18y - 11 = 0$

$(4x^2 - 16x) + (9y^2 + 18y) = 11$

$4(x^2 - 4x) + 9(y^2 + 2y) = 11$

$4(x^2 - 4x + 4) + 9(y^2 + 2y + 1) = 11 + 16 + 9$

$4(x - 2)^2 + 9(y + 1)^2 = 36$

$$\frac{(x - 2)^2}{9} + \frac{(y + 1)^2}{4} = 1$$

$$\frac{(x - 2)^2}{3^2} + \frac{[y - (-1)]^2}{2^2} = 1$$

The center is $(2, -1)$; $a = 3$ and $b = 2$. Since $a > b$, we find c using $c^2 = a^2 - b^2$.

$c^2 = a^2 - b^2 = 9 - 4 = 5$

Thus $c = \sqrt{5}$.

Consider the ellipse $\frac{x^2}{3^2} + \frac{y^2}{2^2} = 1$.

The center is $(0, 0)$; $a = 3$, $b = 2$, and $c = \sqrt{5}$.

The vertices are $(-3, 0)$, $(3, 0)$, $(0, -2)$, and $(0, 2)$.

The foci are $(-\sqrt{5}, 0)$ and $(\sqrt{5}, 0)$.

The vertices and foci of the translated ellipse are found by translation in the same way in which the center has been translated.

The vertices are

$(-3 + 2, 0 - 1)$, $(3 + 2, 0 - 1)$, $(0 + 2, -2 - 1)$, and $(0 + 2, 2 - 1)$

or

$(-1, -1)$, $(5, -1)$, $(2, -3)$, and $(2, 1)$.

The foci are

$(-\sqrt{5} + 2, 0 - 1)$ and $(\sqrt{5} + 2, 0 - 1)$

or

$(2 - \sqrt{5}, -1)$ and $(2 + \sqrt{5}, -1)$.

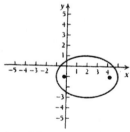

$4x^2 + 9y^2 - 16x + 18y - 11 = 0$

22. C: (5, -2)

V: $(3, -2)$, $(7, -2)$,
    $(5, -2 - \sqrt{2})$ $(5, -2 + \sqrt{2})$

F: $(5 - \sqrt{2}, -2)$, $(5 + \sqrt{2}, -2)$

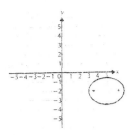

23.
$$4x^2 + y^2 - 8x - 2y + 1 = 0$$
$$(4x^2 - 8x) + (y^2 - 2y) = -1$$
$$4(x^2 - 2x) + (y^2 - 2y) = -1$$
$$4(x^2 - 2x + 1) + (y^2 - 2y + 1) = -1 + 4 + 1$$
$$4(x - 1)^2 + (y - 1)^2 = 4$$
$$\frac{(x - 1)^2}{1^2} + \frac{(y - 1)^2}{2^2} = 1$$

The center is (1, 1); a = 1 and b = 2. Since b > a, we find c using $c^2 = b^2 - a^2$.

$$c^2 = b^2 - a^2 = 4 - 1 = 3$$

Thus $c = \sqrt{3}$.

Consider the ellipse $\frac{x^2}{1^2} + \frac{y^2}{2^2} = 1$.

The center is (0, 0); a = 1, b = 2, and $c = \sqrt{3}$.
The vertices are (-1, 0), (1, 0), (0, -2), and (0, 2).
The foci are $(0, -\sqrt{3})$ and $(0, \sqrt{3})$.

The vertices and foci of the translated ellipse are found by translation in the same way in which the center has been translated.

The vertices are
(-1 + 1, 0 + 1), (1 + 1, 0 + 1), (0 + 1, -2 + 1), and (0 + 1, 2 + 1)
or
(0, 1), (2, 1), (1, -1), and (1, 3).

The foci are
$(0 + 1, -\sqrt{3} + 1)$ and $(0 + 1, \sqrt{3} + 1)$
or
$(1, 1 - \sqrt{3})$ and $(1, 1 + \sqrt{3})$.

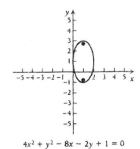

$$4x^2 + y^2 - 8x - 2y + 1 = 0$$

24. C: (-3, 1)

V: (-5, 1),   (-1, 1)
   (-3, -2),   (-3, 4)

F: $(-3, 1 - \sqrt{5})$,
   $(-3, 1 + \sqrt{5})$

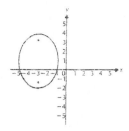

25. Graph the vertices and sketch the axes of the ellipse.

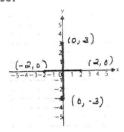

The intersection of the axes, which is (0, 0), is the center of the ellipse. Note that a = 2 and b = 3. Now find an equation of the ellipse (in standard form).

$$\frac{(x - h)^2}{a^2} + \frac{(y - k)^2}{b^2} = 1$$
$$\frac{(x - 0)^2}{2^2} + \frac{(y - 0)^2}{3^2} = 1 \qquad \text{(Substituting)}$$
$$\frac{x^2}{4} + \frac{y^2}{9} = 1$$

26. $x^2 + \frac{y^2}{16} = 1$

27. Graph the vertices and sketch the axes of the ellipse.

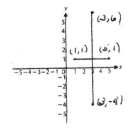

The intersection of the axes, which is (3, 1), is the center of the ellipse. Note that a = 2 and b = 5. Now find an equation of the ellipse (in standard form).

$$\frac{(x - h)^2}{a^2} + \frac{(y - k)^2}{b^2} = 1$$
$$\frac{(x - 3)^2}{2^2} + \frac{(y - 1)^2}{5^2} = 1 \qquad \text{(Substituting)}$$
$$\frac{(x - 3)^2}{4} + \frac{(y - 1)^2}{25} = 1$$

28. $\dfrac{(x+1)^2}{4} + \dfrac{(y-2)^2}{9} = 1$

29. Graph the center, (-2, 3), and sketch the axes of the ellipse. The major axis of length 4 is parallel to the y-axis. The minor axis of length 1 is parallel to the x-axis.

Note that $a = \frac{1}{2}$ and $b = 2$. Find an equation (in standard form) of the ellipse.

$$\dfrac{(x-h)^2}{a^2} + \dfrac{(y-k)^2}{b^2} = 1$$

$$\dfrac{[x-(-2)]^2}{\left(\frac{1}{2}\right)^2} + \dfrac{(y-3)^2}{2^2} = 1 \qquad \text{(Substituting)}$$

$$\dfrac{(x+2)^2}{\frac{1}{4}} + \dfrac{(y-3)^2}{4} = 1$$

30. $\dfrac{x^2}{9} + \dfrac{y^2}{\frac{484}{5}} = 1$, or $\dfrac{x^2}{9} + \dfrac{5y^2}{484} = 1$

31. a)

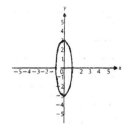

The relation is not a function. It does not pass the vertical line test.

b) $9x^2 + y^2 = 9$

$y^2 = 9 - 9x^2$

$y = \pm\sqrt{9 - 9x^2}$

$y = \pm 3\sqrt{1 - x^2}$

c)

The relation is a function. It does pass the vertical line test.

Domain: $\{x \mid -1 \leqslant x \leqslant 1\}$

Range: $\{y \mid 0 \leqslant y \leqslant 3\}$

d)

The relation is a function. It does pass the vertical line test.

Domain: $\{x \mid -1 \leqslant x \leqslant 1\}$

Range: $\{y \mid -3 \leqslant y \leqslant 0\}$

32. Circle with center (0, 0) and radius a.

33. Draw the graph as instructed.

34. 2×10^6 mi

35. The vertices are (-50, 0), (50, 0), (0, -12), and (0, 12). The equation of the ellipse is $\dfrac{x^2}{50^2} + \dfrac{y^2}{12^2} = 1$, or $\dfrac{x^2}{2500} + \dfrac{y^2}{144} = 1$.

36. $\dfrac{x^2}{524^2} + \dfrac{y^2}{449^2} = 1$, or $\dfrac{x^2}{274,576} + \dfrac{y^2}{201,601} = 1$

37.

Stretching and shrinking, the area of the square, $r \cdot r$, becomes the area of the rectangle, $a \cdot b$. Hence, the area of the circle, $\pi \cdot r \cdot r$, becomes the area of the ellipse, $\pi \cdot a \cdot b$.

b) $a = 4$, $b = 5$
Area $= \pi \cdot 4 \cdot 5 = 20\pi$

c) $a = 524$, $b = 449$
Area $= \pi \cdot 524 \cdot 449 \approx 739,141.4$ ft^2

38. $C = C(x, y)$

θ is the angle which BC makes with the positive x-axis. Let $AC = a$, $BC = b$. Then, in all quadrants, $\dfrac{x}{a} = \cos\theta$ and $\dfrac{y}{b} = \sin\theta$.

Hence $\dfrac{x^2}{a^2} + \dfrac{y^2}{b^2} = 1$ which is the equation of an ellipse.

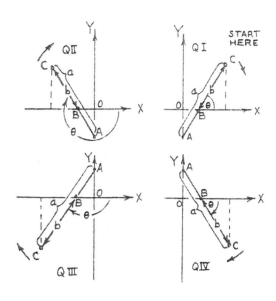

39. C: (2.003125, -1.00515)

    V: (-1.01718, -1.00515), (5.02343, -1.00515),
       (2.003125, -3.018687), (2.003125, 1.008387).

40. C: (-3.0035, 1.002)

    V: (-3.0035, -1.97008), (-3.0035, 3.97408)
       (-1.02211, 1.002), (-4.98489, 1.002)

Exercise Set 7.2

1. $\dfrac{x^2}{9} - \dfrac{y^2}{1} = 1$

   $\dfrac{(x - 0)^2}{3^2} - \dfrac{(y - 0)^2}{1^2} = 1$

   The center is (0, 0); a = 3 and b = 1.
   The vertices are (-3, 0) and (3, 0).
   Since $c^2 = a^2 + b^2$, $c = \sqrt{a^2 + b^2} = \sqrt{9 + 1}$
   $\qquad\qquad\qquad\qquad\qquad = \sqrt{10}.$
   The foci are $(-\sqrt{10}, 0)$ and $(\sqrt{10}, 0)$.
   The asymptotes are $y = -\dfrac{1}{3} x$ and $y = \dfrac{1}{3} x$.

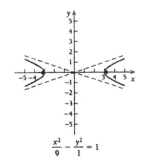

$$\dfrac{x^2}{9} - \dfrac{y^2}{1} = 1$$

2. C: (0, 0)

   V: (-1, 0), (1, 0)

   F: $(-\sqrt{10}, 0)$, $(\sqrt{10}, 0)$

   A: y = 3x, y = -3x

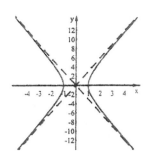

3. $\dfrac{(x - 2)^2}{9} - \dfrac{(y + 5)^2}{1} = 1$

   $\dfrac{(x - 2)^2}{3^2} - \dfrac{[y - (-5)]^2}{1^2} = 1$

   The center is (2, -5); a = 3 and b = 1.
   The transverse axis is parallel to the x-axis.
   Since $c^2 = a^2 + b^2$, $c = \sqrt{a^2 + b^2}$ $= \sqrt{9 + 1}$
   $\qquad\qquad\qquad\qquad\qquad\qquad = \sqrt{10}.$

   Consider the hyperbola $\dfrac{x^2}{3^2} - \dfrac{y^2}{1^2} = 1$.

   The center is (0, 0); a = 3, b = 1, and $c = \sqrt{10}$.
   The vertices are (-3, 0) and (3, 0).
   The foci are $(-\sqrt{10}, 0)$ and $(\sqrt{10}, 0)$.
   The asymptotes are $y = -\dfrac{1}{3} x$ and $y = \dfrac{1}{3} x$.

   The vertices, foci, and asymptotes of the translated hyperbola are found by translation in the
   same way in which the center has been translated.

   The vertices are (-3 + 2, 0 - 5) and
   (3 + 2, 0 - 5), or (-1, -5) and (5, -5).

   The foci are $(-\sqrt{10} + 2, 0 - 5)$ and
   $(\sqrt{10} + 2, 0 - 5)$, or $(2 - \sqrt{10}, -5)$ and
   $(2 + \sqrt{10}, -5)$.

   The asymptotes are

   $y - (-5) = -\dfrac{1}{3} (x - 2)$ and $y - (-5) = \dfrac{1}{3} (x - 2)$

   or $y = -\dfrac{1}{3} x - \dfrac{13}{3}$ and $y = \dfrac{1}{3} x - \dfrac{17}{3}$.

   Graph the hyperpola.

   First draw the rectangle which has
   (-3 + 2, 0 - 5), (3 + 2, 0 - 5), (0 + 2, -1 - 5),
   and (0 + 2, 1 - 5) or (-1, -5), (5, -5), (2, -6),
   and (2, -4) as midpoints of its four sides. Then
   draw the asymptotes and finally the branches of
   the hyperbola outward from the vertices toward the
   asymptotes.

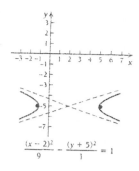

$$\frac{(x-2)^2}{9} - \frac{(y+5)^2}{1} = 1$$

<u>4.</u> C: (2, -5)

V: (1, -5), (3, -5)

F: $(2 - \sqrt{10}, -5)$,
   $(2 + \sqrt{10}, -5)$

A: $y = -3x + 1$,
   $y = 3x - 11$

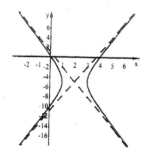

$$\frac{(x-2)^2}{1} - \frac{(y+5)^2}{9} = 1$$

<u>5.</u> $$\frac{(y+3)^2}{4} - \frac{(x+1)^2}{16} = 1$$
$$\frac{[y-(-3)]^2}{2^2} - \frac{[x-(-1)]^2}{4^2} = 1$$

The center is (-1, -3); a = 4 and b = 2.

The transverse axis is parallel to the y-axis.

Since $c^2 = a^2 + b^2$, $c = \sqrt{a^2 + b^2} = \sqrt{16 + 4}$
$= 2\sqrt{5}$.

Consider the hyperbola $\frac{y^2}{2^2} - \frac{x^2}{4^2} = 1$.

The center is (0, 0); a = 4, b = 2, and c = $2\sqrt{5}$.
The vertices are (0, -2) and (0, 2).

The foci are $(0, -2\sqrt{5})$ and $(0, 2\sqrt{5})$.

The asymptotes are $y = -\frac{1}{2}x$ and $y = \frac{1}{2}x$.

The vertices, foci, and asymptotes of the translated hyperbola are found by translation in the same way in which the center has been translated.

The vertices are (0 - 1, -2 - 3) and
(0 - 1, 2 - 3) or (-1, -5) and (-1, -1).

The foci are $(0 - 1, -2\sqrt{5} - 3)$ and
$(0 - 1, 2\sqrt{5} - 3)$ or $(-1, -3 - 2\sqrt{5})$ and
$(-1, -3 + 2\sqrt{5})$.

The asymptotes are

$y - (-3) = -\frac{1}{2}[x - (-1)]$ and

$y - (-3) = \frac{1}{2}[x - (-1)]$ or $y = -\frac{1}{2}x - \frac{7}{2}$

and $y = \frac{1}{2}x - \frac{5}{2}$.

Graph the hyperbola.

First draw the rectangle which has
(0 - 1, -2 - 3), (0 - 1, 2 - 3), (4 - 1, 0 - 3),
and (-4 - 1, 0 - 3) or (-1, -5), (-1, -1),
(3, -3), and (-5, -3) as midpoints of its four
sides. Then draw the asymptotes and finally the
branches of the hyperbola outward from the
vertices toward the asymptotes.

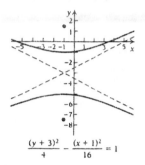

$$\frac{(y+3)^2}{4} - \frac{(x+1)^2}{16} = 1$$

<u>6.</u> C: (-1, -3)

V: (-1, -8), (-1, 2)

F: $(-1, -3 - \sqrt{41})$,
   $(-1, -3 + \sqrt{41})$

A: $y = \frac{5}{4}x - \frac{17}{4}$,

   $y = \frac{5}{4}x - \frac{7}{4}$

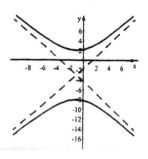

$$\frac{(y+3)^2}{25} - \frac{(x+1)^2}{16} = 1$$

7.
$$x^2 - 4y^2 = 4$$
$$\frac{x^2}{4} - \frac{y^2}{1} = 1$$
$$\frac{(x - 0)^2}{2^2} - \frac{(y - 0)^2}{1^2} = 1$$

The center is (0, 0); a = 2 and b = 1.
The vertices are (-2, 0) and (2, 0).
Since $c^2 = a^2 + b^2$, $c = \sqrt{a^2 + b^2} = \sqrt{4 + 1} = \sqrt{5}$.
The foci are $(-\sqrt{5}, 0)$ and $(\sqrt{5}, 0)$.
The asymptotes are $y = -\frac{1}{2}x$ and $y = \frac{1}{2}x$.

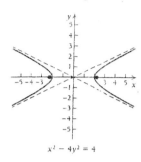

$x^2 - 4y^2 = 4$

8. C: (0, 0)
   V: (-1, 0), (1, 0)
   F: $(-\sqrt{5}, 0)$, $(\sqrt{5}, 0)$
   A: y = -2x, y = 2x

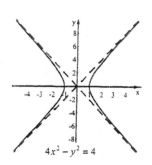

$4x^2 - y^2 = 4$

9.
$$4y^2 - x^2 = 4$$
$$y^2 - \frac{x^2}{4} = 1$$
$$\frac{(y - 0)^2}{1^2} - \frac{(x - 0)^2}{2^2} = 1$$

The center is (0, 0); a = 2 and b = 1.
The vertices are (0, -1) and (0, 1).
Since $c^2 = a^2 + b^2$, $c = \sqrt{a^2 + b^2} = \sqrt{4 + 1} = \sqrt{5}$.
The foci are $(0, -\sqrt{5})$ and $(0, \sqrt{5})$.
The asymptotes are $y = -\frac{1}{2}x$ and $y = \frac{1}{2}x$.

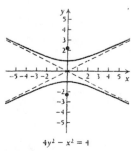

$4y^2 - x^2 = 4$

10.
$$y^2 - 4x^2 = 4$$
$$\frac{y^2}{4} - x^2 = 1$$
C: (0, 0)
V: (0, -2), (0, 2)
$c = \sqrt{1 + 4} = \sqrt{5}$
F: $(0, -\sqrt{5})$, $(0, \sqrt{5})$
A: y = -2x, y = 2x

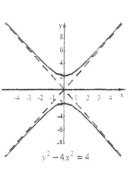

$y^2 - 4x^2 = 4$

11.
$$x^2 - y^2 = 2$$
$$\frac{x^2}{2} - \frac{y^2}{2} = 1$$
$$\frac{(x - 0)^2}{(\sqrt{2})^2} - \frac{(y - 0)^2}{(\sqrt{2})^2} = 1$$

The center is (0, 0); $a = \sqrt{2}$ and $b = \sqrt{2}$.
The vertices are $(-\sqrt{2}, 0)$ and $(\sqrt{2}, 0)$.
Since $c^2 = a^2 + b^2$, $c = \sqrt{a^2 + b^2} = \sqrt{2 + 2} = \sqrt{4} = 2$.
The foci are (-2, 0) and (2, 0).
The asymptotes are y = -x and y = x.

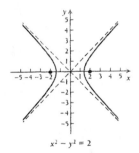

$x^2 - y^2 = 2$

12. $x^2 - y^2 = 3$
$$\frac{x^2}{3} - \frac{y^2}{3} = 1$$
C: (0, 0)
V: $(-\sqrt{3}, 0)$, $(\sqrt{3}, 0)$
$c = \sqrt{3 + 3} = \sqrt{6}$
F: $(-\sqrt{6}, 0)$, $(\sqrt{6}, 0)$
A: y = -x, y = x

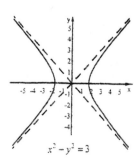

$x^2 - y^2 = 3$

13.
$$x^2 - y^2 = \frac{1}{4}$$
$$\frac{x^2}{\frac{1}{4}} - \frac{y^2}{\frac{1}{4}} = 1$$
$$\frac{(x - 0)^2}{\left(\frac{1}{2}\right)^2} - \frac{(y - 0)^2}{\left(\frac{1}{2}\right)^2} = 1$$

The center is (0, 0); $a = \frac{1}{2}$ and $b = \frac{1}{2}$.

The vertices are $\left[-\frac{1}{2}, 0\right]$ and $\left[\frac{1}{2}, 0\right]$.

Since $c^2 = a^2 + b^2$, $c = \sqrt{a^2 + b^2} = \sqrt{\frac{1}{4} + \frac{1}{4}}$

$$= \sqrt{\frac{1}{2}} = \frac{\sqrt{2}}{2}.$$

The foci are $\left[-\frac{\sqrt{2}}{2}, 0\right]$ and $\left[\frac{\sqrt{2}}{2}, 0\right]$.

The asymptotes are $y = -x$ and $y = x$.

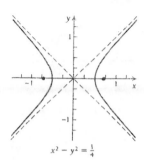

$$x^2 - y^2 = \tfrac{1}{4}$$

<u>14.</u> C: (0, 0)

V: $\left[-\frac{1}{3}, 0\right]$, $\left[\frac{1}{3}, 0\right]$

F: $\left[-\frac{\sqrt{2}}{3}, 0\right]$, $\left[\frac{\sqrt{2}}{3}, 0\right]$

A: $y = -x$, $y = x$

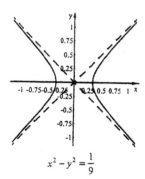

$$x^2 - y^2 = \tfrac{1}{9}$$

<u>15.</u>
$$x^2 - y^2 - 2x - 4y - 4 = 0$$
$$(x^2 - 2x) - (y^2 + 4y) = 4$$
$$(x^2 - 2x + 1) - (y^2 + 4y + 4) = 4 + 1 - 4$$
$$(x - 1)^2 - (y + 2)^2 = 1$$
$$\frac{(x - 1)^2}{1^2} - \frac{[y - (-2)]^2}{1^2} = 1$$

The center of the hyperbola is (1, -2); $a = 1$ and $b = 1$.

The transverse axis is parallel to the x-axis.

Since $c^2 = a^2 + b^2$, $c = \sqrt{a^2 + b^2} = \sqrt{1 + 1}$
$$= \sqrt{2}.$$

Consider the hyperbola $\frac{x^2}{1^2} - \frac{y^2}{1^2} = 1$

The center is (0, 0); $a = 1$, $b = 1$, and $c = \sqrt{2}$.

The vertices are (-1, 0) and (1, 0).
The foci are $(-\sqrt{2}, 0)$ and $(\sqrt{2}, 0)$.
The asymptotes are $y = -x$ and $y = x$.

The vertices, foci, and asymptotes of the translated hyperbola are found in the same way in which the center has been translated.

The vertices are
(-1 + 1, 0 - 2) and (1 + 1, 0 - 2)
or
(0, -2) and (2, -2).

The foci are
$(-\sqrt{2} + 1, 0 - 2)$ and $(\sqrt{2} + 1, 0 - 2)$
or
$(1 - \sqrt{2}, -2)$ and $(1 + \sqrt{2}, -2)$.

The asymptotes are
$y - (-2) = -(x - 1)$ and $y - (-2) = x - 1$
or
$y = -x - 1$ and $y = x - 3$.

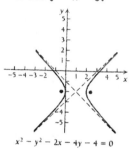

$$x^2 - y^2 - 2x - 4y - 4 = 0$$

<u>16.</u> C: (-1, -2)

V: (-2, -2), (0, -2)

F: $(-1 - \sqrt{5}, -2)$,
$(-1 + \sqrt{5}, -2)$

A: $y = -2x - 4$,
$y = 2x$

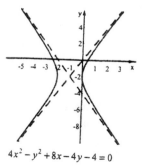

$$4x^2 - y^2 + 8x - 4y - 4 = 0$$

<u>17.</u>
$$36x^2 - y^2 - 24x + 6y - 41 = 0$$
$$(36x^2 - 24x) - (y^2 - 6y) = 41$$
$$36\left[x^2 - \frac{2}{3}x\right] - (y^2 - 6y) = 41$$
$$36\left[x^2 - \frac{2}{3}x + \frac{1}{9}\right] - (y^2 - 6y + 9) = 41 + 4 - 9$$
$$36\left[x - \frac{1}{3}\right]^2 - (y - 3)^2 = 36$$
$$\frac{\left[x - \frac{1}{3}\right]^2}{1} - \frac{(y - 3)^2}{36} = 1$$
$$\frac{\left[x - \frac{1}{3}\right]^2}{1^2} - \frac{(y - 3)^2}{6^2} = 1$$

The center of the hyperbola is $\left(\frac{1}{3}, 3\right)$; a = 1 and b = 6.

The transverse axis is parallel to the x-axis.

Since $c^2 = a^2 + b^2$, $c = \sqrt{a^2 + b^2} = \sqrt{1 + 36}$
$$= \sqrt{37}.$$

Consider the hyperbola $\frac{x^2}{1^2} - \frac{y^2}{6^2} = 1$.

The center is (0, 0); a = 1, b = 6, and $c = \sqrt{37}$.
The vertices are (-1, 0) and (1, 0).
The foci are $(-\sqrt{37}, 0)$ and $(\sqrt{37}, 0)$.
The asymptotes are y = -6x and y = 6x.

The vertices, foci, and asymptotes of the translated hyperbola are found in the same way in which the center has been translated.

The vertices are
$$\left(-1 + \frac{1}{3}, 0 + 3\right) \text{ and } \left(1 + \frac{1}{3}, 0 + 3\right)$$
or
$$\left(-\frac{2}{3}, 3\right) \text{ and } \left(\frac{4}{3}, 3\right).$$

The foci are
$$\left(-\sqrt{37} + \frac{1}{3}, 0 + 3\right) \text{ and } \left(\sqrt{37} + \frac{1}{3}, 0 + 3\right)$$
or
$$\left(\frac{1}{3} - \sqrt{37}, 3\right) \text{ and } \left(\frac{1}{3} + \sqrt{37}, 3\right).$$

The asymptotes are
$$y - 3 = -6\left(x - \frac{1}{3}\right) \text{ and } y - 3 = 6\left(x - \frac{1}{3}\right)$$
or
$$y = -6x + 5 \text{ and } y = 6x + 1.$$

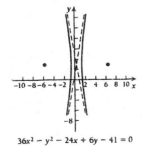

$$36x^2 - y^2 - 24x + 6y - 41 = 0$$

18. C: (-3, 1)

V: $\left(-3 + \frac{4\sqrt{2}}{3}, 1\right)$

$\left(-3 - \frac{4\sqrt{2}}{3}, 1\right)$

F: $\left(-3 + \frac{2\sqrt{26}}{3}, 1\right)$,

$\left(-3 - \frac{2\sqrt{26}}{3}, 1\right)$

A: $y = -\frac{3}{2} - \frac{7}{2}$,

$y = \frac{3}{2}x + \frac{11}{2}$

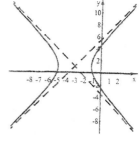

$$9x^2 - 4y^2 + 54x + 8y + 45 = 0$$

19.
$$9y^2 - 4x^2 - 18y + 24x - 63 = 0$$
$$9(y^2 - 2y) - 4(x^2 - 6x) = 63$$
$$9(y^2 - 2y + 1) - 4(x^2 - 6x + 9) = 63 + 9 - 36$$
$$9(y - 1)^2 - 4(x - 3)^2 = 36$$
$$\frac{(y - 1)^2}{4} - \frac{(x - 3)^2}{9} = 1$$

The center of the hyperbola is (3, 1); a = 3 and b = 2.

The transverse axis is parallel to the y-axis.
Since $c^2 = a^2 + b^2$, $c = \sqrt{a^2 + b^2} = \sqrt{9 + 4}$
$$= \sqrt{13}.$$

Consider the hyperbola $\frac{y^2}{4} - \frac{x^2}{9} = 1$.

The center is (0, 0); a = 3, b = 2, and $c = \sqrt{13}$.
The vertices are (0, -2) and (0, 2). The foci are $(0, -\sqrt{13})$ and $(0, \sqrt{13})$. The asymptotes are $y = -\frac{2}{3}x$ and $y = \frac{2}{3}x$.

The vertices, foci, and asymptotes of the translated hyperbola are found in the same way in which the center has been translated.

The vertices are (0 + 3, -2 + 1) and (0 + 3, 2 + 1) or (3, -1) and (3, 3).

The foci are $(0 + 3, -\sqrt{13} + 1)$ and $(0 + 3, \sqrt{13} + 1)$ or $(3, 1 - \sqrt{13})$ and $(3, 1 + \sqrt{13})$.

The asymptotes are
$$y - 1 = -\frac{2}{3}(x - 3) \text{ and } y - 1 = \frac{2}{3}(x - 3) \text{ or}$$
$$y = -\frac{2}{3}x + 3 \text{ and } y = \frac{2}{3}x - 1.$$

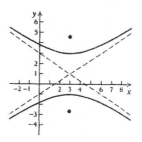

$$9y^2 - 4x^2 - 18y + 24x - 63 = 0$$

20. C: (-3, -1)

V: (2, -1), (-8, -1)

F: $(-3 + \sqrt{26}, -1)$, $(-3 - \sqrt{26}, -1)$

A: $y = -\frac{1}{5}x - \frac{8}{5}$,

$y = \frac{1}{5}x - \frac{2}{5}$

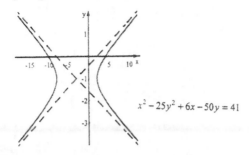

$x^2 - 25y^2 + 6x - 50y = 41$

22. C: (-1, 3)

V: (-1, 5), (-1, 1)

F: $(-1, 3 + \sqrt{13})$, $(-1, 3 - \sqrt{13})$

A: $y = -\frac{2}{3}x + \frac{7}{3}$,

$y = \frac{2}{3}x + \frac{11}{3}$

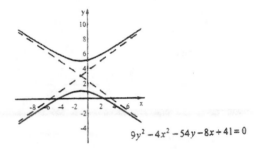

$9y^2 - 4x^2 - 54y - 8x + 41 = 0$

21.

$$x^2 - y^2 - 2x - 4y = 4$$
$$(x^2 - 2x + 1) - (y^2 + 4y + 4) = 4 + 1 - 4$$
$$(x - 1)^2 - (y + 2)^2 = 1, \text{ or}$$
$$\frac{(x - 1)^2}{1} - \frac{(y + 2)^2}{1} = 1$$

The center of the hyperbola is (1, -2), a = 1 and b = 1.

The transverse axis is parallel to the x-axis. Since $c^2 = a^2 + b^2$, $c = \sqrt{a^2 + b^2} = \sqrt{1 + 1} = \sqrt{2}$.

Consider the hyperbola $\frac{x^2}{1} - \frac{y^2}{1} = 1$. The center is (0, 0); a = 1, b = 1, and c = $\sqrt{2}$. The vertices are (-1, 0) and (1, 0). The foci are (-$\sqrt{2}$, 0) and ($\sqrt{2}$, 0). The asymptotes are y = -x and y = x.

The vertices, foci, and asymptotes of the translated hyperbola are found in the same way in which the center has been translated.

The vertices are (-1 + 1, 0 - 2) and (1 + 1, 0 - 2) or (0, -2) and (2, -2).

The foci are (-$\sqrt{2}$ + 1, 0 - 2) and ($\sqrt{2}$ + 1, 0 - 2) or (1 - $\sqrt{2}$, -2) and (1 + $\sqrt{2}$, -2).

The asymptotes are

y + 2 = -(x - 1) and y + 2 = x - 1, or

y = -x - 1 and y = x - 3.

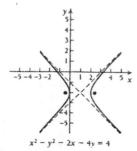

$x^2 - y^2 - 2x - 4y = 4$

23.

$$y^2 - x^2 - 6x - 8y - 29 = 0$$
$$(y^2 - 8y) - (x^2 + 6x) = 29$$
$$(y^2 - 8y + 16) - (x^2 + 6x + 9) = 29 + 16 - 9$$
$$(y - 4)^2 - (x + 3)^2 = 36$$
$$\frac{(y - 4)^2}{36} - \frac{(x + 3)^2}{36} = 1$$

The center of the hyperbola is (-3, 4); a = 6 and b = 6.

The transverse axis is parallel to the y-axis. Since $c^2 = a^2 + b^2$, $c = \sqrt{a^2 + b^2} = \sqrt{36 + 36}$
$= 6\sqrt{2}$.

Consider the hyperbola $\frac{y^2}{36} - \frac{x^2}{36} = 1$. The center is is (0, 0); a = 6, b = 6, and c = 6$\sqrt{2}$. The vertices are (0, -6) and (0, 6). The foci are (0, -6$\sqrt{2}$) and (0, 6$\sqrt{2}$). The asymptotes are y = -x and y = x.

The vertices, foci, and asymptotes of the translated hyperbola are found in the same way in which the center was translated.

The vertices are (0 - 3, 6 + 4) and (0 - 3, -6 + 4) or (-3, 10) and (-3, -2).

The foci are (0 - 3, 6$\sqrt{2}$ + 4) and (0 - 3, -6$\sqrt{2}$ + 4) or (-3, 4 + 6$\sqrt{2}$) and (-3, 4 - 6$\sqrt{2}$).

The asymptotes are

y - 4 = -(x + 3) and y - 4 = x + 3 or

y = -x + 1 and y = x + 7.

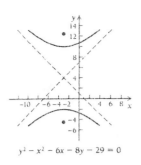

$y^2 - x^2 - 6x - 8y - 29 = 0$

24. C: (4, 1)

    V: $(4 + \sqrt{2}, 1)$, $(4 - \sqrt{2}, 1)$

    F: (6, 1), (2, 1)

    A: y = -x + 5,
        y = x - 3

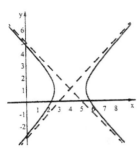

$x^2 - y^2 = 8x - 2y - 13$

25. xy = 1

Since 1 > 0, the branches of the hyperbola lie in the first and third quadrants. The coordinate axes are its asymptotes.

x	y		x	y
$\frac{1}{4}$	4		$-\frac{1}{4}$	-4
$\frac{1}{2}$	2		$-\frac{1}{2}$	-2
1	1		-1	-1
$\frac{3}{2}$	$\frac{2}{3}$		$-\frac{3}{2}$	$-\frac{2}{3}$
3	$\frac{1}{3}$		-3	$-\frac{1}{3}$
4	$\frac{1}{4}$		-4	$-\frac{1}{4}$

$xy = 1$

26.

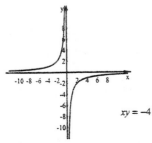

$xy = -4$

27. xy = -8

Since -8 < 0, the branches of the hyperbola lie in the second and fourth quadrants. The coordinates axes are its asymptotes.

x	y		x	y
1	-8		-1	8
2	-4		-2	4
4	-2		-4	2
8	-1		-8	1

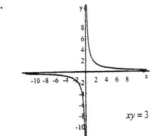

$xy = -8$

28.

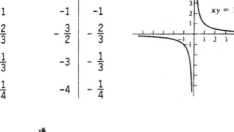

$xy = 3$

29. If the asymptotes are $y = \pm \frac{3}{2} x$, then a = 2 and b = 3.

The center of the hyperbola is (0, 0), the intersection of the asymptotes.

The vertices are (-2, 0) and (2, 0); thus the transverse axis is on the x-axis.

The standard form of the equation of the hyperbola must be in the form $\frac{x^2}{a^2} - \frac{y^2}{b^2} = 1$.

The equation is $\frac{x^2}{4} - \frac{y^2}{9} = 1$.

30. $x^2 - \frac{y^2}{3} = 1$

31. $a = \frac{6}{2} = 3$; $b = \frac{11}{2}$

$\frac{[y - (-8)]^2}{\left[\frac{11}{2}\right]^2} - \frac{(x - 3)^2}{3^2} = 1$, or

$\frac{(y + 8)^2}{\frac{121}{4}} - \frac{(x - 3)^2}{9} = 1$

32. $\frac{(x + 7)^2}{4} - \frac{(y - 4)^2}{36} = 1$

33. The relation in Exercise 32 is

$$\frac{(x + 7)^2}{4} - \frac{(y - 4)^2}{36} = 1.$$

Interchange x and y to find the inverse of the relation:

$$\frac{(y + 7)^2}{4} - \frac{(x - 4)^2}{36} = 1$$

The center is (4, -7).

The vertices are (0 + 4, 2 - 7) and (0 + 4, -2 - 7), or (4, -5) and (4, -9).

34. $\left(\dfrac{x - h}{a}\right)\left(\dfrac{x - h}{a}\right) - \left(\dfrac{y - k}{b}\right)\left(\dfrac{y - k}{b}\right) = 1$, so

$$\frac{(x - h)^2}{a^2} - \frac{(y - k)^2}{b^2} = 1$$

35. a) Graph: $x^2 - 4y^2 = 4$, or $\dfrac{x^2}{4} - \dfrac{y^2}{1} = 1$

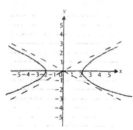

The relation is not a function, because it fails the vertical line test.

b) $x^2 - 4y^2 = 4$

$$x^2 - 4 = 4y^2$$

$$\frac{x^2 - 4}{4} = y^2$$

$$\pm\frac{1}{2}\sqrt{x^2 - 4} = y$$

c) Graph: $y = \frac{1}{2}\sqrt{x^2 - 4}$

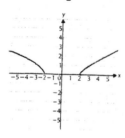

The relation is a function because it passes the vertical line test. The domain is $\{x \mid x \leqslant -2 \text{ or } x \geqslant 2\}$. The range is $\{y \mid y \geqslant 0\}$.

d) Graph: $y = -\frac{1}{2}\sqrt{x^2 - 4}$

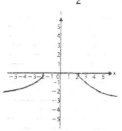

The relation is a function, because it passes the vertical line test. The domain is $\{x \mid x \leqslant -2 \text{ or } x \geqslant 2\}$. The range is $\{y \mid y \leqslant 0\}$.

36. Hyperbola

37.

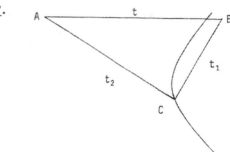

Let the constant b = the velocity of the bullet; let the constant s = the velocity of sound.

Then, by hypothesis,

$t_2 = t + t_1$. Now use: $\text{time} = \dfrac{\text{distance}}{\text{velocity}}$; then

$$\frac{\overline{AC}}{s} = \frac{\overline{AB}}{b} + \frac{\overline{BC}}{s} \text{ or}$$

$\overline{AC} - \overline{BC} = \dfrac{s}{b}\,\overline{AB}$, a constant.

Thus the (absolute value of the) difference of the distances from C to two fixed points A and B is a constant. By definition such a locus is a hyperbola.

38. C: (-1.46, -0.96)

V: (-2.36, -0.96), (-0.56, -0.96)

F: (y = -1.429x - 3.043, y = 1.429x + 1.129

39. C: (1.023, -2.044)

V: (-0.02418, -2.044) and (2.07018, -2.044)

A: y = -x - 1.021 and y = x - 3.067

Exercise Set 7.3

1. $x^2 = 20y$

    $x^2 = 4 \cdot 5 \cdot y$ (Writing $x^2 = 4py$)

    Vertex: $(0, 0)$

    Focus: $(0, 5)$ [(0, p)]

    Directrix: $y = -5$ ($y = -p$)

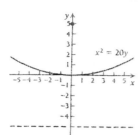

5. $x^2 - 4y = 0$

    $x^2 = 4y$

    $x^2 = 4 \cdot 1 \cdot y$ (Writing $x^2 = 4py$)

    Vertex: $(0, 0)$

    Focus: $(0, 1)$ [(0, p)]

    Directrix: $y = -1$ ($y = -p$)

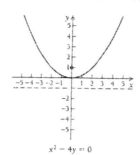

2. V: $(0, 0)$

    F: $(0, 4)$

    D: $y = -4$

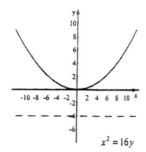

6. V: $(0, 0)$

    F: $(-1, 0)$

    D: $x = 1$

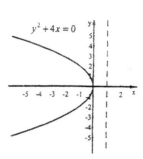

3. $y^2 = -6x$

    $y^2 = 4\left(-\frac{3}{2}\right)x$ (Writing $y^2 = 4px$)

    Vertex: $(0, 0)$

    Focus: $\left(-\frac{3}{2}, 0\right)$ [(p, 0)]

    Directrix: $x = -\left(-\frac{3}{2}\right) = \frac{3}{2}$ ($x = -p$)

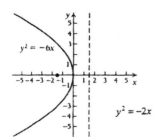

7. $y = 2x^2$

    $x^2 = \frac{1}{2}y$

    $x^2 = 4 \cdot \frac{1}{8} \cdot y$ (Writing $x^2 = 4py$)

    Vertex: $(0, 0)$

    Focus: $\left(0, \frac{1}{8}\right)$ [(0, p)]

    Directrix: $y = -\frac{1}{8}$ ($y = -p$)

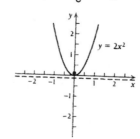

4. V: $(0, 0)$

    F: $\left(-\frac{1}{2}, 0\right)$

    D: $x = \frac{1}{2}$

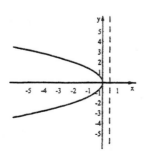

8. V: $(0, 0)$

    F: $\left(0, \frac{1}{2}\right)$

    D: $y = -\frac{1}{2}$

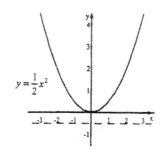

309

9. It helps to sketch a graph of the focus and the directrix.

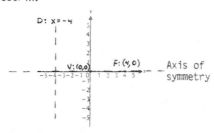

The focus, (4, 0), is on the axis of symmetry, the x-axis. The vertex is (0, 0), and p is 4 - 0, or 4. The equation is of the type

$y^2 = 4px$

$y^2 = 4 \cdot 4 \cdot x$ (Substituting 4 for p)

$y^2 = 16x$

10. $x^2 = y$

11. It is helpful to sketch a graph of the focus and the directrix.

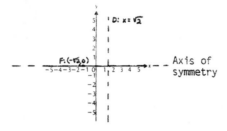

The focus, $(-\sqrt{2}, 0)$, is on the axis of symmetry, the x-axis. The vertex is (0, 0), and $p = -\sqrt{2} - 0$, or $-\sqrt{2}$. The equation is of the type

$y^2 = 4px$

$y^2 = 4(-\sqrt{2})x$ (Substituting $-\sqrt{2}$ for p)

$y^2 = -4\sqrt{2}\, x$

12. $x^2 = -4\pi y$

13. It is helpful to sketch a graph of the focus and the directrix.

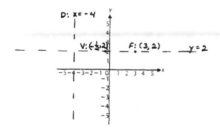

The focus, (3, 2), is on the axis of symmetry, y = 2, which is parallel to the x-axis. The vertex is $\left(-\frac{1}{2}, 2\right)$, and $p = 3 - \left(-\frac{1}{2}\right)$, or $\frac{7}{2}$. The equation is of the type

$(y - k)^2 = 4p(x - h)$

$(y - 2)^2 = 4 \cdot \frac{7}{2}\left[x - \left(-\frac{1}{2}\right)\right]$ (Substituting)

$(y - 2)^2 = 14\left(x + \frac{1}{2}\right)$

14. $(x + 2)^2 = 12y$

15. $(x + 2)^2 = -6(y - 1)$

$[x - (-2)]^2 = 4\left(-\frac{3}{2}\right)(y - 1)$ $[(x - h)^2 = 4p(y-k)]$

Vertex: (-2, 1)   [(h, k)]

Focus: $\left(-2, 1 + \left(-\frac{3}{2}\right)\right)$, or $\left(-2, -\frac{1}{2}\right)$

                         [(h, k + p)]

Directrix: $y = 1 - \left(-\frac{3}{2}\right) = \frac{5}{2}$ (y = k - p)

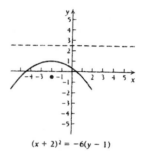

$(x + 2)^2 = -6(y - 1)$

16. V: (-2, 3)

     F: (-7, 3)

     D: x = 3

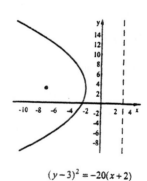

$(y-3)^2 = -20(x+2)$

17. $x^2 + 2x + 2y + 7 = 0$

$x^2 + 2x = -2y - 7$

$(x^2 + 2x + 1) = -2y - 7 + 1 = -2y - 6$

$(x + 1)^2 = -2(y + 3)$

$[x - (-1)]^2 = 4\left(-\frac{1}{2}\right)[y - (-3)]$

                    $[(x - h)^2 = 4p(y - k)]$

Vertex: $(-1, -3)$ [(h, k)]

Focus: $\left[-1, -3 + \left(-\frac{1}{2}\right)\right]$, or $\left[-1, -\frac{7}{2}\right]$ [(h, k+p)]

Directrix: $y = -3 - \left(-\frac{1}{2}\right) = -\frac{5}{2}$ (y = k - p)

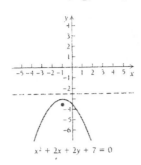

$x^2 + 2x + 2y + 7 = 0$

<u>18.</u> V: $(7, -3)$

F: $\left(\frac{29}{4}, -3\right)$

D: $x = \frac{27}{4}$

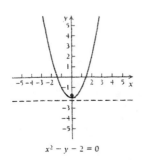

$y^2 + 6y - x + 16 = 0$

<u>19.</u> $x^2 - y - 2 = 0$

$x^2 = y + 2$

$(x - 0)^2 = 4 \cdot \frac{1}{4} \cdot [y - (-2)]$

$[(x - h)^2 = 4p(y - k)]$

Vertex: $(0, -2)$ [(h, k)]

Focus: $\left[0, -2 + \frac{1}{4}\right]$, or $\left[0, -\frac{7}{4}\right]$ [(h, k + p)]

Directrix: $y = -2 - \frac{1}{4} = -\frac{9}{4}$ (y = k - p)

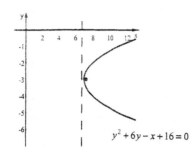

$x^2 - y - 2 = 0$

<u>20.</u> V: $(2, -2)$

F: $\left[2, -\frac{3}{2}\right]$

D: $y = -\frac{5}{2}$

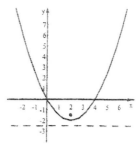

$x^2 - 4x - 2y = 0$

<u>21.</u>

$y = x^2 + 4x + 3$

$y - 3 = x^2 + 4x$

$y - 3 + 4 = x^2 + 4x + 4$

$y + 1 = (x + 2)^2$

$4 \cdot \frac{1}{4} \cdot [y - (-1)] = [x - (-2)]^2$

$[(x - h)^2 = 4p(y - k)]$

Vertex: $(-2, -1)$ [(h, k)]

Focus: $\left[-2, -1 + \frac{1}{4}\right]$, or $\left[-2, -\frac{3}{4}\right]$

[(h, k + p)]

Directrix: $y = -1 - \frac{1}{4} = -\frac{5}{4}$ (y = k - p)

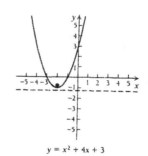

$y = x^2 + 4x + 3$

<u>22.</u> V: $(-3, 1)$

F: $\left[-3, \frac{5}{4}\right]$

D: $y = \frac{3}{4}$

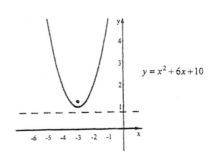

$y = x^2 + 6x + 10$

23. $4y^2 - 4y - 4x + 24 = 0$

$y^2 - y - x + 6 = 0$ $\left[\text{Multiplying by } \frac{1}{4}\right]$

$y^2 - y = x - 6$

$y^2 - y + \frac{1}{4} = x - 6 + \frac{1}{4}$

$\left(y - \frac{1}{2}\right)^2 = x - \frac{23}{4}$

$\left(y - \frac{1}{2}\right)^2 = 4 \cdot \frac{1}{4}\left(x - \frac{23}{4}\right)$

$[(y - k)^2 = 4p(x - h)]$

V: $\left(\frac{23}{4}, \frac{1}{2}\right)$ $[(h, k)]$

F: $\left(\frac{23}{4} + \frac{1}{4}, \frac{1}{2}\right)$, or $\left(6, \frac{1}{2}\right)$ $[(h + p, k)]$

D: $x = \frac{23}{4} - \frac{1}{4} = \frac{22}{4}$, or $\frac{11}{2}$ $(x = h - p)$

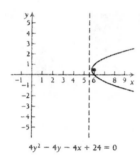

$4y^2 - 4y - 4x + 24 = 0$

24. V: $\left(-\frac{17}{4}, -\frac{1}{2}\right)$

F: $\left(-4, -\frac{1}{2}\right)$

D: $x = -\frac{9}{2}$

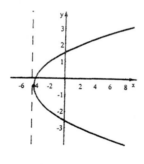

$4y^2 + 4y - 4x - 16 = 0$

25. $x + 1 = 2y^2$

Only one of the variables is squared, so the graph is not a circle, an ellipse, or a hyperbola. We find an equivalent equation:

$(y - 0)^2 = 4\left(\frac{1}{8}\right)(x + 1)$

We have a parabola.

26. Circle

27. $4y^2 + 25x^2 + 4 = 8y + 100x$

$25x^2 - 100x + 4y^2 - 8y = -4$

$25(x^2 - 4x + 4) + 4(y^2 - 2y + 1) = -4 + 100 + 4$

$25(x - 2)^2 + 4(y - 1)^2 = 100$

$\frac{(x - 2)^2}{4} + \frac{(y - 1)^2}{25} = 1$

We have an ellipse.

28. Hyperbola

29. $2x + 13 + y^2 = 8y - x^2$

$(x^2 + 2x + 1) + (y^2 - 8y + 16) = -13 + 1 + 16$

$(x + 1)^2 + (y - 4)^2 = 4$

We have a circle.

30. Hyperbola

31. $x = -16y + y^2 + 7$

$x - 7 + 64 = y^2 - 16y + 64$

$x + 57 = (y - 8)^2$

$4\left(\frac{1}{4}\right)(x + 57) = (y - 8)^2$

We have a parabola.

32. Ellipse

33. $xy + 5x^2 = 9 + 7x^2 - 2x^2$

$xy = 9$

We have a hyperbola.

34. Ellipse

35. 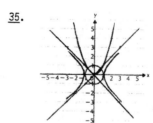   The graph of $x^2 - y^2 = 0$ is the union of the lines $x = y$ and $x = -y$. The others are respectively, a hyperbola, a circle, and a parabola.

36. 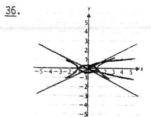   The graph of $x^2 - 4y^2 = 0$ is the union of the lines $x = 2y$ and $x = -2y$. The others are respectively, a hyperbola, an ellipse, and a parabola.

37. If the line of symmetry is parallel to the y-axis and the vertex (h, k) is (-1, 2), then the equation is of the type

$(x - h)^2 = 4p(y - k)$.

Solve for p substituting (-1, 2) for (h, k) and (-3, 1) for (x, y).

$[-3 - (-1)]^2 = 4p(1 - 2)$

$4 = -4p$

$-1 = p$

The equation of the parabola is

$[x - (-1)]^2 = 4(-1)(y - 2)$

or

$(x + 1)^2 = -4(y - 2)$.

38. a)

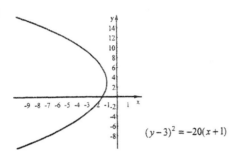

$(y - 3)^2 = -20(x + 1)$

Not a function

b) Not a function unless $p = 0$

39. If the line of symmetry is horizontal and the vertex (h, k) is (-2, 1), then the equation is of the type $(y - k)^2 = 4p(x - h)$.

Find p by substituting (-2, 1) for (h, k) and (-3, 5) for (x, y).

$(5 - 1)^2 = 4p[-3 - (-2)]$

$16 = 4p(-1)$

$16 = -4p$

$-4 = p$

The equation of the parabola is

$(y - 1)^2 = 4(-4)[x - (-2)]$

or $(y - 1)^2 = -16(x + 2)$.

40. 10 ft, 11.6 ft, 16.4 ft, 24.4 ft, 35.6 ft, 50 ft

41. See the answer section in the text.

42.

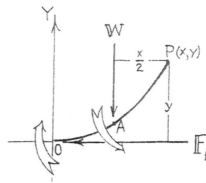

If a load W is distributed uniformly _horizontally_, this means that W = kx; that is, if the horizontal dimension of the cable is doubled, then the load is doubled, etc. We think of the entire load W as being concentrated at point A [whose abscissa is $\frac{x}{2}$], roughly the midpoint of arc OP.

A lever is in rotational equilibrium when the TORQUE(S) [or _moment of force_ defined to be (FORCE) × (⊥ DISTANCE FROM FULCRUM)] tending to turn the lever clockwise about its fulcrum equals (in magnitude) the TORQUE(S) tending to turn the lever counter-clockwise.

Thinking of OP as a (rigid) lever with fulcrum at fixed point P, the torque due to F_1 (applied at O) tends to turn OP clockwise about P; the torque due to W (applied at A) tends to turn the lever counter-clockwise about P. These two torques are equal (in magnitude):

$F_1 y = W \cdot \frac{x}{2}$ or

$F_1 y = (kx) \cdot \frac{x}{2}$.   Thus

$y = \frac{k}{2F_1} \cdot x^2$, a parabola.

[Since F_2 is applied at the fulcrum P, its torque is zero.]

43. V: (0.87, 0.35)

F: (0.87, -0.19)

D: y = 0.89

44. V: (6.90, 1.18)

F: (6.82, 1.18)

D: x = 6.98

Exercise Set 7.4

1. $x^2 + y^2 = 25$,   (1)

   $y - x = 1$ (2)

   First solve Eq. (2) for y.

   $y = x + 1$ (3)

   Then substitute $x + 1$ for y in Eq. (1) and solve for x.

   $$x^2 + y^2 = 25$$
   $$x^2 + (x + 1)^2 = 25$$
   $$x^2 + x^2 + 2x + 1 = 25$$
   $$2x^2 + 2x - 24 = 0$$
   $$x^2 + x - 12 = 0 \quad \text{Multiplying by } \tfrac{1}{2}$$
   $$(x + 4)(x - 3) = 0 \quad \text{Factoring}$$

   $x + 4 = 0$ or $x - 3 = 0$ Principle of zero products

   $x = -4$ or $x = 3$

   Now substitute these numbers into Eq. (3) and solve for y.

   $y = -4 + 1 = -3$

   $y = 3 + 1 = 4$

   The pairs $(-4,-3)$ and $(3,4)$ check, so they are the solutions.

2. $(-8,-6)$, $(6,8)$

3. $4x^2 + 9y^2 = 36$,   (1)

   $3y + 2x = 6$ (2)

   First solve Eq. (2) for y.

   $3y = -2x + 6$

   $y = -\tfrac{2}{3}x + 2$ (3)

   Then substitute $-\tfrac{2}{3}x + 2$ for y in Eq. (1) and solve for x.

   $$4x^2 + 9y^2 = 36$$
   $$4x^2 + 9\left(-\tfrac{2}{3}x + 2\right)^2 = 36$$
   $$4x^2 + 9\left(\tfrac{4}{9}x^2 - \tfrac{8}{3}x + 4\right) = 36$$
   $$4x^2 + 4x^2 - 24x + 36 = 36$$
   $$8x^2 - 24x = 0$$
   $$x^2 - 3x = 0$$
   $$x(x - 3) = 0$$

   $x = 0$ or $x = 3$

   Now substitute these numbers in Eq. (3) and solve for y.

   $y = -\tfrac{2}{3} \cdot 0 + 2 = 2$

   $y = -\tfrac{2}{3} \cdot 3 + 2 = 0$

   The pairs $(0,2)$ and $(3,0)$ check, so they are the solutions.

4. $(2,0)$, $(0,3)$

5. $y^2 - x^2 = 9$ (1)

   $2x - 3 = y$ (2)

   Substitute $2x - 3$ for y in Eq. (1) and solve for x.

   $$y^2 - x^2 = 9$$
   $$(2x - 3)^2 - x^2 = 9$$
   $$4x^2 - 12x + 9 - x^2 = 9$$
   $$3x^2 - 12x = 0$$
   $$x^2 - 4x = 0$$
   $$x(x - 4) = 0$$

   $x = 0$ or $x = 4$

   Now substitute these numbers into the linear Eq. (2) and solve for y.

   If $x = 0$, $y = 2 \cdot 0 - 3 = -3$.

   If $x = 4$, $y = 2 \cdot 4 - 3 = 5$.

   The pairs $(0, -3)$ and $(4, 5)$ check, hence are solutions.

6. $(-7, 1)$, $(1, -7)$

7. $y^2 = x + 3$,   (1)

   $2y = x + 4$ (2)

   First solve Eq. (2) for x.

   $2y - 4 = x$ (3)

   Then substitute $2y - 4$ for x in Eq. (1) and solve for y.

   $$y^2 = x + 3$$
   $$y^2 = (2y - 4) + 3$$
   $$y^2 = 2y - 1$$
   $$y^2 - 2y + 1 = 0$$
   $$(y - 1)(y - 1) = 0$$

   $y = 1$ or $y = 1$

   Now substitute 1 for y in Eq. (3) and solve for x.

   $2 \cdot 1 - 4 = x$

   $-2 = x$

   The pair $(-2,1)$ checks. It is the solution.

8. $(2,4)$, $(1,1)$

9. $x^2 + 4y^2 = 25$,   (1)

   $x + 2y = 7$ (2)

   First solve Eq. (2) for x.

   $x = -2y + 7$ (3)

   Then substitute $-2y + 7$ for x in Eq. (1) and solve for y.

$$x^2 + 4y^2 = 25$$
$$(-2y + 7)^2 + 4y^2 = 25$$
$$4y^2 - 28y + 49 + 4y^2 = 25$$
$$8y^2 - 28y + 24 = 0$$
$$2y^2 - 7y + 6 = 0$$
$$(2y - 3)(y - 2) = 0$$

$$y = \frac{3}{2} \quad \text{or} \quad y = 2$$

Now substitute these numbers in Eq. (3) and solve for x.

$$x = -2 \cdot \frac{3}{2} + 7 = 4$$
$$x = -2 \cdot 2 + 7 = 3$$

The pairs $\left(4, \frac{3}{2}\right)$ and $(3,2)$ check, so they are the solutions.

10. $\left(-\frac{5}{3}, -\frac{13}{3}\right)$, $(3,5)$

11. $x^2 - xy + 3y^2 = 27$,   (1)
    $x - y = 2$ (2)

First solve Eq. (2) for y.
$$x - 2 = y \qquad (3)$$
Then substitute $x - 2$ for y in Eq. (1) and solve for x.
$$x^2 - xy + 3y^2 = 27$$
$$x^2 - x(x - 2) + 3(x - 2)^2 = 27$$
$$x^2 - x^2 + 2x + 3x^2 - 12x + 12 = 27$$
$$3x^2 - 10x - 15 = 0$$

$$x = \frac{-(-10) \pm \sqrt{(-10)^2 - 4(3)(-15)}}{2 \cdot 3}$$
$$= \frac{10 \pm \sqrt{100 + 180}}{6}$$
$$= \frac{10 \pm \sqrt{280}}{6}$$
$$= \frac{10 \pm 2\sqrt{70}}{6}$$
$$= \frac{5 \pm \sqrt{70}}{3}$$

Now substitute these numbers in Eq. (3) and solve for y.
$$y = \frac{5 + \sqrt{70}}{3} - 2 = \frac{-1 + \sqrt{70}}{3}$$
$$y = \frac{5 - \sqrt{70}}{3} - 2 = \frac{-1 - \sqrt{70}}{3}$$

The pairs $\left(\frac{5 + \sqrt{70}}{3}, \frac{-1 + \sqrt{70}}{3}\right)$ and $\left(\frac{5 - \sqrt{70}}{3}, \frac{-1 - \sqrt{70}}{3}\right)$ check, so they are the solutions.

12. $\left(\frac{11}{4}, -\frac{9}{8}\right)$, $(1,-2)$

13. $3x + y = 7$ (1)
    $4x^2 + 5y = 56$ (2)

First solve Eq. (1) for y.
$$3x + y = 7$$
$$y = 7 - 3x \qquad (3)$$

Next substitute $7 - 3x$ for y in Eq. (2) and solve for x.
$$4x^2 + 5y = 56$$
$$4x^2 + 5(7 - 3x) = 56$$
$$4x^2 + 35 - 15x = 56$$
$$4x^2 - 15x - 21 = 0$$

Using the quadratic formula, we find that
$$x = \frac{15 - \sqrt{561}}{8} \quad \text{or} \quad x = \frac{15 + \sqrt{561}}{8}.$$

Now substitute these numbers into Eq. (3) and solve for y.

If $x = \frac{15 - \sqrt{561}}{8}$, $y = 7 - 3\left(\frac{15 - \sqrt{561}}{8}\right)$, or $\frac{11 + 3\sqrt{561}}{8}$.

If $x = \frac{15 + \sqrt{561}}{8}$, $y = 7 - 3\left(\frac{15 + \sqrt{561}}{8}\right)$, or $\frac{11 - 3\sqrt{561}}{8}$.

The pairs $\left(\frac{15 - \sqrt{561}}{8}, \frac{11 + 3\sqrt{561}}{8}\right)$ and $\left(\frac{15 + \sqrt{561}}{8}, \frac{11 - 3\sqrt{561}}{8}\right)$ check and are the solutions.

14. $\left(-3, \frac{5}{2}\right)$, $(3, 1)$

15. $a + b = 7$,   (1)
    $ab = 4$ (2)

First solve Eq. (1) for a.
$$a = -b + 7 \qquad (3)$$
Then substitute $-b + 7$ for a in Eq. (2) and solve for b.
$$(-b + 7)b = 4$$
$$-b^2 + 7b = 4$$
$$0 = b^2 - 7b + 4$$
$$b = \frac{-(-7) \pm \sqrt{(-7)^2 - 4 \cdot 1 \cdot 4}}{2 \cdot 1}$$
$$b = \frac{7 \pm \sqrt{33}}{2}$$

Now substitute these numbers in Eq. (3) and solve for a.
$$a = -\left(\frac{7 + \sqrt{33}}{2}\right) + 7 = \frac{7 - \sqrt{33}}{2}$$
$$a = -\left(\frac{7 - \sqrt{33}}{2}\right) + 7 = \frac{7 + \sqrt{33}}{2}$$

The pairs $\left(\dfrac{7 - \sqrt{33}}{2}, \dfrac{7 + \sqrt{33}}{2}\right)$ and $\left(\dfrac{7 + \sqrt{33}}{2}, \dfrac{7 - \sqrt{33}}{2}\right)$ check, so they are the solutions.

16. $(1,-5)$, $(-5,1)$

17. $2a + b = 1$,   (1)
$b = 4 - a^2$ (2)

Eq. (2) is already solved for b. Substitute $4 - a^2$ for b in Eq. (1) and solve for a.
$2a + 4 - a^2 = 1$
$$0 = a^2 - 2a - 3$$
$$0 = (a - 3)(a + 1)$$
$a = 3$ or $a = -1$
Substitute these numbers in Eq. (2) and solve for b.
$b = 4 - 3^2 = -5$
$b = 4 - (-1)^2 = 3$
The pairs $(3,-5)$ and $(-1,3)$ check.

18. $(3,0)$, $\left(-\dfrac{9}{5}, \dfrac{8}{5}\right)$

19. $a^2 + b^2 = 89$,   (1)
$a - b = 3$ (2)

First solve Eq. (2) for a.
$a = b + 3$ (3)
Then substitute $b + 3$ for a in Eq. (1) and solve for b.
$(b + 3)^2 + b^2 = 89$
$b^2 + 6b + 9 + b^2 = 89$
$2b^2 + 6b - 80 = 0$
$b^2 + 3b - 40 = 0$
$(b + 8)(b - 5) = 0$
$b = -8$ or $b = 5$
Substitute these numbers in Eq. (3) and solve for a.
$a = -8 + 3 = -5$
$a = 5 + 3 = 8$
The pairs $(-5,-8)$ and $(8,5)$ check.

20. $(1,4)$, $(4,1)$

21. $x^2 + y^2 = 5$,   (1)
$x - y = 8$ (2)

First solve Eq. (2) for x.
$x = y + 8$ (3)
Then substitute $y + 8$ for x in Eq. (1) and solve for y.

$(y + 8)^2 + y^2 = 5$
$y^2 + 16y + 64 + y^2 = 5$
$2y^2 + 16y + 59 = 0$
$$y = \dfrac{-16 \pm \sqrt{(16)^2 - 4(2)(59)}}{2\cdot 2}$$
$$y = \dfrac{-16 \pm \sqrt{-216}}{4}$$
$$y = \dfrac{-16 \pm 6i\sqrt{6}}{4}$$
$$y = -4 \pm \dfrac{3}{2}i\sqrt{6}$$

Now substitute these numbers in Eq. (3) and solve for x.

$x = -4 + \dfrac{3}{2}i\sqrt{6} + 8 = 4 + \dfrac{3}{2}i\sqrt{6}$

$x = -4 - \dfrac{3}{2}i\sqrt{6} + 8 = 4 - \dfrac{3}{2}i\sqrt{6}$

The pairs $\left(4 + \dfrac{3}{2}i\sqrt{6},\ -4 + \dfrac{3}{2}i\sqrt{6}\right)$ and $\left(4 - \dfrac{3}{2}i\sqrt{6},\ -4 - \dfrac{3}{2}i\sqrt{6}\right)$ check.

22. $\left(-\dfrac{72}{13} + \dfrac{6}{13}i\sqrt{51},\ \dfrac{32}{13} + \dfrac{6}{13}i\sqrt{51}\right)$,
$\left(-\dfrac{72}{13} - \dfrac{6}{13}i\sqrt{51},\ \dfrac{32}{13} - \dfrac{6}{13}i\sqrt{51}\right)$

23. Familiarize: Let x = one number and y = the other number.
Translate: We translate to a system of equations.
The sum of two numbers is 12.
$x + y = 12$
The sum of their squares is 90.
$x^2 + y^2 = 90$
Carry out: We solve the system:
$x + y = 12$,   (1)
$x^2 + y^2 = 90$ (2)

First solve Eq. (1) for y.
$y = 12 - x$ (3)
Then substitute $12 - x$ for y in Eq. (2) and solve for x.
$x^2 + y^2 = 90$
$x^2 + (12 - x)^2 = 90$
$x^2 + 144 - 24x + x^2 = 90$
$2x^2 - 24x + 54 = 0$
$x^2 - 12x + 27 = 0$
$(x - 9)(x - 3) = 0$
$x = 9$ or $x = 3$
Now substitute these numbers in Eq. (3) and solve for y.
$y = 12 - 9 = 3$
$y = 12 - 3 = 9$

Check: If the numbers are 9 and 3, the sum is 9 + 3, or 12. The sum of their squares is 81 + 9, or 90. The numbers check. The pair (3,9) does not give us another solution.

State: The numbers are 9 and 3.

24. 8, 7

25. Familiarize: We first make a drawing. We let ℓ and w represent the length and width, respectively.

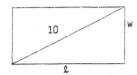

Translate: The perimeter is 28 cm.

$2\ell + 2w = 28$, or $\ell + w = 14$

Using the Pythagorean property we have another equation.

$\ell^2 + w^2 = 10^2$, or $\ell^2 + w^2 = 100$

Carry out: We solve the system:

$\ell + w = 14,$ (1)

$\ell^2 + w^2 = 100$ (2)

First solve Eq. (1) for w.

$w = 14 - \ell$ (3)

Then substitute $14 - \ell$ for w in Eq. (2) and solve for ℓ.

$$\ell^2 + w^2 = 100$$
$$\ell^2 + (14 - \ell)^2 = 100$$
$$\ell^2 + 196 - 28\ell + \ell^2 = 100$$
$$2\ell^2 - 28\ell + 96 = 0$$
$$\ell^2 - 14\ell + 48 = 0$$
$$(\ell - 8)(\ell - 6) = 0$$

$\ell = 8$ or $\ell = 6$

If $\ell = 8$, then $w = 14 - 8$, or 6. If $\ell = 6$, then $w = 14 - 6$, or 8. Since the length is usually considered to be longer than the width, we have the solution $\ell = 8$ and $w = 6$, or (8,6).

Check: If $\ell = 8$ and $w = 6$, then the perimeter is $2 \cdot 8 + 2 \cdot 6$, or 28. The length of a diagonal is $\sqrt{8^2 + 6^2}$, or $\sqrt{100}$, or 10. The numbers check.

State: The length is 8 cm, and the width is 6 cm.

26. 2 m, 1 m

27. Familiarize: We first make a drawing. Let ℓ = the length and w = the width of the rectangle.

Translate:

    Area: $\ell w = 20$

    Perimeter: $2\ell + 2w = 18$, or $\ell + w = 9$

Carry out: We solve the system:

Solve the second equation for ℓ: $\ell = 9 - w$

Substitute $9 - w$ for ℓ in the first equation and solve for w.

$(9 - w)w = 20$

$9w - w^2 = 20$

$0 = w^2 - 9w + 20$

$0 = (w - 5)(w - 4)$

$w = 5$ or $w = 4$

If $w = 5$, then $\ell = 9 - w$, or 4. If $w = 4$, then $\ell = 9 - 4$, or 5. Since length is usually considered to be longer than width, we have the solution $\ell = 5$ and $w = 4$, or (5,4).

Check: If $\ell = 5$ and $w = 4$, the area is $5 \cdot 4$, or 20. The perimeter is $2 \cdot 5 + 2 \cdot 4$, or 18. The numbers check.

State: The length is 5 in. and the width is 4 in.

28. 2 yd, 1 yd

29. Familiarize: We make a drawing of the field. Let ℓ = the length and w = the width.

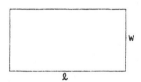

Since it takes 210 yd of fencing to enclose the field, we know that the perimeter is 210 yd.

Translate:

    Perimeter: $2\ell + 2w = 210$, or $\ell + w = 105$

    Area: $\ell w = 2250$

Carry out: We solve the system:

Solve the first equation for ℓ: $\ell = 105 - w$

Substitute $105 - w$ for ℓ in the second equation and solve for w.

$(105 - w)w = 2250$

$105w - w^2 = 2250$

$0 = w^2 - 105w + 2250$

$0 = (w - 30)(w - 75)$

$w = 30$ or $w = 75$

If $w = 30$, then $\ell = 105 - 30$, or 75. If $w = 75$, then $\ell = 105 - 75$, or 30. Since length is usually considered to be longer than width, we have the solution $\ell = 75$ and $w = 30$, or (75,30).

Check: If $\ell = 75$ and $w = 30$, the perimeter is $2 \cdot 75 + 2 \cdot 30$, or 210. The area is 75(30), or 2250. The numbers check.

State: The length is 75 yd and the width is 30 yd.

30. 12 ft, 5 ft

31. Familiarize: Let x and y represent the numbers.
    Translate: The product of the numbers is 2.

    $$xy = 2$$

    The sum of the reciprocals is $\frac{33}{8}$.

    $$\frac{1}{x} + \frac{1}{y} = \frac{33}{8}$$

    Carry out: We solve the system.
    Solve the first equation for y: $y = \frac{2}{x}$
    Substitute $\frac{2}{x}$ for y in the second equation and solve for x.

    $$\frac{1}{x} + \frac{1}{\frac{2}{x}} = \frac{33}{8}$$

    $$\frac{1}{x} + \frac{x}{2} = \frac{33}{8}$$

    $$8 + 4x^2 = 33x \qquad \text{(Multiplying by 8x)}$$

    $$4x^2 - 33x + 8 = 0$$

    $$(4x - 1)(x - 8) = 0$$

    $$x = \frac{1}{4} \quad \text{or} \quad x = 8$$

    If $x = \frac{1}{4}$, then $y = \frac{2}{\frac{1}{4}} = 8$. If x = 8, then
    $y = \frac{2}{8} = \frac{1}{4}$. In either case we get the pair of
    numbers $\frac{1}{4}$ and 8.

    Check: $\frac{1}{4} \cdot 8 = 2$ and $\frac{1}{\frac{1}{4}} + \frac{1}{8} = 4\frac{1}{8} = \frac{33}{8}$. The
    numbers check.
    State: The numbers are $\frac{1}{4}$ and 8.

32. $(x + 2)^2 + (y - 1)^2 = 4$

33. Familiarize: Let x = the length of the longer
    piece and y = the length of the shorter piece.
    Then the lengths of the sides of the squares
    are $\frac{x}{4}$ and $\frac{y}{4}$.

    Translate: The total length of the wire is
    100 cm.

    $$x + y = 100$$

    The area of one square is 144 cm² greater than
    that of the other square.

    $$\left(\frac{x}{4}\right)^2 = \left(\frac{y}{4}\right)^2 + 144$$

    Carry out: We solve the system.
    Solve the first equation for y: $y = 100 - x$

    Substitute for y in the second equation and
    solve for x.

    $$\left(\frac{x}{4}\right)^2 = \left(\frac{100 - x}{4}\right)^2 + 144$$

    $$\frac{x^2}{16} = \frac{10,000 - 200x + x^2}{16} + 144$$

    $$x^2 = 10,000 - 200x + x^2 + 2304$$

    $$200x = 12,304$$

    $$x = 61.52$$

    If x = 61.52, y = 100 - 61.52 = 38.48.

    Check: The total length of the wire is
    61.52 + 38.48 = 100 cm. The area of the larger
    square is $\left(\frac{61.52}{4}\right)^2$, or 236.5444 cm². This is
    144 cm² greater than the area of the smaller
    square, $\left(\frac{38.48}{4}\right)^2$, or 92.5444 cm².

    State: The wire should be cut into 61.52 cm
    and 38.48 cm lengths.

34. Factor: $x^3 + y^3 = (x + y)(x^2 - xy + y^2)$. We know
    that x + y = 1, so $(x + y)^2 = x^2 + 2xy + y^2 = 1$,
    or $x^2 + y^2 = 1 - 2xy$. We also know that xy = 1,
    so $x^2 + y^2 = 1 - 2 \cdot 1 = -1$. Then
    $x^3 + y^3 = 1 \cdot (-1 - 1) = -2$.

35. The equation of the ellipse is of the form

    $$\frac{x^2}{a^2} + \frac{y^2}{b^2} = 1.$$

    Substitute $\left(1, \frac{\sqrt{3}}{2}\right)$ and $\left(\sqrt{3}, \frac{1}{2}\right)$ for (x, y) to
    get two equations.

    $$\frac{1^2}{a^2} + \frac{\left(\frac{\sqrt{3}}{2}\right)^2}{b^2} = 1, \text{ or } \frac{1}{a^2} + \frac{3}{4b^2} = 1$$

    $$\frac{(\sqrt{3})^2}{a^2} + \frac{\left(\frac{1}{2}\right)^2}{b^2} = 1, \text{ or } \frac{3}{a^2} + \frac{1}{4b^2} = 1$$

    Substitute u for $\frac{1}{a^2}$ and v for $\frac{1}{b^2}$.

$u + \frac{3}{4}v = 1,$	$4u + 3v = 4,$
or	
$3u + \frac{1}{4}v = 1$	$12u + v = 4$

    Solving for u and v, we get $u = \frac{1}{4}$, v = 1.
    Then $u = \frac{1}{a^2} = \frac{1}{4}$, so a² = 4; $v = \frac{1}{b^2} = 1$,
    so b² = 1.

    Then the equation of the ellipse is

    $$\frac{x^2}{4} + \frac{y^2}{1} = 1, \text{ or } \frac{x^2}{4} + y^2 = 1.$$

36. $\frac{x^2}{9/4} - \frac{y^2}{15/4} = 1$

37.
$$x - y = a + 2b, \qquad (1)$$
$$x^2 - y^2 = a^2 + 2ab + b^2 \qquad (2)$$

Solve Eq. (1) for x and substitute in Eq. (2). Then solve for y.

$$x = y + a + 2b \qquad (3)$$

$$(y + a + 2b)^2 - y^2 = a^2 + 2ab + b^2$$
$$y^2 + 2ay + 4by + a^2 + 4ab + 4b^2 - y^2 = a^2 + 2ab + b^2$$
$$2ay + 4by = -2ab - 3b^2$$
$$y(2a + 4b) = -2ab - 3b^2$$
$$y = \frac{-2ab - 3b^2}{2a + 4b}$$

Substitute for y in Eq. (3) and solve for x.

$$x = \frac{-2ab - 3b^2}{2a + 4b} + a + 2b$$
$$x = \frac{-2ab - 3b^2 + 2a^2 + 8ab + 8b^2}{2a + 4b}$$
$$x = \frac{2a^2 + 6ab + 5b^2}{2a + 4b}$$

The solution is $\left(\dfrac{2a^2 + 6ab + 5b^2}{2a + 4b}, \dfrac{-2ab - 3b^2}{2a + 4b} \right)$.

38. $(a - b, 0), \left(\dfrac{(a - b)(a^2 + b^2)}{2ab}, -\dfrac{(a^2 - b^2)(a - b)}{2ab} \right)$

39. Solve the system: $2L + 2W = P,$
$$LW = A$$

See the answer section in the text.

40. There is no number x such that

$\dfrac{x^2}{a^2} - \dfrac{\left[\frac{b}{a} x \right]^2}{b^2} = 1$, because the left side

simplifies to $\dfrac{x^2}{a^2} - \dfrac{x^2}{a^2}$ which is 0.

41. $(x - h)^2 + (y - k)^2 = r^2$

If (2, 4) is a point on the circle, then
$(2 - h)^2 + (4 - k)^2 = r^2$.

If (3, 3) is a point on the circle, then
$(3 - h)^2 + (3 - k)^2 = r^2$.

Thus
$$(2 - h)^2 + (4 - k)^2 = (3 - h)^2 + (3 - k)^2$$
$$4 - 4h + h^2 + 16 - 8k + k^2 =$$
$$9 - 6h + h^2 + 9 - 6k + k^2$$
$$-4h - 8k + 20 = -6h - 6k + 18$$
$$2h - 2k = -2$$
$$h - k = -1$$

If the center (h, k) is on the line $3x - y = 3$, then $3h - k = 3$.

Solving the system
$h - k = -1,$
$3h - k = 3$
we find that $(h, k) = (2, 3)$.

Find r, substituting (2, 3) for (h, k) and (2, 4) for (x, y). We could also use (3, 3) for (x, y).

$$(x - h)^2 + (y - k)^2 = r^2$$
$$(2 - 2)^2 + (4 - 3)^2 = r^2$$
$$0 + 1 = r^2$$
$$1 = r^2$$
$$1 = r$$

The equation of the circle whose center is (2, 3) and whose radius is 1 is $(x - 2)^2 + (y - 3)^2 = 1$.

42. $(x + 1)^2 + (y + 3)^2 = 10^2$

43.
$$x^3 + y^3 = 72, \qquad (1)$$
$$x + y = 6 \qquad (2)$$

Solve Eq. (2) for y: $y = 6 - x$
Substitute for y in Eq. (1) and solve for x.
$$x^3 + (6 - x)^3 = 72$$
$$x^3 + 216 - 108x + 18x^2 - x^3 = 72$$
$$18x^2 - 108x + 144 = 0$$
$$x^2 - 6x + 8 = 0$$
$$\left[\text{Multiplying by } \tfrac{1}{18} \right]$$
$$(x - 4)(x - 2) = 0$$

$x = 4$ or $x = 2$

If $x = 4$, then $y = 6 - 4 = 2$.
If $x = 2$, then $y = 6 - 2 = 4$.

The pairs (4, 2) and (2, 4) check.

44. $h \approx 217.386816, k \approx 54.2592, \lambda \approx 0.000016956$

45. (0.965, 4402.33), (-0.965, -4402.33)

46. (785, 45), (45, 785)

47. (2.11, -0.11), (-13.04, -13.34)

48. (7.37, -0.16), (0.19, -6.16)

Exercise Set 7.5

1.
$$x^2 + y^2 = 25, \qquad (1)$$
$$y^2 = x + 5 \qquad (2)$$

We substitute x + 5 for y^2 in Eq. (1) and solve for x.
$$x^2 + y^2 = 25$$
$$x^2 + (x + 5) = 25$$
$$x^2 + x - 20 = 0$$
$$(x + 5)(x - 4) = 0$$

$x + 5 = 0$ or $x - 4 = 0$
$\quad x = -5$ or $\quad\quad x = 4$

Next we substitute these numbers for x in either Eq. (1) or Eq. (2) and solve for y. Here we use Eq. (2).

$y^2 = -5 + 5 = 0$ and $y = 0$.

$y^2 = 4 + 5 = 9$ and $y = \pm 3$.

The possible solutions are $(-5,0)$, $(4,3)$, and $(4,-3)$.

Check:

For $(-5,0)$:

$x^2 + y^2 = 25$		$y^2 = x + 5$	
$(-5)^2 + 0^2$	25	0^2	$-5 + 5$
$25 + 0$		0	0
25			

For $(4,3)$:

$x^2 + y^2 = 25$		$y^2 = x + 5$	
$4^2 + 3^2$	25	3^2	$4 + 5$
$16 + 9$		9	9
25			

For $(4,-3)$:

$x^2 + y^2 = 25$		$y^2 = x + 5$	
$4^2 + (-3)^2$	25	$(-3)^2$	$4 + 5$
$16 + 9$		9	9
25			

The solutions are $(-5,0)$, $(4,3)$, and $(4,-3)$.

2. $(0,0)$, $(1,1)$

3. $x^2 + y^2 = 9$,   (1)
    $x^2 - y^2 = 9$ (2)

Here we use the addition method.

$$\begin{aligned} x^2 + y^2 &= 9 \\ \underline{x^2 - y^2} &= \underline{9} \\ 2x^2 \quad\quad &= 18 \quad \text{Adding} \\ x^2 &= 9 \\ x &= \pm 3 \end{aligned}$$

If $x = 3$, $x^2 = 9$, and if $x = -3$, $x^2 = 9$, so substituting 3 or -3 in Eq. (1) give us

$$\begin{aligned} x^2 + y^2 &= 9 \\ 9 + y^2 &= 9 \\ y^2 &= 0 \\ y &= 0. \end{aligned}$$

The possible solutions are $(3,0)$ and $(-3,0)$.

Check:

$x^2 + y^2 = 9$		$x^2 - y^2 = 9$	
$(\pm 3)^2 + (0)^2$	9	$(\pm 3)^2 - (0)^2$	9
$9 + 0$		$9 - 0$	
9		9	

The solutions are $(3,0)$ and $(-3,0)$.

4. $(0,2)$, $(0,-2)$

5. $x^2 + y^2 = 25$,   (1)
    $xy = 12$ (2)

First we solve Eq. (2) for y.

$xy = 12$

$y = \dfrac{12}{x}$

Then we substitute $\dfrac{12}{x}$ for y in Eq. (1) and solve for x.

$$\begin{aligned} x^2 + y^2 &= 25 \\ x^2 + \left[\frac{12}{x}\right]^2 &= 25 \\ x^2 + \frac{144}{x^2} &= 25 \\ x^4 + 144 &= 25x^2 \quad \text{Multiplying by } x^2 \\ x^4 - 25x^2 + 144 &= 0 \\ u^2 - 25u + 144 &= 0 \quad \text{Letting } u = x^2 \\ (u - 9)(u - 16) &= 0 \end{aligned}$$

$u = 9$ or $u = 16$

We now substitute x^2 for u and solve for x.

$x^2 = 9$ or $x^2 = 16$

$x = \pm 3$ or $x = \pm 4$

Since $y = 12/x$, if $x = 3$, $y = 4$; if $x = -3$, $y = -4$; if $x = 4$, $y = 3$; and if $x = -4$, $y = -3$. The pairs $(3,4)$, $(-3,-4)$, $(4,3)$, $(-4,-3)$ check. They are the solutions.

6. $(-5,3)$, $(-5,-3)$, $(4,0)$

7. $x^2 + y^2 = 4$,   (1)
    $16x^2 + 9y^2 = 144$ (2)

$$\begin{aligned} -9x^2 - 9y^2 &= -36 \quad \text{Multiplying (1) by } -9 \\ \underline{16x^2 + 9y^2} &= \underline{144} \\ 7x^2 \quad\quad &= 108 \quad \text{Adding} \\ x^2 &= \frac{108}{7} \\ x &= \pm\sqrt{\frac{108}{7}} = \pm 6\sqrt{\frac{3}{7}} \\ x &= \pm\frac{6\sqrt{21}}{7} \quad \text{Rationalizing the} \\ &\quad\quad\quad\quad\quad\ \text{denominator} \end{aligned}$$

Substituting $\dfrac{6\sqrt{21}}{7}$ or $-\dfrac{6\sqrt{21}}{7}$ for x in Eq. (1) gives us

$$\begin{aligned} \frac{36 \cdot 21}{49} + y^2 &= 4 \\ y^2 &= 4 - \frac{108}{7} \\ y^2 &= -\frac{80}{7} \\ y &= \pm\sqrt{-\frac{80}{7}} = \pm 4i\sqrt{\frac{5}{7}} \\ y &= \pm\frac{4i\sqrt{35}}{7}. \quad \text{Rationalizing the} \\ &\quad\quad\quad\quad\quad\quad\ \text{denominator} \end{aligned}$$

The pairs $\left(\frac{6\sqrt{21}}{7}, \frac{4i\sqrt{35}}{7}\right)$, $\left(\frac{6\sqrt{21}}{7}, -\frac{4i\sqrt{35}}{7}\right)$,

$\left(-\frac{6\sqrt{21}}{7}, \frac{4i\sqrt{35}}{7}\right)$, and $\left(-\frac{6\sqrt{21}}{7}, -\frac{4i\sqrt{35}}{7}\right)$ check.

They are the solutions.

8. (0,5), (0,-5)

9. $x^2 + y^2 = 16$, $x^2 + y^2 = 16$,   (1)
    $\quad\quad\quad\quad$ or
    $y^2 - 2x^2 = 10$ $-2x^2 + y^2 = 10$ (2)

Here we use the addition method.
    $2x^2 + 2y^2 = 32$ Multiplying (1) by 2
    $\underline{-2x^2 + y^2 = 10}$
    $\quad\quad\quad 3y^2 = 42$ Adding
    $\quad\quad\quady^2 = 14$
    $\quad\quad\quady = \pm\sqrt{14}$

Substituting $\sqrt{14}$ or $-\sqrt{14}$ for y in Eq. (1) gives us
    $x^2 + 14 = 16$
    $\quadx^2 = 2$
    $\quadx = \pm\sqrt{2}$

The pairs $(-\sqrt{2},-\sqrt{14})$, $(-\sqrt{2},\sqrt{14})$, $(\sqrt{2},-\sqrt{14})$, and $(\sqrt{2},\sqrt{14})$ check. They are the solutions.

10. $(-3,-\sqrt{5})$, $(-3,\sqrt{5})$, $(3,-\sqrt{5})$, $(3,\sqrt{5})$

11. $x^2 + y^2 = 5$,   (1)
    $xy = 2$ (2)

First we solve Eq. (2) for y.
    $xy = 2$
    $y = \frac{2}{x}$

Then we substitute $\frac{2}{x}$ for y in Eq. (1) and solve for x.
    $x^2 + y^2 = 5$
    $x^2 + \left(\frac{2}{x}\right)^2 = 5$
    $x^2 + \frac{4}{x^2} = 5$
    $x^4 + 4 = 5x^2$ Multiplying by x^2
    $x^4 - 5x^2 + 4 = 0$
    $u^2 - 5u + 4 = 0$ Letting $u = x^2$
    $(u - 4)(u - 1) = 0$
    $u = 4$ or $u = 1$
We now substitute x^2 for u and solve for x.
    $x^2 = 4$ or $x^2 = 1$
    $x = \pm 2$ $x = \pm 1$
Since $y = 2/x$, if $x = 2$, $y = 1$; if $x = -2$, $y = -1$; if $x = 1$, $y = 2$; and if $x = -1$, $y = -2$. The pairs (2,1), (-2,-1), (1,2), (-1,-2) check. They are the solutions.

12. (4,2), (-4,-2), (2,4), (-2,-4)

13. $x^2 + y^2 = 13$,   (1)
    $xy = 6$ (2)

First we solve Eq. (2) for y.
    $xy = 6$
    $y = \frac{6}{x}$

Then we substitute $\frac{6}{x}$ for y in Eq. (1) and solve for x.
    $x^2 + y^2 = 13$
    $x^2 + \left(\frac{6}{x}\right)^2 = 13$
    $x^2 + \frac{36}{x^2} = 13$
    $x^4 + 36 = 13x^2$ Multiplying by x^2
    $x^4 - 13x^2 + 36 = 0$
    $u^2 - 13u + 36 = 0$ Letting $u = x^2$
    $(u - 9)(u - 4) = 0$
    $u = 9$ or $u = 4$

We now substitute x^2 for u and solve for x.
    $x^2 = 9$ or $x^2 = 4$
    $x = \pm 3$ $x = \pm 2$
Since $y = 6/x$, if $x = 3$, $y = 2$; if $x = -3$, $y = -2$; if $x = 2$, $y = 3$; and if $x = -2$, $y = -3$. The pairs (3,2), (-3,-2), (2,3), (-2,-3) check. They are the solutions.

14. (4,1), (-4,-1), (2,2), (-2,-2)

15. $x^2 + y^2 + 6y + 5 = 0$ (1)
    $x^2 + y^2 - 2x - 8 = 0$ (2)

Using the addition method, multiply Eq. (2) by -1 and add the result to Eq. (1).
    $x^2 + y^2 + 6y + 5 = 0$ (1)
    $\underline{-x^2 - y^2 + 2x + 8 = 0}$ (2)
    $\quad\quad\quad\quad 2x + 6y + 13 = 0$ (3)

Solve Eq. (3) for x.
    $2x + 6y + 13 = 0$
    $\quad\quad\quad 2x = -6y - 13$
    $\quad\quad\quadx = \frac{-6y - 13}{2}$

Substitute $\frac{-6y - 13}{2}$ for x in Eq. (1) and solve for y.

    $x^2 + y^2 + 6y + 5 = 0$
    $\left(\frac{-6y - 13}{2}\right)^2 + y^2 + 6y + 5 = 0$
    $\frac{36y^2 + 156y + 169}{4} + y^2 + 6y + 5 = 0$
    $36y^2 + 156y + 169 + 4y^2 + 24y + 20 = 0$
    $\quad\quad\quad\quad 40y^2 + 180y + 189 = 0$

Using the quadratic formula, we find that

$y = \dfrac{-45 \pm 3\sqrt{15}}{20}$. Substitute $\dfrac{-45 \pm 3\sqrt{15}}{20}$

for y in $x = \dfrac{-6y - 13}{2}$ and solve for x.

If $y = \dfrac{-45 + 3\sqrt{15}}{20}$, then $x = \dfrac{-6\left(\dfrac{-45 + 3\sqrt{15}}{20}\right) - 13}{2}$

$= \dfrac{5 - 9\sqrt{15}}{20}$.

If $y = \dfrac{-45 - 3\sqrt{15}}{20}$, then $x = \dfrac{-6\left(\dfrac{-45 - 3\sqrt{15}}{20}\right) - 13}{2}$

$= \dfrac{5 + 9\sqrt{15}}{20}$.

The pairs $\left(\dfrac{5 + 9\sqrt{15}}{20}, \dfrac{-45 - 3\sqrt{15}}{20}\right)$ and

$\left(\dfrac{5 - 9\sqrt{15}}{20}, \dfrac{-45 + 3\sqrt{15}}{20}\right)$ check and are the
solutions.

16. (2,1), (-2,-1)

17. $xy - y^2 = 2$,   (1)
     $2xy - 3y^2 = 0$ (2)

  $-2xy + 2y^2 = -4$ Multiplying (1) by -2
  $\underline{2xy - 3y^2 = 0}$
       $-y^2 = -4$
        $y^2 = 4$
        $y = \pm 2$

We substitute for y in Eq. (1) and solve for x.
When y = 2: $x \cdot 2 - 2^2 = 2$
              $2x - 4 = 2$
               $2x = 6$
                $x = 3$
When y = -2: $x(-2) - (-2)^2 = 2$
               $-2x - 4 = 2$
                $-2x = 6$
                  $x = -3$

The pairs (3,2) and (-3,-2) check. They are the
solutions.

18. $\left(2, -\dfrac{4}{5}\right)$, $\left(-2, -\dfrac{4}{5}\right)$, (5,2), (-5,2)

19. $m^2 - 3mn + n^2 + 1 = 0$,   (1)
     $3m^2 - mn + 3n^2 = 13$ (2)

     $m^2 - 3mn + n^2 = -1$,   (3)   Rewriting (1)
     $3m^2 - mn + 3n^2 = 13$ (2)

  $-3m^2 + 9mn - 3n^2 = 3$ Multiplying (3) by -3
  $\underline{3m^2 - mn + 3n^2 = 13}$
            $8mn = 16$
             $mn = 2$
              $n = \dfrac{2}{m}$ (4)

Substitute $\dfrac{2}{m}$ for n in Eq. (1) and solve for m.

$m^2 - 3m\left(\dfrac{2}{m}\right) + \left(\dfrac{2}{m}\right)^2 + 1 = 0$

$m^2 - 6 + \dfrac{4}{m^2} + 1 = 0$

$m^2 - 5 + \dfrac{4}{m^2} = 0$

$m^4 - 5m^2 + 4 = 0$ Multiplying by m^2

Substitute u for m^2.
   $u^2 - 5u + 4 = 0$
  $(u - 4)(u - 1) = 0$

   $u = 4$ or $u = 1$
  $m^2 = 4$ or $m^2 = 1$
   $m = \pm 2$ or $m = \pm 1$

Substitute for m in Eq. (4) and solve for n.

When m = 2, $n = \dfrac{2}{2} = 1$.

When m = -2, $n = \dfrac{2}{-2} = -1$.

When m = 1, $n = \dfrac{2}{1} = 2$.

When m = -1, $n = \dfrac{2}{-1} = -2$.

The pairs (2, 1), (-2, -1), (1, 2), and (-1, -2)
check. They are the solutions.

20. $(-\sqrt{2}, \sqrt{2})$, $(\sqrt{2}, -\sqrt{2})$

21. $a^2 + b^2 = 14$,   (1)
     $ab = 3\sqrt{5}$ (2)

     Solve Eq. (1) for b.
        $b = \dfrac{3\sqrt{5}}{a}$

Substitute $\dfrac{3\sqrt{5}}{a}$ for b in Eq. (1) and solve for a.

   $a^2 + \left(\dfrac{3\sqrt{5}}{a}\right)^2 = 14$

     $a^2 + \dfrac{45}{a^2} = 14$

      $a^4 + 45 = 14a^2$

   $a^4 - 14a^2 + 45 = 0$
   $u^2 - 14u + 45 = 0$ Letting $u = a^2$
   $(u - 9)(u - 5) = 0$

    $u = 9$ or $u = 5$
   $a^2 = 9$ or $a^2 = 5$
    $a = \pm 3$ or $a = \pm\sqrt{5}$

Since $b = 3\sqrt{5}/a$, if a = 3, $b = \sqrt{5}$; if a = -3,
$b = -\sqrt{5}$; if $a = \sqrt{5}$, b = 3; and if $a = -\sqrt{5}$,
b = -3. The pairs $(3, \sqrt{5})$, $(-3, -\sqrt{5})$, $(\sqrt{5}, 3)$,
$(-\sqrt{5}, -3)$ check. They are the solutions.

22. $\left(i\sqrt{3}, -\dfrac{8i\sqrt{3}}{3}\right), \left(-i\sqrt{3}, \dfrac{8i\sqrt{3}}{3}\right)$

23. $x^2 + y^2 = 25,$ (1)

    $9x^2 + 4y^2 = 36$ (2)

    $-4x^2 - 4y^2 = -100$ Multiplying (1) by -4

$$\underline{\quad 9x^2 + 4y^2 = \quad\ 36 \quad}$$
$$\quad 5x^2 \qquad\quad = -64$$

$$x^2 = -\frac{64}{5}$$

$$x = \pm\sqrt{\frac{-64}{5}} = \pm\frac{8i}{\sqrt{5}}$$

$$x = \pm\frac{8i\sqrt{5}}{5} \quad \text{Rationalizing the denominator}$$

Substituting $\dfrac{8i\sqrt{5}}{5}$ or $-\dfrac{8i\sqrt{5}}{5}$ for x in Eq. (1) and solving for y gives us

$$-\frac{64}{5} + y^2 = 25$$

$$y^2 = \frac{189}{5}$$

$$y = \pm\sqrt{\frac{189}{5}} = \pm 3\sqrt{\frac{21}{5}}$$

$$y = \pm\frac{3\sqrt{105}}{5}. \quad \text{Rationalizing the denominator}$$

The pairs $\left(\dfrac{8i\sqrt{5}}{5}, \dfrac{3\sqrt{105}}{5}\right), \left(-\dfrac{8i\sqrt{5}}{5}, \dfrac{3\sqrt{105}}{5}\right),$ $\left(\dfrac{8i\sqrt{5}}{5}, -\dfrac{3\sqrt{105}}{5}\right),$ and $\left(-\dfrac{8i\sqrt{5}}{5}, -\dfrac{3\sqrt{105}}{5}\right)$ check. They are the solutions.

24. $\left(\dfrac{4\sqrt{10}}{5}, \dfrac{3i\sqrt{15}}{5}\right), \left(\dfrac{4\sqrt{10}}{5}, -\dfrac{3i\sqrt{15}}{5}\right),$ $\left(-\dfrac{4\sqrt{10}}{5}, \dfrac{3i\sqrt{15}}{5}\right), \left(-\dfrac{4\sqrt{10}}{5}, -\dfrac{3i\sqrt{15}}{5}\right)$

25. Familiarize: Let x and y represent the numbers.
Translate:

    The product of two numbers is 156.

$$xy = 156 \qquad (1)$$

    The sum of their squares is 313.

$$x^2 + y^2 = 313 \qquad (2)$$

Carry out: We solve the system of equations.
First solve Eq. (1) for y.

$$xy = 156$$

$$y = \frac{156}{x}$$

Then we substitute $\dfrac{156}{x}$ for y in Eq. (2) and solve for x.

$$x^2 + y^2 = 313 \qquad (2)$$

$$x^2 + \left(\frac{156}{x}\right)^2 = 313$$

$$x^2 + \frac{24{,}336}{x^2} = 313$$

$$x^4 + 24{,}336 = 313x^2$$

$$x^4 - 313x^2 + 24{,}336 = 0$$

$$u^2 - 313u + 24{,}336 = 0 \quad \text{Letting } u = x^2$$

$$(u - 169)(u - 144) = 0$$

$$u = 169 \quad \text{or} \quad u = 144$$

We now substitute x^2 for u and solve for x.

$$x^2 = 169 \quad \text{or} \quad x^2 = 144$$

$$x = \pm 13 \quad \text{or} \quad x = \pm 12$$

Since $y = 156/x$, if $x = 13$, $y = 12$; if $x = -13$, $y = -12$; if $x = 12$, $y = 13$; and if $x = -12$, $y = -13$. The possible solutions are $(13,12)$, $(-13,-12)$, $(12,13)$, and $(-12,-13)$.

Check: If $x = 13$ and $y = 12$, their product is 156. If $x = -13$ and $y = -12$, their product is 156. The sum of the squares in either case is $(\pm 13)^2 + (\pm 12)^2 = 169 + 144 = 313$. The pairs $(12,13)$ and $(-12,-13)$ do not give us any other solutions.

State: The numbers are 13 and 12 or -13 and -12.

26. 6 and 10 or -6 and -10

27. Familiarize: We first make a drawing. Let ℓ = the length and w = the width.

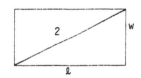

Translate:
Area: $\ell w = \sqrt{3}$ (1)
From the Pythagorean theorem: $\ell^2 + w^2 = 2^2$ (2)
Carry out: We solve the system of equations.
We first solve Eq. (1) for w.

$$\ell w = \sqrt{3}$$

$$w = \frac{\sqrt{3}}{\ell}$$

Then we substitute $\dfrac{\sqrt{3}}{\ell}$ for w in Eq. (2) and solve for ℓ.

$$\ell^2 + \left(\frac{\sqrt{3}}{\ell}\right)^2 = 4$$

$$\ell^2 + \frac{3}{\ell^2} = 4$$

$$\ell^4 + 3 = 4\ell^2$$

$$\ell^4 - 4\ell^2 + 3 = 0$$

$$u^2 - 4u + 3 = 0 \quad \text{Letting } u = \ell^2$$

$$(u - 3)(u - 1) = 0$$

$$u = 3 \quad \text{or} \quad u = 1$$

We now substitute ℓ^2 for u and solve for ℓ.

$$\ell^2 = 3 \quad \text{or} \quad \ell^2 = 1$$

$$\ell = \pm\sqrt{3} \quad \text{or} \quad \ell = \pm 1$$

Length cannot be negative, so we only need to consider $\ell = \sqrt{3}$ and $\ell = 1$. Since $w = \sqrt{3}/\ell$, if $\ell = \sqrt{3}$, $w = 1$ and if $\ell = 1$, $w = \sqrt{3}$. Length is usually considered to be longer than width, so we have the solution $\ell = \sqrt{3}$ and $w = 1$, or $(\sqrt{3}, 1)$.

Check: If $\ell = \sqrt{3}$ and $w = 1$, the area is $\sqrt{3} \cdot 1 = \sqrt{3}$. Also $(\sqrt{3})^2 + 1^2 = 3 + 1 = 4 = 2^2$. The numbers check.

State: The length is $\sqrt{3}$ m, and the width is 1 m.

<u>28.</u> $\sqrt{2}$ m, 1 m

<u>29.</u> Familiarize: We let x = the length of a side of one peanut bed and y = the length of a side of the other peanut bed. Make a drawing.

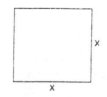
Area: x^2

Area: y^2

Translate:

The sum of the areas is 832 ft².

$$x^2 + y^2 = 832$$

The difference of the areas is 320 ft².

$$x^2 - y^2 = 320$$

Carry out: We solve the system of equations.

$$
\begin{array}{rl}
x^2 + y^2 = & 832 \\
x^2 - y^2 = & 320 \\
\hline
2x^2 \quad\;\; = & 1152 \qquad \text{Adding} \\
x^2 = & 576 \\
x = & \pm 24
\end{array}
$$

Since length cannot be negative, we consider only $x = 24$. Substitute 24 for x in the first equation and solve for y.

$$
\begin{array}{rl}
24^2 + y^2 = & 832 \\
576 + y^2 = & 832 \\
y^2 = & 256 \\
y = & \pm 16
\end{array}
$$

Again, we consider only the positive value, 16. The possible solution is (24,16).

Check: The areas of the peanut beds are 24^2, or 576, and 16^2, or 256. The sum of the areas is $576 + 256$, or 832. The difference of the areas is $576 - 256$, or 320. The values check.

State: The lengths of the beds are 24 ft and 16 ft.

<u>30.</u> $125, 6%

<u>31.</u> $(x - h)^2 + (y - k)^2 = r^2$ (Standard form)

Substitute (4, 6), (-6, 2), and (1, -3) for (x, y).

$$
\begin{array}{ll}
(4 - h)^2 + (6 - k)^2 = r^2 & \qquad (1) \\
(-6 - h)^2 + (2 - k)^2 = r^2 & \qquad (2) \\
(1 - h)^2 + (-3 - k)^2 = r^2 & \qquad (3)
\end{array}
$$

Thus

$(4 - h)^2 + (6 - k)^2 = (-6 - h)^2 + (2 - k)^2$

or $5h + 2k = 3$

$(4 - h)^2 + (6 - k)^2 = (1 - h)^2 + (-3 - k)^2$

or $h + 3k = 7$

We solve the system

$$
\begin{array}{l}
5h + 2k = 3, \\
h + 3k = 7.
\end{array}
$$

Solving we get $h = -\frac{5}{13}$ and $k = \frac{32}{13}$. Substituting these values in equation (1), (2), or (3), we find that $r^2 = \frac{5365}{169}$.

The equation of the circle is

$$\left(x + \frac{5}{13}\right)^2 + \left(y - \frac{32}{13}\right)^2 = \frac{5365}{169}.$$

<u>32.</u> $(x - 10)^2 + (y + 3)^2 = 10^2$

<u>33.</u> Familiarize: Let x and y represent the numbers. Translate:

The square of a certain number exceeds twice the square of another number by $\frac{1}{8}$.

$$x^2 = 2y^2 + \frac{1}{8}$$

The sum of the squares is $\frac{5}{16}$.

$$x^2 + y^2 = \frac{5}{16}$$

Carry out: We solve the system.

$$
\begin{array}{ll}
x^2 - 2y^2 = \frac{1}{8}, & \qquad (1) \\
x^2 + \;\; y^2 = \frac{5}{16} & \qquad (2)
\end{array}
$$

$$
\begin{array}{ll}
x^2 - 2y^2 = \frac{1}{8}, & \\
2x^2 + 2y^2 = \frac{5}{8} & \qquad \text{Multiplying (2) by 2} \\
\hline
3x^2 \qquad\;\; = \frac{6}{8} & \\
x^2 = \frac{1}{4} & \\
x = \pm\frac{1}{2} &
\end{array}
$$

Substitute $\pm\frac{1}{2}$ for x in (2) and solve for y.

$$\left(\pm\frac{1}{2}\right)^2 + y^2 = \frac{5}{16}$$

$$\frac{1}{4} + y^2 = \frac{5}{16}$$

$$y^2 = \frac{1}{16}$$

$$y = \pm\frac{1}{4}$$

We get $\left(\frac{1}{2}, \frac{1}{4}\right)$, $\left(-\frac{1}{2}, \frac{1}{4}\right)$, $\left(\frac{1}{2}, -\frac{1}{4}\right)$ and $\left(-\frac{1}{2}, -\frac{1}{4}\right)$.

Check: It is true that $\left(\pm\frac{1}{2}\right)^2$ exceeds twice $\left(\pm\frac{1}{4}\right)^2$ by $\frac{1}{8}$: $\frac{1}{4} = 2\left(\frac{1}{16}\right) + \frac{1}{8}$

Also $\left(\pm\frac{1}{2}\right)^2 + \left(\pm\frac{1}{4}\right)^2 = \frac{5}{16}$. The pairs check.

State: The numbers are $\frac{1}{2}$ and $\frac{1}{4}$ or $-\frac{1}{2}$ and $\frac{1}{4}$ or $\frac{1}{2}$ and $-\frac{1}{4}$ or $-\frac{1}{2}$ and $-\frac{1}{4}$.

<u>34.</u> 10 in. by 7 in. by 5 in.

<u>35.</u> $x^2 + xy = a$,   (1)

$y^2 + xy = b$ (2)

Solve Eq. (1) for y.

$$xy = a - x^2$$

$$y = \frac{a - x^2}{x} \quad (3)$$

Substitute for y in Eq. (2) and solve for x.

$$\left(\frac{a - x^2}{x}\right)^2 + x\left(\frac{a - x^2}{x}\right) = b$$

$$\frac{a^2 - 2ax^2 + x^4}{x^2} + a - x^2 = b$$

$a^2 - 2ax^2 + x^4 + ax^2 - x^4 = bx^2$ Clearing the fraction

$$a^2 = ax^2 + bx^2$$

$$a^2 = x^2(a + b)$$

$$\frac{a^2}{a + b} = x^2$$

$$\pm\frac{a}{\sqrt{a + b}} = x$$

$$\pm\frac{a\sqrt{a + b}}{a + b} = x \quad \text{Rationalizing the denominator}$$

Substitute for x in Eq. (3) and solve for y.

When $x = \frac{a\sqrt{a + b}}{a + b}$: $y = \dfrac{a - \left(\frac{a\sqrt{a + b}}{a + b}\right)^2}{\frac{a\sqrt{a + b}}{a + b}} =$

$$\frac{a - \frac{a^2}{a + b}}{\frac{a\sqrt{a + b}}{a + b}} = \frac{a^2 + ab - a^2}{a\sqrt{a + b}} = \frac{ab}{a\sqrt{a + b}} = \frac{b}{\sqrt{a + b}} =$$

$$\frac{b\sqrt{a + b}}{a + b}$$

When $x = -\frac{a\sqrt{a + b}}{a + b}$: $y = \dfrac{a - \left(-\frac{a\sqrt{a + b}}{a + b}\right)^2}{-\frac{a\sqrt{a + b}}{a + b}} =$

$$\frac{a - \frac{a^2}{a + b}}{-\frac{a\sqrt{a + b}}{a + b}} = \frac{a^2 + ab - a^2}{-a\sqrt{a + b}} = \frac{ab}{-a\sqrt{a + b}} = -\frac{b}{\sqrt{a + b}} =$$

$$-\frac{b\sqrt{a + b}}{a + b}$$

The pairs $\left(\frac{a\sqrt{a + b}}{a + b}, \frac{b\sqrt{a + b}}{a + b}\right)$ and $\left(-\frac{a\sqrt{a + b}}{a + b}, -\frac{b\sqrt{a + b}}{a + b}\right)$ check. They are the solutions.

<u>36.</u> (a,b)

<u>37.</u> $p^2 + q^2 = 13$ (1)

$\frac{1}{pq} = -\frac{1}{6}$ (2)

Solve Eq. (2) for p.

$$\frac{1}{q} = -\frac{p}{6}$$

$$-\frac{6}{q} = p$$

Substitute -6/q for p in Eq. (1) and solve for q.

$$\left(-\frac{6}{q}\right)^2 + q^2 = 13$$

$$\frac{36}{q^2} + q^2 = 13$$

$$36 + q^4 = 13q^2$$

$q^4 - 13q^2 + 36 = 0$

$u^2 - 13u + 36 = 0$ Letting $u = q^2$

$(u - 9)(u - 4) = 0$

$u = 9$ or $u = 4$

$q^2 = 9$ or $q^2 = 4$

$q = \pm 3$ or $q = \pm 2$

Since p = -6/q, if q = 3, p = -2; if q = -3, p = 2; if q = 2, p = -3, and if q = -2, p = 3. The pairs (-2,3), (2,-3), (-3,2), and (3,-2) check. They are the solutions.

<u>38.</u> $\left(\frac{1}{3}, \frac{1}{2}\right)$, $\left(\frac{1}{2}, \frac{1}{3}\right)$

39. $x^2 + y^2 = 4,$ (1)

$(x - 1)^2 + y^2 = 4$ (2)

Solve Eq. (1) for y^2:

$y^2 = 4 - x^2$ (3)

Substitute $4 - x^2$ for y^2 in Eq. (2) and solve for x.

$(x - 1)^2 + (4 - x^2) = 4$

$x^2 - 2x + 1 + 4 - x^2 = 4$

$-2x = -1$

$x = \frac{1}{2}$

Substitute $\frac{1}{2}$ for x in Eq. (3) and solve for y.

$y^2 = 4 - \left(\frac{1}{2}\right)^2$

$y^2 = 4 - \frac{1}{4}$

$y^2 = \frac{15}{4}$

$y = \pm \frac{\sqrt{15}}{2}$

The pairs $\left(\frac{1}{2}, \frac{\sqrt{15}}{2}\right)$ and $\left(\frac{1}{2}, -\frac{\sqrt{15}}{2}\right)$ check. They are the solutions.

40. $\left(\frac{a\sqrt{3}}{3}, \frac{b\sqrt{3}}{3}\right)$, $\left(\frac{a\sqrt{3}}{3}, -\frac{b\sqrt{3}}{3}\right)$,

$\left(-\frac{a\sqrt{3}}{3}, \frac{b\sqrt{3}}{3}\right)$, $\left(-\frac{a\sqrt{3}}{3}, -\frac{b\sqrt{3}}{3}\right)$ where $a > 0$, $b > 0$

41. $(x - 2)^4 - (x - 2) = 0,$ (1)

$x^2 - kx + k = 0$ (2)

In Eq. (1) let $(x - 2) = z$. Then

$z^4 - z = 0$, or $z(z - 1)(z^2 + z + 1) = 0$;

so that $z = 0, 1, \frac{-1 \pm i\sqrt{3}}{2}$ and then

$x = 2, 3, \frac{3 \pm i\sqrt{3}}{2}.$

The roots of Eq. (2) are $\frac{k \pm \sqrt{k^2 - 4k}}{2}$, and

comparing this with $\frac{3 \pm i\sqrt{3}}{2}$ we have $k = 3$.

[If 2 were a root of Eq. (2) there would be no other common root; likewise for 3.]

42. $\left(\frac{2\sqrt{6}}{3}, \frac{\sqrt{6}}{3}\right)$, $\left(-\frac{2\sqrt{6}}{3}, -\frac{\sqrt{6}}{3}\right)$, $(1, -2)$, $(-1, 2)$

43. $5^{x+y} = 100,$

$3^{2x-y} = 1000$

$(x + y) \log 5 = 2,$ Taking logarithms and

$(2x - y) \log 3 = 3$ simplifying

$x \log 5 + y \log 5 = 2,$ (1)

$2x \log 3 - y \log 3 = 3$ (2)

Multiply Eq. (1) by $\log 3$ and Eq. (2) by $\log 5$.

$x \log 3 \cdot \log 5 + y \log 3 \cdot \log 5 = 2 \log 3$

$\underline{2x \log 3 \cdot \log 5 - y \log 3 \cdot \log 5 = 3 \log 5}$

$3x \log 3 \cdot \log 5 \qquad\qquad = 2 \log 3 +$

$3 \log 5$

$x = \frac{2 \log 3 + 3 \log 5}{3 \log 3 \cdot \log 5}$

Substitute in (1) to find y.

$\frac{2 \log 3 + 3 \log 5}{3 \log 3 \cdot \log 5} \cdot \log 5 + y \log 5 = 2$

$y \log 5 = 2 - \frac{2 \log 3 + 3 \log 5}{3 \log 3}$

$y \log 5 = \frac{6 \log 3 - 2 \log 3 - 3 \log 5}{3 \log 3}$

$y \log 5 = \frac{4 \log 3 - 3 \log 5}{3 \log 3}$

$y = \frac{4 \log 3 - 3 \log 5}{3 \log 3 \cdot \log 5}$

The pair $\left(\frac{2 \log 3 + 3 \log 5}{3 \log 3 \cdot \log 5}, \frac{4 \log 3 - 3 \log 5}{3 \log 3 \cdot \log 5}\right)$

checks. It is the solution.

44. $(0, 0)$

45. Familiarize: Let x and y represent the numbers. Let n represent the sum, product and sum of the squares of the numbers.

Translate: We have a system of equations.

$x + y = n,$ (1)

$xy = n,$ (2)

$x^2 + y^2 = n$ (3)

Carry out: We solve the system.

Solve Eq. (2) for y: $y = \frac{n}{x}$

Substitute in Eq. (1) and solve for n.

$x + \frac{n}{x} = n$

$x^2 + n = nx$

$x^2 = nx - n$

$x^2 = n(x - 1)$

$\frac{x^2}{x - 1} = n$

Then $y = \frac{n}{x} = \frac{\frac{x^2}{x - 1}}{x} = \frac{x}{x - 1}.$

Substitute for y and n in Eq. (3) and solve for x.

$$x^2 + \left[\frac{x}{x-1}\right]^2 = \frac{x^2}{x-1}$$

$$x^2 + \frac{x^2}{(x-1)^2} = \frac{x^2}{x-1}$$

$$x^2(x-1)^2 + x^2 = x^2(x-1) \qquad \text{Multiplying by} \atop (x-1)^2$$

$$x^4 - 2x^3 + x^2 + x^2 = x^3 - x^2$$

$$x^4 - 3x^3 + 3x^2 = 0$$

$$x^2(x^2 - 3x + 3) = 0$$

$$x^2 = 0 \quad \text{or} \quad x^2 - 3x + 3 = 0$$

Solving $x^2 = 0$, we get $x = 0$.

Solving $x^2 - 3x + 3 = 0$, we get

$$x = \frac{3 \pm \sqrt{9-12}}{2} = \frac{3 \pm i\sqrt{3}}{2}.$$

When $x = 0$, $y = \frac{0}{0-1} = 0$.

When $x = \frac{3 + i\sqrt{3}}{2}$, $y = \dfrac{\dfrac{3 + i\sqrt{3}}{2}}{\dfrac{3 + i\sqrt{3}}{2} - 1} = \frac{3 - i\sqrt{3}}{2}.$

When $x = \frac{3 - i\sqrt{3}}{2}$, $y = \dfrac{\dfrac{3 - i\sqrt{3}}{2}}{\dfrac{3 - i\sqrt{3}}{2} - 1} = \frac{3 + i\sqrt{3}}{2}.$

The pairs $(0, 0)$, $\left[\dfrac{3 + i\sqrt{3}}{2}, \dfrac{3 - i\sqrt{3}}{2}\right]$, and

$\left[\dfrac{3 - i\sqrt{3}}{2}, \dfrac{3 + i\sqrt{3}}{2}\right]$ check. They are the

solutions.

<u>46.</u> (400, 1.43), (400, -1.43), (-400, 1.43),
(-400, -1.43)

<u>47.</u> (-8.53, -2.53), (-8.53, 2.53), (8.53, -2.53),
(8.53, 2.53)

Exercise Set 8.1

1. $a_n = 4n - 1$

   $a_1 = 4 \cdot 1 - 1 = 3$, $a_4 = 4 \cdot 4 - 1 = 15$;

   $a_2 = 4 \cdot 2 - 1 = 7$, $a_{10} = 4 \cdot 10 - 1 = 39$;

   $a_3 = 4 \cdot 3 - 1 = 11$, $a_{15} = 4 \cdot 15 - 1 = 59$

2. 0, 0, 0, 6; 504; 2184

3. $a_n = \dfrac{n}{n-1}$, $n \geqslant 2$

   The first 4 terms are a_2, a_3, a_4, and a_5:

   $a_2 = \dfrac{2}{2-1} = 2$, $a_4 = \dfrac{4}{4-1} = \dfrac{4}{3}$,

   $a_3 = \dfrac{3}{3-1} = \dfrac{3}{2}$, $a_5 = \dfrac{5}{5-1} = \dfrac{5}{4}$.

   $a_{10} = \dfrac{10}{10-1} = \dfrac{10}{9}$, $a_{15} = \dfrac{15}{15-1} = \dfrac{15}{14}$

4. 0, 3, 8, 15; 99; 224

5. $a_n = n^2 + 2n$

   $a_1 = 1^2 + 2 \cdot 1 = 3$, $a_4 = 4^2 + 2 \cdot 4 = 24$;

   $a_2 = 2^2 + 2 \cdot 2 = 8$, $a_{10} = 10^2 + 2 \cdot 10 = 120$;

   $a_3 = 3^2 + 2 \cdot 3 = 15$, $a_{15} = 15^2 + 2 \cdot 15 = 255$

6. $0, \dfrac{3}{5}, \dfrac{4}{5}, \dfrac{15}{17}; \dfrac{99}{101}; \dfrac{112}{113}$

7. $a_n = n + \dfrac{1}{n}$

   $a_1 = 1 + \dfrac{1}{1} = 2$, $a_4 = 4 + \dfrac{1}{4} = 4\dfrac{1}{4}$;

   $a_2 = 2 + \dfrac{1}{2} = 2\dfrac{1}{2}$, $a_{10} = 10 + \dfrac{1}{10} = 10\dfrac{1}{10}$;

   $a_3 = 3 + \dfrac{1}{3} = 3\dfrac{1}{3}$, $a_{15} = 15 + \dfrac{1}{15} = 15\dfrac{1}{15}$

8. $1, -\dfrac{1}{2}, \dfrac{1}{4}, -\dfrac{1}{8}; -\dfrac{1}{512}; \dfrac{1}{16,384}$

9. $a_n = (-1)^n n^2$

   $a_1 = (-1)^1 1^2 = -1$, $a_4 = (-1)^4 4^2 = 16$;

   $a_2 = (-1)^2 2^2 = 4$, $a_{10} = (-1)^{10} 10^2 = 100$;

   $a_3 = (-1)^3 3^2 = -9$, $a_{15} = (-1)^{15} 15^2 = -225$

10. -4, 5, -6, 7; 13; -18

11. $a_n = (-1)^{n+1}(3n - 5)$

    $a_1 = (-1)^{1+1}(3 \cdot 1 - 5) = -2$,

    $a_2 = (-1)^{2+1}(3 \cdot 2 - 5) = -1$,

    $a_3 = (-1)^{3+1}(3 \cdot 3 - 5) = 4$,

    $a_4 = (-1)^{4+1}(3 \cdot 4 - 5) = -7$;

    $a_{10} = (-1)^{10+1}(3 \cdot 10 - 5) = -25$;

    $a_{15} = (-1)^{15+1}(3 \cdot 15 - 5) = 40$

12. 0, 7, -26, 63; 999; -3374

13. $a_n = \dfrac{n+2}{n+5}$

    $a_1 = \dfrac{1+2}{1+5} = \dfrac{3}{6} = \dfrac{1}{2}$, $a_4 = \dfrac{4+2}{4+5} = \dfrac{6}{9} = \dfrac{2}{3}$;

    $a_2 = \dfrac{2+2}{2+5} = \dfrac{4}{7}$, $a_{10} = \dfrac{10+2}{10+5} = \dfrac{12}{15} = \dfrac{4}{5}$;

    $a_3 = \dfrac{3+2}{3+5} = \dfrac{5}{8}$, $a_{15} = \dfrac{15+2}{15+5} = \dfrac{17}{20}$

14. $-1, \dfrac{3}{2}, 1, \dfrac{7}{8}; \dfrac{19}{26}; \dfrac{29}{41}$

15. $a_n = 5n - 6$

    $a_8 = 5 \cdot 8 - 6 = 40 - 6 = 34$

16. 37

17. $a_n = (3n - 4)(2n + 5)$

    $a_7 = (3 \cdot 7 - 4)(2 \cdot 7 + 5) = 17 \cdot 19 = 323$

18. 81

19. $a_n = (-1)^{n-1}(4.6n - 18.3)$

    $a_{12} = (-1)^{12-1}[4.6(12) - 18.3] = -36.9$

20. -52,135,198.72

21. $a_n = 5n^2(4n - 100)$

    $a_{11} = 5(11)^2(4 \cdot 11 - 100) = 5(121)(-56) = -33,880$

22. 528,528

23. $a_n = \left(1 + \dfrac{1}{n}\right)^2$

    $a_{20} = \left(1 + \dfrac{1}{20}\right)^2 = \left(\dfrac{21}{20}\right)^2 = \dfrac{441}{400}$

24. $\dfrac{2744}{3375}$

25. $a_n = \log 10^n$

    $a_{43} = \log 10^{43} = 43$

26. 67

27. $a_n = 1 + \dfrac{1}{n^2}$

    $a_{38} = 1 + \dfrac{1}{38^2} = 1\dfrac{1}{1444}$

28. -8

29. 1, 3, 5, 7, 9, . . .

    These are odd integers, so the general term may be $2n - 1$.

30. 3^n

31. $-2, 6, -18, 54, \ldots$

    We can see a pattern if we write the sequence as
    $-1 \cdot 2 \cdot 1, \ 1 \cdot 2 \cdot 3, \ -1 \cdot 2 \cdot 9, \ 1 \cdot 2 \cdot 27, \ldots$
    The general term may be $(-1)^n 2(3)^{n-1}$.

32. $5n - 7$

33. $\dfrac{2}{3}, \dfrac{3}{4}, \dfrac{4}{5}, \dfrac{5}{6}, \dfrac{6}{7}, \ldots$

    These are fractions in which the denominator is 1
    greater than the numerator. Also, each numerator
    is 1 greater than the preceding numerator. The
    general term may be $\dfrac{n+1}{n+2}$.

34. $\sqrt{2n}$

35. $\sqrt{3}, \ 3, \ 3\sqrt{3}, \ 9, \ 9\sqrt{3}, \ldots$

    These are powers of $\sqrt{3}$. The general term may be
    $(\sqrt{3})^n$, or $3^{n/2}$.

36. $n(n + 1)$

37. $-1, -4, -7, -10, -13, \ldots$

    Each term is 3 less than the preceding term. The
    general term may be $-1 - 3(n - 1)$. After removing
    parentheses and simplifying, we can express the
    general term as $-3n + 2$, or $-(3n - 2)$.

38. $\log 10^{n-1}$, or $n - 1$

39. $1, 2, 3, 4, 5, 6, 7, \ldots$
    $S_7 = 1 + 2 + 3 + 4 + 5 + 6 + 7 = 28$

40. -8

41. $2, 4, 6, 8, \ldots$
    $S_5 = 2 + 4 + 6 + 8 + 10 = 30$

42. $1 + \dfrac{1}{4} + \dfrac{1}{9} + \dfrac{1}{16} + \dfrac{1}{25} = \dfrac{5269}{3600}$

43. $\displaystyle\sum_{k=1}^{5} \dfrac{1}{2k} = \dfrac{1}{2 \cdot 1} + \dfrac{1}{2 \cdot 2} + \dfrac{1}{2 \cdot 3} + \dfrac{1}{2 \cdot 4} + \dfrac{1}{2 \cdot 5}$

    $= \dfrac{1}{2} + \dfrac{1}{4} + \dfrac{1}{6} + \dfrac{1}{8} + \dfrac{1}{10}$

    $= \dfrac{60}{120} + \dfrac{30}{120} + \dfrac{20}{120} + \dfrac{15}{120} + \dfrac{12}{120}$

    $= \dfrac{137}{120}$

44. $\dfrac{1}{3} + \dfrac{1}{5} + \dfrac{1}{7} + \dfrac{1}{9} + \dfrac{1}{11} + \dfrac{1}{13} = \dfrac{43,024}{45,045}$

45. $\displaystyle\sum_{k=0}^{5} 2^k = 2^0 + 2^1 + 2^2 + 2^3 + 2^4 + 2^5$

    $= 1 + 2 + 4 + 8 + 16 + 32$

    $= 63$

46. $\sqrt{7} + \sqrt{9} + \sqrt{11} + \sqrt{13} \approx 12.5679$

47. $\displaystyle\sum_{k=7}^{10} \log k = \log 7 + \log 8 + \log 9 + \log 10,$

    or $\log (7 \cdot 8 \cdot 9 \cdot 10)$, or $\log 5040$

48. $0 + \pi + 2\pi + 3\pi + 4\pi \approx 31.4159$

49. $\displaystyle\sum_{k=1}^{8} \dfrac{k}{k + 1} = \dfrac{1}{1 + 1} + \dfrac{2}{2 + 1} + \dfrac{3}{3 + 1} + \dfrac{4}{4 + 1} +$

    $\dfrac{5}{5 + 1} + \dfrac{6}{6 + 1} + \dfrac{7}{7 + 1} + \dfrac{8}{8 + 1}$

    $= \dfrac{1}{2} + \dfrac{2}{3} + \dfrac{3}{4} + \dfrac{4}{5} + \dfrac{5}{6} + \dfrac{6}{7} + \dfrac{7}{8} + \dfrac{8}{9}$

    $= \dfrac{15,551}{2520}$

50. $0 + \dfrac{1}{5} + \dfrac{1}{3} + \dfrac{3}{7} = \dfrac{101}{105}$

51. $\displaystyle\sum_{k=1}^{5} (-1)^k = (-1)^1 + (-1)^2 + (-1)^3 + (-1)^4 + (-1)^5$

    $= -1 + 1 - 1 + 1 - 1$

    $= -1$

52. $1 - 1 + 1 - 1 + 1 = 1$

53. $\displaystyle\sum_{k=1}^{8} (-1)^{k+1} 3k = (-1)^2 3 \cdot 1 + (-1)^3 3 \cdot 2 + (-1)^4 3 \cdot 3 +$

    $(-1)^5 3 \cdot 4 + (-1)^6 3 \cdot 5 + (-1)^7 3 \cdot 6 +$

    $(-1)^8 3 \cdot 7 + (-1)^9 3 \cdot 8$

    $= 3 - 6 + 9 - 12 + 15 - 18 + 21 - 24$

    $= -12$

54. $-4^2 + 4^3 - 4^4 + 4^5 - 4^6 + 4^7 - 4^8 = -52,432$

55. $\displaystyle\sum_{k=1}^{6} = \dfrac{2}{k^2 + 1} = \dfrac{2}{1^2 + 1} + \dfrac{2}{2^2 + 1} + \dfrac{2}{3^2 + 1} + \dfrac{2}{4^2 + 1} +$

    $\dfrac{2}{5^2 + 1} + \dfrac{2}{6^2 + 1}$

    $= \dfrac{2}{2} + \dfrac{2}{5} + \dfrac{2}{10} + \dfrac{2}{17} + \dfrac{2}{26} + \dfrac{2}{37}$

    $= 1 + \dfrac{2}{5} + \dfrac{1}{5} + \dfrac{2}{17} + \dfrac{1}{13} + \dfrac{2}{37}$

    $= \dfrac{75,581}{40,885}$

56. $1 \cdot 2 + 2 \cdot 3 + 3 \cdot 4 + 4 \cdot 5 + 5 \cdot 6 + 6 \cdot 7 + 7 \cdot 8 + 8 \cdot 9 +$
    $9 \cdot 10 + 10 \cdot 11 = 440$

57. $\displaystyle\sum_{k=0}^{5} (k^2 - 2k + 3) = (0^2 - 2 \cdot 0 + 3) +$

    $(1^2 - 2 \cdot 1 + 3) + (2^2 - 2 \cdot 2 + 3) +$

    $(3^2 - 2 \cdot 3 + 3) + (4^2 - 2 \cdot 4 + 3) +$

    $(5^2 - 2 \cdot 5 + 3)$

    $= 3 + 2 + 3 + 6 + 11 + 18$

    $= 43$

58. $4 + 2 + 2 + 4 + 8 + 14 = 34$

59. $\displaystyle\sum_{k=1}^{10} \frac{1}{k(k+1)} = \frac{1}{1(1+1)} + \frac{1}{2(2+1)} + \frac{1}{3(3+1)} +$

$\qquad\qquad \frac{1}{4(4+1)} + \frac{1}{5(5+1)} + \frac{1}{6(6+1)} +$

$\qquad\qquad \frac{1}{7(7+1)} + \frac{1}{8(8+1)} + \frac{1}{9(9+1)} +$

$\qquad\qquad \frac{1}{10(10+1)}$

$\qquad = \frac{1}{2} + \frac{1}{6} + \frac{1}{12} + \frac{1}{20} + \frac{1}{30} + \frac{1}{42} + \frac{1}{56} +$

$\qquad\quad \frac{1}{72} + \frac{1}{90} + \frac{1}{110}$

$\qquad = \frac{10}{11}$

60. $\frac{2}{3} + \frac{4}{5} + \frac{8}{9} + \frac{16}{17} + \frac{32}{33} + \frac{64}{65} + \frac{128}{129} + \frac{256}{257} + \frac{512}{513} +$

$\frac{1024}{1025}$

61. $\frac{1}{2} + \frac{2}{3} + \frac{3}{4} + \frac{4}{5} + \frac{5}{6} + \frac{6}{7}$

This is a sum of fractions in which the denominator is one greater than the numerator. Also, each numerator is 1 greater than the preceding numerator. Sigma notation is

$\displaystyle\sum_{k=1}^{6} \frac{k}{k+1}.$

62. $\displaystyle\sum_{k=1}^{5} 3k$

63. $-2 + 4 - 8 + 16 - 32 + 64$

This is a sum of powers of 2 with alternating signs. Sigma notation is

$\displaystyle\sum_{k=1}^{6} (-1)^k 2^k$, or $\displaystyle\sum_{k=1}^{6} (-2)^k.$

64. $\displaystyle\sum_{k=1}^{5} \frac{1}{k^2}$

65. $4 - 9 + 16 - 25 + \ldots + (-1)^n n^2$

This is a sum of terms of the form $(-1)^k k^2$, beginning with $k = 2$ and continuing through $k = n$. Sigma notation is

$\displaystyle\sum_{k=2}^{n} (-1)^k k^2.$

66. $\displaystyle\sum_{k=3}^{n} (-1)^{k+1} k^2$

67. $5 + 10 + 15 + 20 + 25 + \ldots$

This is a sum of multiples of 5, and it is an infinite series. Sigma notation is

$\displaystyle\sum_{k=1}^{\infty} 5k.$

68. $\displaystyle\sum_{k=1}^{\infty} 7k$

69. $\frac{1}{1 \cdot 2} + \frac{1}{2 \cdot 3} + \frac{1}{3 \cdot 4} + \frac{1}{4 \cdot 5} + \ldots$

This is a sum of fractions in which the numerator is 1 and the denominator is a product of two consecutive integers. The larger integer in each product is the smaller integer in the succeeding product. It is an infinite series. Sigma notation is

$\displaystyle\sum_{k=1}^{\infty} \frac{1}{k(k+1)}.$

70. $\displaystyle\sum_{k=1}^{\infty} \frac{1}{k(k+1)^2}$

71. $a_1 = 4$ $\qquad\qquad a_{k+1} = 1 + \frac{1}{a_k}$

$a_2 = 1 + \frac{1}{4} = 1\frac{1}{4}$

$a_3 = 1 + \frac{1}{\frac{5}{4}} = 1 + \frac{4}{5} = 1\frac{4}{5}$

$a_4 = 1 + \frac{1}{\frac{9}{5}} = 1 + \frac{5}{9} = 1\frac{5}{9}$

72. $256, 16, 4, 2$

73. $a_1 = 6561$ $\qquad\qquad a_{k+1} = (-1)^k \sqrt{a_k}$

$a_2 = (-1)^1 \sqrt{6561} = -81$

$a_3 = (-1)^2 \sqrt{-81} = 9i$

$a_4 = (-1)^3 \sqrt{9i} = -3\sqrt{i}$

74. $e^Q, Q, \ln Q, \ln(\ln Q)$

75. $a_1 = 2$ $\qquad\qquad a_{k+1} = a_k + a_{k-1}$

$a_2 = 3$

$a_3 = 3 + 2 = 5$

$a_4 = 5 + 3 = 8$

76. $-10, 8, 18, 10$

77. $a_n = \frac{1}{2^n} \log 1000^n$

$a_1 = \frac{1}{2} \log 1000 = \frac{1}{2} \cdot 3 = \frac{3}{2}$

$a_2 = \frac{1}{2^2} \log 1000^2 = \frac{1}{4} \cdot 6 = \frac{3}{2}$

$a_3 = \frac{1}{2^3} \log 1000^3 = \frac{1}{8} \cdot 9 = \frac{9}{8}$

$a_4 = \frac{1}{2^4} \log 1000^4 = \frac{1}{16} \cdot 12 = \frac{3}{4}$

$a_5 = \frac{1}{2^5} \log 1000^5 = \frac{1}{32} \cdot 15 = \frac{15}{32}$

$S_5 = \frac{3}{2} + \frac{3}{2} + \frac{9}{8} + \frac{3}{4} + \frac{15}{32}$

$= \frac{48}{32} + \frac{48}{32} + \frac{36}{32} + \frac{24}{32} + \frac{15}{32}$

$= \frac{171}{32}$

78. i, -1, $-i$, 1, i; i

79. $a_n = \ln(1 \cdot 2 \cdot 3 \cdots n)$

$a_1 = \ln 1 = 0$
$a_2 = \ln(1 \cdot 2) = \ln 2 \approx 0.693$
$a_3 = \ln(1 \cdot 2 \cdot 3) = \ln 6 \approx 1.792$
$a_4 = \ln(1 \cdot 2 \cdot 3 \cdot 4) = \ln 24 \approx 3.178$
$a_5 = \ln(1 \cdot 2 \cdot 3 \cdot 4 \cdot 5) = \ln 120 \approx 4.787$

$S_5 \approx 0 + 0.693 + 1.792 + 3.178 + 4.787 = 10.450$

80. 1, 4, 9, 16, 25; 55

81. a) $a_n = n^2 - n + 41$

$a_1 = 1 - 1 + 41 = 41$,
$a_2 = 2^2 - 2 + 41 = 43$,
$a_3 = 3^2 - 3 + 41 = 47$,
$a_4 = 4^2 - 4 + 41 = 53$,
$a_5 = 5^2 - 5 + 41 = 61$,
$a_6 = 6^2 - 6 + 41 = 71$

b) All the terms are prime numbers.

c) $a_{41} = 41^2 - 41 + 41 = 41^2 = 1681$
The pattern does not hold since 1681 is not prime ($1681 = 41 \cdot 41$).

82. 2, 2.25, 2.370370, 2.441406, 2.488320, 2.521626

83. $a_n = \sqrt{n+1} - \sqrt{n}$

$a_1 = \sqrt{1+1} - \sqrt{1} = \sqrt{2} - 1 \approx 0.414214$

$a_2 = \sqrt{2+1} - \sqrt{2} = \sqrt{3} - \sqrt{2} \approx 0.317837$

$a_3 = \sqrt{3+1} - \sqrt{3} = 2 - \sqrt{3} \approx 0.267949$

$a_4 = \sqrt{4+1} - \sqrt{4} = \sqrt{5} - 2 \approx 0.236068$

$a_5 = \sqrt{5+1} - \sqrt{5} = \sqrt{6} - \sqrt{5} \approx 0.213422$

$a_6 = \sqrt{6+1} - \sqrt{6} = \sqrt{7} - \sqrt{6} \approx 0.196262$

84. 2, 1.553774, 1.498834, 1.491398, 1.490378, 1.490238

85. $a_1 = 2$, $\qquad a_{k+1} = \frac{1}{2}\left(a_k + \frac{2}{a_k}\right)$

$a_2 = \frac{1}{2}\left(2 + \frac{2}{2}\right) = 1.5$

$a_3 = \frac{1}{2}\left(1.5 + \frac{2}{1.5}\right) \approx 1.416667$

$a_4 = \frac{1}{2}\left(1.416667 + \frac{2}{1.416667}\right) \approx 1.414216$

$a_5 = \frac{1}{2}\left(1.414216 + \frac{2}{1.414216}\right) \approx 1.414214$

$a_6 = \frac{1}{2}\left(1.414214 + \frac{2}{1.414214}\right) \approx 1.414214$

86. 2, 4, 8, 16, 32, 64, 128, 256, 512, 1024, 2048, 4096, 8192, 16,384, 32,768, 65,536

87. Find each term by multiplying the preceding term by 0.75:
$3900, $2925, $2193.75, $1645.31, $1233.98, $925.49, $694.12, $520.59, $390.44, $292.83

88. $4.20, $4.35, $4.50, $4.65, $4.80, $4.95, $5.10, $5.25, $5.40, $5.55

89. $S_n = \ln 1 + \ln 2 + \ln 3 + \cdots + \ln n$
$= \ln(1 \cdot 2 \cdot 3 \cdots n) = \ln(n!)$.

90. $S_n = \frac{n}{n+1}$

Exercise Set 8.2

1. 3, 8, 13, 18, . . .

$a_1 = 3$
$d = 5$ ($8 - 3 = 5$, $13 - 8 = 5$, $18 - 13 = 5$)

2. $a_1 = 1.08$, $d = 0.08$

3. 9, 5, 1, -3, . . .

  $a_1 = 9$

  $d = -4$ $(5 - 9 = -4, \ 1 - 5 = -4, \ -3 - 1 = -4)$

4. $a_1 = -8$, $d = 3$

5. $\dfrac{3}{2}, \dfrac{9}{4}, 3, \dfrac{15}{4}, \ . . .$

  $a_1 = \dfrac{3}{2}$

  $d = \dfrac{3}{4}$ $\left(\dfrac{9}{4} - \dfrac{3}{2} = \dfrac{3}{4}, \ 3 - \dfrac{9}{4} = \dfrac{3}{4} \right)$

6. $a_1 = \dfrac{3}{5}$, $d = -\dfrac{1}{2}$

7. $1.07, $1.14, $1.21, $1.28, . . .

  $a_1 = \$1.07$

  $d = \$0.07$ ($1.14 - $1.07 = $0.07,
           $1.21 - $1.14 = $0.07)

8. $a_1 = \$316$, $d = -\$3$

9. 2, 6, 10, . . .

  $a_1 = 2$, $d = 4$, and $n = 12$

  $a_n = a_1 + (n - 1)d$

  $a_{12} = 2 + (12 - 1)4 = 2 + 11 \cdot 4 = 2 + 44 = 46$

10. 0.57

11. 7, 4, 1, . . .

  $a_1 = 7$, $d = -3$, and $n = 17$

  $a_n = a_1 + (n - 1)d$

  $a_{17} = 7 + (17 - 1)(-3) = 7 + 16(-3) = 7 - 48 = -41$

12. $-\dfrac{17}{3}$

13. $1200, $964.32, $728.64, . . .

  $a_1 = \$1200$, $d = \$964.32 - \$1200 = -\$235.68$,
    and $n = 13$

  $a_n = a_1 + (n - 1)d$

  $a_{13} = \$1200 + (13 - 1)(-\$235.68) =$
      $\$1200 + 12(-\$235.68) = \$1200 - \$2828.16 =$
      $-\$1628.16$

14. $7941.62

15. $a_1 = 2$, $d = 4$

  $a_n = a_1 + (n - 1)d$

  Let $a_n = 106$, and solve for n.

  $106 = 2 + (n - 1)(4)$

  $106 = 2 + 4n - 4$

  $108 = 4n$

   $27 = n$

  The 27th term is 106.

16. 33rd term

17. $a_1 = 7$, $d = -3$

  $a_n = a_1 + (n - 1)d$

  $-296 = 7 + (n - 1)(-3)$

  $-296 = 7 - 3n + 3$

  $-306 = -3n$

   $102 = n$

  The 102nd term is -296.

18. 46th term

19. $a_n = a_1 + (n - 1)d$

  $a_{17} = 5 + (17 - 1)6$ Substituting 17 for n,
                  5 for a_1, and 6 for d

     $= 5 + 16 \cdot 6$

     $= 5 + 96$

     $= 101$

20. -43

21. $a_n = a_1 + (n - 1)d$

  $33 = a_1 + (8 - 1)4$ Substituting 33 for a_8, 8 for
                  n, and 4 for d

  $33 = a_1 + 28$

   $5 = a_1$

  (Note that this procedure is equivalent to
  subtracting d from a_8 seven times to get a_1:
  $33 - 7(4) = 33 - 28 = 5$)

22. 1.8

23. $a_n = a_1 + (n - 1)d$

  $-76 = 5 + (n - 1)(-3)$ Substituting -76 for a_n,
                  5 for a_1 and -3 for d

  $-76 = 5 - 3n + 3$

  $-76 = 8 - 3n$

  $-84 = -3n$

   $28 = n$

24. 39

25. We know that $a_{17} = -40$ and $a_{28} = -73$. We would have to add d eleven times to get from a_{17} to a_{28}. That is,

$$-40 + 11d = -73$$
$$11d = -33$$
$$d = -3.$$

Since $a_{17} = -40$, we subtract d sixteen times to get to a_1.

$$a_1 = -40 - 16(-3) = -40 + 48 = 8$$

We write the first five terms of the sequence:

$$8, 5, 2, -1, -4$$

26. $\frac{1}{3}, \frac{5}{6}, \frac{4}{3}, \frac{11}{6}, \frac{7}{3}$

27. $5 + 8 + 11 + 14 + \ldots$

Note that $a_1 = 5$, $d = 3$, and $n = 20$. We use the formula of Theorem 3.

$$S_n = \frac{n}{2}\left[2a_1 + (n - 1)d\right]$$

$$S_{20} = \frac{20}{2}[2 \cdot 5 + (20 - 1)3] = 10[10 + 57] = 10 \cdot 67 = 670$$

28. -210

29. The sum is $1 + 2 + 3 + \ldots + 299 + 300$. This is the sum of the arithmetic sequence for which $a_1 = 1$, $a_n = 300$, and $n = 300$. We use the formula of Theorem 2.

$$S_n = \frac{n}{2}(a_1 + a_n)$$

$$S_{300} = \frac{300}{2}(1 + 300) = 150(301) = 45,150$$

30. $80,200$

31. The sum is $2 + 4 + 6 + \ldots + 98 + 100$. This is the sum of the arithmetic sequence for which $a_1 = 2$, $a_n = 100$, and $n = 50$. We use the formula of Theorem 2.

$$S_n = \frac{n}{2}(a_1 + a_n)$$

$$S_{50} = \frac{50}{2}(2 + 100) = 25(102) = 2550$$

32. 2500

33. The sum is $7 + 14 + 21 + \ldots + 91 + 98$. This is the sum of the arithmetic sequence for which $a_1 = 7$, $a_n = 98$, and $n = 14$. We use the formula of Theorem 2.

$$S_n = \frac{n}{2}(a_1 + a_n)$$

$$S_{14} = \frac{14}{2}(7 + 98) = 7(105) = 735$$

34. $34,036$

35. $S_n = \frac{n}{2}\left[2a_1 + (n - 1)d\right]$

$S_{20} = \frac{20}{2}[2 \cdot 2 + (20 - 1)5]$ Substituting 20 for n, 2 for a_1, and 5 for d

$= 10[4 + 19 \cdot 5]$

$= 10[4 + 95]$

$= 10 \cdot 99$

$= 990$

36. -1264

37. We must find the smallest positive term of the sequence $35, 31, 27, \ldots$. It is an arithmetic sequence with $a_1 = 35$ and $d = -4$. We find the largest integer x for which $35 + x(-4) > 0$. Then we evaluate the expression $35 - 4x$ for that value of x.

$$35 - 4x > 0$$
$$35 > 4x$$
$$\frac{35}{4} > x$$
$$8\frac{3}{4} > x$$

The integer we are looking for is 8. Then $35 - 4x = 35 - 4(8) = 3$.

There will be 3 plants in the last row.

38. $62, 950$

39. We go from 50 poles in a row to one pole in the top row, so there must be 50 rows. We want the sum $50 + 49 + 48 + \ldots + 1$. Thus we want the sum of an arithmetic sequence with $a_1 = 50$, $a_n = 1$, and $n = 50$. We will use the formula $S_n = \frac{n}{2}(a_1 + a_n)$.

$$S_{50} = \frac{50}{2}(50 + 1).$$

$$S_{50} = 25(51) = 1275$$

There will be 1275 poles in the pile.

40. 4960¢, or $\$49.60$

41. We want to find the sum of an arithmetic sequence with $a_1 = \$600$, $d = \$100$, and $n = 20$. We will use the formula

$$S_n = \frac{n}{2}[2a_1 + (n - 1)d].$$

$$S_{20} = \frac{20}{2}[2 \cdot \$600 + (20 - 1)\$100].$$

$$S_{20} = 10(\$3100) = \$31,000$$

They save $\$31,000$ (disregarding interest).

42. $\$10,230$

43. We want to find the sum of an arithmetic sequence with $a_1 = 28$, $d = 4$, and $n = 50$. We will use the formula

$$S_n = \frac{n}{2}[2a_1 + (n-1)d].$$

$$S_{50} = \frac{50}{2}[2 \cdot 28 + (50-1)4].$$

$$S_{50} = 25(252) = 6300$$

There are 6300 seats.

44. $462,500

45. 4, m_1, m_2, m_3, m_4, 13

We look for m_1, m_2, m_3, and m_4 such that 4, m_1, m_2, m_3, m_4, 13 is an arithmetic sequence. In this case $a_1 = 4$, $n = 6$, and $a_6 = 13$. We use Theorem 1.

$$a_n = a_1 + (n-1)d$$
$$13 = 4 + (6-1)d$$
$$9 = 5d$$
$$1\frac{4}{5} = d$$

Thus,

$$m_1 = a_1 + d = 4 + 1\frac{4}{5} = 5\frac{4}{5}$$

$$m_2 = m_1 + d = 5\frac{4}{5} + 1\frac{4}{5} = 6\frac{8}{5} = 7\frac{3}{5}$$

$$m_3 = m_2 + d = 7\frac{3}{5} + 1\frac{4}{5} = 8\frac{7}{5} = 9\frac{2}{5}$$

$$m_4 = m_3 + d = 9\frac{2}{5} + 1\frac{4}{5} = 10\frac{6}{5} = 11\frac{1}{5}$$

46. -1, 1, 3

47. $1 + 3 + 5 + 7 + \ldots + n$

$$S_n = \frac{n}{2}[2a_1 + (n-1)d]$$

Note that $a_1 = 1$ and $d = 2$.

$$S_n = \frac{n}{2}[2\cdot1 + (n-1)2] = \frac{n}{2}(2 + 2n - 2)$$

$$= \frac{n}{2} \cdot 2n$$

$$= n^2$$

48. $\dfrac{n(n+1)}{2}$

49. Let x represent the first number in the sequence, and let d represent the common difference. Then the three numbers in the sequence are x, $x + d$, and $x + 2d$.

We write a system of equations:

The sum of the first and third numbers is 10.

$$x + x + 2d = 10$$

The product of the first and second numbers is 15.

$$x(x + d) = 15$$

Solving the system of equations we get $x = 3$ and $d = 2$. Thus the numbers are 3, 5, and 7.

50. $a_1 = -5p - 5q + 60$, $d = 5p + 2q - 20$

51. a_1 = $8760
   a_2 = $8760 + (-$798.23) = $7961.77
   a_3 = $8760 + 2(-$798.23) = $7163.54
   a_4 = $8760 + 3(-$798.23) = $6365.31
   a_5 = $8760 + 4(-$798.23) = $5567.08
   a_6 = $8760 + 5(-$798.23) = $4768.85
   a_7 = $8760 + 6(-$798.23) = $3970.62
   a_8 = $8760 + 7(-$798.23) = $3172.39
   a_9 = $8760 + 8(-$798.23) = $2374.16
   a_{10} = $8760 + 9(-$798.23) = $1575.93

52. $51,679.65

53. $P(x) = x^4 + 4x^3 - 84x^2 - 176x + 640$ has at most 4 zeros because $P(x)$ is of degree 4. By the rational roots theorem, the possible zeros are ±1, ±2, ±4, ±5, ±8, ±10, ±16, ±20, ±32, ±40, ±64, ±80, ±128, ±160, ±320, ±640

We find two of the zeros using synthetic division.

```
2 | 1    4    -84    -176    640
  |      2     12    -144   -640
    1    6    -72    -320 |    0

-4 | 1    6    -72    -320
   |      -4     -8     320
     1    2    -80 |    0
```

Also by synthetic division we determine that ±1 and -2 are not zeros. Therefore we determine that $d = 6$ in the arithmetic sequence.

Possible arithmetic sequences:

a)   -4   2   8   14
b)   -10   -4   2   8
c) -16 -10   -4   2

The solution cannot be a) because 14 is not a possible zero. Checking -16 by synthetic division we find that -16 is not a zero. Thus b) is the only arithmetic sequence which contains all four zeros. The zeros are -10, -4, 2, and 8.

54. Insert 16 arithmetic means with $d = \dfrac{49}{17}$.

55. The sides are a, $a + d$, and $a + 2d$. By the Pythagorean theorem, we get $a^2 + (a+d)^2 = (a + 2d)^2$. Solving for d, we get $d = \frac{a}{3}$. Thus the sides are a, $a + \frac{a}{3}$, and $a + \frac{2a}{3}$, or a, $\frac{4a}{3}$, and $\frac{5a}{3}$ which are in the ratio 3:4:5.

56. a) $a_t = \$5200 - \$512.50t$

  b) $5200, $4687.50, $4175, $3662.50, $3150, $1612.50, $1100

57. $a_n = a_1 + (n - 1)d$

  $a_n = f(n) = d \cdot n + (a_1 - d)$ where
  $f(n)$ is a linear function of n with slope
  d and y-intercept $(a_1 - d)$.

58.

  $m = p + d$
  $\underline{m = q - d}$
  $2m = p + q$ (Adding)
  $m = \dfrac{p + q}{2}$

Exercise Set 8.3

1. 2, 4, 8, 16, . . .

  $\dfrac{4}{2} = 2, \quad \dfrac{8}{4} = 2, \quad \dfrac{16}{8} = 2$

  $r = 2$

2. $-\dfrac{1}{3}$

3. 1, -1, 1, -1, . . .

  $\dfrac{-1}{1} = -1, \quad \dfrac{1}{-1} = -1, \quad \dfrac{-1}{1} = -1$

  $r = -1$

4. 0.1

5. $\dfrac{1}{2}, -\dfrac{1}{4}, \dfrac{1}{8}, -\dfrac{1}{16}, \ldots$

  $\dfrac{-\frac{1}{4}}{\frac{1}{2}} = -\dfrac{1}{4} \cdot \dfrac{2}{1} = -\dfrac{2}{4} = -\dfrac{1}{2}$

  $\dfrac{\frac{1}{8}}{-\frac{1}{4}} = \dfrac{1}{8} \cdot \left(-\dfrac{4}{1}\right) = -\dfrac{4}{8} = -\dfrac{1}{2}$

  $r = -\dfrac{1}{2}$

6. -2

7. 75, 15, 3, $\dfrac{3}{5}$, . . .

  $\dfrac{15}{75} = \dfrac{1}{5}, \quad \dfrac{3}{15} = \dfrac{1}{5}, \quad \dfrac{\frac{3}{5}}{3} = \dfrac{3}{5} \cdot \dfrac{1}{3} = \dfrac{1}{5}$

  $r = \dfrac{1}{5}$

8. 0.1

9. $\dfrac{1}{x}, \dfrac{1}{x^2}, \dfrac{1}{x^3}, \ldots$

  $\dfrac{\frac{1}{x^2}}{\frac{1}{x}} = \dfrac{1}{x^2} \cdot \dfrac{x}{1} = \dfrac{x}{x^2} = \dfrac{1}{x}$

  $\dfrac{\frac{1}{x^3}}{\frac{1}{x^2}} = \dfrac{1}{x^3} \cdot \dfrac{x^2}{1} = \dfrac{x^2}{x^3} = \dfrac{1}{x}$

  $r = \dfrac{1}{x}$

10. $\dfrac{m}{2}$

11. $780, $858, $943.80, $1038.18, . . .

  $\dfrac{\$858}{\$780} = 1.1, \quad \dfrac{\$943.80}{\$858} = 1.1,$

  $\dfrac{\$1038.18}{\$943.80} = 1.1$

  $r = 1.1$

12. 0.95

13. 2, 4, 8, 16, . . .

  $a_1 = 2$, n = 6, and $r = \dfrac{4}{2}$, or 2.

  We use the formula $a_n = a_1 r^{n-1}$.

  $a_6 = 2(2)^{6-1} = 2 \cdot 2^5 = 2 \cdot 32 = 64$

14. 781,250

15. 2, $2\sqrt{3}$, 6, . . .

  $a_1 = 2$, n = 9, and $r = \dfrac{2\sqrt{3}}{2}$, or $\sqrt{3}$

  $a_n = a_1 r^{n-1}$

  $a_9 = 2(\sqrt{3})^{9-1} = 2(\sqrt{3})^8 = 2 \cdot 81 = 162$

16. 1

17. $\dfrac{8}{243}, \dfrac{8}{81}, \dfrac{8}{27}, \ldots$

  $a_1 = \dfrac{8}{243}$, n = 10, and $r = \dfrac{\frac{8}{81}}{\frac{8}{243}} = \dfrac{8}{81} \cdot \dfrac{243}{8} = 3$

  $a_n = a_1 r^{n-1}$

  $a_{10} = \dfrac{8}{243}(3)^{10-1} = \dfrac{8}{243}(3)^9 = \dfrac{8}{243} \cdot 19,683 = 648$

18. $7(25)^{20}$

19. $1000, $1080, $1166.40, . . .

  $a_1 = \$1000$, n = 5, and $r = \dfrac{\$1080}{\$1000} = 1.08$

  $a_n = a_1 r^{n-1}$

  $a_5 = \$1000(1.08)^{5-1} \approx \$1360.49$

20. $1402.55

21. 1, 3, 9, . . .

$a_1 = 1$ and $r = \frac{3}{1}$, or 3

$a_n = a_1 r^{n-1}$

$a_n = 1(3)^{n-1} = 3^{n-1}$

22. 5^{3-n}

23. 1, -1, 1, -1, . . .

$a_1 = 1$ and $r = \frac{-1}{1} = -1$

$a_n = a_1 r^{n-1}$

$a_n = 1(-1)^{n-1} = (-1)^{n-1}$

24. 2^n

25. $\frac{1}{x}, \frac{1}{x^2}, \frac{1}{x^3}, . . .$

$a_1 = \frac{1}{x}$ and $r = \frac{1}{x}$ (see Exercise 9)

$a_n = a_1 r^{n-1}$

$a_n = \frac{1}{x}\left(\frac{1}{x}\right)^{n-1} = \frac{1}{x} \cdot \frac{1}{x^{n-1}} = \frac{1}{x^{1+n-1}} = \frac{1}{x^n}$

26. $5\left(\frac{m}{2}\right)^{n-1}$

27. 6 + 12 + 24 + . . .

$a_1 = 6$, $n = 7$, and $r = \frac{12}{6}$, or 2

$S_n = \frac{a_1(r^n - 1)}{r - 1}$

$S_7 = \frac{6(2^7 - 1)}{2 - 1} = \frac{6(128 - 1)}{1} = 6 \cdot 127 = 762$

28. $\frac{21}{2}$

29. $\frac{1}{18} - \frac{1}{6} + \frac{1}{2} - . . .$

$a_1 = \frac{1}{18}$, $n = 7$, and $r = \frac{-\frac{1}{6}}{\frac{1}{18}} = -\frac{1}{6} \cdot \frac{18}{1} = -3$

$S_n = \frac{a_1(r^n - 1)}{r - 1}$

$S_7 = \frac{\frac{1}{18}[(-3)^7 - 1]}{-3 - 1} = \frac{\frac{1}{18}[-2187 - 1]}{-4} = \frac{\frac{1}{18}(-2188)}{-4} =$

$\frac{1}{18}(-2188)\left(-\frac{1}{4}\right) = \frac{547}{18}$

30. $-\frac{171}{32}$

31. $1 + x + x^2 + x^3 + . . .$

$a_1 = 1$, $n = 8$, and $r = \frac{x}{1}$, or x

$S_n = \frac{a_1(r^n - 1)}{r - 1}$

$S_8 = \frac{1(x^8 - 1)}{x - 1} = \frac{(x^4 + 1)(x^4 - 1)}{x - 1}$

$= \frac{(x^4 + 1)(x^2 + 1)(x^2 - 1)}{x - 1}$

$= \frac{(x^4 + 1)(x^2 + 1)(x + 1)(x - 1)}{x - 1}$

$= (x^4 + 1)(x^2 + 1)(x + 1)$

32. $\frac{x^{20} - 1}{x^2 - 1}$

33. $200, $200(1.06), $200(1.06)^2, . . .

$a_1 = \$200$, $n = 16$, and $r = \frac{\$200(1.06)}{\$200} = 1.06$

$S_n = \frac{a_1(r^n - 1)}{r - 1}$

$S_{16} = \frac{\$200(1.06^{16} - 1)}{1.06 - 1} \approx \$5134.51$

34. $60,893.30

35. $\sum\limits_{k=1}^{\infty} \left(\frac{1}{2}\right)^{k-1}$

$a_1 = 1$, $|r| = \left|\frac{1}{2}\right| = \frac{1}{2}$

$\sum\limits_{k=1}^{\infty} \left(\frac{1}{2}\right)^{k-1} = \frac{a_1}{1 - r} = \frac{1}{1 - \frac{1}{2}} = \frac{1}{\frac{1}{2}} = 2$

36. No

37. 4 + 2 + 1 + . . .

$|r| = \left|\frac{2}{4}\right| = \left|\frac{1}{2}\right| = \frac{1}{2}$, and since $|r| < 1$, the series does have a sum.

$S_\infty = \frac{a_1}{1 - r} = \frac{4}{1 - \frac{1}{2}} = \frac{4}{\frac{1}{2}} = 4 \cdot \frac{2}{1} = 8$

38. Yes, $\frac{49}{4}$

39. 25 + 20 + 16 + . . .

$|r| = \left|\frac{20}{25}\right| = \left|\frac{4}{5}\right| = \frac{4}{5}$, and since $|r| < 1$, the series does have a sum.

$S_\infty = \frac{a_1}{1 - r} = \frac{25}{1 - \frac{4}{5}} = \frac{25}{\frac{1}{5}} = 25 \cdot \frac{5}{1} = 125$

40. Yes, 48

41. $100 - 10 + 1 - \frac{1}{10} + \ldots$

$|r| = \left|\frac{-10}{100}\right| = \left|-\frac{1}{10}\right| = \frac{1}{10}$, and since $|r| < 1$, the series does have a sum.

$S_\infty = \frac{a_1}{1-r} = \frac{100}{1 - \left(-\frac{1}{10}\right)} = \frac{100}{\frac{11}{10}} = 100 \cdot \frac{10}{11} = \frac{1000}{11}$

42. No

43. $8 + 40 + 200 + \ldots$

$|r| = \left|\frac{40}{8}\right| = |5| = 5$, and since $|r| \not< 1$ the series does not have a sum.

44. Yes, -4

45. $0.6 + 0.06 + 0.006 + \ldots$

$|r| = \left|\frac{0.06}{0.6}\right| = |0.1| = 0.1$, and since $|r| < 1$, the series does have a sum.

$S_\infty = \frac{a_1}{1-r} = \frac{0.6}{1 - 0.1} = \frac{0.6}{0.9} = \frac{6}{9} = \frac{2}{3}$

46. Yes, $\frac{37}{99}$

47. $\$500(1.11)^{-1} + \$500(1.11)^{-2} + \$500(1.11)^{-3} + \ldots$

$|r| = \left|\frac{\$500(1.11)^{-2}}{\$500(1.11)^{-1}}\right| = |(1.11)^{-1}| = (1.11)^{-1}$, or $\frac{1}{1.11}$, and since $|r| < 1$, the series does have a sum.

$S_\infty = \frac{a_1}{1-r} = \frac{\$500(1.11)^{-1}}{1 - \left(\frac{1}{1.11}\right)} = \frac{\frac{\$500}{1.11}}{\frac{0.11}{1.11}} =$

$\frac{\$500}{1.11} \cdot \frac{1.11}{0.11} \approx \$4545.45$

48. Yes, $\$12,500$

49. $\sum\limits_{k=1}^{\infty} 16(0.1)^{k-1}$

$a_1 = 16(0.1)^{1-1} = 16 \cdot 1 = 16$

$a_2 = 16(0.1)^{2-1} = 16(0.1) = 1.6$

$|r| = \left|\frac{1.6}{16}\right| = |0.1| = 0.1 < 1$, so the series has a sum.

$S_\infty = \frac{16}{1 - 0.1} = \frac{16}{0.9} = \frac{160}{9}$

50. Yes, 10

51. $\sum\limits_{k=1}^{\infty} \frac{1}{2^{k-1}}$

$a_1 = \frac{1}{2^{1-1}} = \frac{1}{1} = 1$

$a_2 = \frac{1}{2^{2-1}} = \frac{1}{2^1} = \frac{1}{2}$

$|r| = \left|\frac{\frac{1}{2}}{1}\right| = \frac{1}{2} < 1$, so the series has a sum.

$S_\infty = \frac{1}{1 - \frac{1}{2}} = \frac{1}{\frac{1}{2}} = 2$

52. Yes, $\frac{16}{3}$

53. $0.7\overline{77} = 0.7 + 0.07 + 0.007 + 0.0007 + \ldots$

Note that $a_1 = 0.7$ and $r = 0.1$

$S_\infty = \frac{a_1}{1-r}$

$S_\infty = \frac{0.7}{1 - 0.1} = \frac{0.7}{0.9} = \frac{7}{9}$

54. 9

55. $0.53\overline{33} = 0.5 + 0.03\overline{33}$

First we consider the repeating part:
$0.03\overline{33} = 0.03 + 0.003 + 0.0003 + \cdots$

$a_1 = 0.03$ and $r = 0.1$

$S_\infty = \frac{a_1}{1-r} = \frac{0.03}{1 - 0.1} = \frac{0.03}{0.9} = \frac{3}{90} = \frac{1}{30}$

Thus

$0.5333 = 0.5 + 0.03\overline{33} = \frac{1}{2} + \frac{1}{30} = \frac{16}{30} = \frac{8}{15}$

56. $\frac{29}{45}$

57. $5.15\overline{15} = 5 + 0.15\overline{15}$

First we consider the repeating part.
$0.15\overline{15} = 0.15 + 0.0015 + 0.000015 + \ldots$
$a_1 = 0.15$ and $r = 0.01$

$S_\infty = \frac{a_1}{1-r} = \frac{0.15}{1 - 0.01} = \frac{0.15}{0.99} = \frac{15}{99} = \frac{5}{33}$

Thus

$5.15\overline{15} = 5 + \frac{5}{33} = 5\frac{5}{33}$, or $\frac{170}{33}$

58. $\frac{4121}{9990}$

59. The rebound distances form a geometric sequence:

$\frac{1}{4} \times 16$, $\left(\frac{1}{4}\right)^2 \times 16$, $\left(\frac{1}{4}\right)^3 \times 16$, $\ldots$,

or 4, $\frac{1}{4} \times 4$, $\left(\frac{1}{4}\right)^2 \times 4$, $\ldots$

The height of the 6th rebound is the 6th term of the sequence.

We will use the formula $a_n = a_1 r^{n-1}$, with $a_1 = 4$, $r = \frac{1}{4}$, and $n = 6$:

$$a_6 = 4\left(\frac{1}{4}\right)^{6-1}$$

$$a_6 = \frac{4}{4^5} = \frac{1}{4^4} = \frac{1}{256}$$

The ball rebounds $\frac{1}{256}$ ft the 6th time.

60. $5\frac{1}{3}$ ft

61. In one year, the population will be
100,000 + 0.03(100,000), or (1.03)100,000. In
two years, the population will be (1.03)100,000 +
0.03(1.03)100,000, or $(1.03)^2$100,000. Thus the
populations form a geometric sequence:

100,000,   (1.03)100,000, $(1.03)^2$100,000, . . .

The population in 15 years will be the 16th term
of the sequence.

We will use the formula $a_n = a_1 r^{n-1}$ with
$a_1 = 100,000$, $r = 1.03$, and $n = 16$:

$$a_{16} = 100,000(1.03)^{16-1}$$

$$a_{16} \approx 155,797$$

In 15 years the population will be about 155,797.

62. About 24 years

63. The amounts owed at the beginning of successive
years form a geometric sequence:

$1200, (1.12)$1200, $(1.12)^2$1200,

$(1.12)^3$1200, . . .

The amount to be repaid at the end of 13 years is
the amount owed at the beginning of the 14th year.

We use the formula $a_n = a_1 r^{n-1}$ with $a_1 = \$1200$,
$r = 1.12$, and $n = 14$:

$$a_{14} = \$1200(1.12)^{14-1}$$

$$a_{14} \approx \$5236.19$$

At the end of 13 years, $5236.19 will be repaid.

64. 10,485.76 in.

65. The lengths of the falls form a geometric
sequence:

556, $\left(\frac{3}{4}\right)$556, $\left(\frac{3}{4}\right)^2$556, $\left(\frac{3}{4}\right)^3$556, . . .

The total length of the first 6 falls is the sum
of the first six terms of this sequence. The
heights of the rebounds also form a geometric
sequence:

$\left(\frac{3}{4}\right)$556, $\left(\frac{3}{4}\right)^2$556, $\left(\frac{3}{4}\right)^3$556, . . . , or

417, $\left(\frac{3}{4}\right)$417, $\left(\frac{3}{4}\right)^2$417, . . .

When the ball hits the ground for the 6th time,
it will have rebounded 5 times. Thus the total
length of the rebounds is the sum of the first
five terms of this sequence.

We use the formula $S_n = \frac{a_1(r^n - 1)}{r - 1}$ twice,
once with $a_1 = 556$, $r = \frac{3}{4}$, and $n = 6$ and a
second time with $a_1 = 417$, $r = \frac{3}{4}$, and $n = 5$.

D = Length of falls + length of rebounds

$$= \frac{556\left[\left(\frac{3}{4}\right)^6 - 1\right]}{\frac{3}{4} - 1} + \frac{417\left[\left(\frac{3}{4}\right)^5 - 1\right]}{\frac{3}{4} - 1}.$$

We use a calculator to obtain $D \approx 3100$.

The ball will have traveled about 3100 ft.

66. 3892 ft

67. The amounts form a geometric series:
$0.01, $0.01(2), $0.01(2)^2, $0.01(2^3) + . . . +
($0.01)(2)^{27}$

We use the formula $S_n = \frac{a_1(r^n - 1)}{r - 1}$ to find
the sum of the geometric series with $a_1 = 0.01$,
$r = 2$, and $n = 28$:

$$S_{28} = \frac{0.01(2^{28} - 1)}{2 - 1}$$

We use a calculator to obtain
$S_{28} \approx \$2,684,355$.

You would earn about $2,684,355.

68. $645,826.93

69. The total effect on the economy is
$13,000,000,000 + $13,000,000,000(0.85) +
$13,000,000,000$(0.85)^2$+$13,000,000,000$(0.85)^3$+...
which is a geometric series with
$a_1 = \$13,000,000,000$ and $r = 0.85$.

$$S_\infty = \frac{a_1}{1 - r}$$

$$S_\infty = \frac{\$13,000,000,000}{1 - 0.85} = \frac{\$13,000,000,000}{0.15}$$

$$\approx \$86,666,666,667$$

70. 3,333,333, $66\frac{2}{3}$%

71. See the answer section in the text.

72. a) $\frac{13}{3}$, $\frac{22}{3}$, $\frac{34}{3}$, $\frac{46}{3}$, $\frac{58}{3}$

     b) $-\frac{11}{3}$; $-\frac{2}{3}$, $\frac{10}{3}$, $-\frac{50}{3}$, $\frac{250}{3}$ or 5; 8, 12, 18, 27

73. $1 + x + x^2 + . . .$
This is a geometric series with $a_1 = 1$ and $r = x$.

$$S_n = \frac{a_1(r^n - 1)}{r - 1} = \frac{1(x^n - 1)}{x - 1} = \frac{x^n - 1}{x - 1}$$

74. $\dfrac{x^2[1 - (-x)^n]}{x + 1}$

75. See the answer section in the text.

76.
$$\frac{a_{n+1}}{a_n} = r$$

$$\ln \frac{a_{n+1}}{a_n} = \ln r$$

$$\ln a_{n+1} - \ln a_n = \ln r$$

Thus, $\ln a_1$, $\ln a_2$, $\ln a_3$, ... is an arithmetic sequence.

77. $a_{n+1} - a_n = d$ (a_1, a_2, a_3, ... is arithmetic)

$$\frac{5^{a_{n+1}}}{5^{a_n}} = 5^{a_{n+1}-a_n} = 5^d \text{, a constant}$$

Hence 5^{a_1}, 5^{a_2}, 5^{a_3}, ... is a geometric sequence.

78. 512 cm²

Exercise Set 8.4

---

1. $n^2 < n^3$

   $1^2 < 1^3$, $2^2 < 2^3$, $3^2 < 3^3$, $4^2 < 4^3$, $5^2 < 5^3$

2. $1^2 - 1 + 41$ is prime, $2^2 - 2 + 41$ is prime, etc.

3. A polygon of n sides has $\frac{n(n-3)}{2}$ diagonals.

   A polygon of 3 sides has $\frac{3(3-3)}{2}$ diagonals.

   A polygon of 4 sides has $\frac{4(4-3)}{2}$ diagonals., etc.

4. The sum of the angles of a polygon of 3 sides is $(3-2)\cdot 180°$., etc.

5. See the answer section in the text.

6. S_n: $4 + 8 + 12 + \ldots + 4n = 2n(n+1)$

   S_1: $4 = 2\cdot 1\cdot(1+1)$

   S_k: $4 + 8 + 12 + \ldots + 4k = 2k(k+1)$

   S_{k+1}: $4 + 8 + 12 + \ldots + 4k + 4(k+1) = 2(k+1)(k+2)$

   1) **Basis step**: Since $2\cdot 1\cdot(1+1) = 2\cdot 2 = 4$, S_1 is true.

   2) **Induction step**: Let k be any natural number. Assume S_k. Deduce S_{k+1}. Starting with the left side of S_{k+1}, we have

   $4 + 8 + 12 + \ldots + 4k + 4(k+1)$

   $= 2k(k+1) + 4(k+1)$ (By S_k)

   $= (k+1)(2k+4)$

   $= 2(k+1)(k+2)$

7. See the answer section in the text.

8. S_n: $3 + 6 + 9 + \ldots + 3n = \frac{3n(n+1)}{2}$

   S_1: $3 = \frac{3(1+1)}{2}$

   S_k: $3 + 6 + 9 + \ldots + 3k = \frac{3k(k+1)}{2}$

   S_{k+1}: $3 + 6 + 9 + \ldots + 3k + 3(k+1) = \frac{3(k+1)(k+2)}{2}$

   1) **Basis step**: Since $\frac{3(1+1)}{2} = \frac{3\cdot 2}{2} = 3$, S_1 is true.

   2) **Induction step**: Let k be any natural number. Assume S_k. Deduce S_{k+1}. Starting with the left side of S_{k+1}, we have

   $3 + 6 + 9 + \ldots + 3k + 3(k+1)$

   $= \frac{3k(k+1)}{2} + 3(k+1)$ (By S_k)

   $= \frac{3k(k+1) + 6(k+1)}{2}$

   $= \frac{(k+1)(3k+6)}{2}$

   $= \frac{3(k+1)(k+2)}{2}$

9. See the answer section in the text for S_n, S_1, S_k, and S_{k+1}.

   1) **Basis step**: Since $\frac{1}{1+1} = \frac{1}{2} = \frac{1}{1\cdot 2}$, S_1 is true.

   2) **Induction step**: Let k be any natural number. Assume S_k. Deduce S_{k+1}.

   $\frac{1}{1\cdot 2} + \frac{1}{2\cdot 3} + \ldots + \frac{1}{k(k+1)} = \frac{k}{k+1}$ (S_k)

   $\frac{1}{1\cdot 2} + \frac{1}{2\cdot 3} + \ldots + \frac{1}{k(k+1)} + \frac{1}{(k+1)(k+2)} =$

   $\frac{k}{k+1} + \frac{1}{(k+1)(k+2)}$

   $\left[\text{Adding } \frac{1}{(k+1)(k+2)} \text{ on both sides}\right]$

   $= \frac{k(k+2) + 1}{(k+1)(k+2)}$

   $= \frac{k^2 + 2k + 1}{(k+1)(k+2)}$

   $= \frac{(k+1)(k+1)}{(k+1)(k+2)}$

   $= \frac{k+1}{k+2}$

10. S_n: $2 + 4 + 8 + \ldots + 2^n = 2(2^n - 1)$

    S_1: $2 = 2(2-1)$

    S_k: $2 + 4 + 8 + \ldots + 2^k = 2(2^k - 1)$

    S_{k+1}: $2 + 4 + 8 + \ldots + 2^k + 2^{k+1} = 2(2^{k+1} - 1)$

1) <u>Basis step</u>: Since $2(2 - 1) = 2 \cdot 1 = 2$, S_1 is true by substitution.

2) <u>Induction step</u>: Let k be any natural number. Assume S_k. Deduce S_{k+1}. Starting with the left side of S_{k+1}, we have

$$2 + 4 + 8 + \ldots + \underbrace{2^k + 2^{k+1}}$$

$$= \quad 2(2^k - 1) \quad + 2^{k+1} \qquad \text{(By } S_k\text{)}$$
$$= 2^{k+1} - 2 + 2^{k+1}$$
$$= 2 \cdot 2^{k+1} - 2$$
$$= 2(2^{k+1} - 1)$$

<u>11.</u> S_n: $1^3 + 2^3 + 3^3 + \ldots + n^3 = \dfrac{n^2(n + 1)^2}{4}$

S_1: $1^3 = \dfrac{1^2(1 + 1)^2}{4}$

S_k: $1^3 + 2^3 + 3^3 + \ldots + k^3 = \dfrac{k^2(k + 1)^2}{4}$

S_{k+1}: $1^3 + 2^3 + 3^3 + \ldots + k^3 + (k + 1)^3 =$

$$\dfrac{(k + 1)^2(k + 2)^2}{4}$$

1) <u>Basis step</u>: Since $\dfrac{1^2(1 + 1)^2}{4} = \dfrac{1 \cdot 2^2}{4} = \dfrac{1 \cdot 4}{4} = 1 = 1^3$, S_1 is true.

2) <u>Induction step</u>: Let k be any natural number. Assume S_k. Deduce S_{k+1}. Add $(k + 1)^3$ on both sides of S_k. Then simplify the right side.

$$1^3 + 2^2 + 3^3 + \ldots + k^3 + (k + 1)^3$$
$$= \dfrac{k^2(k + 1)^2}{4} + (k + 1)^3$$
$$= \dfrac{k^2(k + 1)^2 + 4(k + 1)^3}{4}$$
$$= \dfrac{(k + 1)^2[k^2 + 4(k + 1)]}{4}$$
$$= \dfrac{(k + 1)^2(k^2 + 4k + 4)}{4}$$
$$= \dfrac{(k + 1)^2(k + 2)^2}{4}$$

<u>12.</u> S_n: $\dfrac{1}{1 \cdot 2 \cdot 3} + \dfrac{1}{2 \cdot 3 \cdot 4} + \dfrac{1}{3 \cdot 4 \cdot 5} + \cdots + \dfrac{1}{n(n + 1)(n + 2)}$

$$= \dfrac{n(n + 3)}{4(n + 1)(n + 2)}$$

S_1: $\dfrac{1}{1 \cdot 2 \cdot 3} = \dfrac{1(1 + 3)}{4 \cdot 2 \cdot 3}$

S_k: $\dfrac{1}{1 \cdot 2 \cdot 3} + \dfrac{1}{2 \cdot 3 \cdot 4} + \cdots + \dfrac{1}{k(k + 1)(k + 2)}$

$$= \dfrac{k(k + 3)}{4(k + 1)(k + 2)}$$

S_{k+1}: $\dfrac{1}{1 \cdot 2 \cdot 3} + \dfrac{1}{2 \cdot 3 \cdot 4} + \cdots + \dfrac{1}{k(k + 1)(k + 2)} +$

$$\dfrac{1}{(k + 1)(k + 2)(k + 3)}$$

$$= \dfrac{(k + 1)(k + 1 + 3)}{4(k + 1 + 1)(k + 1 + 2)} = \dfrac{(k + 1)(k + 4)}{4(k + 2)(k + 3)}$$

1) <u>Basis step</u>: Since $\dfrac{1}{1 \cdot 2 \cdot 3} = \dfrac{1}{6}$ and $\dfrac{1(1 + 3)}{4 \cdot 2 \cdot 3} = \dfrac{1 \cdot 4}{4 \cdot 2 \cdot 3} = \dfrac{1}{6}$, S_1 is true.

2) <u>Induction step</u>: Let k be any natural number. Assume S_k. Deduce S_{k+1}.

Add $\dfrac{1}{(k + 1)(k + 2)(k + 3)}$ on both sides of S_k and simplify the right side.

Only the right side is shown here.

$$\dfrac{k(k + 3)}{4(k + 1)(k + 2)} + \dfrac{1}{(k + 1)(k + 2)(k + 3)}$$
$$= \dfrac{k(k + 3)(k + 3) + 4}{4(k + 1)(k + 2)(k + 3)}$$
$$= \dfrac{k^3 + 6k^2 + 9k + 4}{4(k + 1)(k + 2)(k + 3)}$$
$$= \dfrac{(k + 1)^2(k + 4)}{4(k + 1)(k + 2)(k + 3)}$$
$$= \dfrac{(k + 1)(k + 4)}{4(k + 2)(k + 3)}$$

<u>13.</u> S_n: $n < n + 1$

S_1: $1 < 1 + 1$

S_k: $k < k + 1$

S_{k+1}: $k + 1 < k + 2$

1) <u>Basis step</u>: Since $1 + 1 = 2$ and $1 < 2$, S_1 is true.

2) <u>Induction step</u>: Let k be any natural number. Assume S_k. Deduce S_{k+1}.

$$k < k + 1 \qquad (S_k)$$
$$k + 1 < k + 1 + 1 \quad \text{(Adding 1)}$$
$$k + 1 < k + 2$$

<u>14.</u> S_n: $2 \leqslant 2^n$

S_1: $2 \leqslant 2^1$

S_k: $2 \leqslant 2^k$

S_{k+1}: $2 \leqslant 2^{k+1}$

1) <u>Basis step</u>: Since $2 \leqslant 2$, S_1 is true.

2) <u>Induction step</u>: Let k be any natural number. Assume S_k. Deduce S_{k+1}.

$$2 \leqslant 2^k \qquad (S_k)$$
$$2 \cdot 2 \leqslant 2^k \cdot 2 \qquad \text{(Multiplying by 2)}$$
$$2 < 2 \cdot 2 \leqslant 2^{k+1} \qquad (2 < 2 \cdot 2)$$
$$2 \leqslant 2^{k+1}$$

15. S_n: $3^n < 3^{n+1}$

   S_1: $3^1 < 3^{1+1}$

   S_k: $3^k < 3^{k+1}$

   S_{k+1}: $3^{k+1} < 3^{k+2}$

   1) Basis step: Since $1 + 1 = 2$ and $3^1 < 3^2$, S_1 is true.

   2) Induction step: Let k be any natural number. Assume S_k. Deduce S_{k+1}.

   $\begin{array}{ll} 3^k < 3^{k+1} & (S_k) \\ 3^k \cdot 3 < 3^{k+1} \cdot 3 & \text{(Multiplying by 3)} \\ 3^{k+1} < 3^{k+1+1} & \\ 3^{k+1} < 3^{k+2} & \end{array}$

16. S_n: $2n \leqslant 2^n$

   S_1: $2 \cdot 1 \leqslant 2^1$

   S_k: $2k \leqslant 2^k$

   S_{k+1}: $2(k + 1) \leqslant 2^{k+1}$

   1) Basis step: Since $2 = 2$, S_1 is true.

   2) Induction step: Let k be any natural number. Assume S_k. Deduce S_{k+1}.

   $\begin{array}{ll} 2k \leqslant 2^k & (S_k) \\ 2 \cdot 2k \leqslant 2 \cdot 2^k & \text{(Multiplying by 2)} \\ 4k \leqslant 2^{k+1} & \end{array}$

   Since $1 \leqslant k$, $k + 1 \leqslant k + k$, or $k + 1 \leqslant 2k$. Then $2(k + 1) \leqslant 4k$.

   Thus $2(k + 1) \leqslant 4k \leqslant 2^{k+1}$, so $2(k + 1) \leqslant 2^{k+1}$.

17. See the answer section in the text.

18. S_n: $a_1 + (a_1 + d) + \ldots + [a_1 + (n - 1)d] = \frac{n}{2}[2a_1 + (n - 1)d]$

   S_1: $a_1 = \frac{1}{2}[2a_1 + (1 - 1)d]$

   S_k: $a_1 + (a_1 + d) + \ldots + [a_1 + (k - 1)d] = \frac{k}{2}[2a_1 + (k - 1)d]$

   S_{k+1}: $a_1 + (a_1 + d) + \ldots + [a_1 + (k - 1)d] + [a_1 + kd] = \frac{k + 1}{2}[2a_1 + kd]$

   1) Basis step: Since $\frac{1}{2}\left[2a_1 + (1 - 1)d\right] = \frac{1}{2} \cdot 2a_1 = a_1$, S_1 is true.

   2) Induction step: Let k be any natural number. Assume S_k. Deduce S_{k+1}. Starting with the left side of S_{k+1}, we have

$\underbrace{a_1 + (a_1 + d) + \ldots + [a_1 + (k-1)d]} + [a_1 + kd]$

$= \frac{k}{2}[2a_1 + (k - 1)d] + [a_1 + kd] \quad \text{(By } S_k\text{)}$

$= \frac{k[2a_1 + (k - 1)d]}{2} + \frac{2[a_1 + kd]}{2}$

$= \frac{2ka_1 + k(k - 1)d + 2a_1 + 2kd}{2}$

$= \frac{2a_1(k + 1) + k(k - 1)d + 2kd}{2}$

$= \frac{2a_1(k + 1) + (k - 1 + 2)kd}{2}$

$= \frac{2a_1(k + 1) + (k + 1)kd}{2}$

$= \frac{k + 1}{2}[2a_1 + kd]$

19. See the answer section in the text.

20. S_n: $x + y$ is a factor of $x^{2n} - y^{2n}$

   S_1: $x + y$ is a factor of $x^{2 \cdot 1} - y^{2 \cdot 1} = x^2 - y^2$.

   S_k: $x + y$ is a factor of $x^{2k} - y^{2k}$.

   S_{k+1}: $x + y$ is a factor of $x^{2(k+1)} - y^{2(k+1)} = x^{2k+2} - y^{2k+2}$.

   1) Basis step: Since $x^2 - y^2 = (x + y)(x - y)$, $x + y$ is a factor of $x^2 - y^2$.

   2) Induction step: Let k be any natural number. Assume S_k.

   $\begin{aligned} x^{2k+2} - y^{2k+2} &= x^{2k}x^2 - y^{2k}y^2 \\ &= x^{2k}x^2 - x^2y^{2k} + x^2y^{2k} - y^{2k}y^2 \\ &= x^2(x^{2k} - y^{2k}) + y^{2k}(x^2 - y^2) \end{aligned}$

   $x + y$ is a factor of $x^{2k} - y^{2k}$ by S_k, and $x + y$ is a factor of $x^2 - y^2$ by S_1, so $x + y$ is a factor of $x^{2k+2} - y^{2k+2}$.

21. See the answer section in the text.

22. S_2: $\log_a (b_1 b_2) = \log_a b_1 + \log_a b_2$

   S_k: $\log_a (b_1 b_2 \ldots b_k) = \log_a b_1 + \log_a b_2 + \ldots + \log_a b_k$

   S_{k+1}: $\log_a (b_1 b_2 \ldots b_{k+1}) = \log_a b_1 + \log_a b_2 + \ldots + \log_a b_{k+1}$

   1) Basis step: S_2 is true by the properties of logarithms.

   2) Induction step: Let k be a natural number $k \geqslant 2$. Assume S_k. Deduce S_{k+1}.

   $\log_a (b_1 b_2 \ldots b_{k+1}) \quad \text{(Left side of } S_{k+1}\text{)}$
   $= \log_a (b_1 b_2 \ldots b_k) + \log_a b_{k+1} \quad \text{(By } S_2\text{)}$
   $= \log_a b_1 + \log_a b_2 + \ldots + \log_a b_k + \log_a b_{k+1}$

23. See the answer section in the text.

24. $S_1 = \overline{z^1} = \overline{z}^1$

$S_k = \overline{z^k} = \overline{z}^k$

$\overline{z^k} \cdot \overline{z} = \overline{z}^k \cdot \overline{z}$ (Multiplying both sides of S_k by $\overline{z}$)

$\overline{z^k} \cdot \overline{z} = \overline{z}^{k+1}$

$\overline{z^{k+1}} = \overline{z}^{k+1}$

25. See the answer section in the text.

26. S_2: $\overline{z_1 z_2} = \overline{z}_1 \cdot \overline{z}_2$

S_k: $\overline{z_1 z_2 \ldots z_k} = \overline{z}_1 \overline{z}_2 \ldots \overline{z}_k$

Starting with the left side of S_{k+1}, we have

$\overline{z_1 z_2 \ldots z_k z_{k+1}} = \overline{z_1 z_2 \ldots z_k} \cdot \overline{z_{k+1}}$ (By S_2)

$= \overline{z}_1 \overline{z}_2 \ldots \overline{z}_k \cdot \overline{z}_{k+1}$ (By S_k)

27. See the answer section in the text.

28. S_1: 3 is a factor of $1^3 + 2 \cdot 1$

S_k: 3 is a factor of $k^3 + 2k$, i.e. $k^3 + 2k = 3 \cdot m$

S_{k+1}: 3 is a factor of $(k+1)^3 + 2(k+1)$

Consider:

$(k+1)^3 + 2(k+1)$

$= k^3 + 3k^2 + 5k + 3$

$= (k^3 + 2k) + 3k^2 + 3k + 3$

$= 3m + 3(k^2 + k + 1)$ (A multiple of 3)

29. See the answer section in the text.

30. S_1: 5 is a factor of $1^5 - 1$.

S_k: 5 is a factor of $k^5 - k$, i.e., $k^5 - k = 5 \cdot m$

S_{k+1}: 5 is a factor of $(k+1)^5 - (k+1)$.

Consider:

$(k+1)^5 - (k+1)$

$= (k^5 + 5k^4 + 10k^3 + 10k^2 + 5k + 1) - (k+1)$

$= (k^5 - k) + 5(k^4 + 2k^3 + 2k^2 + k)$

$= 5 \cdot m + 5(k^4 + 2k^3 + 2k^2 + k)$ (A multiple of 5)

31. See the answer section in the text.

32. S_n: $\dfrac{1}{\sqrt{1}} + \dfrac{1}{\sqrt{2}} + \dfrac{1}{\sqrt{3}} + \cdots + \dfrac{1}{\sqrt{n}} > \sqrt{n}$

S_2: $\dfrac{1}{\sqrt{1}} + \dfrac{1}{\sqrt{2}} > \sqrt{2}$

S_k: $\dfrac{1}{\sqrt{1}} + \dfrac{1}{\sqrt{2}} + \dfrac{1}{\sqrt{3}} + \cdots + \dfrac{1}{\sqrt{k}} > \sqrt{k}$

S_{k+1}: $\dfrac{1}{\sqrt{1}} + \dfrac{1}{\sqrt{2}} + \dfrac{1}{\sqrt{3}} + \cdots + \dfrac{1}{\sqrt{k}} + \dfrac{1}{\sqrt{k+1}}$

$> \sqrt{k+1}$

1) Basis Step: $\dfrac{1}{\sqrt{1}} + \dfrac{1}{\sqrt{2}} = 1 + \dfrac{1}{\sqrt{2}}$

$= \dfrac{\sqrt{2}+1}{\sqrt{2}} > \dfrac{2}{\sqrt{2}} = \sqrt{2}$

(Since $\sqrt{2} > 1$)

2) Induction Step: Let k be a natural number, $k \geq 2$. Assume S_k. Deduce S_{k+1}.

$\dfrac{1}{\sqrt{1}} + \dfrac{1}{\sqrt{2}} + \dfrac{1}{\sqrt{3}} + \cdots + \dfrac{1}{\sqrt{k}} > \sqrt{k}$, (by S_k).

Therefore, adding $\dfrac{1}{\sqrt{k+1}}$,

$\dfrac{1}{\sqrt{1}} + \dfrac{1}{\sqrt{2}} + \dfrac{1}{\sqrt{3}} + \cdots + \dfrac{1}{\sqrt{k}} + \dfrac{1}{\sqrt{k+1}} >$

$\sqrt{k} + \dfrac{1}{\sqrt{k+1}}$

Then

$\sqrt{k} + \dfrac{1}{\sqrt{k+1}} = \dfrac{\sqrt{k} \cdot \sqrt{k+1} + 1}{\sqrt{k+1}} > \dfrac{k+1}{\sqrt{k+1}}$

$= \sqrt{k+1}.$

33. See the answer section in the text.

34. S_n: $n = n + 1$

S_1: $1 = 1 + 1$

S_k: $k = k + 1$

S_{k+1}: $k + 1 = (k + 1) + 1$

a) Basis step: $1 = 1 + 1$ cannot be proved.

b) Induction step: Let k be any natural number. Assume S_k. Deduce S_{k+1}.

$k = k + 1$ (S_k)

$k + 1 = (k + 1) + 1$ (Adding 1)

The induction step can be proved.

c) The statement is not true.

35. See the answer section in the text.

36. a) Figure (2): Number of sides: 12,
    Perimeter: $4a$, Area: $A_1\left[1 + \frac{1}{3}\right]$, where A_1 is the area of Figure (1)

    Figure (3): Number of sides: 48,
    Perimeter: $\frac{16a}{3}$, Area: $A_1\left[1 + \frac{1}{3} + \frac{1}{3} \cdot \frac{4}{9}\right]$,
    where A_1 is the area of Figure (1)

    Figure (4): Number of sides: 192,
    Perimeter: $\frac{64a}{9}$,

    Area: $A_1\left[1 + \frac{1}{3} + \frac{1}{3} \cdot \frac{4}{9} + \frac{1}{3}\left(\frac{4}{9}\right)^2\right]$,

    where A_1 is the area of Figure (1)

    b) Let N_n represent the number of sides of the nth figure.

    Conjecture: $N_n = 3(4^{n-1})$

    S_1: $N_1 = 3(4^{1-1}) = 3$
    S_k: $N_k = 3(4^{k-1})$
    S_{k+1}: $N_{k+1} = 3(4^k)$

    1) Basis step: S_1 is true by part (a).

    2) Induction step: Let k be any natural number. Assume S_k. Deduce S_{k+1}.
    $N_k = 3(4^{k-1})$

    Adding another equilateral triangle to each side creates 4 sides where each side originally was. Then
    $N_{k+1} = 3(4^{k-1})(4) = 3(4^k)$.

    c) Let P_n represent the perimeter of the nth figure.

    Conjecture: $P_n = 3a\left(\frac{4}{3}\right)^{n-1}$ $\left[\text{or } \frac{4^{n-1}\, a}{3^{n-2}}\right]$

    S_1: $P_1 = 3a\left(\frac{4}{3}\right)^{1-1} = 3a$

    S_k: $P_k = 3a\left(\frac{4}{3}\right)^{k-1}$

    S_{k+1}: $P_{k+1} = 3a\left(\frac{4}{3}\right)^k$

    1) Basis step: S_1 is true by part (a).

    2) Induction step: Let k be any natural number. Assume S_k. Deduce S_{k+1}.
    $P_k = 3a\left(\frac{4}{3}\right)^{k-1}$

    Adding another equilateral triangle to each side increases the length of each side by $\frac{1}{3}$, so the new length is $\frac{4}{3}$ of the former length. Then
    $P_{k+1} = 3a\left(\frac{4}{3}\right)^{k-1}\left(\frac{4}{3}\right) = 3a\left(\frac{4}{3}\right)^k$.

    d) Since Figure (2) has 12 sides, we observe that in going from Figure (2) to Figure (3) we gain 12 additional triangles of side $\frac{a}{3^2}$.

    Hence, the additional area is $12 \cdot \dfrac{\left(\frac{a}{9}\right)^2}{4}\sqrt{3}$,
    or $\frac{1}{3}\, r \cdot A_1$, where $r = \frac{4}{9}$. In going from Figure (3) to Figure (4) we gain
    $48 \cdot \dfrac{\left(\frac{a}{27}\right)^2}{4}\sqrt{3}$, or $\frac{1}{3}\, r^2 \cdot A_1$, and so on.

    Let A_n represent the area of the nth figure. Recall that N_n represents the number of sides of the nth figure.

    Conjecture:

    $A_n = A_1\left[1 + \left(\frac{1}{3} + \frac{1}{3}\, r + \cdots + \frac{1}{3}\, r^{n-2}\right)\right]$

    $A_n = A_1\left[1 + \left(\frac{1}{3} \cdot \frac{1 - r^{n-1}}{1 - r}\right)\right]$

    $A_n = \frac{a^2\sqrt{3}}{20}(8 - 3r^{n-1})$, $n \geqslant 1$

    S_1: $A_1 = \frac{a^2\sqrt{3}}{20}(8 - 3 \cdot 1) = \frac{a^2\sqrt{3}}{4}$

    S_k: $A_k = \frac{a^2\sqrt{3}}{20}(8 - 3r^{k-1})$, or
    $\frac{1}{5}\, A_1(8 - 3r^{k-1})$

    S_{k+1}: $A_{k+1} = \frac{a^2\sqrt{3}}{20}(8 - 3r^k)$

    1) Basis step: S_1 is true by part (a).
    2) Induction step: Let k be any natural number. Assume S_k. Deduce S_{k+1}.

    Adding another equilateral triangle to each side of the kth figure adds the area of N_k equilateral triangles with side of length $\frac{a}{3^k}$. Thus

    $A_{k+1} = A_k + N_k \cdot \dfrac{\left(\frac{a}{3^k}\right)^2}{4}\sqrt{3}$

    $= \frac{1}{5}\, A_1(8 - 3r^{k-1}) + N_k \cdot \frac{1}{9^k} \cdot A_1$

    $= A_1\left[\frac{8}{5} - \frac{3}{5}\, r^{k-1} + 3 \cdot 4^{k-1} \cdot \frac{1}{9^k}\right]$

    $= A_1\left[\frac{8}{5} - \frac{3}{5}\, r^{k-1} + \frac{3}{4}\, r^k\right]$

    $= \frac{a^2\sqrt{3}}{20}(8 - 3r^k)$

    e) $0.4\, a^2\sqrt{3}$

Exercise Set 8.5

1. $_4P_3 = 4\cdot3\cdot2 = 24$ (Theorem 8, Formula 1)

2. $7\cdot6\cdot5\cdot4\cdot3 = 2520$

3. $_{10}P_7 = 10\cdot9\cdot8\cdot7\cdot6\cdot5\cdot4 = 604{,}800$ (Theorem 8, Formula 1)

4. $10\cdot9\cdot8 = 720$

5. Without repetition: $_5P_5 = 5\cdot4\cdot3\cdot2\cdot1 = 120$
    (Theorem 7)

    With repetition: $5^5 = 3125$ (Theorem 10)

6. a) $_4P_4 = 24$

    b) $4^4 = 256$

7. $_5P_5 = 5\cdot4\cdot3\cdot2\cdot1 = 120$ (Theorem 7)

8. $7! = 5040$

9. DIGIT

    There are 2 I's, 1 D, 1 G, and 1 T for a total of 5.

    $P = \dfrac{5!}{2!\cdot1!\cdot1!\cdot1!}$ (Theorem 9)

    $= \dfrac{5!}{2!} = \dfrac{5\cdot4\cdot3\cdot2!}{2!} = 5\cdot4\cdot3 = 60$

10. $\dfrac{6!}{2!} = 360$

11. There are only 9 choices for the first digit since 0 is excluded. There are also 9 choices for the second digit since 0 can be included and the first digit cannot be repeated. Because no digit is used more than once there are only 8 choices for the third digit, 7 for the fourth, 6 for the fifth, 5 for the sixth and 4 for the seventh. By the Fundamental Counting Principle the total number of permutations is

     $9\cdot9\cdot8\cdot7\cdot6\cdot5\cdot4$, or 544,320.

     Thus 544,320 7-digit phone numbers can be formed.

12. $_5P_5 \cdot {}_4P_4 = 2880$

13. $a^2b^3c^4 = a\cdot a\cdot b\cdot b\cdot b\cdot c\cdot c\cdot c\cdot c$

     There are 2 a's, 3 b's, and 4 c's, for a total of 9.

     Thus $P = \dfrac{9!}{2!\cdot3!\cdot4!}$ (Theorem 9)

     $= \dfrac{9\cdot8\cdot7\cdot6\cdot5\cdot4!}{2\cdot1\cdot3\cdot2\cdot1\cdot4!} = \dfrac{9\cdot8\cdot7\cdot6\cdot5}{2\cdot3\cdot2} = 1260$

14. $\dfrac{6!}{3!2!} = 60$

15. a) $_5P_5 = 5!$ (Theorem 7)
        $= 5\cdot4\cdot3\cdot2\cdot1 = 120$

    b) There are 5 choices for the first coin and 2 possibilities (head or tail) for each choice. This results in a total of 10 choices for the first selection.

        There are 4 choices (no coin can be used more than once) for the second coin and 2 possibilities (head or tail) for each choice. This results in a total of 8 choices for the second selection.

        There are 3 choices for the third coin and 2 possibilities (head or tail) for each choice. This results in a total of 6 choices for the third selection.

        Likewise there are 4 choices for the fourth selection and 2 choices for the fifth selection.

        Using the Fundamental Counting Principle we know there are

        $10\cdot8\cdot6\cdot4\cdot2$, or 3840

        ways the coins can be lined up.

16. a) $4\cdot3\cdot2\cdot1 = 24$

    b) $(4!)\cdot2^4 = 384$

17. $_{52}P_4 = 52\cdot51\cdot50\cdot49$ (Theorem 8, Formula 1)
        $= 6{,}497{,}400$

18. $50\cdot49\cdot48\cdot47\cdot46 = 254{,}251{,}200$

19. $P = \dfrac{24!}{3!\cdot5!\cdot9!\cdot4!\cdot3!}$ (Theorem 9)

     $= \dfrac{24\cdot23\cdot22\cdot21\cdot20\cdot19\cdot18\cdot17\cdot16\cdot15\cdot14\cdot13\cdot12\cdot11\cdot10\cdot9!}{3\cdot2\cdot1\cdot5\cdot4\cdot3\cdot2\cdot1\cdot4\cdot3\cdot2\cdot1\cdot3\cdot2\cdot1\cdot9!}$

     $= 23\cdot11\cdot7\cdot19\cdot17\cdot8\cdot15\cdot14\cdot13\cdot12\cdot11\cdot10$

                                (Simplifying)
     $= 16{,}491{,}024{,}950{,}400$

20. $\dfrac{20!}{2!\ 5!\ 8!\ 3!\ 2!} = 20{,}951{,}330{,}400$

21. MATH:

     $_4P_4 = 4\cdot3\cdot2\cdot1 = 24$ (Theorem 7)

     BUSINESS: 1 B, 1 U, 3 S's, 1 I, 1 N, 1 E, a total of 8.

     $P = \dfrac{8!}{1!\cdot1!\cdot3!\cdot1!\cdot1!\cdot1!}$ (Theorem 9)

     $= \dfrac{8!}{3!} = \dfrac{8\cdot7\cdot6\cdot5\cdot4\cdot3!}{3!} = 8\cdot7\cdot6\cdot5\cdot4 = 6720$

PHILOSOPHICAL: 2 P's, 2 H's, 2 I's, 2 L's, 2 O's,
1 S, 1 C, and 1 A, a total of 13

$$P = \frac{13!}{2! \cdot 2! \cdot 2! \cdot 2! \cdot 2! \cdot 1! \cdot 1! \cdot 1!} \qquad \text{(Theorem 9)}$$

$$= \frac{13 \cdot 12 \cdot 11 \cdot 10 \cdot 9 \cdot 8 \cdot 7 \cdot 6 \cdot 5 \cdot 4 \cdot 3 \cdot 2 \cdot 1}{2 \cdot 1 \cdot 2 \cdot 1 \cdot 2 \cdot 1 \cdot 2 \cdot 1 \cdot 2 \cdot 1 \cdot 1 \cdot 1 \cdot 1}$$

$$= 13 \cdot 12 \cdot 11 \cdot 5 \cdot 9 \cdot 4 \cdot 7 \cdot 3 \cdot 5 \cdot 2 \cdot 3 \qquad \text{(Simplifying)}$$

$$= 194,594,400$$

<u>22.</u> a) $6! = 720$

b) $\frac{7!}{2!} = 2520$

c) $\frac{11!}{2! \, 2! \, 2!} = 4,989,600$

<u>23.</u> There are 80 choices for the number of the county, 26 choices for the letter of the alphabet, and 9999 choices for the number that follows the letter. By the Fundamental Counting Principle we know there are $80 \cdot 26 \cdot 9999$, or 20,797,920 possible license plates.

<u>24.</u> a) $_5P_4 = 120$

b) $5 \cdot 5 \cdot 5 \cdot 5 = 625$

c) $1 \cdot {}_4P_3 = 24$

d) $_3P_2 \cdot 1 = 6$

<u>25.</u> a) We want to find the number of permutations of 10 objects taken 5 at a time with repetition. This is 10^5, or 100,000.

b) Since there are 100,000 possible zip-codes, there could be 100,000 post offices.

<u>26.</u> a) 10^9, or 1,000,000,000

b) Yes

<u>27.</u> a) We want to find the number of permutations of 10 objects taken 9 at a time with repetition. This is 10^9, or 1,000,000,000.

b) Since there are more than 257 million social security numbers, each person can have one.

<u>28.</u> $\frac{15!}{31,557,600} \approx 41,437.7$ yr (using 1 yr = 365.25 days)

<u>29.</u> $$_nP_5 = 7 \cdot {}_nP_4$$

$$\frac{n!}{(n-5)!} = 7 \cdot \frac{n!}{(n-4)!} \qquad \text{(Theorem 8)}$$

$$\frac{n!}{7(n-5)!} = \frac{n!}{(n-4)!} \qquad \left[\text{Multiplying by } \tfrac{1}{7}\right]$$

$$7(n-5)! = (n-4)! \qquad \begin{array}{l}\text{(The denominators} \\ \text{must be the same.)}\end{array}$$

$$7(n-5)! = (n-4)(n-5)!$$

$$7 = n - 4 \qquad \left[\text{Multiplying by } \tfrac{1}{(n-5)!}\right]$$

$$11 = n$$

<u>30.</u> 8

<u>31.</u> $$_nP_5 = 9 \cdot {}_{n-1}P_4$$

$$\frac{n!}{(n-5)!} = 9 \cdot \frac{(n-1)!}{(n-5)!} \qquad \text{(Theorem 8)}$$

$$n! = 9(n-1)! \qquad [\text{Multiplying by } (n-5)!]$$

$$\frac{n!}{(n-1)!} = 9 \qquad \left[\text{Multiplying by } \tfrac{1}{(n-1)!}\right]$$

$$\frac{n(n-1)!}{(n-1)!} = 9$$

$$n = 9$$

<u>32.</u> 11

<u>33.</u> There is one losing team per game. In order to leave one tournament winner there must be $n - 1$ losers produced in $n - 1$ games.

<u>34.</u> $2n - 1$

Exercise Set 8.6

<u>1.</u> The 13 tells where to start.
$_{13}C_2 = \frac{13 \cdot 12}{2 \cdot 1} = 78$
The 2 tells how many factors there are in both numerator and denominator and where to start the denominator.

We used formula (2) from Theorem 11.

<u>2.</u> 84

<u>3.</u> $\binom{13}{11} = \frac{13!}{11!(13-11)!} = \frac{13!}{11!2!} = \frac{13 \cdot 12 \cdot 11!}{11!2!} = \frac{13 \cdot 12}{2 \cdot 1} = 78$

We used formula (1) from Theorem 11. Using Theorem 12 we could also have observed that $\binom{13}{11} = \binom{13}{2}$, giving us 78 as in Exercise 1.

<u>4.</u> 84

<u>5.</u> $\binom{7}{1} = \frac{7}{1} = 7$ Formula (2)

6. 1

7. $\dfrac{_5P_3}{3!} = {_5C_3} = \dfrac{5 \cdot 4 \cdot 3}{3 \cdot 2 \cdot 1} = 10$ Formula (2)

8. 252

9. $\begin{pmatrix} 6 \\ 0 \end{pmatrix} = \dfrac{6!}{0!(6-0)!} = \dfrac{6!}{1 \cdot 6!} = 1$ Formula (1)

10. 6

11. $\begin{pmatrix} 6 \\ 2 \end{pmatrix} = \dfrac{6 \cdot 5}{2 \cdot 1} = 15$ Formula (2)

12. 20

13. $_{12}C_{11} = {_{12}C_1}$ Theorem 12

       $= \dfrac{12}{1}$ Formula (2)

       $= 12$

14. 66

15. $_{12}C_9 = {_{12}C_3}$ Theorem 12

       $= \dfrac{12 \cdot 11 \cdot 10}{3 \cdot 2 \cdot 1}$ Formula (2)

       $= 220$

16. 495

17. $\begin{pmatrix} m \\ 2 \end{pmatrix} = \dfrac{m(m-1)}{2 \cdot 1}$ Formula (2)

       $= \dfrac{m(m-1)}{2}$, or $\dfrac{m^2 - m}{2}$

Also $\begin{pmatrix} m \\ 2 \end{pmatrix} = \dfrac{m!}{2!(m-2)!}$.   Formula (1)

This is equivalent to the result using Formula (2).

18. $\dfrac{t!}{4!(t-4)!}$

19. $\begin{pmatrix} p \\ 3 \end{pmatrix} = \dfrac{p(p-1)(p-2)}{3 \cdot 2 \cdot 1}$ Formula (2)

       $= \dfrac{p(p-1)(p-2)}{6}$, or $\dfrac{p^3 - 3p^2 + 2p}{6}$

Also $\begin{pmatrix} p \\ 3 \end{pmatrix} = \dfrac{p!}{3!(p-3)!}$.   Formula (1)

This is equivalent to the result using Formula (2).

20. 1

21. By Theorem 12, $\begin{pmatrix} n \\ r \end{pmatrix} = \begin{pmatrix} n \\ n-r \end{pmatrix}$, so

$\begin{pmatrix} 7 \\ 0 \end{pmatrix} + \begin{pmatrix} 7 \\ 1 \end{pmatrix} + \begin{pmatrix} 7 \\ 2 \end{pmatrix} + \begin{pmatrix} 7 \\ 3 \end{pmatrix} + \begin{pmatrix} 7 \\ 4 \end{pmatrix} + \begin{pmatrix} 7 \\ 5 \end{pmatrix} + \begin{pmatrix} 7 \\ 6 \end{pmatrix} + \begin{pmatrix} 7 \\ 7 \end{pmatrix}$

$= 2\left[\begin{pmatrix} 7 \\ 0 \end{pmatrix} + \begin{pmatrix} 7 \\ 1 \end{pmatrix} + \begin{pmatrix} 7 \\ 2 \end{pmatrix} + \begin{pmatrix} 7 \\ 3 \end{pmatrix} \right]$

$= 2\left[\dfrac{7!}{7!0!} + \dfrac{7!}{6!1!} + \dfrac{7!}{5!2!} + \dfrac{7!}{4!3!} \right]$

$= 2(1 + 7 + 21 + 35) = 2 \cdot 64 = 128$

22. 64

23. $_{23}C_4 = \begin{pmatrix} 23 \\ 4 \end{pmatrix} = \dfrac{23!}{4!(23-4)!}$ (Theorem 11)

       $= \dfrac{23!}{4! \cdot 19!} = \dfrac{23 \cdot 22 \cdot 21 \cdot 20 \cdot 19!}{4! \cdot 19!}$

       $= \dfrac{23 \cdot 22 \cdot 21 \cdot 20}{4 \cdot 3 \cdot 2 \cdot 1} = 8855$

24. 36, 72

25. $_{10}C_6 = \begin{pmatrix} 10 \\ 6 \end{pmatrix} = \dfrac{10!}{6(10-6)!}$ (Theorem 11)

       $= \dfrac{10!}{6! \cdot 4!} = \dfrac{10 \cdot 9 \cdot 8 \cdot 7 \cdot 6!}{6! \cdot 4!}$

       $= \dfrac{10 \cdot 9 \cdot 8 \cdot 7}{4 \cdot 3 \cdot 2 \cdot 1} = 210$

26. $_{11}C_7 = 330$

27. Since two points determine a line and no three of these 8 points are collinear, we need to find out the number of combinations of 8 points taken 2 at a time, $_8C_2$.

$_8C_2 = \begin{pmatrix} 8 \\ 2 \end{pmatrix} = \dfrac{8!}{2!(8-2)!}$ (Theorem 11)

       $= \dfrac{8 \cdot 7 \cdot 6!}{2 \cdot 1 \cdot 6!}$

       $= \dfrac{8 \cdot 7}{2} = 28$

Thus 28 lines are determined.

Since three noncollinear points determine a triangle and no four of these 8 points are coplanar, we need to find out the number of combinations of 8 points taken 3 at a time, $_8C_3$.

$_8C_3 = \begin{pmatrix} 8 \\ 3 \end{pmatrix} = \dfrac{8!}{3!(8-3)!}$ (Theorem 11)

       $= \dfrac{8 \cdot 7 \cdot 6 \cdot 5!}{3 \cdot 2 \cdot 1 \cdot 5!}$

       $= \dfrac{8 \cdot 7 \cdot 6}{3 \cdot 2} = 56$

Thus 56 triangles are determined.

28. $_7C_2 = 21$, $_7C_3 = 35$

29. $_{10}C_7 \cdot {}_5C_3 = \binom{10}{7} \cdot \binom{5}{3}$ (Using the Fundamental Counting Principle)

$= \dfrac{10!}{7!(10-7)!} \cdot \dfrac{5!}{3!(5-3)!}$ (Theorem 11)

$= \dfrac{10 \cdot 9 \cdot 8 \cdot 7!}{7! \cdot 3!} \cdot \dfrac{5 \cdot 4 \cdot 3!}{3! \cdot 2!}$

$= \dfrac{10 \cdot 9 \cdot 8}{3 \cdot 2 \cdot 1} \cdot \dfrac{5 \cdot 4}{2 \cdot 1} = 120 \cdot 10 = 1200$

30. $_8C_6 \cdot {}_4C_3 = 112$

31. $_{58}C_6 \cdot {}_{42}C_4$ (Using the Fundamental Counting Principle)

$= \binom{58}{6} \cdot \binom{42}{4}$

32. $\binom{63}{8} \cdot \binom{37}{12}$

33. In a 52-card deck there are 4 aces and 48 cards that are not aces.

$_4C_3 \cdot {}_{48}C_2 = \binom{4}{3}\binom{48}{2}$ (Using the Fundamental Counting Principle)

$= \dfrac{4!}{3!(4-3)!} \cdot \dfrac{48!}{2!(48-2)!}$ (Theorem 11)

$= \dfrac{4!}{3! \cdot 1!} \cdot \dfrac{48!}{2! \cdot 46!}$

$= \dfrac{4 \cdot 3!}{3! \cdot 1!} \cdot \dfrac{48 \cdot 47 \cdot 46!}{2! \cdot 46!}$

$= \dfrac{4}{1} \cdot \dfrac{48 \cdot 47}{2}$

$= 4512$

34. $_4C_2 \cdot {}_{48}C_3 = 103,776$

35. a) If order is considered and repetition is not allowed, the number of possible cones is
$_{33}P_3 = 33 \cdot 32 \cdot 31 = 32,736.$

b) If order is considered and repetition is allowed, the number of possibilities is
$33^3 = 35,937.$

c) If order is not considered and there is no repetition, the number of possibilities is
$_{33}C_3 = \dfrac{33 \cdot 32 \cdot 31}{3 \cdot 2 \cdot 1} = 5456.$

36. a) $_{31}P_2 = 930$
b) $31^2 = 961$
c) $_{31}C_2 = 465$

37. The number of plain pizzas (pizzas with no toppings) is $\binom{12}{0}$. The number of pizzas with 1 topping is $\binom{12}{1}$. The number of pizzas with 2 toppings is $\binom{12}{2}$, and so on. The number of pizzas with all 12 toppings is $\binom{12}{12}$. The total number of pizzas is $\binom{12}{0} + \binom{12}{1} + \binom{12}{2} + \cdots + \binom{12}{12} = 4096.$

38. $3 \cdot 2 \cdot 4096 = 24,576$

39. $_{52}C_5 = \binom{52}{5} = \dfrac{52!}{5!(52-5)!}$ (Theorem 11)

$= \dfrac{52!}{5! \cdot 47!}$

$= \dfrac{52 \cdot 51 \cdot 50 \cdot 49 \cdot 48 \cdot 47!}{5! \cdot 47!}$

$= \dfrac{52 \cdot 51 \cdot 50 \cdot 49 \cdot 48}{5 \cdot 4 \cdot 3 \cdot 2 \cdot 1}$

$= 13 \cdot 17 \cdot 10 \cdot 49 \cdot 24$

$= 2,598,960$

40. $_{52}C_{13} = 635,013,559,600$

41. $_8C_3 = \binom{8}{3} = \dfrac{8!}{3!(8-3)!} = \dfrac{8!}{3! \cdot 5!}$

$= \dfrac{8 \cdot 7 \cdot 6 \cdot 5!}{3 \cdot 2 \cdot 1 \cdot 5!}$

$= 8 \cdot 7 = 56$

42. $\binom{n}{4} = \dfrac{n(n-1)(n-2)(n-3)}{24}$

43. For each parallelogram 2 lines from each group of parallel lines are needed.

Two lines can be selected from the group of 5 parallel lines in $_5C_2$ ways and the other two lines can be selected from the group of 8 parallel lines in $_8C_2$ ways.

Using the Fundamental Counting Principle, it follows that the number of possible parallelograms is

$_5C_2 \cdot {}_8C_2 = \binom{5}{2}\binom{8}{2}$ (Using the Fundamental Counting Principle)

$= \dfrac{5!}{2!(5-2)!} \cdot \dfrac{8!}{2!(8-2)!}$ (Theorem 11)

$= \dfrac{5 \cdot 4 \cdot 3!}{2 \cdot 1 \cdot 3!} \cdot \dfrac{8 \cdot 7 \cdot 6!}{2 \cdot 1 \cdot 6!}$

$= \dfrac{5 \cdot 4}{2} \cdot \dfrac{8 \cdot 7}{2}$

$= 10 \cdot 28 = 280$

Thus 280 parallelograms can be formed.

44. $\binom{n}{r} = \dfrac{n!}{r!(n-r)!} = \dfrac{n!}{(n-r)!r!}$

$= \dfrac{n!}{(n-r)![n-(n-r)]!}$

$= \binom{n}{n-r}$

45. If each team plays each other team once, we have
$\binom{8}{2} = \dfrac{8!}{2!6!} = \dfrac{8 \cdot 7 \cdot 6!}{2 \cdot 1 \cdot 6!} = 28.$

If each team plays each other team twice, we have
$2\binom{8}{2} = 2 \cdot 28 = 56.$

46. $\begin{bmatrix} n \\ 2 \end{bmatrix} = \frac{n(n-1)}{2}$; $2\begin{bmatrix} n \\ 2 \end{bmatrix} = n(n-1)$

47.
$$\begin{bmatrix} n+1 \\ 3 \end{bmatrix} = 2 \cdot \begin{bmatrix} n \\ 2 \end{bmatrix}$$

$$\frac{(n+1)!}{3!(n+1-3)!} = 2 \cdot \frac{n!}{2!(n-2)!}$$
(Theorem 11)

$$\frac{(n+1)(n)(n-1)(n-2)!}{3!(n-2)!} = \frac{n(n-1)(n-2)!}{(n-2)!}$$

$$\frac{(n+1)(n)(n-1)}{6} = n(n-1)$$

$$(n+1)(n)(n-1) = 6n(n-1)$$

$$n + 1 = 6$$

$$n = 5$$

48. 4

49.
$$\begin{bmatrix} n+2 \\ 4 \end{bmatrix} = 6 \cdot \begin{bmatrix} n \\ 2 \end{bmatrix}$$

$$\frac{(n+2)!}{4!(n+2-4)!} = 6 \cdot \frac{n!}{2!(n-2)!}$$
(Theorem 11)

$$\frac{(n+2)!}{4!(n-2)!} = 6 \cdot \frac{n!}{2!(n-2)!}$$

$$\frac{(n+2)!}{4!} = 6 \cdot \frac{n!}{2!}$$

$$4! \cdot \frac{(n+2)!}{4!} = 4! \cdot 6 \cdot \frac{n!}{2!}$$

$$(n+2)! = 72 \cdot n!$$

$$(n+2)(n+1)n! = 72 \cdot n!$$

$$(n+2)(n+1) = 72$$

$$n^2 + 3n + 2 = 72$$

$$n^2 + 3n - 70 = 0$$

$$(n+10)(n-7) = 0$$

$$n + 10 = 0 \quad \text{or} \quad n - 7 = 0$$

$$n = -10 \quad \text{or} \quad n = 7$$

The only solution is 7 since we cannot have a set of -10 objects.

50. 6

51. a) $_5C_2 = 10$

b) $_5C_2 - 5 = 5$

52. $_6C_2 = 15$, $_6C_2 - 6 = 9$

53. a) $_nC_2 = \frac{n(n-1)}{2}$

b) $_nC_2 - n = \frac{n(n-1)}{2} - \frac{2n}{2} = \frac{n(n-3)}{2}$,

where $n = 4, 5, 6, \ldots$

c) See the answer section in the text.

54. $\begin{bmatrix} n \\ r-1 \end{bmatrix} + \begin{bmatrix} n \\ r \end{bmatrix}$

$$= \frac{n!}{(r-1)!(n-r+1)!} \cdot \frac{r}{r} + \frac{n!}{r!(n-r)!} \cdot \frac{(n-r+1)}{(n-r+1)}$$

$$= \frac{n![r + (n-r+1)]}{r!(n-r+1)!}$$

$$= \frac{(n+1)!}{r!(n-r+1)!}$$

$$= \begin{bmatrix} n+1 \\ r \end{bmatrix}$$

Exercise Set 8.7

1. Pascal Method: Expand $(m+n)^5$.

The expansion of $(m+n)^5$ has $5 + 1$, or 6, terms. The sum of the exponents in each term is 5. The exponents of m start with 5 and decrease to 0. The last term has no factor of m. The first term has no factor of n. The exponents of n start in the second term with 1 and increase to 5. We get the coefficients from the 6th row of Pascal's Triangle.

```
            1
         1     1
       1    2    1
     1    3    3    1
   1    4    6    4    1
 1    5   10   10    5    1
```

$(m+n)^5 = 1 \cdot m^5 + 5 \cdot m^4 n^1 + 10 \cdot m^3 \cdot n^2 + 10 \cdot m^2 \cdot n^3 +$
$\qquad 5 \cdot m \cdot n^4 + 1 \cdot n^5$

$\qquad = m^5 + 5m^4 n + 10m^3 n^2 + 10m^2 n^3 + 5mn^4 + n^5$

2. $a^4 - 4a^3 b + 6a^2 b^2 - 4ab^3 + b^4$

3. Pascal Method: Expand $(x-y)^6$.

The expansion of $(x-y)^6$ has $6 + 1$, or 7 terms. The sum of the exponents in each term is 6. The exponents of x start with 6 and decrease to 0. The last term has no factor of x. The first term has no factor of -y. The exponents of -y start in the second term with 1 and increase to 6. We get the coefficients from the 7th row of Pascal's Triangle.

```
              1
           1     1
         1    2    1
       1    3    3    1
     1    4    6    4    1
   1    5   10   10    5    1
 1    6   15   20   15    6    1
```

$(x-y)^6 = 1 \cdot x^6 + 6 \cdot x^5 \cdot (-y) + 15 \cdot x^4 \cdot (-y)^2 +$
$\qquad 20 \cdot x^3 \cdot (-y)^3 + 15 \cdot x^2 \cdot (-y)^4 +$
$\qquad 6 \cdot x \cdot (-y)^5 + 1 \cdot (-y)^6$

$\qquad = x^6 - 6x^5 y + 15x^4 y^2 - 20x^3 y^3 + 15x^2 y^4 -$
$\qquad 6xy^5 + y^6$

4. $p^7 + 7p^6q + 21p^5q^2 + 35p^4q^3 + 35p^3q^4 + 21p^2q^5 + 7pq^6 + q^7$

5. Pascal Method: Expand $(x^2 - 3y)^5$.

   The expansion of $(x^2 - 3y)^5$ has $5 + 1$, or 6, terms. The sum of the exponents in each term is 5. The exponents of x^2 start with 5 and decrease to 0. The last term has no factor of x^2. The first term has no factor of $-3y$. The exponents of $-3y$ start in the second term with 1 and increase to 5. We get the coefficients from the 6th row of Pascal's Triangle.

$$\begin{array}{ccccccccccc} & & & & & 1 & & & & & \\ & & & & 1 & & 1 & & & & \\ & & & 1 & & 2 & & 1 & & & \\ & & 1 & & 3 & & 3 & & 1 & & \\ & 1 & & 4 & & 6 & & 4 & & 1 & \\ 1 & & 5 & & 10 & & 10 & & 5 & & 1 \end{array}$$

$$\begin{aligned} (x^2 - 3y)^5 &= 1\cdot(x^2)^5 + 5\cdot(x^2)^4\cdot(-3y) + \\ &\quad 10\cdot(x^2)^3\cdot(-3y)^2 + 10\cdot(x^2)^2\cdot(-3y)^3 + \\ &\quad 5\cdot(x^2)\cdot(-3y)^4 + 1\cdot(-3y)^5 \\ &= x^{10} - 15x^8y + 90x^6y^2 - 270x^4y^3 + \\ &\quad 405x^2y^4 - 243y^5 \end{aligned}$$

6. $2187c^7 - 5103c^6d + 5103c^5d^2 - 2835c^4d^3 + 945c^3d^4 - 189c^2d^5 + 21cd^6 - d^7$

7. Pascal Method: Expand $(3c - d)^6$.

   The sum of the exponents in each term is 6. The exponents of $3c$ start with 6 and decrease to 0. The last term has no factor of $3c$. The first term has no factor of $-d$. The exponents of $-d$ start in the second term with 1 and increase to 6. We get the coefficients from the 7th row of Pascal's Triangle. The expansion has $6 + 1$, or 7 terms.

$$\begin{array}{ccccccccccccc} & & & & & & 1 & & & & & & \\ & & & & & 1 & & 1 & & & & & \\ & & & & 1 & & 2 & & 1 & & & & \\ & & & 1 & & 3 & & 3 & & 1 & & & \\ & & 1 & & 4 & & 6 & & 4 & & 1 & & \\ & 1 & & 5 & & 10 & & 10 & & 5 & & 1 & \\ 1 & & 6 & & 15 & & 20 & & 15 & & 6 & & 1 \end{array}$$

$$\begin{aligned} (3c - d)^6 &= 1\cdot(3c)^6 + 6\cdot(3c)^5\cdot(-d) + \\ &\quad 15\cdot(3c)^4\cdot(-d)^2 + 20\cdot(3c)^3\cdot(-d)^3 + \\ &\quad 15\cdot(3c)^2\cdot(-d)^4 + 6\cdot(3c)\cdot(-d)^5 + \\ &\quad 1\cdot(-d)^6 \\ &= 3^6c^6 - 6\cdot3^5c^5d + 15\cdot3^4c^4d^2 - \\ &\quad 20\cdot3^3c^3d^3 + 15\cdot3^2c^2d^4 - 6\cdot3cd^5 + d^6 \\ &= 729c^6 - 1458c^5d + 1215c^4d^2 - 540c^3d^3 + \\ &\quad 135c^2d^4 - 18cd^5 + d^6 \end{aligned}$$

8. $t^{-12} + 12t^{-10} + 60t^{-8} + 160t^{-6} + 240t^{-4} + 192t^{-2} + 64$

9. Pascal Method: Expand $(x - y)^3$.

   The expansion of $(x - y)^3$ has $3 + 1$, or 4 terms. The sum of the exponents in each term is 3. The exponents of x start with 3 and decrease to 0. The last term has no factor of x. The first term has no factor of $-y$. The exponents of $-y$ start in the second term with 1 and increase to 3. We get the coefficients from the 4th row of Pascal's Triangle.

$$\begin{array}{ccccccc} & & & 1 & & & \\ & & 1 & & 1 & & \\ & 1 & & 2 & & 1 & \\ 1 & & 3 & & 3 & & 1 \end{array}$$

$$\begin{aligned} (x - y)^3 &= 1\cdot x^3 + 3\cdot x^2\cdot(-y) + 3\cdot x\cdot(-y)^2 + 1\cdot(-y)^3 \\ &= x^3 - 3x^2y + 3xy^2 - y^3 \end{aligned}$$

10. $x^5 - 5x^4y + 10x^3y^2 - 10x^2y^3 + 5xy^4 - y^5$

11. Binomial Theorem Method: Expand $\left[\frac{1}{x} + y\right]^7$.

    Note that $a = \frac{1}{x}$, $b = y$, and $n = 7$.

$$\begin{aligned} \left[\frac{1}{x} + y\right]^7 &= \binom{7}{0}\left[\frac{1}{x}\right]^7 + \binom{7}{1}\left[\frac{1}{x}\right]^6 y + \binom{7}{2}\left[\frac{1}{x}\right]^5 y^2 + \\ &\quad \binom{7}{3}\left[\frac{1}{x}\right]^4 y^3 + \binom{7}{4}\left[\frac{1}{x}\right]^3 y^4 + \binom{7}{5}\left[\frac{1}{x}\right]^2 y^5 + \\ &\quad \binom{7}{6}\left[\frac{1}{x}\right] y^6 + \binom{7}{7} y^7 \\ &= \frac{7!}{7!0!}\left[\frac{1}{x}\right]^7 + \frac{7!}{6!1!}\left[\frac{1}{x}\right]^6 y + \frac{7!}{5!2!}\left[\frac{1}{x}\right]^5 y^2 + \\ &\quad \frac{7!}{4!3!}\left[\frac{1}{x}\right]^4 y^3 + \frac{7!}{3!4!}\left[\frac{1}{x}\right]^3 y^4 + \\ &\quad \frac{7!}{2!5!}\left[\frac{1}{x}\right]^2 y^5 + \frac{7!}{1!6!}\left[\frac{1}{x}\right] y^6 + \frac{7!}{0!7!} y^7 \\ &= x^{-7} + 7x^{-6}y + 21x^{-5}y^2 + 35x^{-4}y^3 + \\ &\quad 35x^{-3}y^4 + 21x^{-2}y^5 + 7x^{-1}y^6 + y^7 \end{aligned}$$

12. $8s^3 - 36s^2t^2 + 54st^4 - 27t^6$

13. $\left[a - \frac{2}{a}\right]^9$

    Let $a = a$, $b = -\frac{2}{a}$, and $n = 9$.

    Expand using the Binomial Theorem Method.

$$\begin{aligned} \left[a - \frac{2}{a}\right]^9 &= \binom{9}{0}a^9 + \binom{9}{1}a^8\left[-\frac{2}{a}\right] + \binom{9}{2}a^7\left[-\frac{2}{a}\right]^2 + \\ &\quad \binom{9}{3}a^6\left[-\frac{2}{a}\right]^3 + \binom{9}{4}a^5\left[-\frac{2}{a}\right]^4 + \\ &\quad \binom{9}{5}a^4\left[-\frac{2}{a}\right]^5 + \binom{9}{6}a^3\left[-\frac{2}{a}\right]^6 + \\ &\quad \binom{9}{7}a^2\left[-\frac{2}{a}\right]^7 + \binom{9}{8}a\left[-\frac{2}{a}\right]^8 + \\ &\quad \binom{9}{9}\left[-\frac{2}{a}\right]^9 \end{aligned}$$

$$= \frac{9!}{9!0!} a^9 + \frac{9!}{8!1!} a^8 \left(-\frac{2}{a}\right) + \frac{9!}{7!2!} a^7 \left(\frac{4}{a^2}\right) +$$

$$\frac{9!}{6!3!} a^6 \left(-\frac{8}{a^3}\right) + \frac{9!}{5!4!} a^5 \left(\frac{16}{a^4}\right) +$$

$$\frac{9!}{4!5!} a^4 \left(-\frac{32}{a^5}\right) + \frac{9!}{3!6!} a^3 \left(\frac{64}{a^6}\right) +$$

$$\frac{9!}{2!7!} a^2 \left(-\frac{128}{a^7}\right) + \frac{9!}{1!8!} a \left(\frac{256}{a^8}\right) +$$

$$\frac{9!}{0!9!} \left(-\frac{512}{a^9}\right)$$

$$= a^9 - 9(2a^7) + 36(4a^5) - 84(8a^3) +$$
$$126(16a) - 126(32a^{-1}) + 84(64a^{-3}) -$$
$$36(128a^{-5}) + 9(256a^{-7}) - 512a^{-9}$$

$$= a^9 - 18a^7 + 144a^5 - 672a^3 + 2016a -$$
$$4032a^{-1} + 5376a^{-3} - 4608a^{-5} + 2304a^{-7} -$$
$$512a^{-9}$$

14. $512x^9 + 2304x^7 + 4608x^5 + 5376x^3 + 4032x +$
$2016x^{-1} + 672x^{-3} + 144x^{-5} + 18x^{-7} + x^{-9}$

15. $(1 - 1)^n$

Let $a = 1$ and $b = -1$.

Expand using the Binomial Theorem Method:

$$(1 - 1)^n = \binom{n}{0} 1^n + \binom{n}{1} 1^{n-1}(-1) + \binom{n}{2} 1^{n-2}(-1)^2 +$$

$$\binom{n}{3} 1^{n-3}(-1)^3 + \ldots + \binom{n}{n}(-1)^n$$

$$= 1 \cdot 1 + n \cdot 1 \cdot (-1) + \binom{n}{2} \cdot 1 \cdot 1 +$$

$$\binom{n}{3} \cdot 1 \cdot (-1) + \ldots + \binom{n}{n}(-1)^n$$

$$= 1 - n + \binom{n}{2} - \binom{n}{3} + \ldots + \binom{n}{n}(-1)^n$$

16. $1 + 3n + \binom{n}{2} 3^2 + \binom{n}{3} 3^3 + \ldots + \binom{n}{n-2} 3^{n-2} +$
$\binom{n}{n-1} 3^{n-1} + 3^n$

17. $(\sqrt{3} - t)^4$

Let $a = \sqrt{3}$, $b = -t$, and $n = 4$.

Expand using the Binomial Theorem Method.

$$(\sqrt{3} - t)^4 = \binom{4}{0}(\sqrt{3})^4 + \binom{4}{1}(\sqrt{3})^3(-t) +$$

$$\binom{4}{2}(\sqrt{3})^2(-t)^2 + \binom{4}{3}(\sqrt{3})(-t)^3 +$$

$$\binom{4}{4}(-t)^4$$

$$= 1 \cdot 9 + 4 \cdot 3\sqrt{3} \cdot (-t) + 6 \cdot 3 \cdot t^2 +$$
$$4 \cdot \sqrt{3} \cdot (-t^3) + 1 \cdot t^4$$

$$= 9 - 12\sqrt{3}\, t + 18t^2 - 4\sqrt{3}\, t^3 + t^4$$

18. $125 + 150\sqrt{5}\, t + 375t^2 + 100\sqrt{5}\, t^3 + 75t^4 +$
$6\sqrt{5}\, t^5 + t^6$

19. $(\sqrt{2} + 1)^6 - (\sqrt{2} - 1)^6$

First, expand $(\sqrt{2} + 1)^6$.

$$(\sqrt{2} + 1)^6 = \binom{6}{0}(\sqrt{2})^6 + \binom{6}{1}(\sqrt{2})^5(1) +$$

$$\binom{6}{2}(\sqrt{2})^4(1)^2 + \binom{6}{3}(\sqrt{2})^3(1)^3 +$$

$$\binom{6}{4}(\sqrt{2})^2(1)^4 + \binom{6}{5}(\sqrt{2})(1)^5 +$$

$$\binom{6}{6}(1)^6$$

$$= \frac{6!}{6!0!} \cdot 8 + \frac{6!}{5!1!} \cdot 4\sqrt{2} + \frac{6!}{4!2!} \cdot 4 +$$

$$\frac{6!}{3!3!} \cdot 2\sqrt{2} + \frac{6!}{2!4!} \cdot 2 + \frac{6!}{1!5!} \cdot \sqrt{2} +$$

$$\frac{6!}{0!6!}$$

$$= 8 + 24\sqrt{2} + 60 + 40\sqrt{2} + 30 +$$
$$6\sqrt{2} + 1$$

$$= 99 + 70\sqrt{2}$$

Next, expand $(\sqrt{2} - 1)^6$.

$$(\sqrt{2} - 1)^6 = \binom{6}{0}(\sqrt{2})^6 + \binom{6}{1}(\sqrt{2})^5(-1) +$$

$$\binom{6}{2}(\sqrt{2})^4(-1)^2 + \binom{6}{3}(\sqrt{2})^3(-1)^3 +$$

$$\binom{6}{4}(\sqrt{2})^2(-1)^4 + \binom{6}{5}(\sqrt{2})(-1)^5 +$$

$$\binom{6}{6}(-1)^6$$

$$= \frac{6!}{6!0!} \cdot 8 - \frac{6!}{5!1!} \cdot 4\sqrt{2} + \frac{6!}{4!2!} \cdot 4 -$$

$$\frac{6!}{3!3!} \cdot 2\sqrt{2} + \frac{6!}{2!4!} \cdot 2 - \frac{6!}{1!5!} \cdot \sqrt{2} +$$

$$\frac{6!}{0!6!}$$

$$= 8 - 24\sqrt{2} + 60 - 40\sqrt{2} + 30 -$$
$$6\sqrt{2} + 1$$

$$= 99 - 70\sqrt{2}$$

$$(\sqrt{2} + 1)^6 - (\sqrt{2} - 1)^6 = (99 + 70\sqrt{2}) - (99 - 70\sqrt{2})$$
$$= 99 + 70\sqrt{2} - 99 + 70\sqrt{2}$$
$$= 140\sqrt{2}$$

20. 34

21. Expand $(x^{-2} + x^2)^4$ using the binomial theorem. Note that $a = x^{-2}$, $b = x^2$, and $n = 4$.

$$(x^{-2} + x^2)^4 = \begin{bmatrix} 4 \\ 0 \end{bmatrix}(x^{-2})^4 + \begin{bmatrix} 4 \\ 1 \end{bmatrix}(x^{-2})^3(x^2) +$$

$$\begin{bmatrix} 4 \\ 2 \end{bmatrix}(x^{-2})^2(x^2)^2 + \begin{bmatrix} 4 \\ 3 \end{bmatrix}(x^{-2})(x^2)^3 +$$

$$\begin{bmatrix} 4 \\ 4 \end{bmatrix}(x^2)^4$$

$$= \frac{4!}{4!0!}(x^{-8}) + \frac{4!}{3!1!}(x^{-6})(x^2) +$$

$$\frac{4!}{2!2!}(x^{-4})(x^4) + \frac{4!}{1!3!}(x^{-2})(x^6) +$$

$$\frac{4!}{0!4!}(x^8)$$

$$= x^{-8} + 4x^{-4} + 6x^0 + 4x^4 + x^8$$
$$= x^{-8} + 4x^{-4} + 6 + 4x^4 + x^8$$

22. $x^{-3} - 6x^{-2} + 15x^{-1} - 20 + 15x - 6x^2 + x^3$

23. Find the 3rd term of $(a + b)^6$.

First, we note that $3 = 2 + 1$, $a = a$, $b = b$, and $n = 6$. Then the 3rd term of the expansion of $(a + b)^6$ is

$$\begin{bmatrix} 6 \\ 2 \end{bmatrix}a^{6-2}b^2, \text{ or } \frac{6!}{4!2!}\,a^4b^2, \text{ or } 15a^4b^2.$$

24. $21x^2y^5$

25. Find the 12th term of $(a - 2)^{14}$.

First, we note that $12 = 11 + 1$, $a = a$, $b = -2$, and $n = 14$. Then the 12th term of the expansion of $(a - 2)^{14}$ is

$$\begin{bmatrix} 14 \\ 11 \end{bmatrix}a^{14-11}\cdot(-2)^{11} = \frac{14!}{3!11!}\,a^3(-2048)$$
$$= 364a^3(-2048)$$
$$= -745,472a^3$$

26. $3,897,234x^2$

27. Find the 5th term of $(2x^3 - \sqrt{y})^8$.

First, we note that $5 = 4 + 1$, $a = 2x^3$, $b = -\sqrt{y}$, and $n = 8$. Then the 5th term of the expansion of $(2x^3 - \sqrt{y})^8$ is

$$\begin{bmatrix} 8 \\ 4 \end{bmatrix}(2x^3)^{8-4}(-\sqrt{y})^4$$

$$= \frac{8!}{4!4!}(2x^3)^4(-\sqrt{y})^4$$

$$= 70(16x^{12})(y^2)$$

$$= 1120x^{12}y^2$$

28. $\frac{35}{27}\,b^{-5}$

29. The middle term of the expansion of $(2u - 3v^2)^{10}$ is the 6th term. Note that $6 = 5 + 1$, $a = 2u$, $b = -3v^2$, and $n = 10$. Then the 6th term of the expansion of $(2u - 3v^2)^{10}$ is

$$\begin{bmatrix} 10 \\ 5 \end{bmatrix}(2u)^{10-5}(-3v^2)^5$$

$$= \frac{10!}{5!5!}(2u)^5(-3v^2)^5$$

$$= 252(32u^5)(-243v^{10})$$

$$= -1,959,552u^5v^{10}$$

30. $30x\sqrt{x}$, $30x\sqrt{3}$

31. The number of subsets is 2^7, or 128 (Theorem 16).

32. 2^6, or 64

33. The number of subsets is 2^{26}, or 67,108,864 (Theorem 16).

34. 2^{24}, or 16,777,216

35. We use the binomial theorem. Note that $a = \sqrt{2}$, $b = -i$, and $n = 4$.

$$(\sqrt{2} - i)^4 = \begin{bmatrix} 4 \\ 0 \end{bmatrix}(\sqrt{2})^4 + \begin{bmatrix} 4 \\ 1 \end{bmatrix}(\sqrt{2})^3(-i) +$$

$$\begin{bmatrix} 4 \\ 2 \end{bmatrix}(\sqrt{2})^2(-i)^2 + \begin{bmatrix} 4 \\ 3 \end{bmatrix}(\sqrt{2})(-i)^3 +$$

$$\begin{bmatrix} 4 \\ 4 \end{bmatrix}(-i)^4$$

$$= \frac{4!}{0!4!}(4) + \frac{4!}{1!3!}(2\sqrt{2})(-i) +$$

$$\frac{4!}{2!2!}(2)(-1) + \frac{4!}{3!1!}(\sqrt{2})(i) +$$

$$\frac{4!}{4!0!}(1)$$

$$= 4 - 8\sqrt{2}\,i - 12 + 4\sqrt{2}\,i + 1$$

$$= -7 - 4\sqrt{2}\,i$$

36. $-8i$

37. $(a - b)^n = \begin{bmatrix} n \\ 0 \end{bmatrix}a^n + \begin{bmatrix} n \\ 1 \end{bmatrix}a^{n-1}(-b) +$

$$\begin{bmatrix} n \\ 2 \end{bmatrix}a^{n-2}(-b)^2 + \cdots + \begin{bmatrix} n \\ n \end{bmatrix}(-b)^n$$

$$= \begin{bmatrix} n \\ 0 \end{bmatrix}a^n + \begin{bmatrix} n \\ 1 \end{bmatrix}(-1)a^{n-1}b +$$

$$\begin{bmatrix} n \\ 2 \end{bmatrix}(-1)^2a^{n-2}b^2 + \cdots + \begin{bmatrix} n \\ n \end{bmatrix}(-1)^nb^n$$

$$= \sum_{r=0}^{n} \begin{bmatrix} n \\ r \end{bmatrix}(-1)^r a^{n-r}b^r$$

38. $\dfrac{(x + h)^n - x^n}{h} = \displaystyle\sum_{r=1}^{n} \begin{bmatrix} n \\ r \end{bmatrix}x^{n-r}h^{r-1}$

39. $\sum\limits_{r=0}^{8} \binom{8}{r} x^{8-r}3^r = 0$

The left side of the equation is sigma notation for the binomial series $(x + 3)^8$, so we have:

$(x + 3)^8 = 0$

$x + 3 = 0$ (Taking the 8th root on both sides)

$x = -3$

40. $x = -5 \pm 2\sqrt{2}$

41. $\sum\limits_{r=0}^{5} \binom{5}{r}(-1)^r x^{5-r}3^r = \sum\limits_{r=0}^{5} \binom{5}{r} x^{5-r}(-3)^r$, so the

left side of the equation is sigma notation for the binomial series $(x - 3)^5$. We have:

$(x - 3)^5 = 32$

$x - 3 = 2$ (Taking the 5th root on both sides)

$x = 5$

42. 9

43. Find the third term of $(0.313 + 0.687)^5$:

$\binom{5}{2}(0.313)^{5-2}(0.687)^2 = \dfrac{5!}{3!2!}(0.313)^3(0.687)^2 \approx$

0.14473

44. $\binom{8}{5}(0.15)^3(0.85)^5 \approx 0.08386$

45. Find and add the 3rd through 6th terms of $(0.313 + 0.687)^5$:

$\binom{5}{2}(0.313)^3(0.687)^2 + \binom{5}{3}(0.313)^2(0.687)^3 +$

$\binom{5}{4}(0.313)(0.687)^4 + \binom{5}{5}(0.687)^5 \approx 0.96403$

46. $\binom{8}{6}(0.15)^2(0.85)^6 + \binom{8}{7}(0.15)(0.85)^7 +$

$\binom{8}{8}(0.85)^8 \approx 0.89479$

47. There are 11 terms in the expression of $(8u + 3v^2)^{10}$, so the 6th term is the middle term.

$\binom{10}{5}(8u)^5(3v^2)^5 = \dfrac{10!}{5!5!}(32,768u^5)(243v^{10}) =$

$2,006,581,248u^5v^{10}$

48. $30x\sqrt{x}, -30x\sqrt{3}$

49. The $(r + 1)$st term of $\left(\dfrac{3x^2}{2} - \dfrac{1}{3x}\right)^{12}$ is

$\binom{12}{r}\left(\dfrac{3x^2}{2}\right)^{12-r}\left(-\dfrac{1}{3x}\right)^r$. In the term which does not contain x, the exponent of x in the numerator is equal to the exponent of x in the denominator.

$2(12 - r) = r$

$24 - 2r = r$

$24 = 3r$

$8 = r$

Find the $(8 + 1)$st, or 9th term:

$\binom{12}{8}\left(\dfrac{3x^2}{2}\right)^4\left(-\dfrac{1}{3x}\right)^8 = \dfrac{12!}{4!8!}\left(\dfrac{3^4x^8}{2^4}\right)\left(\dfrac{1}{3^8x^8}\right) = \dfrac{55}{144}$

50. $-4320x^6y^{9/2}$

51. $\dfrac{\binom{5}{3}(p^2)^2\left(-\frac{1}{2}p\sqrt[3]{q}\right)^3}{\binom{5}{2}(p^2)^3\left(-\frac{1}{2}p\sqrt[3]{q}\right)^2} = \dfrac{-\frac{1}{8}p^7q}{\frac{1}{4}p^8\sqrt[3]{q^2}} = -\dfrac{\sqrt[3]{q}}{2p}$

52. $-35x^{-1/6}$, or $-\dfrac{35}{x^{1/6}}$

53. The term of highest degree of $(x^5 + 3)^4$ is the first term, or

$\binom{4}{0}(x^5)^{4-0}3^0 = \dfrac{4!}{4!0!}x^{20} = x^{20}$.

Therefore, the degree of $(x^5 + 3)^4$ is 20.

54. 127

55. Familiarize: Let x, $x + 1$, $x + 2$, and $x + 3$ represent the integers.

Translate: We write an equation.

$x^3 + (x + 1)^3 + (x + 2)^3 = (x + 3)^3$

Carry out: We solve the equation.

$x^3 + x^3 + 3x^2 + 3x + 1 + x^3 + 6x^2 + 12x + 8 =$
$\qquad x^3 + 9x^2 + 27x + 27$

$2x^3 - 12x - 18 = 0$

$x^3 - 6x - 9 = 0$ $\left[\text{Multiplying by } \frac{1}{2}\right]$

Using the rational roots theorem and synthetic division, we see that 3 is a root of the equation.

$(x - 3)(x^2 + 3x + 3) = 0$

The quadratic factor does not have real number zeros, so the only real number solution is 3.

Check: If $x = 3$, then the integers are 3, 4, 5, and 6. It is true that $3^3 + 4^3 + 5^3 = 6^3$. The values check.

State: The integers are 3, 4, 5, and 6.

56. 2^{100}

57. $_nC_0 + {_nC_1} + \ldots + {_nC_n}$ is the total number of subsets of a set with n members, or 2^n (by Theorem 16).

58. $[\log_a (xt)]^{23}$

59.
$$\sum_{r=0}^{15} \binom{15}{r} i^{30-2r}$$

$$= \sum_{r=0}^{15} \binom{15}{r} i^{30-2r}(1)^{2r}$$

$$= \sum_{r=0}^{15} \binom{15}{r} (i^2)^{15-r}(i^2)^r$$

$$= (i^2 + i^2)^{15} = (-1 + 1)^{15} = 0^{15} = 0$$

60. Basis step: Since $a + b = (a + b)^1$, S_1 is true.
Induction step: Let k be any positive integer.

Let S_k be the statement of the binomial theorem with n replaced by k. Multiply both sides of S_k by $(a + b)$ to obtain
$(a + b)^{k+1}$

$$= \left[a^k + \cdots + \binom{k}{r-1} a^{k-(r-1)} b^{r-1} + \binom{k}{r} a^{k-r} b^r + \cdots + b^k \right](a + b)$$

$$= a^{k+1} + \cdots + \left[\binom{k}{r-1} + \binom{k}{r} \right] a^{(k+1)-r} b^r + \cdots + b^{k+1}$$

$$= a^{k+1} + \cdots + \binom{k+1}{r} a^{(k+1)-r} b^r + \cdots + b^{k+1}.$$

This proves S_{k+1}, assuming S_k. Hence S_n is true for n = 1, 2, 3, ...

1. 100 surveyed
   57 wore either glasses or contacts
   43 wore neither glasses nor contacts

   Using Principle P,

   P(wearing either glasses or contacts) = $\frac{57}{100}$, or 0.57

   P(wearing neither glasses nor contacts) = $\frac{43}{100}$, or 0.43

2. 0.18, 0.24, 0.23, 0.23, 0.12; Opinions might vary, but it would seem that people tend not to pick the first or last numbers.

3. 1044 Total letters in the three paragraphs
   78 A's occurred
   140 E's occurred
   60 I's occurred
   74 O's occurred
   31 U's occurred

   Using Principle P,

   P(the occurrence of letter A) = $\frac{78}{1044} \approx 0.075$

   P(the occurrence of letter E) = $\frac{140}{1044} \approx 0.134$

   P(the occurrence of letter I) = $\frac{60}{1044} \approx 0.057$

   P(the occurrence of letter O) = $\frac{74}{1044} \approx 0.071$

   P(the occurrence of letter U) = $\frac{31}{1044} \approx 0.030$

4. 0.367

5. 1044 Total letters
   $-$ 383 Total vowels   (from Exercise 4)
   661 Total consonants

   Using Principle P,

   P(the occurrence of a consonant) = $\frac{661}{1044} \approx 0.633$

6. Z; 0.999

7. There are 52 equally likely outcomes.

8. $\frac{1}{13}$

9. Since there are 52 equally likely outcomes and there are 13 ways to obtain a heart, by Principle P we have

   P(drawing a heart) = $\frac{13}{52} = \frac{1}{4}$.

10. $\frac{1}{4}$

11. Since there are 52 equally likely outcomes and there are 4 ways to obtain a 4, by Principle P we have

$$P(\text{drawing a 4}) = \frac{4}{52} = \frac{1}{13}.$$

12. $\frac{1}{2}$

13. Since there are 52 equally likely outcomes and there are 26 ways to obtain a black card, by Principle P we have

$$P(\text{drawing a black card}) = \frac{26}{52} = \frac{1}{2}.$$

14. $\frac{2}{13}$

15. Since there are 52 equally likely outcomes and there are 8 ways to obtain a 9 or a king (four 9's and four kings), we have, by Principle P,

$$P(\text{drawing a 9 or a king}) = \frac{8}{52} = \frac{2}{13}.$$

16. $\frac{2}{7}$

17. Since there are 14 equally likely ways of selecting a marble from a bag containing 4 red marbles and 10 green marbles, we have, by Principle P,

$$P(\text{selecting a green marble}) = \frac{10}{14} = \frac{5}{7}.$$

18. 0

19. There are 14 equally likely ways of selecting any marble from a bag containing 4 red marbles and 10 green marbles. Since the bag does not contain any white marbles, there are 0 ways of selecting a white marble. By Principle P, we have

$$P(\text{selecting a white marble}) = \frac{0}{14} = 0.$$

20. $\frac{11}{4165}$

21. The number of ways of drawing 4 cards from a deck of 52 cards is $_{52}C_4$. Now 13 of the 52 cards are hearts, so the number of ways of drawing 4 hearts is $_{13}C_4$. Thus,

$$P(\text{getting 4 hearts}) = \frac{_{13}C_4}{_{52}C_4} = \frac{715}{270,725} = \frac{11}{4165}$$

22. $\frac{60}{143}$

23. The number of ways of selecting 4 people from a group of 15 is $_{15}C_4$. Two men can be selected in $_8C_2$ ways, and 2 women can be selected in $_7C_2$ ways. By the Fundamental Counting Principle, the number of ways of selecting 2 men and 2 women is $_8C_2 \cdot {}_7C_2$. Thus,

$$P(\text{2 men and 2 women are chosen}) = \frac{_8C_2 \cdot {}_7C_2}{_{15}C_4}$$
$$= \frac{28 \cdot 21}{1365} = \frac{28}{65}$$

24. $\frac{5}{36}$

25. On each die there are 6 possible outcomes. The outcomes are paired so there are $6 \cdot 6$, or 36 possible ways in which the two can fall. The pairs that total 3 are (1, 2) and (2, 1). Thus there are 2 possible ways of getting a total of 3, so the probability is $\frac{2}{36}$, or $\frac{1}{18}$.

26. $\frac{1}{36}$

27. On each die there are 6 possible outcomes. The outcomes are paired so there are $6 \cdot 6$, or 36 possible ways in which the two can fall. There is only 1 way of getting a total of 12, the pair (6, 6), so the probability is $\frac{1}{36}$.

28. $\frac{245}{1938}$

29. The number of ways of getting 6 coins from a bag containing 20 coins is $_{20}C_6$. Three nickels can be selected in $_6C_3$ ways since the bag contains 6 nickels. Two dimes can be selected in $_{10}C_2$ ways since the bag contains 10 dimes. One quarter can be selected in $_4C_1$ ways since the bag contains 4 quarters. By the Fundamental Counting Principle the number of ways of selecting 3 nickels, 2 dimes, and 1 quarter is $_6C_3 \cdot {}_{10}C_2 \cdot {}_4C_1$. Thus,

$$P(\text{getting 3 nickels, 2 dimes, and 1 quarter})$$
$$= \frac{_6C_3 \cdot {}_{10}C_2 \cdot {}_4C_1}{_{20}C_6} = \frac{20 \cdot 45 \cdot 4}{38,760} = \frac{30}{323}$$

30. $\frac{9}{19}$

31. The roulette wheel contains 38 equally likely slots. Eighteen of the 38 slots are colored red. Thus, by Principle P,

$$P(\text{the ball falls in a red slot}) = \frac{18}{38} = \frac{9}{19}$$

32. $\frac{18}{19}$

33. The roulette wheel contains 38 equally likely slots. Only 1 slot is numbered 00. Then, by Principle P,

P(the ball falls in the 00 slot) = $\frac{1}{38}$.

34. $\frac{1}{38}$

35. The roulette wheel contains 38 equally likely slots, 2 of which are numbered 00 and 0. Thus, using Principle P,

P$\left(\begin{array}{l}\text{the ball falling in either}\\\text{the 00 or the 0 slot}\end{array}\right)$ = $\frac{2}{38}$ = $\frac{1}{19}$.

36. $\frac{9}{19}$

37. $_{52}C_5$ = 2,598,960

38. a) 4

b) $\frac{4}{_{52}C_5}$ = $\frac{4}{2,598,960}$ ≈ 1.54 × 10⁻⁶

39. Consider a suit

A K Q J 10 9 8 7 6 5 4 3 2

A straight flush can be any of the following combinations in the same suit.

K Q J 10 9
Q J 10 9 8
J 10 9 8 7
10 9 8 7 6
9 8 7 6 5
8 7 6 5 4
7 6 5 4 3
6 5 4 3 2
5 4 3 2 A

Remember a straight flush does not include A K Q J 10 which is a royal flush.

a) Since there are 9 straight flushes per suit, there are 36 straight flushes in all 4 suits.

b) Since $_{52}C_5$, or 2,598,960 poker hands can be dealt from a standard 52-card deck and 36 of those hands are straight flushes, the probability of getting a straight flush is $\frac{36}{2,598,960}$, or 0.0000139.

40. a) 624

b) $\frac{13 \cdot 48}{_{52}C_5}$ = $\frac{624}{2,598,960}$ ≈ 2.4 × 10⁻⁴

41. a) $[13 \cdot {}_4C_3] \cdot [12 \cdot {}_4C_2]$ = 3744; that is, there are 13 ways to select a denomination. Then from that denomination there are $_4C_3$ ways to pick 3 of the 4 cards in that denomination. Now there are 12 ways to select any one of the remaining 12 denominations and $_4C_2$ ways to pick 2 cards from the 4 cards in that denomination.

b) $\frac{3744}{_{52}C_5}$ = $\frac{3744}{2,598,960}$ ≈ 0.00144

42. a) $\left[13 \cdot \binom{4}{2}\right] \cdot \left[\binom{12}{3} \cdot \binom{4}{1}\binom{4}{1}\binom{4}{1}\right]$
= 1,098,240

b) $\frac{1,098,240}{_{52}C_5}$ = $\frac{1,098,240}{2,598,960}$ ≈ 0.423

43. a) $\left[13 \cdot \binom{4}{3}\right] \cdot \left[\binom{48}{2}\right]$ - 3744 = 54,912.

There are 13 ways to select a denomination and then $\binom{4}{3}$ ways to pick 3 cards from the 4 in this denomination. Now there are $\binom{48}{2}$ ways to pick 2 cards from the remaining 12 denominations (48 cards); but $13\binom{4}{3}\binom{48}{2}$ includes the 3744 hands (like QQQ-10-10) in a full house. (See Exercise 41.) Subtracting 3744, we have the above result.

b) $\frac{54,912}{_{52}C_5}$ = $\frac{54,912}{2,598,960}$ ≈ 0.0211

44. a) $4 \cdot \binom{13}{5}$ - 4 - 36 = 5108

b) $\frac{5108}{_{52}C_5}$ = $\frac{5108}{2,598,960}$ ≈ 0.00197

45. a) $\binom{13}{2}\binom{4}{2}\binom{4}{2}\binom{44}{1}$ = 123,552

There are $\binom{13}{2}$ ways to select 2 denominations from 13 denominations; then there are $\binom{4}{2}$ ways to pick 2 of the 4 cards in either of these 2 denominations, and $\binom{4}{2}$ ways to pick 2 of the 4 cards in the other of these 2 denominations. Finally, there are $\binom{44}{1}$ ways to select the fifth card from the remaining 44.

b) $\frac{123,552}{_{52}C_5}$ = $\frac{123,552}{2,598,960}$ ≈ 0.0475

46. a) $\binom{10}{1}\binom{4}{1}\binom{4}{1}\binom{4}{1}\binom{4}{1}\binom{4}{1}$ - 4 - 36
= 10,200

b) $\dfrac{10,200}{_{52}C_5} = \dfrac{10,200}{2,598,960} \approx 0.00392$